MW00441851

MATHEMATICAL STATISTICS

A Decision Theoretic Approach

Probability and Mathematical Statistics

A Series of Monographs and Textbooks

Editors **Z. W. Birnbaum** *University of Washington Seattle, Washington*

E. Lukacs *Bowling Green State University Bowling Green, Ohio*

---

1. Thomas Ferguson. Mathematical Statistics: A Decision Theoretic Approach. 1967
2. Howard Tucker. A Graduate Course in Probability. 1967
3. K. R. Parthasarathy. Probability Measures on Metric Spaces. 1967
4. P. Révész. The Laws of Large Numbers. 1968
5. H. P. McKean, Jr. Stochastic Integrals. 1969
6. B. V. Gnedenko, Yu. K. Belyayev, and A. D. Solovyev. Mathematical Methods of Reliability Theory. 1969
7. Demetrios A. Kappos. Probability Algebras and Stochastic Spaces. 1969
8. Ivan N. Pesin. Classical and Modern Integration Theories. 1970
9. S. Vajda. Probabilistic Programming. 1972
10. Sheldon M. Ross. Introduction to Probability Models. 1972
11. Robert B. Ash. Real Analysis and Probability. 1972
12. V. V. Fedorov. Theory of Optimal Experiments. 1972
13. K. V. Mardia. Statistics of Directional Data. 1972
14. H. Dym and H. P. McKean. Fourier Series and Integrals. 1972
15. Tatsuo Kawata. Fourier Analysis in Probability Theory. 1972
16. Fritz Oberhettinger. Fourier Transforms of Distributions and Their Inverses: A Collection of Tables. 1973
17. Paul Erdös and Joel Spencer. Probabilistic Methods in Combinatorics. 1973
18. K. Sarkadi and I. Vincze. Mathematical Methods of Statistical Quality Control. 1973
19. Michael R. Anderberg. Cluster Analysis for Applications. 1973
20. W. Hengartner and R. Theodorescu. Concentration Functions. 1973
21. Kai Lai Chung. A Course in Probability Theory, Second Edition. 1974
22. L. H. Koopmans. The Spectral Analysis of Time Series. 1974
23. L. E. Maistrov. Probability Theory: A Historical Sketch. 1974
24. William F. Stout. Almost Sure Convergence. 1974
25. E. J. McShane. Stochastic Calculus and Stochastic Models. 1974
26. Z. Govindarajulu. Sequential Statistical Procedures. 1975
27. Robert B. Ash and Melvin F. Gardner. Topics in Stochastic Processes. 1975
28. Avner Friedman, Stochastic Differential Equations and Applications, Volume 1, 1975; Volume 2. 1976
29. Roger Cuppens. Decomposition of Multivariate Probabilities. 1975
30. Eugene Lukacs. Stochastic Convergence, Second Edition. 1975
31. Harry Dym and Henry P. McKean. Gaussian Processes, Function Theory, and the Inverse Spectral Problem. 1976

MATHEMATICAL STATISTICS

A DECISION THEORETIC APPROACH

Thomas S. Ferguson

DEPARTMENT OF MATHEMATICS
UNIVERSITY OF CALIFORNIA
LOS ANGELES, CALIFORNIA

1967

ACADEMIC PRESS New York San Francisco London

A Subsidiary of Harcourt Brace Jovanovich, Publishers

COPYRIGHT © 1967, BY ACADEMIC PRESS, INC.
ALL RIGHTS RESERVED.
NO PART OF THIS PUBLICATION MAY BE REPRODUCED OR TRANSMITTED IN ANY FORM OR BY ANY MEANS, ELECTRONIC OR MECHANICAL, INCLUDING PHOTOCOPY, RECORDING, OR ANY INFORMATION STORAGE AND RETRIEVAL SYSTEM, WITHOUT PERMISSION IN WRITING FROM THE PUBLISHER.

ACADEMIC PRESS, INC.
111 Fifth Avenue, New York, New York 10003

United Kingdom Edition published by
ACADEMIC PRESS, INC. (LONDON) LTD.
24/28 Oval Road, London NW1

LIBRARY OF CONGRESS CATALOG CARD NUMBER: 66-30080

PRINTED IN THE UNITED STATES OF AMERICA

Preface

The theory of games is a part of the rich mathematical legacy left by John von Neumann, one of the outstanding mathematicians of our era. Although others—notably Emil Borel—preceded him in formulating a theory of games, it was von Neumann who with the publication in 1927 of a proof of the minimax theorem for finite games laid the foundation for the theory of games as it is known today. Von Neumann's work culminated in a book written in collaboration with Oskar Morgenstern entitled *Theory of Games and Economic Behavior* published in 1944.

At about the same time, statistical theory was being given an increasingly rigorous mathematical foundation in a series of papers by J. Neyman and Egon Pearson. Statistical theory until that time, as developed by Karl Pearson, R. A. Fisher, and others had lacked the precise mathematical formulation, supplied by Neyman and Pearson, that allows the delicate foundational questions involved to be treated rigorously.

Apparently it was Abraham Wald who first appreciated the connections between the theory of games and the statistical theory of Neyman and Pearson, and who recognized the advantages of basing statistical theory on the theory of games. Wald's theory of statistical decisions, as it is called, generalizes and simplifies the Neyman-Pearson theory by unifying, that is, by treating problems considered as distinct in the Neyman-Pearson theory as special cases of the decision theory problem.

In the 1940's, Wald produced a prodigious amount of research that resulted in the publication of his book *Statistical Decision Functions* in 1950, the year of his tragic death in an airplane accident.

It is our objective to present the elements of Wald's decision theory and an investigation of the extent to which problems of mathematical statistics may be treated successfully by this approach. The main viewpoint is developed in the first two chapters and culminates in a rather general complete class theorem (Theorem 2.10.3). The remaining five chapters deal with statistical topics. No separate chapter on estimation is included since estimation is discussed as examples for general decision problems. It was originally intended that only those parts of statistical theory that could be justified from a decision-theoretic viewpoint would be included. Mainly, this entails the omission of those topics whose mathematical justification is given by large sample theory, such as maximum likelihood estimates, minimum χ^2 methods, and likelihood ratio tests. However, one exception is made. Although the theory of confidence sets as treated does not allow a decision-theoretic justification, it was felt that this topic "belongs" in any discourse on statistics wherein tests of hypotheses are treated. For purposes of comparison, the decision-theoretic notion of a set estimate is included in the exercises.

This book is intended for first-year graduate students in mathematics. It has been used in mimeographed form at UCLA in a two-semester or three-quarter course attended mainly by mathematicians, bio-statisticians, and engineers. I have generally finished the first four chapters in the first semester, deleting perhaps Sections 1.4 and 3.7, but I have never succeeded in completing the last three chapters in the second semester.

There are four suggested prerequisites.

(1) The main prerequisite is a *good* undergraduate course in probability. Ideally, this course should pay a little more attention to conditional expectation than the usual course. In particular, the formula $E(E(X \mid Y)) = E(X)$ should be stressed. Although the abstract approach to probability theory through measure theory is not used (except in Section 3.7, which may be omitted), it is assumed that the reader is acquainted with the notions of a σ-field of sets (as the natural domain of definition of a probability) and of a set of probability zero.

(2) An undergraduate course in analysis on Euclidean spaces is strongly recommended. It is assumed that the reader knows the con-

cepts of continuity, uniform continuity, open and closed sets, the Riemann integral, and so forth.

(3) An introductory undergraduate course in statistics is highly desirable as background material. Although the usual notions of test, power function, and so on, are defined as they arise, the discussion and illustration are rather abstract.

(4) A course in the algebra of matrices would be helpful to the student.

Rudimentary notes leading to this book have been in existence for about six years. Each succeeding generation of students has improved the quality of the text and removed errors overlooked by their predecessors. Without the criticism and interest of these students, too numerous to mention individually, this book would not have been written. Early versions of the notes benefitted from comments by Jack Kiefer and Herbert Robbins. The notes were used by Milton Sobel for a course at the University of Minnesota; his criticisms and those of his students were very useful. Further improvements followed when Paul Hoel used the notes in a course at UCLA. Finally, Gus Haggstrom gave the galleys a critical reading and caught several errors that eluded all previous readers. To all these, I express my deep appreciation.

THOMAS S. FERGUSON

Berkeley, California
February, 1967

Contents

CHAPTER 6. Multiple Decision Problems

CHAPTER 7. Sequential Decision Problems

CHAPTER 1

Game Theory and Decision Theory

1.1 Basic Elements

The elements of decision theory are similar to those of the theory of games. In particular, decision theory may be considered as the theory of a two-person game, in which nature takes the role of one of the players. The so-called normal form of a zero-sum two-person game, henceforth to be referred to as a *game*, consists of three basic elements:

1. A nonempty set, Θ, of possible states of nature, sometimes referred to as the parameter space.
2. A nonempty set, $\mathcal{A}$, of actions available to the statistician.
3. A loss function, $L(\theta, a)$, a real-valued function defined on $\Theta \times \mathcal{A}$.

A game in the mathematical sense is just such a triplet $(\Theta, \mathcal{A}, L)$, and any such triplet defines a game, which is interpreted as follows. Nature chooses a point θ in Θ, and the statistician, without being informed of the choice nature has made, chooses an action a in $\mathcal{A}$. As a consequence of these two choices, the statistician loses an amount $L(\theta, a)$. [The function $L(\theta, a)$ may take negative values. A negative loss may be interpreted as a gain, but throughout this book $L(\theta, a)$ represents the loss to the statistician if he takes action a when θ is the "true state of nature".] Simple though this definition may be, its scope is quite broad, as the following examples illustrate.

EXAMPLE 1. ODD OR EVEN. Two contestants simultaneously put up either one or two fingers. One of the players, call him player I, wins if the sum of the digits showing is odd, and the other player, player II, wins if the sum of the digits showing is even. The winner in all cases receives in dollars the sum of the digits showing, this being paid to him by the loser.

To create a triplet $(\Theta, \mathcal{A}, L)$ out of this game we give player I the label "nature" and player II the label "statistician". Each of these players has two possible choices, so that $\Theta = \{1, 2\} = \mathcal{A}$, in which "1" and "2" stand for the decisions to put up one and two fingers, respectively. The loss function is given by Table 1.1. Thus $L(1, 1) = -2$,

Table 1.1

Θ \ $\mathcal{A}$	1	2
1	-2	3
2	3	-4

$L(\theta, a)$

$L(1, 2) = 3$, $L(2, 1) = 3$, and $L(2, 2) = -4$. It is quite clear that this is a game in the sense described in the first paragraph. This example is discussed later in Section 1.7, in which it is shown that one of the players has a distinct advantage over the other. Can you tell which one it is? Which player would you rather be?

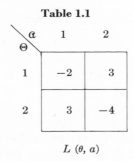

EXAMPLE 2. TIC-TAC-TOE, CHESS. In the game $(\Theta, \mathcal{A}, L)$ an element of the space Θ or $\mathcal{A}$ is sometimes referred to as a *strategy*. In some games strategies are built on a more elementary concept, that of a "move". Many parlor games illustrate this feature; for example, the games tic-tac-toe, chess, checkers, Battleship, Nim, Go, and so forth. A *move* is an action made by a specified player at a specified time during the game. The rules determine at each move the player whose turn it is to move and the choices of move available to that player at that time. For such a game a strategy is a rule that specifies for a given player the exact move he is to make each time it is his turn to move, for all possible histories of the game. The game of tic-tac-toe has at most nine moves, one player making five of them, the other making four. A player's

strategy must tell him exactly what move to make in each possible position that may occur in the game. Because the number of possible games of tic-tac-toe is rather small (less than 9!), it is possible to write down an optimal strategy for each player. In this case each player has a strategy that guarantees its user at least a tie, no matter what his opponent does. Such strategies are called optimal strategies. Naturally, in the game of chess it is physically impossible to describe "all possible histories", for there are too many possible games of chess and many more strategies, in fact, than there are atoms in our solar system. We can write down strategies for the game of chess, but none so far constructed has much of a chance of beating the average amateur. When the two players have written down their strategies, they may be given to a referee who may play through the game and determine the winner. In the triplet $(\Theta, \mathcal{A}, L)$, which describes either tic-tac-toe or chess, the spaces Θ and $\mathcal{A}$ are the sets of all strategies for the two players, and the loss function $L(\theta, a)$ may be $+1$ if the strategy θ beats the strategy a, 0 for a draw, and -1 if a beats θ.

Example 3. A Game with Bluffing. Another feature of many games, and one that is characteristic of card games, is the notion of a *chance move*. The dealing or drawing of cards, the rolling of dice, the spinning of a roulette wheel, and so on, are examples of chance moves. In the theory of games it is assumed that both players are aware of the probabilities of the various outcomes resulting from a chance move. Sometimes, as in card games, one player may be informed of the actual outcome of a chance move, whereas the other player is not. This leads to the possibility of "bluffing". The following example is a loose description of a situation which sometimes occurs in the game of stud poker.

Two players each put an "ante" of a units into a pot ($a > 0$). Player I then draws a card from a deck, which gives him a winning or a losing card. Both players are aware of the probability P that the card drawn is a winning card ($0 < P < 1$). Player I then may bet b units ($b > 0$) by putting b units into the pot or he may check. If player I checks, he wins the pot if he has a winning card and loses the pot if he has a losing card. If player I bets, player II may call and put b units in the pot or he may fold. If player II folds, player I wins the pot whatever card he has drawn. If player II calls, player I wins the pot if he has a winning card and loses it otherwise.

If I receives a winning card, it is clear that he should bet: if he checks, he automatically receives total winnings of a units, whereas if he bets, he will receive at least a units and possibly more. For the purposes of our discussion we assume that the rules of the game enforce this condition: that if I receives a winning card, he must bet. This will eliminate some obviously poor strategies from player I's strategy set. With this restriction, player I has two possible strategies: (a) *the bluff strategy*—bet with a winning card or a losing card; and (b) *the honest strategy*—bet with a winning card, check with a losing card. The two strategies for player II are (a) *the call strategy*—if player I bets, call; and (b) *the fold strategy*—if player I bets, fold. Given a strategy for each player in a game with chance moves, a referee can play the game through as before, playing each chance move with the probability distribution specified, and determining who has won and by how much. The actual payoff in such games is thus a random quantity determined by the chance moves. In writing down a loss function, we replace these random quantities by their expected values in order to obtain a game as defined. (Further discussion of this may be found in Sections 1.3 and 1.4.) Table 1.2

Table 1.2

I \ II	Call	Fold
Bluff	$(2P - 1)(a + b)$	a
Honest	$(2P - 1)a + Pb$	$(2P - 1)a$

shows player I's expected winnings and player II's expected losses. For example, if I uses the honest strategy and II uses the call strategy, player II's loss will be $(a + b)$ with probability P(I receives a winning card) and $-a$ with probability $(1 - P)$(I receives a losing card). The expected loss is

$$(a + b)P - a(1 - P) = (2P - 1)a + Pb,$$

as found in the table. If player I is given the label "nature" and player II the label "statistician," the triplet $(\Theta, \mathcal{A}, L)$, in which Θ = (bluff, honest), $\mathcal{A}$ = (call, fold), and L is given by Table 1.2, defines a game that contains the main aspects of the bluffing game already described. This game is considered in Exercises 1.7.4 and 5.2.8.

1.2 A Comparison of Game Theory and Decision Theory

There are certain differences between game theory and decision theory that arise from the philosophical interpretation of the elements Θ, $\mathcal{A}$, and L. The main differences are these.

1. In a two-person game the two players are trying simultaneously to maximize their winnings (or to minimize their losses), whereas in decision theory nature chooses a state without this view in mind. This difference plays a role mainly in the interpretation of what is considered to be a good decision for the statistician and results in presenting him with a broader dilemma and a correspondingly wider class of what might be called "reasonable" decision rules. This is natural, for one can depend on an intelligent opponent to behave "rationally", that is to say, in a way profitable to him. However, a criterion of "rational" behavior for nature may not exist or, if it does, the statistician may not have knowledge of it. We do not assume that nature wins the amount $L(\theta, a)$ when θ and a are the points chosen by the players. An example will make this clear. Consider the game $(\Theta, \mathcal{A}, L)$ in which $\Theta = \{\theta_1, \theta_2\}$ and $\mathcal{A} = \{a_1, a_2\}$ and in which the loss function L is given by Table 1.3. In game

Table 1.3

	a_1	a_2
θ_1	4	1
θ_2	-3	0

$L(\theta, a)$

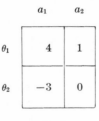

theory, in which the player choosing a point from Θ is assumed to be intelligent and his winnings in the game are given by the function L, the only "rational" choice for him is θ_1. No matter what his opponent does, he will gain more if he chooses θ_1 than if he chooses θ_2. Thus it is clear that the statistician should choose action a_2, instead of a_1, for he will lose only one instead of four. Again, this is the only reasonable thing for him to do. Now, suppose that the function L does not reflect the winnings of nature or that nature chooses a state without any clear objective in mind. Then we can no longer state categorically that the statistician should choose action a_2. If nature happens to choose θ_2, the statistician will prefer to take action a_1. This basic conceptual difference between game theory and decision theory is reflected in the difference between the theorems we have called fundamental for game theory and fundamental for decision theory (Sec. 2.2).

2. It is assumed that nature chooses the "true state" once and for all and that the statistician has at his disposal the possibility of gathering information on this choice by sampling or by performing an experiment. This difference between game theory and decision theory is more apparent than real, for one can easily imagine a game between two intelligent adversaries in which one of the players has an advantage given to him by the rules of the game by which he can get some information on the choice his opponent has made before he himself has to make a decision. It turns out (Sec. 1.3) that the over-all problem which allows the statistician to gain information by sampling may simply be viewed as a more complex game. However, all statistical games have this characteristic feature, and it is the exploitation of the structure which such gathering of information gives to a game that distinguishes decision theory from game theory proper.

For an entertaining introduction to finite games the delightful book *The Compleat Strategyst* by the late J. D. Williams (1954) is highly recommended. The more serious student should also consult the lucid accounts of game theory found in McKinsey (1952), Karlin (1959), and Luce and Raiffa (1957). An elementary text by Chernoff and Moses (1959) provides a good introduction to the main concepts of decision theory. The important book by Blackwell and Girshick (1954), which is a more advanced text, is recommended as collateral reading for this study.

1.3 Decision Function; Risk Function

To give a mathematical structure to this process of information gathering, we suppose that the statistician before making a decision is allowed to look at the observed value of a random variable or vector, X, whose distribution depends on the true state of nature, θ. Throughout most of this book *the sample space*, denoted by $\mathfrak{X}$, is taken to be (a Borel subset of) a finite dimensional Euclidean space, and the probability distributions of X are supposed to be defined on the Borel subsets, $\mathcal{B}$ of $\mathfrak{X}$. Thus for each $\theta \in \Theta$ there is a probability measure P_θ defined on $\mathcal{B}$, and a corresponding cumulative distribution function $F_X(x \mid \theta)$, which represents the distribution of X when θ is the true value of the parameter. [If X is an n-dimensional vector, it is best to consider X as a notation for $(X_1, \cdots, X_n)$ and $F_X(x \mid \theta)$ as a notation for the multivariate cumulative distribution function $F_{X_1,\cdots,X_n}(x_1, \cdots, x_n \mid \theta)$.]

A *statistical decision problem* or a *statistical game* is a game $(\Theta, \mathcal{A}, L)$ coupled with an experiment involving a random observable X whose distribution P_θ depends on the state $\theta \in \Theta$ chosen by nature.

On the basis of the outcome of the experiment $X = x$ (x is the observed value of X), the statistician chooses an action $d(x) \in \mathcal{A}$. Such a function d, which maps the sample space $\mathfrak{X}$ into $\mathcal{A}$, is an elementary strategy for the statistician in this situation. The loss is now the random quantity $L(\theta, d(X))$. The expected value of $L(\theta, d(X))$ when θ is the true state of nature is called *the risk function*

$$R(\theta, d) = E_\theta L(\theta, d(X)) \tag{1.1}$$

and represents the average loss to the statistician when the true state of nature is θ and the statistician uses the function d. Note that for some choices of the function d and some values of the parameter θ the expected value in (1.1) may be $\pm\infty$ or, worse, it may not even exist. As the following definition indicates, we do not bother ourselves about such functions.

Definition 1. Any function $d(x)$ that maps the sample space $\mathfrak{X}$ into $\mathcal{A}$ is called a nonrandomized *decision rule* or a nonrandomized *decision function,* provided the risk function $R(\theta, d)$ exists and is finite for all $\theta \in \Theta$. The class of all nonrandomized decision rules is denoted by D.

Unfortunately, the class D is not well defined unless we specify the sense in which the expectation in (1.1) is to be understood. The reader may take this expectation to be the Lebesgue integral,

$$R(\theta, d) = E_\theta L(\theta, d(X)) = \int L(\theta, d(x))\, dP_\theta(x).$$

With such an understanding, D consists of those functions d for which $L(\theta, d(x))$ is for each $\theta \in \Theta$ a Lebesgue integrable function of x. In particular, D contains all simple functions. (A function d from $\mathfrak{X}$ to $\mathfrak{A}$ is called simple if there is a finite partition of $\mathfrak{X}$ into measurable subsets $B_1, \cdots, B_m \in \mathfrak{B}$, and a finite subset $\{a_1, \cdots, a_m\}$ of $\mathfrak{A}$ such that for $x \in B_i$, $d(x) = a_i$ for $i = 1, \cdots, m$.) On the other hand, the expectation in (1.1) may be taken as the Riemann or the Riemann-Stieltjes integral,

$$R(\theta, d) = E_\theta L(\theta, d(X)) = \int L(\theta, d(x))\, dF_X(x \mid \theta).$$

In that case D would contain only functions d for which $L(\theta, d(x))$ is for each $\theta \in \Theta$ continuous on a set of probability one under $F_X(x \mid \theta)$. For the purposes of understanding what follows, it is not too important which of the various definitions is given to the expectation in (1.1). In most of the proofs of the theorems given later we use only certain linearity $[E(aX + Y) = aEX + EY]$ and ordering $(X > 0 \Rightarrow EX > 0)$ properties of the expectation; such proofs are equally valid for Lebesgue and Riemann integrals. Therefore we let the definition of the expectation be arbitrary (unless otherwise stated) and assume that the class D of decision rules is well defined.

EXAMPLE 1. The game of "odd or even" mentioned in Sec. 1.1 may be extended to a statistical decision problem. Suppose that before the game is played the player called "the statistician" is allowed to ask the player called "nature" how many fingers he intends to put up and that nature must answer truthfully with probability 3/4 (hence untruthfully with probability 1/4). The statistician therefore observes a random variable X (the answer nature gives) taking the values 1 or 2. If $\theta = 1$ is the true state of nature, the probability that $X = 1$ is 3/4; that is, $P_1\{X = 1\} = 3/4$. Similarly, $P_2\{X = 1\} = 1/4$. There are exactly four possible functions from $\mathfrak{X} = \{1, 2\}$ into $\mathfrak{A} = \{1, 2\}$. These are the four decision rules:

$$d_1(1) = 1, \qquad d_1(2) = 1;$$

$$d_2(1) = 1, \qquad d_2(2) = 2;$$

$$d_3(1) = 2, \qquad d_3(2) = 1;$$

$$d_4(1) = 2, \qquad d_4(2) = 2.$$

Rules d_1 and d_4 ignore the value of X. Rule d_2 reflects the belief of the statistician that nature is telling the truth, and rule d_3, that nature is not telling the truth. The risk table (Table 1.4) should be checked by the student as an exercise.

Table 1.4

$\Theta \backslash D$	d_1	d_2	d_3	d_4
1	−2	−3/4	7/4	3
2	3	−9/4	5/4	−4

$R(\theta, d)$

It is a custom, which we steadfastly observe, that the choice of a decision function should depend only on the risk function $R(\theta, d)$ (the smaller in value the better) and not otherwise on the distribution of the random variable $L(\theta, d(X))$. (For example, this would entail the supposition that a poor man would be indifferent when choosing between the offer of $10,000 as an outright gift, and the offer of a gamble that would give him $20,000 with probability one half and $0 with probability one half.) There is a relatively sound mathematical reason for the statistician to behave in this fashion, provided the loss function is measured in utiles rather than in some monetary way. This topic is the subject of the next section.

Notice that the original game $(\Theta, \mathcal{A}, L)$ has been replaced by a new game, (Θ, D, R), in which the space D and the function R have an underlying structure, depending on $\mathcal{A}$, L, and the distribution of X, whose exploitation must be the main objective of decision theory.

Naturally, only a small part of statistics can be contained within such a simple framework. No room has been made for such broad topics as the choice of experiments, the design of experiments, or sequential analysis. In each case a new structure could be added to the framework to include these topics, and the problem would be reduced once again to a simple game. For example, in sequential analysis the statistician may take observations one at a time, paying c units each time he does so. Therefore a decision rule will have to tell him both when to stop taking observations and what action to take once he has stopped. He will try to choose a decision rule that will minimize in some sense his new risk, which is defined now as the expected value of the loss plus the cost.

Nevertheless, even the simple structure of the first paragraph of this section is broad enough to contain the main aspects of three important categories in what might be called "classical" mathematical statistics.

1. $\mathcal{A}$ *consists of two points*, $\mathcal{A} = \{a_1, a_2\}$. Decision theoretic problems in which $\mathcal{A}$ consists of exactly two points are called *problems in testing hypotheses*. Consider the special case in which Θ is the real line and suppose that the loss function is for some fixed number θ_0 given by the formulas

$$L(\theta, a_1) = \begin{cases} l_1 & \text{if} \quad \theta > \theta_0 \\ 0 & \text{if} \quad \theta \leq \theta_0 \end{cases}$$

and

$$L(\theta, a_2) = \begin{cases} 0 & \text{if} \quad \theta > \theta_0 \\ l_2 & \text{if} \quad \theta \leq \theta_0 , \end{cases}$$

where l_1 and l_2 are positive numbers. Here we would like to take action a_1 if $\theta \leq \theta_0$ and action a_2 if $\theta > \theta_0$. The space D of decision rules consists of those functions d from the sample space into $\{a_1, a_2\}$ with the property that $P_\theta\{d(X) = a_1\}$ is well-defined for all values of $\theta \in \Theta$. The risk function in this case is easy to compute

$$R(\theta, d) = \begin{cases} l_1 P_\theta\{d(X) = a_1\} & \text{if} \quad \theta > \theta_0 \\ l_2 P_\theta\{d(X) = a_2\} & \text{if} \quad \theta \leq \theta_0 . \end{cases}$$

In this way probabilities of making two types of error are involved. For $\theta > \theta_0$, $P_\theta\{d(X) = a_1\}$ is the probability of making the error of taking action a_1 when we should take action a_2 and θ is the true state of nature. Similarly, for $\theta \leq \theta_0$,

$$P_\theta\{d(X) = a_2\} = 1 - P_\theta\{d(X) = a_1\}$$

is the probability of making the error of taking action a_2 when we should take action a_1 and θ is the true state of nature.

2. $\mathcal{A}$ *consists of* k *points*, $\{a_1, a_2, \cdots, a_k\}$, $k \geq 3$. These decision theoretic problems are called *multiple decision problems*. As an example, a statistician might be called on to decide which of k worthy students is

to receive a scholarship on the basis of school grades and financial need in which the loss is based on the students expected performance with and without a scholarship. Another typical example occurs when an experimenter is to judge which of two treatments has a greater yield on the basis of an experiment. He may (a) decide treatment 1 is better, (b) decide treatment 2 is better, or (c) withhold judgment until more data are available. In this example $k = 3$.

3. *$\mathcal{A}$ consists of the real line,* $\mathcal{A} = (-\infty, \infty)$. Such decision theoretic problems are referred to in a broad sense as *point estimation of a real parameter*. Consider the special case in which Θ is also the real line and suppose that the loss function is given by the formula

$$L(\theta, a) = c(\theta - a)^2,$$

where c is some positive constant. A decision function, d, in this case a real-valued function defined on the sample space, may be considered as an "estimate" of the true unknown state of nature θ. It is the statistician's desire to choose the function d to minimize the risk function

$$R(\theta, d) = cE_\theta(\theta - d(X))^2,$$

which is c times the mean squared error of the estimate $d(X)$. Note that the criterion arrived at here—that of choosing an estimate with a small mean squared error in some sense—is exactly the criterion most frequently used in classical statistics.

1.4 Utility and Subjective Probability

The method a "rational" man uses in choosing between two alternative actions, a_1 and a_2 in $\mathcal{A}$, is complex. Let us assume for the moment that the payoff or loss function $L(\theta, a)$ is not necessarily numerical but may represent complex entities, such as "you receive a ticket to a ball game tomorrow when there is a good chance of rain and your raincoat is torn" or "you lose five dollars on a bet to someone you dislike and the chances are that he is going to rub it in." Such entities we refer to as *payoffs*, or *prizes*. The "rational" man in choosing between two actions evaluates the value of the various payoffs to himself and balances it with the probabilities with which he thinks the payoffs will occur. He may do, and usually does, such an evaluation subconsciously. We give in this section a mathematical model by which such choices among actions are made. This model is based on the notion that a "rational" man can express his preference between two payoffs in a method that is consistent with certain axioms. We show that the value to himself of a payoff may be expressed as a numerical function of the payoffs, called a *utility*, and that preference between actions giving him a probability distribution over the payoffs is based only on the expected value of the utility of the action.

Let $\mathcal{P}$ denote the set of payoffs of the game and let $\mathcal{P}^*$ denote the set of all probability distributions over $\mathcal{P}$ that give all their mass to only a finite number of points of $\mathcal{P}$. We use P, P_1, P_2, and so on, to denote payoffs (that is, elements of $\mathcal{P}$) and p, q, p_1, p_2, and so on, to denote elements of $\mathcal{P}^*$.

Note. The only properties of $\mathcal{P}^*$ used here are (a) $\mathcal{P}^*$ is closed under convex linear combinations (that is, $p_1 \in \mathcal{P}^*$ and $p_2 \in \mathcal{P}^*$ imply $\pi p_1 + (1 - \pi)p_2 \in \mathcal{P}^*$ for $0 \leq \pi \leq 1$) and (b) all degenerate probability distributions belong to $\mathcal{P}^*$. We can enlarge $\mathcal{P}^*$ to contain continuous distributions, for example, provided that these two conditions are still satisfied.

Definition 1. A preference pattern or a preference relation on $\mathcal{P}^*$ is a linear ordering, $\leq$, of $\mathcal{P}^*$; that is, (a) (*linearity*) if p_1 and $p_2 \in \mathcal{P}^*$, then either $p_1 \leq p_2$ or $p_2 \leq p_1$ (or both), (b) (*transitivity*) if p_1, p_2 and $p_3 \in \mathcal{P}^*$, and if $p_1 \leq p_2$ and $p_2 \leq p_3$, then $p_1 \leq p_3$.

An element $p \in \mathcal{P}^*$ may be considered as a superpayoff that chooses a payoff in $\mathcal{P}$ according to the probability distribution p. It is assumed that for each "rational" man there is a preference pattern that reflects his preference among the superpayoffs. The statement $p_1 \leq p_2$ means that the "rational" man does not prefer p_1 to p_2; that is, he prefers p_2 to p_1 or he is indifferent to which he receives.

Notation. We say that p_2 is preferred to p_1 and write $p_1 < p_2$ if $p_1 \leq p_2$ and not $p_2 \leq p_1$.

We say that p_1 and p_2 are equivalent and write $p_1 \sim p_2$ if $p_1 \leq p_2$ and $p_2 \leq p_1$.

The proofs of the following lemmas are easy and are left to the reader.

Lemma 1. $\sim$ is an equivalence relation [that is, (i) $p \sim p$ for all p, (ii) $p_1 \sim p_2$ implies $p_2 \sim p_1$, and (iii) $p_1 \sim p_2$ and $p_2 \sim p_3$ imply $p_1 \sim p_3$].

Lemma 2. If $p_1 \leq p_2$ and $p_2 < p_3$, then $p_1 < p_3$.

We note that if $p_1, \cdots, p_n \in \mathcal{P}^*$ and $\lambda_1 \geq 0$ with $\sum_1^n \lambda_i = 1$, then $\sum_{i=1}^n \lambda_i p_i$ is a finite distribution on $\mathcal{P}$, and hence is in $\mathcal{P}^*$.

Definition 2. A *utility* on $\mathcal{P}^*$ is a real-valued function, u, defined on $\mathcal{P}^*$, which is linear on $\mathcal{P}^*$; that is, if p_1 and $p_2 \in \mathcal{P}^*$ and $0 \leq \lambda \leq 1$, then

$$u(\lambda p_1 + (1 - \lambda) p_2) = \lambda\, u(p_1) + (1 - \lambda)\, u(p_2). \tag{1.2}$$

If a function u satisfies (1.2), it follows by induction that for $p_i \in \mathcal{P}^*$ and $\lambda_i \geq 0$ for $i = 1, \cdots, n$, with $\sum_1^n \lambda_i = 1$, then

$$u\left(\sum_{i=1}^{n} \lambda_i p_i \right) = \sum_{i=1}^{n} \lambda_i\, u(p_i). \tag{1.3}$$

Definition 3. A preference pattern, $\leq$, and a utility, u, on $\mathcal{P}^*$ are said to agree if for all p_1 and $p_2 \in \mathcal{P}^*$

$$p_1 \leq p_2 \quad \text{if, and only if,} \quad u(p_1) \leq u(p_2).$$

If $P_1, P_2, \cdots, P_k$ are elements of $\mathcal{P}$, and $f_1, f_2, \cdots, f_k$ are nonnegative numbers whose sum is one, we shall use the notation $(f_1P_1, f_2P_2, \cdots, f_kP_k)$ to denote that element of $\mathcal{P}^*$ which chooses payoff P_1 with probability f_1, payoff P_2 with probability f_2, $\cdots$, and payoff P_k with probability f_k. If we define for each $P \in \mathcal{P}$, $h(P) = u(1P)$ for a given utility u, then

$$u((f_1P_1, f_2P_2, \cdots, f_kP_k)) = \sum_{i=1}^{k} f_i\, h(P_i), \tag{1.4}$$

which is simply the expected value of the function h using the distribution $(f_1P_1, f_2P_2, \cdots, f_kP_k)$. Thus, if there exists a utility, u, which agrees with a person's preference pattern, $\leq$, that person will act as if he wished to maximize the expected value of the function h. Not all preference patterns have agreeing utilities. However, there will exist an agreeing utility if the preference pattern satisfies the following two additional reasonable hypotheses.

Hypothesis H_1. If p_1, p_2, and $q \in \mathcal{P}^*$ and $\lambda > 0$, $\lambda \leq 1$, then

$$p_1 \leq p_2 \quad \text{if, and only if,} \quad \lambda p_1 + (1 - \lambda)q \leq \lambda p_2 + (1 - \lambda)q.$$

Hypothesis H_2. If p_1, p_2, and $p_3 \in \mathcal{P}^*$ are such that $p_1 < p_2 < p_3$, then there exist numbers λ and μ with $0 < \lambda < 1$ and $0 < \mu < 1$, such that

$$\lambda p_3 + (1 - \lambda)p_1 < p_2 < \mu p_3 + (1 - \mu)p_1 .$$

Hypothesis H_1 seems reasonable. A minor objection is that we might be indifferent between $\lambda p_1 + (1 - \lambda)q$ and $\lambda p_2 + (1 - \lambda)q$ when λ is sufficiently small, say $\lambda = 10^{-1000}$, even though we prefer p_1 to p_2. Another objection comes from the man who dislikes gambles with random payoffs. He might prefer p_2 that would give him \$2.00 for sure to a gamble p_1 that would give him \$3.10 with probability 1/2 and \$1.00 with probability 1/2; but if q is \$5.00 for sure and $\lambda = 1/2$, he might prefer $\lambda p_1 + (1 - \lambda)q$ to $\lambda p_2 + (1 - \lambda)q$ on the basis of larger expected value, for the payoff is random in either case. Hypothesis H_2 is more debatable. It is safe to assume that death < 10¢ < \$1.00. Yet would there exist a $\mu < 1$ such that 10¢ < μ(\$1.00) + $(1 - \mu)$(death)? Perhaps not. For myself, I would say that $1 - \mu = 10^{-1000}$ would suffice. At any rate, hypothesis H_2 implies that there is no payoff infinitely more desirable or infinitely less desirable than any other payoff. For penetrating critiques of the whole subject of utility and subjective probability, two entertaining and informative books are recommended: Luce and Raiffa (1957) and Savage (1954).

Theorem 1. If a preference pattern $\leq$ on $\mathcal{P}^*$ satisfies H_1 and H_2, then there exists a utility, u, on $\mathcal{P}^*$ which agrees with $\leq$. Furthermore, u is uniquely determined up to a linear transformation.

Note. If u is a utility that agrees with $\leq$, then $\hat{u} = \alpha u + \beta$, where $\alpha > 0$ and β are real numbers, is also a utility that agrees with $\leq$. Thus the uniqueness of u up to a linear transformation is as strong a uniqueness as any that can be obtained.

Proof. We break up the proof into a number of easy steps.

1. If $p_0 < p_1$ and $0 \leq \lambda < \mu \leq 1$, then

$$\lambda p_1 + (1 - \lambda)p_0 < \mu p_1 + (1 - \mu)p_0 .$$

(If p_1 is preferred to p_0, then, between any two linear combinations of p_1 and p_0, the one giving larger weight to p_1 is preferred.)

Proof. Because $\mu - \lambda$ is positive, H_1 implies that

$$\begin{aligned} \lambda p_1 &+ (1 - \lambda)p_0 \\ &= (\mu - \lambda)p_0 + (1 - \mu + \lambda)\left(\frac{\lambda}{1 - \mu + \lambda}p_1 + \frac{1 - \mu}{1 - \mu + \lambda}p_0\right) \\ &< (\mu - \lambda)p_1 + (1 - \mu + \lambda)\left(\frac{\lambda}{1 - \mu + \lambda}p_1 + \frac{1 - \mu}{1 - \mu + \lambda}p_0\right) \\ &= \mu p_1 + (1 - \mu)p_0 . \end{aligned}$$

2. If $p_0 < p_1$ and $p_0 \leq q \leq p_1$, there exists a unique number λ', $0 \leq \lambda' \leq 1$, such that $\lambda' p_1 + (1 - \lambda')p_0 \sim q$.

Proof. If either $p_1 \sim q$ or $p_2 \sim q$, the result is immediate. Hence assume that $p_0 < q < p_1$ and let

$$T = (\lambda : 0 \leq \lambda \leq 1 \quad \text{and} \quad \lambda p_1 + (1 - \lambda)p_0 < q);$$

then $0 \in T$ and $1 \notin T$. By (1), if $\lambda_1 \in T$ and $\lambda_2 < \lambda_1$, then $\lambda_2 \in T$. Thus T is an interval. Let λ' be the least upper bound of T. We will show that λ' satisfies the requirement of statement (2).

(a). $q \leq \lambda' p_1 + (1 - \lambda')p_0$. This statement is obvious if $\lambda' = 1$. Now suppose that $\lambda' < 1$ and that this statement is false. Then

$$\lambda' p_1 + (1 - \lambda')p_0 < q < p_1 ,$$

so that from H_2 there is a λ, $0 < \lambda < 1$, such that

$$\lambda p_1 + (1 - \lambda)[\lambda' p_1 + (1 - \lambda')p_0] < q.$$

This is the same as

$$(\lambda' + \lambda(1 - \lambda'))p_1 + (1 - \lambda)(1 - \lambda')p_0 < q,$$

so that

$$\lambda' + \lambda(1 - \lambda') \in T; \qquad \text{but } \lambda' + \lambda(1 - \lambda') > \lambda',$$

which contradicts the definition of λ' as an upper bound of T.

(b). $\lambda' p_1 + (1 - \lambda')p_0 \leq q$. The proof, similar to that of (a), is left to the reader. Together, (a) and (b) imply that $\lambda' p_1 + (1 - \lambda')p_0 \sim q$. Only unicity remains to be proved.

(c). *Unicity.* If $\lambda' p_1 + (1 - \lambda')p_0 \sim \lambda'' p_1 + (1 - \lambda'')p_0$, then from (1) both $\lambda' \leq \lambda''$ and $\lambda'' \leq \lambda'$ so that $\lambda'' = \lambda'$, completing the proof of (2).

If all $p \in \mathcal{P}^*$ are equivalent, the result is trivial. So we suppose there exist p_0 and $p_1 \in \mathcal{P}^*$ such that $p_0 < p_1$. By the interval $[p_0, p_1]$ we shall mean the set

$$[p_0, p_1] = \{q \in \mathcal{P}^* : p_0 \leq q \leq p_1\}.$$

3. Let p_0 and p_1 be any elements of $\mathcal{P}^*$ for which $p_0 < p_1$. Then there exists a utility function, u, on the interval $[p_0, p_1]$, uniquely determined up to a linear transformation, which agrees with $\leq$ on $[p_0, p_1]$.

Proof. For $q \in [p_0, p_1]$ define $u(q)$ to be that unique number λ' such that $q \sim \lambda' p_1 + (1 - \lambda')p_0$. Note that $u(p_0) = 0$ and $u(p_1) = 1$.

(a). *u agrees with $\leq$.* If $p \leq q$, then

$$u(p)p_1 + (1 - u(p))p_0 \leq u(q)p_1 + (1 - u(q))p_0\,,$$

so that from (1) $u(p) \leq u(q)$. On the other hand, if $u(p) \leq u(q)$, then

$$u(p)p_1 + (1 - u(p))p_0 \leq u(q)p_1 + (1 - u(q))p_0$$

from (1), so that $p \leq q$.

(b). *u is linear.* If $0 \leq \lambda \leq 1$, $p_0 \leq p \leq p_1$, and $p_0 \leq q \leq p_1$, then

$$\lambda p + (1 - \lambda)q \sim u(\lambda p + (1 - \lambda)q)p_1 + (1 - u(\lambda p + (1 - \lambda)q))p_0\,;$$

also from H_1

$$\begin{aligned}\lambda p + (1 - \lambda)q &= \lambda(u(p)p_1 + (1 - u(p))p_0) \\ &\qquad + (1 - \lambda)(u(q)p_1 + (1 - u(q))p_0) \\ &= (\lambda\, u(p) + (1 - \lambda)\, u(q))p_1 \\ &\qquad + (1 - \lambda\, u(p) - (1 - \lambda)\, u(q))p_0\,.\end{aligned}$$

Thus from the unicity part of (2)

$$u(\lambda p + (1 - \lambda)q) = \lambda\, u(p) + (1 - \lambda)\, u(q).$$

(c). *Unique up to linear transformation.* If $\hat{u}$ is also linear and agrees with $\leq$ on $[p_0, p_1]$, we may by a linear transformation take $\hat{u}(p_0) = 0$ and $\hat{u}(p_1) = 1$; but for $p \in [p_0, p_1]$

$$\begin{aligned} \hat{u}(p) &= \hat{u}(u(p)p_1 + (1 - u(p))p_0) \\ &= u(p)\, \hat{u}(p_1) + (1 - u(p))\, \hat{u}(p_0) \\ &= u(p), \end{aligned}$$

completing the proof of (3).

4. To complete the proof of the theorem we consider fixed elements $p_0 < p_1$ and find the utility u on $[p_0, p_1]$ with $u(p_0) = 0$, $u(p_1) = 1$, which agrees with $\leq$. We extend the domain of definition of $u(p)$ to all of $\mathcal{P}^*$ as follows. For any $p \in \mathcal{P}^*$ choose an interval $[p_0', p_1']$ that contains p and $[p_0, p_1]$. Because the utility u' on $[p_0', p_1']$ is determined only up to a linear transformation, we choose that u' such that $u'(p_0) = 0$ and $u'(p_1) = 1$. This u' must agree with u on $[p_0, p_1]$. We define $u(p)$ as $u'(p)$; this is well defined, for if $[p_0'', p_1'']$ is any other interval containing p and $[p_0, p_1]$ and u'' is a utility agreeing with $\leq$ for which $u''(p_0) = 0$ and $u(p_1'') = 1$, then $u''(p) = u'(p)$ because they are both utilities on an interval $[p_0''', p_1''']$ that contains p and $[p_0, p_1]$. It is immediate from this that u is linear and agrees with $\leq$ on all of $\mathcal{P}^*$ and that u is unique up to linear transformation.

This theorem states that if a person has a preference pattern that satisfies certain reasonable restrictions, he will behave as if he had assigned a numerical utility to the set of payoffs $\mathcal{P}$, and his preference in $\mathcal{P}^*$ will coincide with a preference for the element of $\mathcal{P}^*$ with the larger expected utility. In this we have assumed that the person knows the exact distribution of an element $(f_1P_1, \cdots, f_kP_k)$ in $\mathcal{P}^*$, as he would, for example, if he had received the payoffs as a result of the outcome of one roll of the roulette wheel. The more general situation also occurs in which the probability distribution can only be guessed at (for example when the payoffs are distributed according to the outcome of a horse race). If we assume that a person still has a preference pattern on a set of gambles for which prizes are awarded on the basis of the outcome of a horse race, this question arises: When will the person act as if he had assigned probabilities to the various outcomes of a horse race and was trying to maximize his expected utility? Such an assignment of probabilities would be considered as that individual's *subjective probability* or *personal probability*, as it is sometimes called, of the outcomes of the race. The approach to subjective probability we follow here is contained in a paper by Anscombe and Aumann (1963).

The situation we consider is as follows. We start with a set of payoffs, $\mathcal{P}$, the set $\mathcal{P}^*$ of finite probability distributions on $\mathcal{P}$, and a preference pattern $\leq$ on $\mathcal{P}^*$ that satisfies hypotheses H_1 and H_2. From Theorem 1 there is a utility u on $\mathcal{P}^*$ that agrees with $\leq$. We consider an experiment with a finite number of outcomes $\theta_1, \theta_2, \cdots, \theta_m$, one and only one of which is bound to occur, and we consider the gamble whose payoff is p_1 if θ_1 occurs, p_2 if θ_2 occurs, $\cdots$, and p_m if θ_m occurs, in which $p_1, p_2, \cdots, p_m$ are arbitrary elements of $\mathcal{P}^*$. Such a gamble is denoted by $[p_1, p_2, \cdots, p_m]$. The set $\mathcal{G}$ of all such gambles may be considered as a set of payoffs. Therefore, if on the set $\mathcal{G}^*$ of all finite probability measures over $\mathcal{G}$ there is a preference pattern $\leq_g$ that satisfies H_1 and H_2, there will be a utility u_g on $\mathcal{G}^*$ that agrees with $\leq_g$. The following assumptions connect the two preference patterns.

Assumption A_1. For all i, if $p_i \leq p_i'$, then

$$[p_1, \cdots, p_i, \cdots, p_m] \leq_g [p_1, \cdots, p_i', \cdots, p_m].$$

This implies that if $p_i \leq p_i'$ for all i, then $[p_1, \cdots, p_m] \leq_g [p_1', \cdots, p_m']$.

Assumption A_2. If $p < p'$, then $[p, \cdots, p] <_g [p', \cdots, p']$.

Assumption A_3. $(f_1[p_{11}, \cdots, p_{1m}], \cdots, f_k[p_{k1}, \cdots, p_{km}]) \sim_g$

$$[(f_1p_{11}, \cdots, f_kp_{k1}), \cdots, (f_1p_{1m}, \cdots, f_kp_{km})].$$

Assumptions A_1 and A_2 say that the preference pattern $\leq$ on $\mathcal{P}^*$ carries over onto the preference pattern $\leq_g$ on $\mathcal{G}^*$. Assumption A_3 says that every element in $\mathcal{G}^*$ is equivalent$_g$ to the element of $\mathcal{G}$ in which the randomization is performed after the outcome of the experiment is observed. As Anscombe and Aumann express it, if the payoff is to be determined by a roulette wheel and a horse race, the individual is indifferent whether the roulette wheel is spun before or after the race.

Suppose, then, that an individual has preference patterns $\leq$ on $\mathcal{P}^*$ and $\leq_g$ on $\mathcal{G}^*$, both of which satisfy H_1 and H_2, and suppose that these preference patterns are related in the sense that Assumptions A_1, A_2, and A_3 are satisfied. Does it follow that this individual's preferences on $\mathcal{G}^*$ agree with a utility u_g obtained by assigning some probabilities $\pi_1, \cdots, \pi_m$ to the outcomes $\theta_1, \cdots, \theta_m$, respectively, and defining

$$u_g[p_1, \cdots, p_m] = u(p_1)\pi_1 + \cdots + u(p_m)\pi_m\,, \tag{1.5}$$

where u is a utility on $\mathcal{P}^*$ that agrees with $\leq$? If so, this individual is acting as if it were known that the probability of θ_i is π_i, $i = 1, \cdots, m$, and π_i can be considered as his subjective probability or personal probability of the outcome θ_i. The following theorem answers this fundamental question on the existence of the probabilities π_i affirmatively.

Theorem 2. If both preference patterns, $\leq$ on $\mathcal{P}^*$ and $\leq_g$ on $\mathcal{G}^*$, satisfy H_1 and H_2, and if Assumptions A_1, A_2, and A_3 are satisfied, then there exist utilities u on $\mathcal{P}^*$ and u_g on $\mathcal{G}^*$ and there exist nonnegative numbers $\pi_1, \cdots, \pi_m$ whose sum is one such that u agrees with $\leq$, u_g agrees with $\leq_g$, and (1.5) holds.

Proof. If $p \sim q$ for all p and q in $\mathcal{P}^*$, then by A_1 all elements of $\mathcal{G}^*$ are equivalent$_g$ and the theorem is trivial, for both u and u_g are constants.

Let us suppose then that there exist elements q_0 and q_1 in $\mathcal{P}^*$ such that $q_0 < q_1$; then by A_2

$$[q_0, \cdots, q_0] <_g [q_1, \cdots, q_1].$$

Let us choose u so that $u(q_0) = 0$ and $u(q_1) = 1$ and u_g so that

$$u_g[q_0, \cdots, q_0] = 0 \qquad \text{and} \qquad u_g[q_1, \cdots, q_1] = 1.$$

A_1 implies that if

$$p_i \sim p_i' \text{ for } i = 1, \cdots, m$$

then

$$[p_1, \cdots, p_m] \sim_g [p_1', \cdots, p_m'].$$

In other words, $u_g[p_1, \cdots, p_m]$ is determined by the numbers $u(p_1), \cdots, u(p_m)$. This allows us to abuse our notation and write $[r_1, \cdots, r_m]$ in place of $[p_1, \cdots, p_m]$, where $r_i = u(p_i)$.

1. If for some constant $c > 0$, and for $i = 1, \cdots, m$ we have $0 \leq r_i \leq 1$ and $0 \leq cr_i \leq 1$, then

$$u_g[cr_1, \cdots, cr_m] = cu_g[r_1, \cdots, r_m].$$

Proof. First assume that $c \leq 1$; then A_3 implies

$$[cr_1, \cdots, cr_m] = [cr_1 + (1-c)0, \cdots, cr_m + (1-c)0]$$
$$\sim_g (c[r_1, \cdots, r_m], (1-c)[0, \cdots, 0]).$$

Hence

$$u_g[cr_1, \cdots, cr_m] = cu_g[r_1, \cdots, r_m] + (1-c)u_g[0, \cdots, 0]$$
$$= cu_g[r_1, \cdots, r_m].$$

Next, if $c > 1$, it follows that

$$u_g[r_1, \cdots, r_m] = u_g\left[\frac{cr_1}{c}, \cdots, \frac{cr_m}{c}\right]$$
$$= \frac{1}{c}u_g[cr_1, \cdots, cr_m].$$

Multiplying through by c completes the proof of (1).

Now let $\pi_i = u_g[0, \cdots, 1, \cdots, 0]$ with 1 in the ith spot and 0 elsewhere.

2. If $0 \leq r_i \leq 1$, then $u_g[r_1, \cdots, r_m] = \sum_{i=1}^m r_i\pi_i$.

Proof. First note that if all $r_i = 0$, then (2) is obviously true. Suppose then that at least one of the r_i is positive and let $c = \sum r_i$, so that $c > 0$. Then

$$u_g[r_1, \cdots, r_m]$$
$$= u_g\left[\frac{cr_1}{c}, \cdots, \frac{cr_m}{c}\right]$$
$$= cu_g\left[\frac{r_1}{c}, \cdots, \frac{r_m}{c}\right] \qquad \text{from (1)}$$
$$= cu_g\left(\frac{r_1}{c}[1, 0, \cdots, 0], \cdots, \frac{r_m}{c}[0, 0, \cdots, 1]\right) \qquad \text{from } A_3$$
$$= c\left(\frac{r_1}{c}\pi_1 + \cdots + \frac{r_m}{c}\pi_m\right) \qquad \text{from (1.4)}$$
$$= \sum_{i=1}^m r_i\pi_i,$$

proving (2). Note that automatically $\pi_i \geq 0$ and $\sum \pi_i = 1$.

We have proved Theorem 2, provided $p_1, \cdots, p_m$ are all in a fixed interval $[q_0, q_1]$. Now let $p_1, \cdots, p_m$ be arbitrary elements of $\mathcal{P}^*$ and find an interval $[q_0', q_1']$ that contains $[q_0, q_1]$ and $p_1, p_2, \cdots, p_m$. If we let $\hat{u}$ be that representation of the utility on $\mathcal{P}^*$ such that $\hat{u}(q_0') = 0$ and $\hat{u}(q_1') = 1$, then $\hat{u}(p) = \alpha u(p) + \beta$ for some $\alpha > 0$ and β. Similarly, if we let $\hat{u}_g$ be that representation of the utility on $\mathcal{G}^*$ such that

$$\hat{u}_g[q_0', \cdots, q_0'] = 0 \qquad \text{and} \qquad \hat{u}_g[q_1', \cdots, q_1'] = 1,$$

then $\hat{u}_g = \alpha_g u_g + \beta_g$ for some $\alpha_g > 0$ and β_g. Then, by what we have proved in (1) and (2),

$$\hat{u}_g[p_1', \cdots, p_m'] = \sum_{i=1}^{m} \hat{u}(p_i')\hat{\pi}_i \tag{1.6}$$

for all $p_1', \cdots, p_m'$ in $[q_0', q_1']$, where $\hat{\pi}_i = \hat{u}_g[q_0', \cdots, q_1', \cdots, q_0']$ with q_1' in the ith spot and q_0' elsewhere. However,

$$\hat{u}_g[q_0, \cdots, q_0] = \beta_g ,$$

$$\hat{u}_g[q_1, \cdots, q_1] = \alpha_g + \beta_g ,$$

whereas, from (1.6),

$$\hat{u}_g[q_0, \cdots, q_0] = \hat{u}(q_0) = \beta,$$

$$\hat{u}_g[q_1, \cdots, q_1] = \hat{u}(q_1) = \alpha + \beta.$$

Hence $\beta_g = \beta$ and $\alpha_g = \alpha > 0$. Now (1.6) may be written

$$\alpha u_g[p_1', \cdots, p_m'] + \beta = \sum (\alpha\, u(p_i') + \beta)\hat{\pi}_i$$

$$= \alpha \sum (u\, p_i')\hat{\pi}_i + \beta,$$

from which it is easy to see that $\hat{\pi}_i = \pi_i$ by putting $[p_1', \cdots, p_m'] = [q_0, \cdots, q_1, \cdots, q_0]$ with q_1 in the ith spot and q_0 elsewhere. Thus we have

$$u_g[p_1, \cdots, p_m] = \sum_{i=1}^{n} u(p_i)\pi_i .$$

Exercises

1. Prove Lemma 1.
2. Prove Lemma 2.
3. Assume that $\mathcal{P}^*$ is defined to contain all probability distributions that give mass to denumerable sets of points of $\mathcal{P}$. Show that in Theorem 1 it may also be concluded that the utility u is bounded,

provided a utility is defined to satisfy (1.3) with $n = \infty$, and H_1 is replaced by the following: if $p_i \leq p'_i$ for $i = 1, 2, \cdots$, and $\sum \lambda_i = 1$, then $\sum \lambda_i p_1 \leq \sum \lambda_i p'_i$. [See Blackwell and Girshick (1954), p. 105, Eq. (1.2), and Hypothesis H_1; Luce and Raiffa (1957); or Chernoff and Moses (1959).]

4. From Hypothesis H_1 derive the following. If p_1, p_2, and $q \in \mathcal{P}^*$ and $0 < \lambda \leq 1$, then

$$p_1 \sim p_2 \quad \text{if, and only if,} \quad \lambda p_1 + (1 - \lambda)q \sim \lambda p_2 + (1 - \lambda)q.$$

5. Show that if in Theorem 2 there exist elements q_0 and q_1 in $\mathcal{P}^*$ such that $q_0 < q_1$, then the π_i, $i = 1, \cdots, m$ are uniquely determined.

1.5 Randomization

Let us begin with an example. Consider a game $(\Theta, \mathcal{A}, L)$ in which Θ consists of two elements $\{\theta_1, \theta_2\}$, and $\mathcal{A}$ consists of three elements $\{a_1, a_2, a_3\}$, and suppose that the loss function (or the negative of the utility) is given by Table 1.5. The question to be discussed is, "Should

Table 1.5

	a_1	a_2	a_3
θ_1	4	1	3
θ_2	1	4	3

Loss function

the statistician ever take action a_3?" If nature chooses θ_1, action a_3 is preferable to action a_1. If, on the other hand, nature chooses θ_2, action a_3 is preferable to action a_2. Thus a_3 is preferred to either of the other actions under the proper circumstances. However, suppose the statistician flips a fair coin to choose between actions a_1 and a_2; that is, suppose the statistician's decision is to choose a_1 if the coin comes up heads and a_2 if the coin comes up tails. This decision, denoted by δ, is a *randomized decision*; such decisions allow the actual choice of the action in $\mathcal{A}$ to be left to a random mechanism and the statistician chooses only the probabilities of the various outcomes. In game theory δ would be called a *mixed strategy*. The randomized decision δ chooses action a_1 with probability 1/2, action a_2 with probability 1/2, and action a_3 with probability zero. The expected loss in the use of δ (or the negative of the utility of δ) is

$$\tfrac{1}{2}L(\theta_1, a_1) + \tfrac{1}{2}L(\theta_1, a_2) + 0\,L(\theta_1, a_3) = 5/2 \quad \text{if } \theta_1 \text{ is true,}$$

$$\tfrac{1}{2}L(\theta_2, a_1) + \tfrac{1}{2}L(\theta_2, a_2) + 0\,L(\theta_2, a_3) = 5/2 \quad \text{if } \theta_2 \text{ is true.}$$

Because it is understood that the choice between strategies is to be made on the basis of expected loss only, δ is certainly to be preferred to a_3; for no matter what the true state of nature, the expected loss is smaller if we use δ than if we use a_3. Moreover, any randomized decision that gives positive weight to a_3 can be improved by distributing that weight equally between the other two actions. For example, if δ_1 chooses a_1 with probability 1/4, a_2 with probability 1/2, and a_3 with probability 1/4, the rule δ_2, which chooses a_1 with probability 3/8, a_2 with probability 5/8, and a_3 with probability 0, has a smaller expected loss, regardless of the true state of nature. (What are the expected losses?) Thus the answer to the question posed earlier is, "If randomized decisions are allowed and the choice between strategies is based on expected loss only, the statistician should never take action a_3."

Note. In the situation in which the statistician may base his choice of action on a random variable, X, whose distribution depends on θ, again, for the same reasons, a_3 should never be chosen.

This example shows that randomization is valuable—the statistician may disregard a_3 entirely in making his choice, thus simplifying the problem with which he is faced.

More generally, a randomized decision for the statistician in a game $(\Theta, \mathcal{A}, L)$ is a probability distribution over $\mathcal{A}$ (it is understood that a fixed σ-field of subsets of $\mathcal{A}$ containing the individual points of $\mathcal{A}$ is given). If P is a probability distribution over $\mathcal{A}$ and Z is a random variable taking values in $\mathcal{A}$, whose distribution is given by P, the expected or average loss in the use of the randomized decision P is

$$L(\theta, P) = E\,L(\theta, Z), \tag{1.8}$$

provided it exists. This formula is to be regarded as an extension of the domain of definition of the function $L(\theta, \cdot)$ from $\mathcal{A}$ to the space of randomized decisions, for each element $a \in \mathcal{A}$ may, and shall, be regarded as a probability distribution degenerate at a, that is, the distribution giving probability one to point a. *The space of randomized decisions, P, for which $L(\theta, P)$ exists and is finite for all $\theta \in \Theta$ is denoted by $\mathcal{A}^*$.*

With this definition, the game $(\Theta, \mathcal{A}^*, L)$ is to be considered as the game $(\Theta, \mathcal{A}, L)$ in which the statistician is allowed randomization.

For $\mathcal{A}^*$ to be rigorously defined we must specify the σ-field of subsets of $\mathcal{A}$ on which the probabilities are defined and the sense in which the expectation in (1.8) is to be understood. We assume in the following that this has been done. (Theoretically, there is a natural σ-field of subsets of $\mathcal{A}$ to use; namely, the smallest σ-field with respect to which all the functions $L(\theta, \cdot)$ for $\theta \in \Theta$ are measurable.) However, we make one restriction: that the σ-field of subsets of $\mathcal{A}$ contain all the sets consisting of individual points of $\mathcal{A}$. With this restriction, $\mathcal{A}^*$ *contains all probability distributions giving mass one to a finite number of points of* $\mathcal{A}$.

By analogy, we may extend the game (Θ, D, R) to (Θ, D^*, R), where D^* is a space containing probability distributions over D. If δ denotes a probability distribution over D, $R(\theta, \delta)$ is defined analogously to (1.8) as

$$R(\theta, \delta) = E\,R(\theta, Z), \tag{1.9}$$

where Z is a random variable taking values in D, whose distribution is given by δ.

Definition 1. Any probability distribution δ on the space of nonrandomized decision functions, D, is called a *randomized decision function* or a *randomized decision rule*, provided the risk function (1.9) exists and is finite for all $\theta \in \Theta$. The space of all randomized decision rules is denoted by D^*.

As with $\mathcal{A}^*$, we assume that, however the expectation in (1.9) is defined, D^* *contains all probability distributions giving mass one to a finite number of points of* D.

The space D of nonrandomized decision rules may, and shall, be considered as a subset of the space D^* of randomized decision rules, $D \subset D^*$, by identifying a point $d \in D$ with the probability distribution $\delta \in D^*$ degenerate at the point d. Thus, in speaking of randomized decision functions or randomized decision rules, we may drop the adjective "randomized", for this class also contains the nonrandomized rules. When specific reference to nonrandomized decision rules is necessary, the corresponding adjective is not dropped.

One advantage in the extension of the definition of $L(\theta, \cdot)$ from $\mathcal{A}$ to $\mathcal{A}^*$ and the definition of $R(\theta, \cdot)$ from D to D^* is that these functions become linear on $\mathcal{A}^*$ and D^*, respectively. In other words, if $P_1 \in \mathcal{A}^*$,

$P_2 \in \mathcal{A}^*$, and $0 \leq \alpha \leq 1$, then $\alpha P_1 + (1 - \alpha)P_2 \in \mathcal{A}^*$ and

$$L(\theta, \alpha P_1 + (1 - \alpha)P_2) = \alpha L(\theta, P_1) + (1 - \alpha) L(\theta, P_2). \quad (1.10)$$

Similarly, if $\delta_1 \in D^*$, $\delta_2 \in D^*$, and $0 \leq \alpha \leq 1$, then $\alpha\delta_1 + (1 - \alpha)\delta_2 \in D^*$ and

$$R(\theta, \alpha\delta_1 + (1 - \alpha)\delta_2) = \alpha R(\theta, \delta_1) + (1 - \alpha) R(\theta, \delta_2). \quad (1.11)$$

An alternative way of setting up randomization for decision problems leads to a space of decision functions, to be denoted by $\mathfrak{D}$, which is often easier to work with than D^*. A behavioral strategy, to borrow another term from game theory, or a behavioral decision rule, as we shall call it, is a function, δ, which gives for each x in the sample space a probability distribution over $\mathcal{A}$, that is, an element of $\mathcal{A}^*$, $\delta(x) \in \mathcal{A}^*$. For a behavioral decision rule, δ, the risk function is now defined as

$$\hat{R}(\theta, \delta) = E_\theta L(\theta, \delta(X)), \quad (1.12)$$

where $L(\theta, \delta(x))$ is the loss defined by (1.8).

Definition 2. A function, $\delta(x)$, from the sample space into $\mathcal{A}^*$ is called a *behavioral decision function* or a *behavioral decision rule*, provided the risk function defined by (1.12) exists and is finite for all $\theta \in \Theta$. The space of all behavioral decision rules is denoted by $\mathfrak{D}$.

In particular, if the expectation in (1.12) is the Lebesgue integral, $\mathfrak{D}$ contains all simple functions. In any case, $\mathfrak{D}$ contains as a subset the set D of functions from the sample space into $\mathcal{A}$ (considered as a subset of $\mathcal{A}^*$), so that both (Θ, D^*, R) and $(\Theta, \mathfrak{D}, \hat{R})$ may be considered as generalizations of (Θ, D, R). These two methods of randomization correspond to the two routes in the diagram

$$\begin{array}{ccc} & \nearrow & (\Theta, D, R) \to (\Theta, D^*, R) \\ (\Theta, \mathcal{A}, L) & & \\ & \searrow & (\Theta, \mathcal{A}^*, L) \to (\Theta, \mathfrak{D}, \hat{R}). \end{array}$$

In the upper route the set of functions from the sample space is considered first and then the randomization of the decision rule. In the lower route this process is reversed, the randomization being incorporated first and the set of functions from the sample space afterward.

The term "behavioral strategy" in game theory refers to those strategies that tell the player how to randomize at each move, whereas a "mixed strategy" chooses at random a strategy that tells him exactly what to do at each move. Similarly, in decision theory a behavioral decision rule tells the statistician how to randomize after observing the outcome of the experiment, whereas a randomized decision rule chooses at random a decision function that tells him before observing the outcome of the experiment exactly what action to take as a result of the experiment. It is intuitively clear that randomized decision rules are no more general than behavioral decision rules; for the randomization provided by a rule $\delta \in D^*$ may be performed after the statistician looks at the outcome of the experiment, thus providing for each x in the sample space a probability distribution over $\mathcal{A}$. A mapping, $\delta \to \hat{\delta}$, of this sort will embed D^* in $\mathfrak{D}$ in such a way that $R(\theta, \delta) = \hat{R}(\theta, \hat{\delta})$ for all $\theta \in \Theta$. [In a similar way it is clear that the game $(\Theta, \mathfrak{D}^*, \hat{R})$, which allows randomization over the set $\mathfrak{D}$ of behavioral decision rules, is no more general than the game $(\Theta, \mathfrak{D}, \hat{R})$.] However, it is not immediately clear that (Θ, D^*, R) is not less general than $(\Theta, \mathfrak{D}, \hat{R})$. The argument in the following paragraph indicates that in quite general situations the games (Θ, D^*, R) and $(\Theta, \mathfrak{D}, \hat{R})$ are equivalent in the sense that for any $\delta \in D^*$ there is a $\hat{\delta}$ in $\mathfrak{D}$ for which $R(\theta, \delta) = \hat{R}(\theta, \hat{\delta})$ for all θ and conversely. A rigorous proof is beyond the scope of this book. (See Blackwell and Girshick (1954), Theorems 7.2.1 and 8.3.1, for proofs in certain situations, and Wald and Wolfowitz (1951) for a proof in a rather general setting.)

A behavioral decision rule chooses, for each x in the sample space, a random point Y_x in $\mathcal{A}$. A behavioral decision rule merely specifies the distribution of Y_x for each x in the sample space. On the other hand, a randomized decision rule chooses at random a function, Y_x, which is a function from the sample space into $\mathcal{A}$. A randomized decision rule must specify the distribution of the random function Y_x. (This includes, for example, the joint distribution of $Y_{x_1}, Y_{x_2}, \cdots, Y_{x_n}$, when $x_1, x_2, \cdots, x_n$ are points of the sample space.) Thus a behavioral decision rule specifies only the marginal distributions of Y_x for each x—and this is why, in general, behavioral strategies are simpler—whereas a randomized decision rule specifies the entire distribution of Y_x. To say that (Θ, D^*, R) is no more general than $(\Theta, \mathfrak{D}, \hat{R})$ is to say that the marginal distributions can be determined from the joint distributions. To say that (Θ, D^*, R) is no less general than $(\Theta, \mathfrak{D}, \hat{R})$ is to say that, given a set of marginal distribution of Y_x, there exists a distribution of the random function Y_x with these specified marginals; for example, we may take each Y_x to be completely independent of the rest of the variables.

This method can be used to give a rigorous proof when the sample space is discrete (that is, has a countable number of points). However, if the distribution of the observable random variable X has a continuous distribution, choosing the Y_x independent will not do, because the resulting random function Y_x will with probability one choose a non-measurable decision rule (that is, a rule for which the risk function is undefined). Wald and Wolfowitz (1951) have overcome these measurability difficulties when $\mathcal{A}$ is a separable complete metric space (and the Lebesgue integral is used) by a different choice of the distribution of the random function Y_x with the specified marginals.

In the rest of this book we assume that the games (Θ, D^*, R) and $(\Theta, \mathfrak{D}, \hat{R})$ are equivalent in the sense defined, but we try to point out explicitly where this fact is used. In most of the main theorems of decision theory we deal with the game (Θ, D^*, R).

Example. Consider the simplest nontrivial case, namely, $\mathcal{A} = \{a_1, a_2\}$ and $\mathfrak{X} = \{x_1, x_2\}$. The set $\mathcal{A}^*$ of probability distributions on $\mathcal{A}$ may be taken to be the closed interval $[0, 1]$ with the understanding that $\pi \in \mathcal{A}^*$ represents the probability of taking action a_1, whereas $1 - \pi$ is the probability of taking action a_2. As in the example of Sec. 1.3, D consists of four elements,

$$D = \{d_1, d_2, d_3, d_4\}:$$

$$\begin{aligned} d_1(x_1) &= a_1, & d_1(x_2) &= a_1; \\ d_2(x_1) &= a_1, & d_2(x_2) &= a_2; \\ d_3(x_1) &= a_2, & d_3(x_2) &= a_1; \\ d_4(x_1) &= a_2, & d_4(x_2) &= a_2. \end{aligned}$$

Hence we may take

$$D^* = \{(p_1, p_2, p_3, p_4) : p_i \geq 0, \sum_1^4 p_i = 1\}$$

with the understanding that decision rule d_1 is chosen with probability p_1, decision rule d_2 with probability p_2, etc. On the other hand, $\mathfrak{D}$ is the space of maps from $\mathfrak{X}$ into $\mathcal{A}^*$ and can be represented as the unit square

$$\mathfrak{D} = \{(\pi_1, \pi_2) : 0 \leq \pi_1 \leq 1, 0 \leq \pi_2 \leq 1\}$$

with the understanding that if x_1 is observed $\pi_1 \in \mathcal{A}^*$ is used, whereas if x_2 is observed, then $\pi_2 \in \mathcal{A}^*$ is used; $\mathfrak{D}$ is two-dimensional and D^* is essentially three-dimensional. In general, the dimension of $\mathfrak{D}$ is less than the dimension of D^*.

Exercises

1. Given $(1/10, 1/2, 1/10, 3/10) \in D^*$ in the preceding example, find an equivalent element of $\mathfrak{D}$.
2. Given $(1/3, 3/4) \in \mathfrak{D}$ in the preceding example, find an equivalent element of D^*.
3. If $\mathcal{A} = \{a_1, \cdots, a_m\}$ and $\mathfrak{X} = \{x_1, \cdots, x_n\}$, show that D^* is essentially $(m^n - 1)$-dimensional, whereas $\mathfrak{D}$ is $[n(m - 1)]$-dimensional.

1.6 Optimal Decision Rules

The fundamental problem of decision theory can be stated quite simply. Given a game $(\Theta, \mathcal{A}, L)$ and a random observable X whose distribution depends on $\theta \in \Theta$, what decision rule δ should the statistician use? The reason that this question, so easy to pose, is so difficult to answer, if, indeed, it is not meaningless, is discussed in the following paragraphs.

It is a natural reaction to search for a "best" decision rule, a rule that has the smallest risk no matter what the true state of nature. Unfortunately, *situations in which a best decision rule exists are rare and uninteresting*. For each fixed state of nature there may be a best action for the statistician to take. However, this best action will differ, in general, for different states of nature, so that no one action can be presumed best over all. As an example, consider the problem of estimating a real parameter θ with quadratic loss function, $L(\theta, a) = (\theta - a)^2$. If the true state of nature is θ_0, the best action the statistician can take is $a = \theta_0$ and the best decision rule is the nonrandomized decision rule $d_0(x) \equiv \theta_0$. There is, in general, no other significantly different decision rule as good as d_0 if θ_0 is the true state of nature. Yet it is clear that d_0 cannot be considered best over all, for it is not so good as $d_1(x) \equiv \theta_1$ if θ_1 is the true state of nature. Thus a best decision rule does not exist. Any given decision rule δ can be improved at any value θ_0 for which $R(\theta_0, \delta) > 0$ by the rule $d_0(x) \equiv \theta_0$.

Frequently, because a problem is too complex for a thorough analysis, the statistician has to be content with finding any "reasonable" rule at all. (In our terminology the statistician is *just guessing* if he uses a rule that does not depend on the observed outcome of the experiment. A *reasonable* rule is one that is better than just guessing.) A general method for obtaining a reasonable estimate in most situations is the well-known maximum likelihood method. Another that leads, in general, to reasonable rules in hypothesis-testing situations is the likelihood ratio technique. Moreover, even if exceptionally good rules do exist, the statistician may prefer one that, although admittedly not very good, at least has the advantage of being easy to apply and evaluate [what Tukey (1953) calls "quick and easy" rules] or perhaps of being easy to explain and defend to one's employer (such as graphs and histograms). Although such problems could be included within decision theory proper by enlarging the framework, the formidable structure thus encountered and the consequent additional cost of analysis are convincing arguments that a theory of reasonable rules has a place in statistics. However, decision theory, as developed here, is devoted to the problem of finding optimal rules, so that we do not refer to maximum likelihood estimates, likelihood ratio techniques, quick and easy rules, and so on, unless they turn out naturally to be optimal in some sense.

To circumvent the fact that a best rule usually does not exist, two general methods, which have been proposed for arriving at a decision rule, are frequently satisfactory.

Method 1. Restricting the Available Rules. The reason a uniformly best rule usually does not exist is that there are too many others available, some of which (like d_0) are not really good because they guard against specific states of nature extremely well and neglect the others. This suggests a restriction of the rules, among which the statistician is to make his choice, to a smaller class of rules, all of which have good over-all properties, in the hope that among these rules there is one that is uniformly best. Two restrictions of this kind are treated in this book.

1. Unbiasedness. An estimate $\hat{\theta}$ of a parameter θ is said to be unbiased if, when θ is the true value of the parameter, the mean of the distribution of $\hat{\theta}$ is θ.

$$E_\theta \hat{\theta} = \theta \quad \text{for all } \theta. \tag{1.13}$$

Thus an unbiased estimate in a very weak sense treats all states of nature equally. The estimate d_0 of the preceding example is obviously not unbiased, for $E_\theta d_0 = \theta_0$ for all θ. If, in this example, we apply the principle of unbiasedness and restrict the available rules to be unbiased, it is then possible (and in some cases factual) that a "uniformly best unbiased estimate" of θ will exist. This theory is discussed briefly in Section 3.6. The theory of unbiasedness in the hypothesis testing situation is presented in Sec. 5.4.

2. INVARIANCE. If the decision problem is symmetric, or invariant, with respect to certain operations, then it may seem reasonable to restrict the available rules to be symmetric, or invariant, with respect to those operations also. For example, consider the problem of estimating a parameter θ, which is a location parameter for the distribution of X, and in which the loss function is $L(\theta, a) = c(\theta - a)^2$. If the statistician is willing to estimate $a = \theta_0$ when $X = x_0$ is observed ($\hat{\theta}(x_0) = \theta_0$), then he should be willing to estimate $\theta_0 + 5$ when $X = x_0 + 5$ is observed; for his original problem is the same as the problem of estimating an unknown parameter φ ($\varphi = \theta - 5$), which is a location parameter for the distribution of Y ($Y = X - 5$), when the loss function is $L(\varphi, b) = c(\varphi - b)^2$ ($b = a - 5$). This problem is the same as the original, so that the statistician should be willing to estimate $b = \theta_0$ when $Y = x_0$ is observed (that is, to estimate $a = \theta_0 + 5$ when $X = x_0 + 5$ is observed). Any estimate $\hat{\theta}(x)$ for which $\hat{\theta}(x + c) = \hat{\theta}(x) + c$ is called an "invariant estimate" of θ for this problem. Such an estimate is completely determined when the value of the estimate is given at one point. It is not surprising, then, that in general a "uniformly best invariant estimate" of θ will exist in this situation. The invariance principle is the topic of Chapter 4.

Method 2. Ordering the Decision Rules. The statistician may, if he likes, invent a principle by which to act in choosing a decision rule. Such a principle will, in general, lead to an ordering of the available decision rules, and any such ordering may be considered a principle. Two important and useful principles are basic to the study of decision theory.

1. THE BAYES PRINCIPLE. The Bayes principle involves the notion of a

distribution on the parameter space Θ called a *prior distribution*. Two things are needed of a prior distribution τ on Θ. First, we must be able to speak of the *Bayes risk of a decision rule* δ *with respect to a prior distribution* τ, *namely*,

$$r(\tau, \delta) = ER(T, \delta), \tag{1.14}$$

where T is a random variable over Θ having distribution τ. Second, we need to be able to speak of the joint distribution of T and X and of the conditional distribution of T, given X, the latter being called *the posterior distribution of the parameter given the observations*. It is clear that with the usual definition of expectation any finite distribution τ on Θ (that is, any distribution τ that gives all its mass to a finite number of points of Θ) satisfies these two conditions. For the main theorems we intend to prove it is sufficient to take the space of prior distributions, to be denoted by Θ^*, as this set of finite distributions on Θ. (In fact, Theorems 2.9.2 and 2.10.3 are stated in strongest form when Θ^* is taken thus.) However, it is useful to speak of continuous prior distributions also. Hence we shall take Θ^* as any space of distributions τ on Θ that satisfies these two conditions and that (a) contains all finite distributions and (b) is linear (that is, $\tau_1 \in \Theta^*$ and $\tau_2 \in \Theta^*$ implies $\alpha\tau_1 + (1 - \alpha)\tau_2 \in \Theta^*$ for all $0 \leq \alpha \leq 1$).

In using the Bayes principle, the statistician acts as if the parameter were actually a random variable whose distribution he knows. For a fixed distribution $\tau \in \Theta^*$ the statistician prefers a rule δ_1 to a rule δ_2 if δ_1 has smaller Bayes risk. This sets up a linear ordering on the space of decision rules. A Bayes decision rule is one that is best with respect to this ordering (that is, which has the smallest Bayes risk).

Definition 1. A decision rule δ_0 is said to be Bayes with respect to the prior distribution $\tau \in \Theta^*$ if

$$r(\tau, \delta_0) = \inf_{\delta \in D^*} r(\tau, \delta). \tag{1.15}$$

The value on the right side of (1.15) is known as the *minimum Bayes risk*. Bayes rules may not exist even if the minimum Bayes risk is defined and finite for the same reason a smallest positive number does not exist. In such a case the statistician has to be satisfied with a rule whose Bayes risk is close to the minimum value.

Definition 2. Let $\epsilon > 0$. A decision rule δ_0 is said to be ϵ-Bayes with respect to the prior distribution $\tau \in \Theta^*$ if

$$r(\tau, \delta_0) \le \inf_{\delta \epsilon D^*} r(\tau, \delta) + \epsilon. \tag{1.16}$$

This procedure for choosing a decision rule derives its name from the eighteenth-century philosopher Thomas Bayes, who first suggested and investigated methods sometimes referred to as "inverse probability methods," which are basic to the study of what is now termed Bayes methods and with which every present-day probability student is acquainted. In the third and fourth decades of this century, when mathematical statistics as it is known today was being conceived and formulated with ever-increasing rigor, Bayes procedures were somewhat neglected and even frowned on. In fact, one of the few points of agreement between the antagonists R. A. Fisher and J. Neyman was that one would seldom have enough information about the unknown state of nature to assume a prior distribution for it and that lacking such prior probabilities the formula of Bayes would not be applicable. [See Neyman (1952), pg. 193.] Nowadays, Bayes procedures are treated with great respect. This is partly because of the large sample optimality of some of these rules and partly because of the basic theorem of decision theory, presented in Sec. 2.10, which states (I trust the reader will excuse the imprecise terms) that, in general, every really good decision rule is practically Bayes with respect to some prior distribution.

Another perhaps more direct cause of the renaissance in the use of Bayes procedures is the attention drawn to these methods by a still small but extremely vocal new school of statisticians, formed in the middle of this century and led by the Italian B. de Finetti and the American L. J. Savage, which has a somewhat different point of view. Members of this school, who may be called "subjectivists" in contrast to the "objectivists" of the old school, hold that it is not necessary to believe that nature actually chooses a state according to the prior distribution, but rather, and more generally, the prior distribution is viewed merely as a reflection of the belief of the statistician about where the true state of nature lies. In Sec. 1.4 it was shown that such prior distributions exist when Θ is finite, provided that the preference patterns of the statistician satisfy certain reasonable assumptions. The statistician naturally changes his belief after he acquires new information on the true value of the parameter by means of an experiment. His new belief should correspond to the posterior distribution of the parameter, given the outcome of the experiment.

2. THE MINIMAX PRINCIPLE. An essentially different type of ordering of the decision rules may be obtained by ordering the rules according to the worst that could happen to the statistician. In other words, a rule δ_1 is preferred to a rule δ_2 if $\sup_\theta R(\theta, \delta_1) < \sup_\theta R(\theta, \delta_2)$. This relation, "is preferred to," leads to a linear ordering of the space D^* of decision rules. A rule that is most preferred in this ordering is called a *minimax* decision rule.

Definition 3. A decision rule δ_0 is said to be *minimax* if

$$\sup_{\theta \in \Theta} R(\theta, \delta_0) = \inf_{\delta \in D^*} \sup_{\theta \in \Theta} R(\theta, \delta). \tag{1.17}$$

The value on the right side of (1.17) is called the *minimax value* or *upper value* of the game.

If "inf" is replaced by "min" and if "sup" is replaced by "max," the reader can easily see how the word "minimax" was coined. The reader may check that a decision rule δ_0 is minimax if, and only if,

$$R(\theta', \delta_0) \leq \sup_{\theta \in \Theta} R(\theta, \delta) \tag{1.18}$$

for all $\theta' \in \Theta$ and $\delta \in D^*$.

Even if the minimax value is finite, there may not be a minimax decision rule, so that the statistician may have to be satisfied with a rule whose maximum risk is within ϵ of the minimax value.

Definition 4. Let $\epsilon > 0$. A decision rule δ_0 is said to be *ϵ-minimax* if

$$\sup_{\theta} R(\theta, \delta_0) \leq \inf_{\delta} \sup_{\theta} R(\theta, \delta) + \epsilon. \tag{1.19}$$

More simply, δ_0 is ϵ-minimax if for all $\theta' \in \Theta$ and $\delta \in D^*$

$$R(\theta', \delta_0) \leq \sup_{\theta} R(\theta, \delta) + \epsilon. \tag{1.20}$$

Considering a statistical game (Θ, D, R) symmetrically, we can also define a minimax rule for the other player, nature. Such a rule is an element of the space Θ^* of prior distributions. In the terminology of decision theory a minimax strategy for nature is called a least favorable distribution.

Definition 5. A distribution $\tau_0 \in \Theta^*$ is said to be *least favorable* if

$$\inf_{\delta} r(\tau_0, \delta) = \sup_{\tau} \inf_{\delta} r(\tau, \delta). \tag{1.21}$$

The value on the right side of (1.21) is called the *maximin* or *lower value* of the game.

Again, there may not be a least favorable distribution. One of the advantages of allowing Θ^* to contain some continuous distributions is that a continuous distribution may then be least favorable. The name "least favorable" derives from the fact that if the statistician were told which prior distribution nature was using he would like least to be told a distribution τ_0 satisfying (1.21).

Exercises

1. Show that a decision rule δ_0 is minimax if, and only if, (1.18) holds for all $\theta' \in \Theta$ and $\delta \in D^*$.
2. Show that a decision rule δ_0 is ϵ-minimax if, and only if, (1.20) holds for all $\theta' \in \Theta$ and $\delta \in D^*$.
3. Show that a prior distribution τ_0 is least favorable if, and only if,

$$r(\tau_0, \delta') \geq \inf_{\delta} r(\tau, \delta)$$

for all $\delta' \in D^*$ and all $\tau \in \Theta^*$.

1.7 Geometric Interpretation for Finite Θ

We give a geometric interpretation of the fundamental problem of decision theory in the case in which the parameter space Θ is finite. This interpretation proves extremely useful in visualizing Bayes and minimax decision rules and in formulating the proofs of the main theorems. It can also be useful in situations in which Θ is infinite.

Suppose that Θ consists of k points, $\Theta = \{\theta_1, \theta_2, \cdots, \theta_k\}$ and consider the set S, to be called the *risk set*, contained in k-dimensional Euclidean

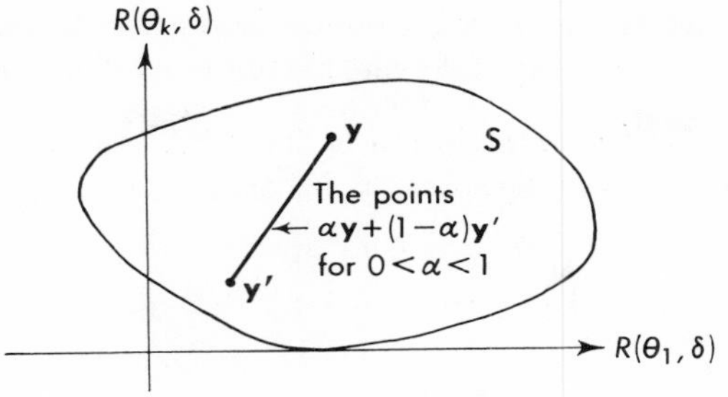

Fig. 1.1

space, E_k, of points of the form $(R(\theta_1, \delta), R(\theta_2, \delta), \cdots, R(\theta_k, \delta))$, where δ ranges through D^*. Formally, the risk set is

$$S = \{(y_1, \cdots, y_k): \quad \text{for some } \delta \in D^*, y_j = R(\theta_j, \delta) \quad \text{for } j = 1, \cdots, k\}. \tag{1.22}$$

If $k = 2$, this set may easily be plotted in the plane. If $k > 2$, we graph the set from the point of view of two coordinates only, always imagining the other coordinates and remembering that they may have an effect on what may be visually interpreted in two dimensions.

The first step in obtaining a picture of the fundamental problem is to note that a risk set must be convex. Recall that a subset A of Euclidean k-dimensional space is said to be convex if, whenever $\mathbf{y} = (y_1, \cdots, y_k)$ and $\mathbf{y}' = (y_1', \cdots, y_k')$ are elements of A, the points

$$\alpha\mathbf{y} + (1 - \alpha)\mathbf{y}' = (\alpha y_1 + (1 - \alpha)y_1', \cdots, \alpha y_k + (1 - \alpha)y_k'),$$

as α varies between zero and one, are also elements of A.

Lemma 1. The risk set S is a convex subset of E_k.

Proof. Let $\mathbf{y}$ and $\mathbf{y}'$ be arbitrary points of S. According to the definition of S, there exist decision rules δ and δ' in D^* for which $y_j = R(\theta_j, \delta)$ and $y_j' = R(\theta_j, \delta')$ for $j = 1, \cdots, k$. Let α be an arbitrary number between zero and one and consider the decision rule δ_α, which chooses a nonrandomized decision rule in D according to the distribution mixing δ with probability α and δ' with probability $1 - \alpha$. Clearly, $\delta_\alpha \in D^*$ and [see (1.11)]

$$R(\theta_j, \delta_\alpha) = \alpha R(\theta_j, \delta) + (1 - \alpha)R(\theta_j, \delta') \tag{1.23}$$

for $j = 1, 2, \cdots, k$. If $\mathbf{z}$ denotes the point whose jth coordinate is $R(\theta_j, \delta_\alpha)$, then $\mathbf{z} = \alpha\mathbf{y} + (1 - \alpha)\mathbf{y}' \in S$, completing the proof.

Later, a more precise description of the set S may be given; namely, that S is the convex hull of the set S_0, where S_0 is the *nonrandomized risk set*

$$S_0 = \{(y_1, \cdots, y_k)\colon \text{ for some } d \in D, y_j = R(\theta_j, d) \quad \text{for } j = 1, \cdots, k\}. \tag{1.24}$$

The convex hull of a set S_0 is defined as the smallest convex set containing S_0 or, alternatively, as the intersection of all convex sets containing S_0 (see Exercise 2). Lemma 1 immediately shows that S contains the convex hull of S_0. To show that the convex hull of S_0 contains S, the supporting hyperplane theorem is used. (See Corollary 2.7.1.)

Because the risk function contains all the pertinent information about a decision rule as far as we are concerned, the risk set S contains all the information about the decision problem. For a given decision problem (Θ, D^*, R) for Θ finite the risk set S is convex; conversely, for any convex set S in k-dimensions there is a decision problem, (Θ, D^*, R), in which Θ consists of k points, whose risk set is the set S. (For a proof of this

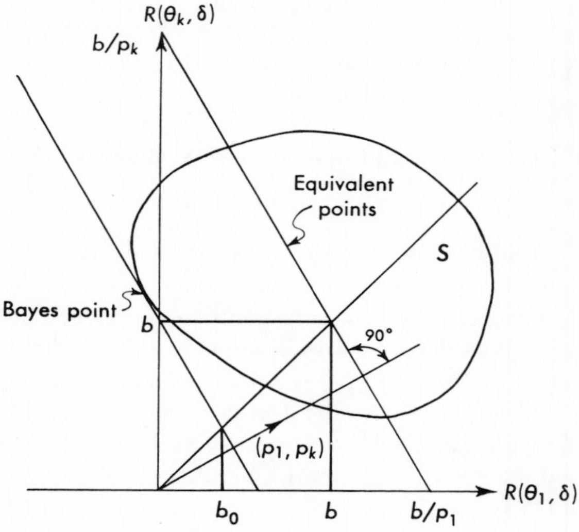

Fig. 1.2

statement see Exercise 2.7.1.) Consequently, we may discuss minimax and Bayes rules for the set S without reference to the associated decision problem.

Bayes Rules. A prior distribution for nature when Θ consists of k points is merely a k-tuple of nonnegative numbers $(p_1, p_2, \cdots, p_k)$ whose sum is one, with the understanding that p_j represents the probability that nature chooses θ_j. All points that yield the same expected risk,

$$\sum p_j R(\theta_j, \delta) = \sum p_j y_j,$$

are equivalent in the ordering given by the Bayes principle for the prior distribution $(p_1, p_2, \cdots, p_k)$. Thus all points on the plane $\sum p_j y_j = b$ for any real number b are equivalent. Every such plane is perpendicular to the vector from the origin to the point $(p_1, \cdots, p_k)$, and because each p_j is nonnegative the slope of the line of intersection of the plane $\sum p_i y_i = b$ with the coordinate planes cannot be positive (see Fig. 1.2). The quantity b can best be visualized by noting that the point of intersection of the diagonal line $y_1 = y_2 = \cdots = y_k$ with the plane $\sum p_j y_j = b$ must occur at $(b, b, \cdots, b)$. To find the Bayes rules we find the infimum of those values of b, call it b_0, for which the plane $\sum p_j y_j = b$ intersects the set S. Decision rules corresponding to points in this intersection are Bayes rules with respect to the prior distribution $(p_1, \cdots, p_k)$. Of course, Bayes rules do not exist when the set S does not contain its boundary points. On the other hand, there may be many Bayes points with respect

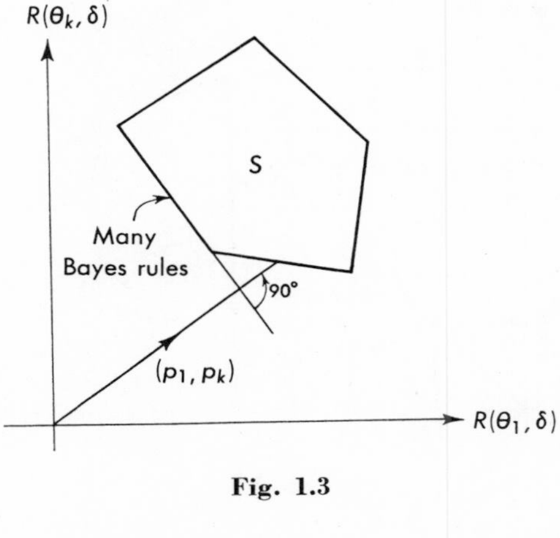

Fig. 1.3

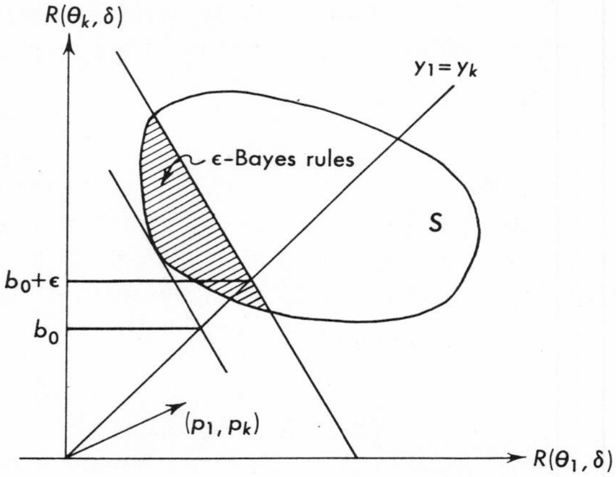

Fig. 1.4

to a given prior distribution, for example, in Fig. 1.3 or Fig. 1.6 if the prior distribution is given by $p_1 = 0, \cdots, p_k = 1$. For a fixed $\epsilon > 0$, the ϵ-Bayes rules correspond to points in S which are on or below the plane $\sum p_j y_j = b_0 + \epsilon$ (Fig. 1.4).

Minimax Rules. The maximum risk for a fixed δ is $\max_j R(\theta_j, \delta) = \max_j y_j$. Any points $\mathbf{y} \in S$ that give rise to the same value of $\max_j y_j$ are

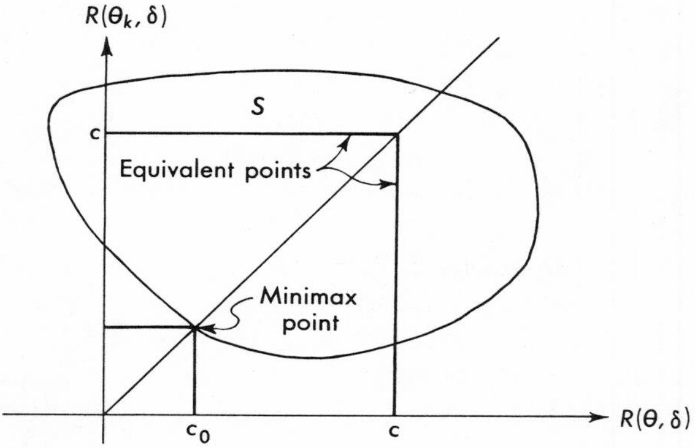

Fig. 1.5

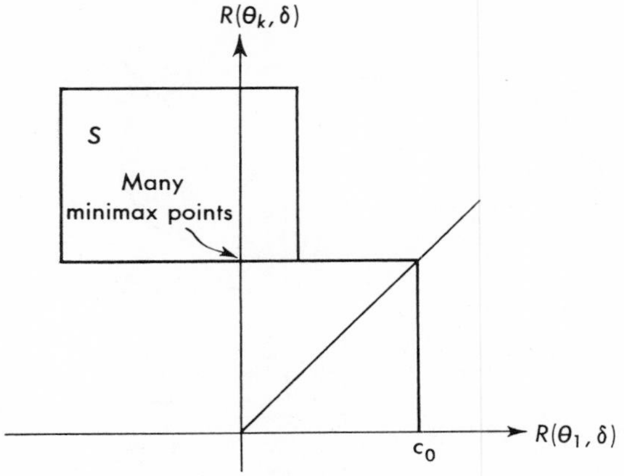

Fig. 1.6

equivalent in the ordering given by the minimax principle. Thus all points on the boundary of the set

$$Q_c = \{(y_1, \cdots, y_k) : y_i \leq c \quad \text{for } i = 1, \cdots, k\}$$

for any real number c are equivalent. To find the minimax rules we find the infimum of those values of c, call it c_0, such that the set Q_c intersects S. Any decision rule δ, whose associated risk point is an element of the intersection $Q_{c_0} \cap S$, is a minimax decision rule. Of course, minimax decision rules do not exist when the set S does not contain its boundary points. We note in Fig. 1.6 that there may be many minimax points for a given decision problem and that minimax points do not have to lie on the diagonal line $y_1 = y_2 = \cdots = y_k$. For a fixed $\epsilon > 0$ the rules corresponding to points in the intersection $Q_{c_0+\epsilon} \cap S$ are ϵ-minimax rules. The number c_0 is the minimax risk (Fig. 1.7).

A minimax strategy for nature, which is otherwise called a "least favorable distribution", may also be visualized geometrically. A strategy for nature is a prior distribution, $\tau = (p_1, \cdots, p_k)$, which, as we have seen, represents the family of planes perpendicular to $(p_1, \cdots, p_k)$. In using a Bayes rule to obtain $\inf_\delta r(\tau, \delta)$, the statistician finds the plane out of this family that is tangent to and below S. Because the minimum Bayes risk $\inf_\delta r(\tau, \delta)$ is b_0, where $(b_0, b_0, \cdots, b_0)$ is the intersection of the line $y_1 = y_2 = \cdots = y_k$ and the plane, tangent to and below S, and perpendicular to $(p_1, \cdots, p_k)$, a least favorable distribution is the choice

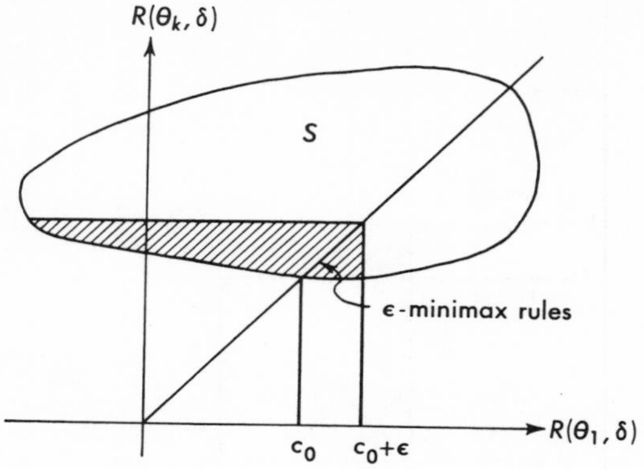

Fig. 1.7

of $(p_1, \cdots, p_k)$ that makes this intersection as far up the line as possible. It is clear that b_0 is not greater than c_0, the minimax risk, so that if we find a prior distribution whose minimum Bayes risk is c_0 this distribution must be least favorable.

As an example, consider the game of "odd or even" mentioned in Sec. 1.1. In this game $\Theta = \{\theta_1, \theta_2\}$, $\mathcal{A} = \{a_1, a_2\}$, and the loss is given by Table 1.6. A randomized strategy $\delta \in \mathcal{A}^*$ may be represented as a

Table 1.6

$L(\theta, a)$		a_1	a_2
	θ_1	-2	3
	θ_2	3	-4

number q, $0 \leq q \leq 1$, with the understanding that a_1 is taken with probability q and a_2, with probability $1 - q$. Thus, the "risk" set for the game $(\Theta, \mathcal{A}^*, L)$, is

$$S = \{(L(\theta_1, \delta), L(\theta_2, \delta)) : \delta \in \mathcal{A}^*\} = \{(3 - 5q, -4 + 7q) : 0 \leq q \leq 1\},$$

which is merely the line segment joining $(3, -4)$ and $(-2, 3)$ (Fig. 1.8). The minimax strategy occurs when $3 - 5q = -4 + 7q$ (that is, when

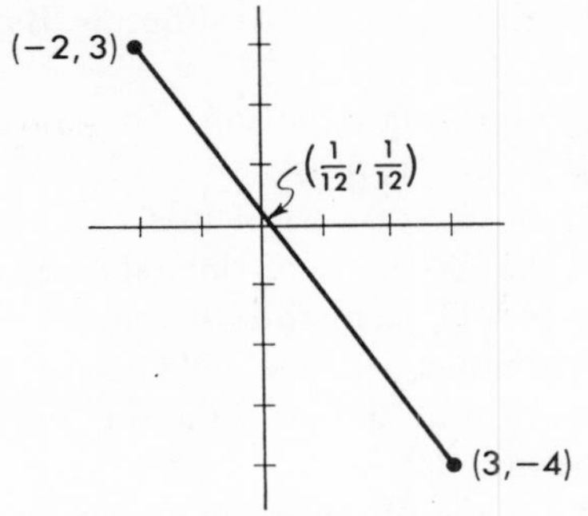

Fig. 1.8

$q = 7/12$) and the minimax risk is $3 - 5(7/12) = 1/12$. Is this minimax rule also a Bayes rule? Yes, with respect to a prior distribution taking θ_1 with probability p and θ_2 with probability $1 - p$ if the vector $(p, 1 - p)$ is perpendicular to the line S. Because the slope of the vector $(p, 1 - p)$ must then be the negative of the reciprocal of the slope of S, we find that $(1 - p)/p = 5/7$ or $p = 7/12$. In fact, every $\delta \in \mathcal{A}^*$ is Bayes with respect to this prior distribution, which is easily seen to be

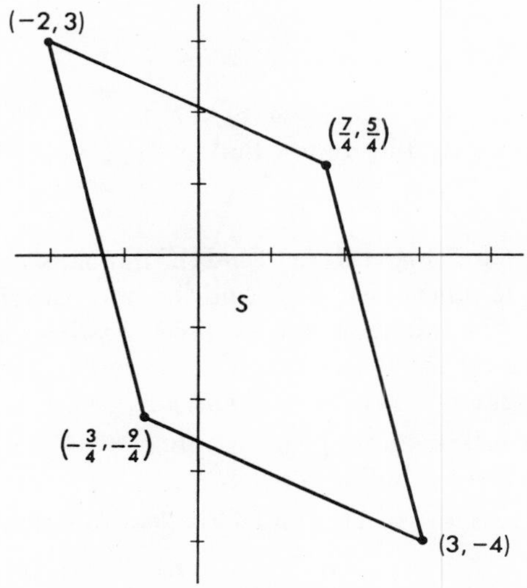

Fig. 1.9

a least favorable distribution, for its minimum Bayes risk is equal to the minimax risk, namely 1/12.

This is a rather interesting situation: The statistician has a rule that restricts his expected loss (at most) to 1/12, and nature has a rule that maintains the statistician's expected loss (at least) at 1/12. It seems reasonable to call 1/12 the value of the game. If a referee were to arbitrate this game, it would seem fair to require the statistician to pay nature 1/12. The minimax theorem (Theorem 2.9.1) shows that this situation occurs in all finite games (Θ and $\mathcal{A}$ finite). Further discussion may be found in Section 2.2.

In Section 1.3 the game of odd or even was extended to a statistical decision problem (Θ, D, R) in which the risk function is given by Table 1.7. As in Fig. 1.9, the risk set S must contain all lines between any two

Table 1.7

	d_1	d_2	d_3	d_4
θ_1	-2	$-3/4$	$7/4$	3
θ_2	3	$-9/4$	$5/4$	-4

$R(\theta, d)$

of the points $(-2, 3)$, $(-3/4, -9/4)$, $(7/4, 5/4)$, $(3, -4)$. Because S is convex, it is exactly the convex hull of these four points.

Exercises

1. Using the risk set of Fig. 1.9: (a) Find the minimax rule for the statistician and the minimax risk; (b) Find the prior distribution with respect to which the minimax rule is Bayes. Is this distribution least favorable? (c) Find the Bayes rule with respect to the prior distribution giving weight 1/2 to each of θ_1 and θ_2.
2. Show that the intersection of any number of convex sets is convex, hence that the convex hull is convex.
3. Let $\Theta = \{0, 1\}$, $\mathcal{A} = \{0, 1\}$, and let the loss function be

$$L(0, 0) = L(1, 1) \doteq 0, \qquad L(1, 0) = L(0, 1) = 1.$$

(The statistician tries to guess what nature has chosen. If he guesses correctly, he loses nothing. If he guesses incorrectly, he loses one.) Suppose the statistician observes the random variable X with the discrete distribution

$$P(X = x \mid \theta) = 2^{-k} \quad \text{if} \quad x = k - \theta \quad \text{for} \quad k = 1, 2, \cdots.$$

Describe the set of all nonrandomized decision rules of the statistician. Plot the risk set S in the plane. Find a minimax decision rule. Find a least favorable distribution. (Can you find a nonrandomized minimax rule?)

4. Consider the game of bluffing defined in Section 1.1. Assume that the statistician can take no observations on the true state of nature (or, equivalently, that X is degenerate at 0 for all θ). Show that if $P \geq (2a + b)/(2a + 2b)$, then I should use the bluff strategy and II should use the fold strategy. Show that if $P < (2a + b)/(2a + 2b)$ then II's minimax strategy is to fold with probability $b/(2a + b)$ [independent of $P(!)$] and I's minimax strategy (the least favorable distribution) is to bluff with probability $Pb/[(1 - P)(2a + b)]$.

1.8 The Form of Bayes Rules for Estimation Problems

In this section we discuss finding Bayes rules with special reference to estimation problems. A corresponding section on minimax rules is postponed to Section 2.11 after the main theorems have been developed.

One definite computational advantage that the Bayes approach has over the minimax approach to decision theory problems is that the search for good decision rules may be restricted to the class of nonrandomized decision rules. More precisely, *if a Bayes rule with respect to a prior distribution τ exists, there exists a nonrandomized Bayes rule with respect to τ; similarly for ϵ-Bayes rules.* This assertion may be given a rough proof. Suppose that $\delta_0 \in D^*$ is Bayes with respect to a distribution τ over Θ. Let Z denote the random variable with values in D whose distribution is given by δ_0. Then $r(\tau, \delta_0) = Er(\tau, Z)$ (after an appropriate change in the order of integration), but, because δ_0 is Bayes with respect to τ, $r(\tau, \delta_0) \leq r(\tau, d)$ for all $d \in D$. This entails $r(\tau, Z) = r(\tau, \delta_0)$ with probability one, so that any $d \in D$ that Z chooses will with probability one satisfy the equality $r(\tau, d) = r(\tau, \delta_0)$, implying that d is Bayes with respect to τ. A similar argument obviously works for ϵ-Bayes rules.

To make such a proof rigorous provision has to be made so that we can tell whether the interchange in the order of integration is legal. If either τ on Θ or δ_0 on D were finite, this interchange would be valid; but the general problem requires more structure on the loss function and distributions than we have given. We assume for the present section that this interchange of order of integration and those that follow are valid.

Given the prior distribution τ, we want to choose a nonrandomized decision rule $d \in D$ that minimizes the Bayes risk

$$r(\tau, d) = \int R(\theta, d)\, d\tau(\theta),$$

where $R(\theta, d)$ is the risk function

$$R(\theta, d) = \int L(\theta, d(x))\, dF_X(x \mid \theta).$$

A choice of θ by the distribution $\tau(\theta)$, followed by a choice of X from the distribution $F_X(x \mid \theta)$, determines (in general) a joint distribution of θ and X, which in turn can be determined (in general) by first choosing X according to its marginal distribution

$$F_X(x) = \int F_X(x \mid \theta)\, d\tau(\theta)$$

and then choosing θ according to the conditional distribution of θ, given $X = x$, $\tau(\theta \mid x)$. Hence by a change in the order of integration we may write

$$r(\tau, d) = \int \left[\int L(\theta, d(x))\, d\tau(\theta \mid x) \right] dF_X(x). \tag{1.25}$$

Given that these operations are legal, it is easy now to describe a Bayes decision rule. To find a function $d(x)$ that minimizes the double integral (1.25), we may minimize the inside integral separately for each x; that is, we may find for each x the action, call it $d(x)$, that minimizes

$$\int L(\theta, d(x))\, d\tau(\theta \mid x).$$

In other words, *the Bayes decision rule minimizes the posterior conditional expected loss, given the observation(s)*. If an action that achieves the infimum of

$$\int L(\theta, a)\, d\tau(\theta \mid x),$$

over $a \in \mathcal{A}$, does not exist, we must find an action that comes within ϵ of achieving this infimum, call this action $d(x)$, and if this forms a decision rule, it will obviously be ϵ-Bayes.

An example will help to clarify these ideas. Consider the estimation problem in which $\Theta = \mathcal{A} = (0, \infty)$ and $L(\theta, a) = c(\theta - a)^2$, where $c > 0$. Suppose the statistician observes the value of a random variable X having a uniform distribution on the interval $(0, \theta)$ with density

$$f(x \mid \theta) = \begin{cases} 1/\theta & \text{if} \quad 0 < x < \theta \\ 0 & \text{otherwise.} \end{cases} \tag{1.26}$$

We are to find a Bayes rule with respect to the prior distribution τ with density

$$g(\theta) = \begin{cases} \theta e^{-\theta} & \text{if} \quad \theta > 0 \\ 0 & \text{if} \quad \theta < 0, \end{cases} \tag{1.27}$$

a particular case of the gamma distribution (see Section 3.1). The joint density of X and θ is therefore

$$h(x, \theta) = g(\theta)\, f(x \mid \theta)$$

and the marginal distribution of X has the density

$$f(x) = \int h(x, \theta)\, d\theta = \begin{cases} e^{-x} & \text{if} \quad x > 0 \\ 0 & \text{if} \quad x < 0. \end{cases}$$

Hence the posterior distribution of θ, given $X = x$, has the density

$$g(\theta \mid x) = \frac{h(x, \theta)}{f(x)} = \begin{cases} e^{x-\theta} & \text{if} \quad \theta > x \\ 0 & \text{if} \quad \theta < x, \end{cases}$$

where $x > 0$. The posterior expected loss given $X = x$ is

$$E\{L(\theta, a) \mid X = x\} = ce^{x} \int_{x}^{\infty} (\theta - a)^2 e^{-\theta}\, d\theta.$$

To find the action a that minimizes this expression we may set the derivative with respect to a equal to zero.

$$\frac{d}{da} E\{L(\theta, a) \mid X = x\} = -2ce^{x} \int_{x}^{\infty} (\theta - a)e^{-\theta}\, d\theta = 0.$$

This implies

$$d(x) = a = \frac{\int_{x}^{\infty} \theta e^{-\theta}\, d\theta}{\int_{x}^{\infty} e^{-\theta}\, d\theta} = \frac{(x+1)e^{-x}}{e^{-x}} = x + 1.$$

This, therefore, is a Bayes decision rule with respect to τ: if $X = x$ is observed, "estimate" θ to be $x + 1$.

The problem of point estimation of a real parameter, using quadratic loss, occurs so frequently that it is worthwhile to make the following observation. The posterior expected loss, given $X = x$, for a quadratic loss function at action a is merely the second moment about a of the posterior distribution of θ given x.

$$E\{L(\theta, a) \mid X = x\} = c \int (\theta - a)^2\, d\tau(\theta \mid x).$$

This quantity is minimized by taking a as the mean of this distribution. Hence the Bayes decision rule is simply

$$d(x) = E(\theta \mid X = x).$$

The reader who is unacquainted with this fact should execute Exercise 1. It is important enough to state as a general rule.

Rule 1. In the problem of estimating a real parameter, θ, with loss proportional to squared error, a Bayes decision rule with respect to a given prior distribution for θ is to estimate θ as the mean of the posterior distribution of θ, given the observations.

This form of the Bayes rule is particular to decision problems involving squared error loss. Other loss functions may be investigated with profit.

A handy generalization of the squared error loss is the weighted squared error loss

$$L(\theta, a) = w(\theta)(\theta - a)^2,$$

where $w(\theta) > 0$ for all $\theta \in \Theta$. This analysis may be carried through eventually to yield, as a Bayes rule, the decision function

$$d(x) = \frac{E\{\theta w(\theta) \mid X = x\}}{E\{w(\theta) \mid X = x\}} = \frac{\int \theta w(\theta)\, d\tau(\theta \mid x)}{\int w(\theta)\, d\tau(\theta \mid x)}.$$

Another popular loss function is the absolute error loss $L(\theta, a) = c \mid \theta - a \mid$. For a given observed value of $X = x$ the Bayes decision rule $d(x)$ is the action a that minimizes

$$E\{L(\theta, a) \mid X = x\} = c \int \mid \theta - a \mid d\tau(\theta \mid x).$$

This quantity is minimized by taking a as the median of the posterior distribution of θ, given $X = x$ (see Exercise 2), which is also worth stating as a rule.

Rule 2. In the problem of estimating a real parameter θ, with loss proportional to absolute error, a Bayes rule with respect to a given prior distribution of θ is to estimate θ as the median of the posterior distribution of θ, given the observations.

These statements are given as "rules" rather than "theorems" because they are obviously not precisely stated. A precise statement would require conditions that would assert the existence of the joint distribution of θ and X, the finiteness of the integrals, and the interchangeability of the integrals.

Limit of Bayes, Generalized Bayes, and Extended Bayes Rules. There are several extensions to the notion of a Bayes rule. These extensions are useful because certain important decision rules are not Bayes rules in the narrow sense we have used so far.

As an example, let $\Theta = \mathcal{A} =$ the real line, let $L(\theta, a) = (\theta - a)^2$, and let the distribution of X, given θ, be normal with mean θ and variance one,

$$f_X(x \mid \theta) = \frac{1}{\sqrt{2\pi}} \exp\{-\tfrac{1}{2}(x - \theta)^2\}.$$

The usual estimate (that is, the maximum likelihood estimate) of θ is $d(X) = X$. A well-known property of this estimate is that it is unbiased; that is, $E_\theta\, d(X) = \theta$ for all $\theta \in \Theta$. This very fact, however, implies that it is not a Bayes rule, as the following argument shows. If $d(X)$ were Bayes with respect to a prior distribution τ, so that $d(X) = E\{\theta \mid X\}$, the two equations

$$E\theta\, d(X) = E(E\{\theta\, d(X) \mid X\}) = E(d(X)\, E\{\theta \mid X\}) = E\, d(X)^2$$

and

$$E\theta\, d(X) = E(E\{\theta\, d(X) \mid \theta\}) = E(\theta E\{d(X) \mid \theta\}) = E\theta^2$$

imply that

$$r(\tau, d) = E(\theta - d(X))^2 = E\theta^2 - 2E\,\theta\, d(X) + E\, d(X)^2 = 0.$$

But this is impossible, for

$$r(\tau, d) = E(\theta - d(X))^2 = E(E\{(\theta - X)^2 \mid \theta\}) = 1.$$

This contradiction implies that d is not a Bayes rule. (This argument may be generalized. See Exercise 1.8.6.)

For this example let us consider the Bayes rule with respect to the prior distribution τ_σ that is normal with mean zero and variance σ^2. The joint distribution of X and θ has density

$$h(\theta, x) = \frac{1}{2\pi\sigma} \exp\left[-\frac{(x - \theta)^2}{2} - \frac{\theta^2}{2\sigma^2}\right].$$

The marginal density of X is therefore

$$f(x) = [2\pi(1 + \sigma^2)]^{-1/2} \exp\left[-\frac{x^2}{2(1 + \sigma^2)}\right],$$

and the posterior density of θ given $X = x$ is

$$g(\theta \mid x) = \left(\frac{1 + \sigma^2}{2\pi\sigma^2}\right)^{1/2} \exp\left[-\frac{(1 + \sigma^2)}{2\sigma^2}\left(\theta - \frac{x\sigma^2}{1 + \sigma^2}\right)^2\right],$$

that is, normal with mean $x\sigma^2/(1 + \sigma^2)$ and variance $\sigma^2/(1 + \sigma^2)$. The Bayes rule with respect to τ_σ is

$$d_\sigma(x) = \frac{x\sigma^2}{1 + \sigma^2},$$

which has Bayes risk

$$r(\tau_\sigma, d_\sigma) = E(E\{(\theta - d_\sigma(X))^2 \mid X\}) = \frac{\sigma^2}{1 + \sigma^2}.$$

Thus, although $d(x) = x$ is not a Bayes rule, it is almost a Bayes rule in that the rules $d_\sigma(x)$ are Bayes and $d_\sigma(x) \to d(x)$ as $\sigma \to \infty$.

The following three extensions in the definition of a Bayes rule are useful in decision theory.

Definition 1. A rule δ is said to be a *limit of Bayes rules* δ_n, if for almost all x [that is, for $x \notin N$, where $P_\theta(N) = 0$ for all θ], $\delta_n(x) \to \delta(x)$ in the sense of distributions [that is, for any open set $A \subset \mathcal{A}$ whose boundary has $\delta(x)$-probability zero, $\delta_n(x)(A) \to \delta(x)(A)$]. In this definition it is assumed that $\mathcal{A}$ has a well-behaved topology and the decision rules are defined on the σ-field generated by the open sets.

For nonrandomized decision rules this definition becomes $d_n \to d$ if $d_n(x) \to d(x)$ for almost all x. For the example given for the normal distribution $d_\sigma \to d$ as $\sigma \to \infty$, so that d is a limit of Bayes rules. We make no essential use of this notion in this book.

A second extension in the notion of a Bayes rule may be obtained by generalizing the notion of a prior distribution to include nonfinite measures on Θ. When this is done, it is no longer easy to keep a probabilistic interpretation of the analysis; in particular, the marginal "distribution" of X may have infinite mass. Nonetheless, we can proceed by analogy and seek a function $\delta(x)$ which minimizes

$$\int L(\theta, \delta) f_X(x \mid \theta)\, d\tau(\theta),$$

where τ is a fixed measure on Θ and $f_X(x \mid \theta)$ is the density or probability mass function of X. Such a minimizing rule δ is called a generalized Bayes rule.

Definition 2. A rule δ_0 is said to be a *generalized Bayes rule* if there exists a measure τ on Θ (or a nondecreasing function on Θ if Θ is real), such that

$$\int L(\theta, \delta)\, f_X(x \mid \theta)\, d\tau(\theta)$$

takes on a finite minimum value when $\delta = \delta_0$.

In our example the rule d turns out to be generalized Bayes with respect to Lebesgue measure on the line, $\tau(\theta) = \theta$ or $d\tau(\theta) = d\theta$. The posterior distribution of θ,

$$f(\theta \mid x)\, d\theta = \frac{1}{\sqrt{2\pi}} \exp\left[-\tfrac{1}{2}(x - \theta)^2\right] d\theta,$$

is a normal distribution with mean x and variance one. The generalized Bayes rule with respect to Lebesgue measure is therefore $d(x) = x$. It might have been possible to guess that the measure with respect to which $d(x) = x$ is generalized Bayes would be Lebesgue measure, for d is a limit of Bayes rules with respect to measures τ_σ tending more and more to be uniform over the whole real line.

Our third extension in the notion of a Bayes rule is perhaps simplest of all.

Definition 3. A rule δ_0 is said to be *extended Bayes* if δ_0 is ϵ-Bayes for every $\epsilon > 0$.

In other words, δ_0 is extended Bayes if for every $\epsilon > 0$ there is a prior distribution τ such that δ_0 is ϵ-Bayes with respect to τ; that is, $r(\tau, \delta_0) \leq \inf_\delta r(\tau, \delta) + \epsilon$.

To show that the rule d of our example is extended Bayes we compute

$$r(\tau_\sigma, d) = E(\theta - X)^2 = E(E\{(\theta - X)^2 \mid \theta\}) = 1;$$

but because

$$\inf_\delta r(\tau_\sigma, \delta) = r(\tau_\sigma, d_\sigma) = \frac{\sigma^2}{1 + \sigma^2}$$

we have

$$r(\tau_\sigma, d) = \inf_\delta r(\tau_\sigma, \delta) + \epsilon \quad \text{for } \epsilon = \frac{1}{1 + \sigma^2}.$$

Thus d is ϵ-Bayes for every $\epsilon > 0$.

When it becomes necessary to use some notion of an "almost" Bayes rule in the sequel, we will choose the extended Bayes rules of Definition 3. This notion seems to be the easiest to deal with mathematically.

Exercises

1. Let Z be a random variable with finite second moment. Show that $f(b) = E(Z - b)^2$ is minimized when $b = EZ$.
2. Let Z be a random variable with finite first moment. Show that $f(b) = E \mid Z - b \mid$ is minimized when $b =$ any median of the distribution of Z.
3. Consider the decision problem with $\Theta = \mathcal{A} =$ real line and loss function

$$L(\theta, a) = \begin{cases} k_1 \mid \theta - a \mid & \text{if} \quad a \leq \theta \\ k_2 \mid \theta - a \mid & \text{if} \quad a > \theta, \end{cases}$$

   where $k_1 > 0$ and $k_2 > 0$. The function $f(b) = EL(Z, b)$ for an arbitrary random variable Z with finite first moment is minimized when b is a pth quantile of the distribution of Z. Find p as a function of k_1 and k_2 and state a rule for finding Bayes rules, using this loss function.
4. Let $\Theta = \mathcal{A} = (0, \infty)$ and $L(\theta, a) = c \mid \theta - a \mid$. Let the distribution of X be given by (1.26) and find a Bayes rule with respect to the prior distribution τ of (1.27). Compare this rule with the Bayes rule found in the text, using squared error loss.
5. Let $\Theta = \mathcal{A}$ be real line and let the loss function be

$$L(\theta, a) = \begin{cases} 0 & \text{if} \quad \mid \theta - a \mid \leq c \\ 1 & \text{if} \quad \mid \theta - a \mid > c, \end{cases}$$

   where $c > 0$. The function $f(b) = EL(Z, b)$ for an arbitrary random variable Z is minimized when b is the midpoint of the modal interval of length $2c$. Define "modal interval of length $2c$" so that this makes sense and state a rule for finding Bayes rules, using this loss function.

6. Let $\Theta = \mathcal{A}$ be a subset of the real line and let $L(\theta, a) = w(\theta)(\theta - a)^2$. Show that if a Bayes rule d with respect to a prior distribution τ is an unbiased estimate of θ and $Ew(\theta) < \infty$, then $r(\tau, d) = 0$.
7. Let $\Theta = (0, \infty)$, let $\mathcal{A}$ be the real line, and let $L(\theta, a) = (\theta - a)^2$. Let the distribution of X be Poisson with parameter $\theta > 0$,

$$f_X(x \mid \theta) = e^{-\theta} \frac{\theta^x}{x!} \qquad x = 0, 1, 2, \cdots.$$

Take as the prior distribution of θ the gamma distribution $\mathcal{G}(\alpha, \beta)$ (see Section 3.1) with density

$$g(\theta) = (\Gamma(\alpha)\beta^\alpha)^{-1} e^{-\theta/\beta} \theta^{\alpha-1} \quad \text{for} \quad \theta > 0,$$

where $\alpha > 0$ and $\beta > 0$.

(a) Show that the posterior distribution of θ given $X = x$ is the gamma distribution $\mathcal{G}(\alpha + x, \beta/(\beta + 1))$.

(b) The first two moments of the gamma distribution $\mathcal{G}(\alpha, \beta)$ are $\alpha\beta$ and $\alpha(\alpha + 1)\beta^2$. Show that the Bayes rule with respect to $\mathcal{G}(\alpha, \beta)$ is $d_{\alpha,\beta}(x) = \beta(\alpha + x)/(\beta + 1)$.

(c) Show that the usual (maximum likelihood) estimate $d(x) = x$ is not a Bayes rule. Note that if $\Theta = [0, \infty)$, $d(x) = x$ will be a Bayes rule.

(d) Show that the rule $d(x) = x$ is a limit of Bayes rules.

(e) Show that the rule $d(x) = x$ is a generalized Bayes rule with respect to the measure $\tau(\theta) = \log \theta$ [or $d\tau(\theta) = (1/\theta)\, d\theta$].

(f) Show that the rule $d(x) = x$ is an extended Bayes rule.

8. Let $\Theta = (0, 1)$ (the open interval from zero to one), let $\mathcal{A} = [0, 1]$ (the closed interval from zero to one), let $L(\theta, a) = (\theta - a)^2$, and let the distribution of X be binomial with n trials and probability θ of success,

$$f_X(x \mid \theta) = \binom{n}{x} \theta^x (1 - \theta)^{n-x}, \qquad x = 0, 1, \cdots, n.$$

Take as a prior distribution of θ the beta distribution $\mathcal{B}e(\alpha, \beta)$ (see Section 3.1) with density

$$g(\theta) = \frac{\Gamma(\alpha + \beta)}{\Gamma(\alpha)\,\Gamma(\beta)} \theta^{\alpha-1}(1 - \theta)^{\beta-1} \quad \text{for} \quad 0 < \theta < 1,$$

where $\alpha > 0$ and $\beta > 0$.

(a) Show that the posterior distribution of θ given $X = x$ is the beta distribution $\mathcal{B}e(\alpha + x, \beta + n - x)$.
(b) The first moment of the beta distribution $\mathcal{B}e(\alpha, \beta)$ is $\alpha/(\alpha + \beta)$ and the second moment is $\alpha(\alpha + 1)/[(\alpha + \beta)(\alpha + \beta + 1)]$. Show that the Bayes rule with respect to $\mathcal{B}e(\alpha, \beta)$ is

$$d_{\alpha,\beta}(x) = \frac{\alpha + x}{\alpha + \beta + n}.$$

(c) Show that the usual (maximum likelihood) estimate of θ, $d(x) = x/n$, is not a Bayes rule.
(d) Show that d is a limit of Bayes rules.
(e) Find a measure τ on $(0, 1)$ with respect to which d is generalized Bayes.
(f) Show that d is extended Bayes.

9. Consider the problem of Exercise 8 in which the loss is changed to $L(\theta, a) = (\theta - a)^2/[\theta(1 - \theta)]$. Show that the usual (maximum likelihood) estimate $d(x) = x/n$ is Bayes with respect to the uniform distribution on $(0, 1)$.

10. Let X and Y be independent with binomial distributions

$$f_{X,Y}(x, y \mid p_1, p_2) = \binom{n}{x}\binom{n}{y} p_1^x p_2^y (1 - p_1)^{n-x} (1 - p_2)^{n-y},$$

$$x = 0, 1, \cdots, n,$$

$$y = 0, 1, \cdots, n.$$

An estimate is required of the difference $p_1 - p_2$, using squared error loss $L((p_1, p_2), a) = (p_1 - p_2 - a)^2$, where $|a| \leq 1$, $0 \leq p_1 \leq 1$ and $0 \leq p_2 \leq 1$. Find the Bayes estimate with respect to the prior distribution which assigns independent uniform distributions on $(0, 1)$ to p_1 and p_2.

CHAPTER 2

The Main Theorems of Decision Theory

2.1 Admissibility and Completeness

The main theorems of decision theory involve two more notions we have not yet defined, namely the admissibility of a decision rule and the completeness of a class of decision rules. It is the object of this section to define these notions and discuss the relations between them.

Definition 1. *Natural ordering.* A decision rule δ_1, is said to be *as good as* a rule δ_2, if $R(\theta, \delta_1) \leq R(\theta, \delta_2)$ for all $\theta \in \Theta$. A rule δ_1, is said to be *better than* a rule δ_2 if $R(\theta, \delta_1) \leq R(\theta, \delta_2)$ for all $\theta \in \Theta$, and $R(\theta, \delta_1) < R(\theta, \delta_2)$ for at least one $\theta \in \Theta$. A rule δ_1, is said to be *equivalent to* a rule δ_2 if $R(\theta, \delta_1) = R(\theta, \delta_2)$ for all $\theta \in \Theta$.

The natural ordering gives a partial ordering of the space D^* (or $\mathfrak{D}$) of decision rules.

In any linear ordering of the decision rules there is a minimal requirement: that the linear ordering shall not disagree with the natural ordering "as good as" (that is, if δ_1 is as good as δ_2, then δ_2 is not to be preferred to δ_1). Both the minimax and Bayes orderings satisfy this requirement.

Definition 2. A rule δ is said to be *admissible* if there exists no rule better than δ. A rule is said to be *inadmissible* if it is not admissible.

Admissibility is an optimum property, although in a very weak sense. Conversely, we could never feel very proud about proving a rule optimum in some sense if it is inadmissible. Thus in a very real sense the word "admissible" is a synonym for the word "optimal." Note, however, that in a given decision problem every rule may be inadmissible, for example, when the risk set S does not contain its boundary points.

Definition 3. A class C of decision rules, $C \subset D^*$, is said to be *complete*, if, given any rule $\delta \in D^*$ not in C, there exists a rule $\delta_0 \in C$ that is better than δ. A class C of decision rules is said to be *essentially complete*, if, given any rule δ (not in C), there exists a rule $\delta_0 \in C$ that is as good as δ.

The difference between complete and essentially complete classes of decision rules may be illuminated by the following lemmas whose proofs are left as exercises.

Lemma 1. If C is a complete class, and A denotes the class of all admissible rules, then $A \subset C$.

Lemma 2. If C is an essentially complete class and there exists an admissible $\delta \notin C$, there exists a $\delta' \in C$ which is equivalent to δ.

Definition 4. A class C of decision rules is said to be *minimal complete* if C is complete and if no proper subclass of C is complete. A class C of decision rules is said to be *minimal essentially complete* if C is essentially complete and if no proper subclass of C is essentially complete.

We note that it is not necessary that minimal complete or minimal essentially complete classes exist.

The use of the notion of complete class is clear. If the statistician is presented with an (essentially) complete class, there is no need for him to look outside this class to find a decision rule, for he can do just as well inside the class. Thus the statistician can simplify his task by finding a small (essentially) complete class from which to make his choice. A smallest class may not exist, but if it does it is called a minimal (essentially) complete class and affords the maximal reduction of this problem. The following theorem clarifies the relationship between admissible rules and minimal complete classes.

Theorem 1. If a minimal complete class exists, it consists of exactly the admissible rules.

Proof. Let C denote a minimal complete class and let A denote the class of all admissible rules. We are to show $C = A$. Lemma 1 implies that $A \subset C$, because a minimal complete class is complete. We must show that $C \subset A$. This is done by assuming it to be false and arriving at a contradiction. Let $\delta_0 \in C$ and suppose that $\delta_0 \notin A$. We assert that there exists a $\delta_1 \in C$ that is better than δ_0. Because δ_0 is inadmissible, there exists a δ better than δ_0. If $\delta \in C$, we may take $\delta_1 = \delta$. If $\delta \notin C$, then, because C is complete, there exists a $\delta_1 \in C$ that is better than δ, hence better than δ_0. In either case our claim is verified. Now let $C_1 = C \sim \{\delta_0\}$. We will show that C_1 is complete, contradicting the fact that C is minimal. Let δ be an arbitrary rule not in C_1. If $\delta = \delta_0$, then $\delta_1 \in C_1$ is better than δ. If $\delta \neq \delta_0$, there exists a $\delta' \in C$ that is better than δ; if $\delta' = \delta_0$, then $\delta_1 \in C_1$ is better than δ; if $\delta' \neq \delta_0$, then $\delta' \in C_1$is better than δ. In any case, there exists an element of C_1 better than δ, which proves the completeness of C_1. This contradiction completes the proof.

Exercises

1. Prove Lemma 1.
2. Prove Lemma 2.
3. Show that every complete class is essentially complete.
4. Show, using the axiom of choice, that every minimal complete class contains a minimal essentially complete subclass.
5. Show that if C is complete and contains no proper essentially complete subclass, then C is minimal complete and minimal essentially complete.
6. Prove the following converse to the theorem of this section: if the class of admissible rules is complete, it is minimal complete.
7. Find a counterexample to the following statement: if C_1 and C_2 are complete, then $C_1 \cap C_2$ is essentially complete.

2.2 Decision Theory

We are now in a position to discuss the problems with which decision theory is concerned. At the most general level there is a large number of theorems that are valid for game theory. These theorems, presented in this chapter, use none of the structure given to the decision problem by the fact that a random quantity has been observed. They are valid in the game $(\Theta, \mathcal{A}, L)$ which may be viewed as (Θ, D, R) when the random variable X is degenerate at zero for all $\theta \in \Theta$. The two main problems at this level are the following:

1. The Minimax Theorem (The Fundamental Theorem of Game Theory). In what circumstances does a minimax rule for the statistician exist? Under what conditions is it true that

$$\sup_{\tau \in \Theta^*} \inf_{\delta \in D^*} r(\tau, \delta) = \inf_{\delta \in D^*} \sup_{\tau \in \Theta^*} r(\tau, \delta)? \tag{2.1}$$

This formula requires some comments. The left side represents the maximin or lower value of the game. If nature were a rational opponent plotting the ruin of the statistician, she could use a least favorable distribution, if it existed, which would guarantee her that the statistician's expected loss would be at least this amount (the lower value) no matter what decision rule he might use. However, even if a least favorable distribution did not exist, there does exist, for any given number less than the lower value, a prior distribution which guarantees that the statistician's expected loss is at least that number. The right side of (2.1) represents the minimax or upper value of the game as given in Definition 1.6.3, since for every $\delta \in D^*$

$$\sup_{\tau \in \Theta^*} r(\tau, \delta) = \sup_{\theta \in \Theta} R(\theta, \delta) \tag{2.2}$$

(the proof is an exercise). The statistician therefore has a rule which ensures him that his expected loss will not be greater than any preassigned number larger than the upper value, no matter what prior distribution nature decides to use. Now, suppose that the equality of (2.1) holds, denote this common value by V, and assume that $|V| < \infty$. Then, for every $\epsilon > 0$, the statistician has a rule δ_0 which keeps his expected loss at most to $V + \epsilon$, regardless of what nature does, and nature has a rule τ_0 which keeps the statistician's expected loss at least to $V - \epsilon$, regardless of what the statistician does. This is the situation that occurred in the game of odd or even analyzed in Section 1.7, except that we could choose $\epsilon = 0$. If we are lucky enough to find δ_0 and τ_0, which work for $\epsilon = 0$, then δ_0 is minimax and τ_0 is least favorable. Thus in the situation in which nature is a thinking opponent, that is to say, in game theory, it is of prime importance to know when (2.1) holds. If it does, we say that the game has a value and call the common quantity in (2.1) the value of the game, denoted by V. That (2.1) holds under quite general conditions is the central result of the theory of zero-sum two-person games.

In decision theory we cannot take the view that nature is dedicated to the ruin of the statistician, and the fact that (2.1) holds under general conditions is not so interesting. Nevertheless, the minimax theorem is useful in helping the statistician to discover a minimax decision rule in particular problems. The minimax theorem is also useful in answering another question it is natural to ask; namely, when are minimax rules also Bayes rules with respect to some prior distribution? If the minimax theorem holds and if a least favorable distribution τ_0 exists, then any minimax rule δ_0 is Bayes with respect to τ_0 (the proof is an exercise).

2. The Complete Class Theorem (The Fundamental Theorem of Decision Theory). When does the collection of Bayes rules (extended Bayes rules) form a complete class? What subset of the class of Bayes rules forms a minimal complete class?

An advantage that the Bayes principle has over most other principles leading to optimal rules (including the minimax principle) is that often it leads to rules that are relatively easy to compute. Thus it is especially important to know whether the statistician can restrict his attention to the class of Bayes rules, as he may if it forms a complete class or an essentially complete class. If the Bayes rules do form a complete class and a minimal complete class exists, it is then of interest to characterize the prior distributions whose corresponding Bayes rules form the minimal complete class. In this connection the theorem of Section 2.1 makes it important to answer another question; namely, when are Bayes rules admissible? This problem is discussed in Section 2.3.

In addition to these problems, there are others that arise at the most general level. Among them are the following. (a) When do Bayes rules exist? (b) Are minimax rules admissible? (c) Are all admissible rules also Bayes rules? (d) When does a minimal complete class exist? We give relatively complete answers to these questions in the case of finite Θ, for which the geometric interpretation of Section 1.7 comes in handy. Partial answers to some of these questions are also given in cases in which Θ is infinite.

Permission to use randomized decision rules is a great mathematical boon in our work, but it also has the disadvantage of greatly complicating the nature of the decision space from which we must choose. It is therefore important to know when it is possible to restrict attention to the class D of nonrandomized decision rules. Thus the question: when is the class D of nonrandomized decision rules essentially complete? This question is discussed in Section 2.8.

At the next most general level are problems concerned with taking advantage of any structure that exists in the distribution of the random quantity X observed by the statistician. Thus these problems are specific to decision theory and do not arise naturally in game theory itself. The most important and useful of these problems is concerned with the reduction obtained in basing the decision rule on sufficient statistics. This problem is discussed in Chapter 3. Related problems connected with the invariance principle and location and scale parameters are discussed in Chapter 4. Problems in which the space of actions of the statistician is given forms that lead to testing hypotheses and multiple decision problems are discussed in Chapters 5 and 6.

The last chapter is devoted to problems for which it is necessary to enlarge the structure of the decision problem beyond that described in the first chapter. Chapter 7 is concerned with sequential decision problems, from which we present a few basic details out of a wealth of known results.

Exercises

1. Show that for every $\delta \in D^*$,

$$\sup_{\tau \in \Theta^*} r(\tau, \delta) = \sup_{\theta \in \Theta} R(\theta, \delta).$$

2. *Minimax rules are Bayes.* Assume that (2.1) holds and that a least favorable distribution, τ_0, for nature exists. Show that any minimax rule, δ_0, for the statistician is also Bayes with respect to τ_0.
3. *Minimax rules are extended Bayes.* Assume that (2.1) holds. Show that any minimax rule, δ_0, for the statistician is an extended Bayes rule.

2.3 Admissibility of Bayes Rules

One of the questions posed in Section 2.2 can be given a satisfactory answer without introducing any further terminology; namely, when are Bayes rules admissible? One type of answer is given in Theorem 1 and another in Theorems 2 and 3.

The first theorem allows Θ, $\mathcal{A}$, L, and X to be quite arbitrary. The only requirement is that the Bayes rule be unique up to equivalence, (by equivalence we mean that of Definition 2.1.1). In other words, we require any two Bayes rules with respect to τ to have the same risk function.

Theorem 1. If for a given prior distribution τ a Bayes rule with respect to τ is unique up to equivalence, this Bayes rule is admissible.

The proof is left as an exercise.

Recalling the argument at the beginning of Section 1.8 that if there exists a randomized Bayes rule with respect to τ there exists a nonrandomized Bayes rule with respect to τ, we see that Theorem 1 applies essentially to nonrandomized rules. We note also a trivial improvement that may be made; namely, that in checking uniqueness we do not have to go outside the class of nonrandomized rules. Thus, if for a given prior distribution τ the Bayes rule d with respect to τ is unique up to equivalence among the nonrandomized rules, then d is admissible.

By contrast, the following theorem implies that certain randomized Bayes rules are admissible, but it is applicable only in special situations. The basic idea occurs when the parameter space is finite; $\Theta = \{\theta_1, \cdots, \theta_k\}$. A prior distribution is denoted by the k-tuple $(p_1, \cdots, p_k)$, where $p_j \geq 0, j = 1, \cdots, k$ and $\sum_1^k p_j = 1$, with the interpretation that θ_j is chosen with probability p_j for $j = 1, \cdots, k$.

Theorem 2. Assume that $\Theta = \{\theta_1, \cdots, \theta_k\}$ and that a Bayes rule, δ_0, with respect to the prior distribution $(p_1, \cdots, p_k)$ exists. If $p_j > 0$ for $j = 1, \cdots, k$, then δ_0 is admissible.

Proof. Suppose that δ_0 is inadmissible; then there exists a $\delta' \in D^*$ which is better than δ_0; that is,

$$R(\theta_j, \delta') \leq R(\theta_j, \delta_0) \quad \text{for all} \quad j$$

and

$$R(\theta_j, \delta') < R(\theta_j, \delta_0) \quad \text{for some} \quad j.$$

Because all p_j are positive,

$$\sum_{j=1}^{k} p_j R(\theta_j, \delta') < \sum_{j=1}^{k} p_j R(\theta_j, \delta_0),$$

the strict inequality showing that δ_0 is not Bayes with respect to $(p_1, \cdots, p_k)$. This contradiction proves the theorem.

The following counterexample shows that δ_0 is not necessarily admissible if the hypothesis $p_j > 0$ for $j = 1, \cdots, k$ is violated. It also provides a counterexample to Theorem 1 if the Bayes rule is not unique. Suppose that $\Theta = \{\theta_1, \theta_2\}$, and that the risk set S of (1.22) is the square, $S = \{(y_1, y_2): 1 \leq y_1 \leq 2, 0 \leq y_2 \leq 1\}$. Consider the decision rules that are Bayes with respect to the prior distribution (p_1, p_2) with $p_1 = 1$ and $p_2 = 0$. Because $\sum p_j R(\theta_j, \delta) = R(\theta_1, \delta)$, it is clear that any decision rule that achieves the minimum value of $y_1 = R(\theta_1, \delta)$, namely the value $R(\theta_1, \delta) = 1$, will be a Bayes rule with respect to the prior distribution $(1, 0)$ (Fig. 2.1). Thus the rule δ_0 for which $R(\theta_1, \delta_0) = 1$, $R(\theta_2, \delta_0) = 1$, is Bayes with respect to $(1, 0)$ (and, incidentally, minimax), yet it is not admissible because the rule δ' for which $R(\theta_1, \delta') = 1$, $R(\theta_2, \delta') = 0$, is better than δ_0. (The set S would result, for example, if the variable X were degenerate at zero for all θ, if $A = \{a_1, a_2, a_3, a_4\}$ and if the loss function is given by Table 2.1. Here, δ_0 chooses a_2 with probability one and δ' chooses a_1 with probability one.)

Table 2.1

	a_1	a_2	a_3	a_4
θ_1	1	1	2	2
θ_2	0	1	0	1

The extension of Theorem 2 to infinite Θ requires a new concept. The principal ideas are contained in the case $\Theta = E_1$, in which E_1 is the real line or Euclidean 1-dimensional space; so attention is restricted to this case. A point $\theta_0 \in E_1$ is said to be in the *support* of a distribution τ on the real line if for every $\epsilon > 0$ the interval $(\theta_0 - \epsilon, \theta_0 + \epsilon)$ has positive probability, $\tau(\theta_0 - \epsilon, \theta_0 + \epsilon) > 0$. For example, the binomial distribution with n trials and probability p, $0 < p < 1$, of success has support consisting of the points $(0, 1, 2, \cdots, n)$, whereas the uniform distribution on the interval $[0, 1]$ has this interval as support. Note that a discrete distribution may also have an interval as support. For example, if r_1,

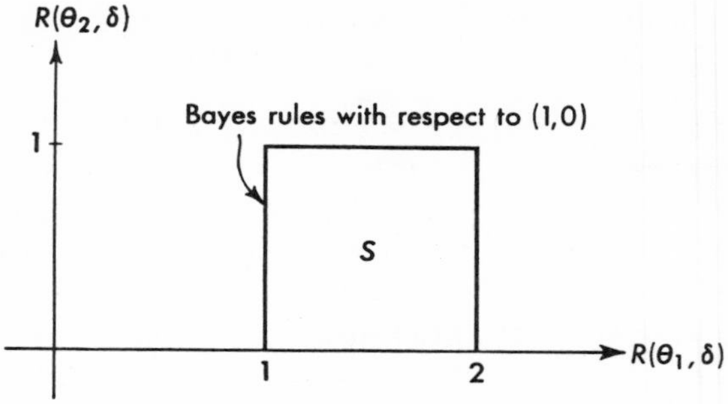

Fig. 2.1

$r_2, r_3, \cdots, r_n, \cdots$ represents an ordering of all the rationals on the interval $[0, 1]$, the distribution which gives probability $1/2^n$ to r_n is a discrete probability distribution whose support is the interval $[0, 1]$.

Theorem 3. Let $\Theta = E_1$ and assume that $R(\theta, \delta)$ is a continuous function of θ for all $\delta \in D^*$. If δ_0 is a Bayes rule with respect to a probability distribution τ on the real line, for which $r(\tau, \delta_0)$ is finite, and if the support of τ is the whole real line, then δ_0 is admissible.

Proof. As before, assume that δ_0 is not admissible and find a $\delta' \in D^*$ for which $R(\theta, \delta') \leq R(\theta, \delta_0)$ for all θ and $R(\theta_0, \delta') < R(\theta_0, \delta_0)$ for some $\theta_0 \in E_1$. Because $R(\theta, \delta)$ is continuous in θ for all δ, there exists an $\epsilon > 0$ for which $R(\theta, \delta') \leq R(\theta, \delta_0) - \eta/2$ (where $\eta = R(\theta_0, \delta_0) - R(\theta_0, \delta') > 0$) whenever $|\theta_0 - \theta| < \epsilon$. Then, letting T denote a random variable whose distribution is τ, we have

$$r(\tau, \delta_0) - r(\tau, \delta') = E(R(T, \delta_0) - R(T, \delta')) \geq \frac{\eta}{2} \tau(\theta_0 - \epsilon, \theta_0 + \epsilon);$$

but, since θ_0 is in the support of τ, the right side of the inequality is strictly greater than zero. This contradicts the assumption that δ_0 is a Bayes rule with respect to τ and completes the proof.

The reader may well be disturbed by the assumption in Theorem 3. After all, that is quite a few functions to expect to be continuous. We return to this question in Section 3.7, in which it is shown that this assumption is satisfied in many important cases.

Exercises

1. Prove Theorem 1.

2. The support of a distribution on E_1 is a closed set.

3. A rule δ_0 is *ϵ-admissible* if there does not exist a rule δ_1 for which $R(\theta, \delta_1) < R(\theta, \delta_0) - \epsilon$ for all $\theta \in \Theta$. Show that if δ_0 is ϵ-Bayes with $\epsilon > 0$ it is ϵ-admissible (no restriction on Θ or on the distribution τ with respect to which δ_0 is ϵ-Bayes).

2.4 Basic Assumptions

Further progress on the solutions to the problems mentioned in Section 2.2 is possible in the case of finite Θ after two important assumptions are made. These assumptions are that the risk set S,

$$S = \{(y_1, \cdots, y_k) : \text{for some } \delta \in D^*, y_j = R(\theta_j, \delta) \text{ for } j = 1, \cdots, k\}, \tag{2.3}$$

be bounded from below and closed from below. We proceed to make these notions precise.

Definition 1. A set, S, in k-dimensional Euclidean space, E_k, is said to be *bounded from below* if there exists a finite number, M, such that for every $\mathbf{y} = (y_1, \cdots, y_k) \in S$

$$y_j > -M \quad \text{for} \quad j = 1, \cdots, k. \tag{2.4}$$

Thus a set S is bounded from below if for each fixed j, $1 \leq j \leq k$, the coordinate y_j is bounded below as $\mathbf{y}$ ranges through S. The set of points in E_2

$$\{(y_1, y_2) : y_1 y_2 = 1, y_2 > 0\}$$

is bounded from below, whereas the set of points on the parabola

$$\{(y_1, y_2) : y_1^2 = y_2, y_2 > 0\}$$

is not bounded from below.

Definition 2. Let $\mathbf{x}$ be a point in E_k. The *lower quantant at* $\mathbf{x}$, denoted by $Q_{\mathbf{x}}$, is defined as the set

$$Q_{\mathbf{x}} = \{\mathbf{y} \in E_k: y_j \le x_j \text{ for } j = 1, \cdots, k\}. \tag{2.5}$$

Thus $Q_{\mathbf{x}}$ is the set of risk points as good as $\mathbf{x}$ and $Q_{\mathbf{x}} \sim \{\mathbf{x}\}$ is the set of risk points better than $\mathbf{x}$.

In the following definition $\bar{S}$ denotes the closure of the set S, so that $\bar{S}$ is the union of S and the set of all limit points of S or, alternatively, $\bar{S}$ is the smallest closed set containing S.

Definition 3. A point $\mathbf{x}$ is said to be a *lower boundary point* of a convex set $S \subset E_k$ if $Q_{\mathbf{x}} \cap \bar{S} = \{\mathbf{x}\}$. The set of lower boundary points of a convex set S is denoted by $\lambda(S)$.

The lower boundary of the unit disk

$$S_1 = \{(y_1, y_2): y_1^2 + y_2^2 \le 1\}$$

is

$$\lambda(S_1) = \{(y_1, y_2): y_1^2 + y_2^2 = 1, y_1 \le 0, y_2 \le 0\}.$$

The lower boundary of the unit square

$$S_2 = \{(y_1, y_2): 0 \le y_1 \le 1, 0 \le y_2 \le 1\}$$

is the set consisting of one point

$$\lambda(S_2) = \{(0, 0)\}.$$

Definition 4. A convex set $S \subset E_k$ is said to be *closed from below* if $\lambda(S) \subset S$.

The unit disk S_1 and the unit square S_2 are closed from below; in fact, it is clear that any closed convex set is closed from below (the proof is an exercise).

The importance of the lower boundary of the convex set S of (2.3) is that elements of $\lambda(S)$ lead to admissible decision rules in the following manner.

Theorem 1. If $\mathbf{x} = (R(\theta_1, \delta_0), \cdots, R(\theta_k, \delta_0))$ is in $\lambda(S)$, then δ_0 is admissible.

The proof is an exercise.

A partial converse is contained in Exercise 3. (See also Exercise 2.5.3.)

In the theorems for finite Θ, which follow in this chapter, we assume that the risk set S of (2.3) is both bounded from below and closed from below. Therefore it is of interest to investigate conditions placed on the space $\mathcal{A}$, the loss function L, and the distribution of the random variable X, which ensure that the risk set S is bounded from below and closed from below. The reader is referred to Blackwell and Girshick (1954), Theorem 6.2.1, for one set of conditions. We mention here that in one very important and frequently occurring case the risk set has the above-mentioned nice properties. If the space $\mathcal{A}$ is finite and the random variable X can assume only a finite number of values, the set D of nonrandomized decision rules is also finite, and Theorem 2 implies that S is bounded and closed, hence bounded from below and closed from below. It is useful here to prove something a little stronger, namely, that when the nonrandomized risk set S_0 of (1.24) is compact (that is, bounded and closed), its convex hull is also compact. The proof is based on the following lemma. Recall that the convex hull of a set S_0 is defined as the smallest convex set containing S_0, that is, the intersection of all convex sets containing S_0.

Lemma 1. The convex hull of a subset S_0 of k-dimensional Euclidean space, E_k, is the set of all convex linear combinations of at most $k + 1$ points of S_0, that is,

$$\{\mathbf{z}:\mathbf{z} = \sum_1^{k+1} \lambda_i \mathbf{y}_i,\ \mathbf{y}_i \in S_0,\ \lambda_i \geq 0,\ \sum \lambda_i = 1\}.$$

Proof. First, we note that the convex hull of a set S_0 may be defined as the set of all finite convex linear combinations of points of S_0, that is,

$$\{\mathbf{z}:\mathbf{z} = \sum_1^{r} \lambda_i \mathbf{y}_i,\ \mathbf{y}_i \in S_0,\ \lambda_i > 0,\ \sum \lambda_i = 1\}.$$

(This set is convex, yet must be contained in every convex set containing S_0.) That the set of convex linear combinations of at most $k + 1$ points of S_0 is contained in the convex hull of S_0 is clear.

To show the reverse inclusion let $\mathbf{z} = \sum_1^r \lambda_i \mathbf{y}_i$, with $\mathbf{y}_i \in S_0$, $\lambda_i > 0$, and $\sum \lambda_i = 1$, and suppose that $r > k + 1$. For any set of $k + 2$ points $\mathbf{y}_1, \cdots, \mathbf{y}_{k+2}$ in k-dimensional space there exist $\beta_1, \cdots, \beta_{k+2}$ not all zero such that $\sum_1^{k+2} \beta_i \mathbf{y}_i = \mathbf{0}$ and $\sum_1^{k+2} \beta_i = 0$. (To see this, note that there are two different nontrivial linear combinations for which $\sum \beta_i' \mathbf{y}_i = \mathbf{0}$ and $\sum \beta_i'' \mathbf{y}_i = \mathbf{0}$ and that $\beta_i = c_1\beta_i' + c_2\beta_i''$, for suitable c_1 and c_2, will do the job.) Let

$$\lambda_i' = \lambda_i + \epsilon\beta_i \qquad (\beta_i = 0 \quad \text{for } i > k + 2),$$

where $\epsilon > 0$ is chosen so that $\lambda_i' \geq 0$ for all i and $\lambda_i' = 0$ for some i. Then, because

$$\sum_1^r \lambda_i' = \sum_1^r \lambda_i + \epsilon \sum_1^r \beta_i = 1$$

and

$$\sum_1^r \lambda_i' \mathbf{y}_i = \sum_1^r \lambda_i \mathbf{y}_i + \epsilon \sum_1^r \beta_i \mathbf{y}_i = \mathbf{z},$$

we see that z can be represented as a convex linear combination of less than r points of S_0. This process may obviously be continued until $\mathbf{z}$ is a convex linear combination of at most $k + 1$ points of S_0, which completes the proof.

That we cannot write every point of a convex hull of a set S_0 in E_k as a convex linear combination of k points is most easily seen when $k = 1$ and $S_0 = \{0, 1\}$. The point $\frac{1}{2}$ in the convex hull of S_0 requires the combination $\frac{1}{2}(0) + \frac{1}{2}(1)$ of *two* points of S_0.

Frequently we will be able to show that the nonrandomized risk set S_0 is compact, as it is, for example, when D is finite. In the corollary of Lemma 2.7.3 it is shown that the risk set S is the convex hull of S_0. In conjunction with the following theorem, this implies that if S_0 is compact then S is also compact.

Theorem 2. The convex hull of a compact subset S_0 of E_k is compact.

Proof. Define $f(\lambda_1, \cdots, \lambda_{k+1}, \mathbf{y}_1, \cdots, \mathbf{y}_{k+1})$ on the compact subset $\{\lambda_i \geq 0, \sum \lambda_i = 1\} \times S_0 \times \cdots \times S_0$ of $E_{(k+1)^2}$ as equal to $\sum_1^{k+1} \lambda_i \mathbf{y}_i$. This is a continuous function defined on a compact set. Its image, the convex hull of S_0 from Lemma 1, therefore is compact.

Exercises

1. Show that any closed convex set is closed from below.

2. Prove Theorem 1.

3. Show that if S is closed and δ_0 admissible, then

$$(R(\theta_1, \delta_0), \cdots, R(\theta_k, \delta_0)) \in \lambda(S).$$

4. Give a counterexample to Exercise 3 if S is not assumed to be closed.

5. The *interior* of a set S, denoted by S^0, is defined as the set of all interior points of S or equivalently as the largest open set contained in S. The *boundary* of a set S, denoted by ∂S, is defined as the closure of S minus the interior of S, $\partial S = \bar{S} - S^0$. Extend Exercise 1 by showing that if S is a convex set the lower boundary of S is contained in the boundary of S, $\lambda(S) \subset \partial S$.

2.5 Existence of Bayes Decision Rules

The first question to be answered by using the basic assumptions of the preceding section concerns the existence of Bayes decision rules for specified prior distributions. Again, we assume that Θ is finite, $\Theta = \{\theta_1, \cdots, \theta_k\}$, and we denote a distribution over Θ by the k-tuple $(p_1, \cdots, p_k)$, where $p_j \geq 0$ for $j = 1, \cdots, k$ and $\sum p_j = 1$, with the interpretation that θ_j is chosen with probability p_j for $j = 1, \cdots, k$.

Theorem 1. Suppose that $\Theta = \{\theta_1, \cdots, \theta_k\}$ and that the risk set S is bounded from below and closed from below. Then, for every prior distribution $(p_1, \cdots, p_k)$ for which $p_j > 0$ for $j = 1, \cdots, k$, a Bayes rule with respect to $(p_1, \cdots, p_k)$ exists.

Proof. Let $(p_1, \cdots, p_k)$ be a distribution over Θ for which $p_j > 0$ for all j and let B denote the set of all numbers of the form $b = \sum p_j y_j$, where $\mathbf{y} = (y_1, \cdots, y_k) \in S$.

$$B = \{b = \sum_{j=1}^{k} p_j y_j \quad \text{for some} \quad \mathbf{y} \in S\}.$$

Because S is bounded from below, so is B; let b_0 be the greatest lower bound of B. In a sequence of points $\mathbf{y}^{(n)} \in S$ for which $\sum p_j y_j^{(n)} \to b_0$, each $p_j > 0$ implies that each sequence $y_j^{(n)}$ is bounded above. Thus there exists a finite limit point $\mathbf{y}^0$ of the sequence $\mathbf{y}^{(n)}$ and $\sum p_j y_j^0 = b_0$. We now show that $\mathbf{y}^0 \in \lambda(S)$. Clearly, because $\mathbf{y}^0$ is a limit point of points of S, $\mathbf{y}^0 \in \bar{S}$ and $\{\mathbf{y}^0\} \subset Q_{\mathbf{y}^0} \cap \bar{S}$. Furthermore, $Q_{\mathbf{y}^0} \cap \bar{S} \subset \{\mathbf{y}^0\}$, for if $\mathbf{y}'$ is any point of $Q_{\mathbf{y}^0}$ other than $\mathbf{y}^0$ itself, $\sum p_j y_j' < b_0$; so that if $\mathbf{y}' \in \bar{S}$, there would exist points $\mathbf{y}$ of S for which $\sum p_j y_j < b_0$. This contradicts the assumption that b_0 is a lower bound of B. Thus $Q_{\mathbf{y}^0} \cap \bar{S} = \{\mathbf{y}^0\}$, implying that $\mathbf{y}^0 \in \lambda(S)$.

Because S is closed from below, $\mathbf{y}^0 \in S$ and the minimum value of $\sum p_j R(\theta_j, \delta)$ is achieved by a point of S. The proof is completed by noting that any $\delta \in D^*$, for which $R(\theta_j, \delta) = y_j^0$ for $j = 1, \cdots, k$, is therefore a Bayes decision rule with respect to $(p_1, \cdots, p_k)$.

The restriction that the prior distribution $(p_1, \cdots, p_k)$ give positive mass to each state of nature cannot be dropped, as is shown in the following counterexample for the case $k = 2$. Suppose that S is the set

$$S = \{(y_1, y_2) : y_1 y_2 \geq 1,\ y_1 > 0\}.$$

This set is convex, bounded from below, and closed from below. Consider the prior distribution (p_1, p_2) with $p_1 = 1$ and $p_2 = 0$. Then $\sum p_j y_j = y_1$ so that the minimum Bayes risk over points of S is zero. Yet this minimum risk is not attained by a point of S, which shows that a Bayes rule with respect to $(1, 0)$ does not exist (Fig. 2.2).

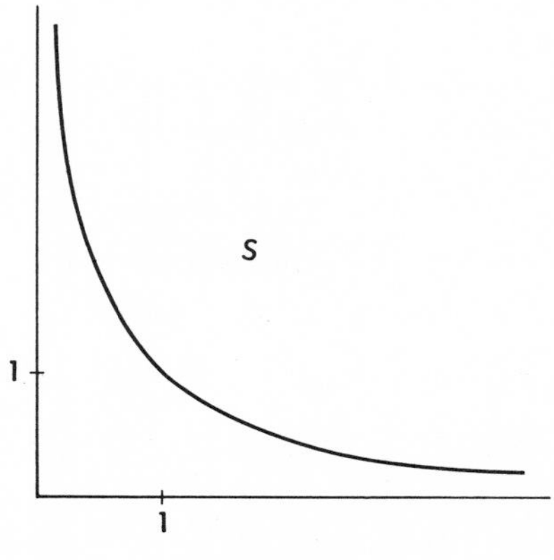

Fig. 2.2

However, if the hypotheses of this theorem are strengthened by requiring S to be bounded (above as well as below), it can be shown that Bayes solutions with respect to all prior distributions exist. The proof is an exercise.

In conclusion we give the following lemma, whose validity is implied by the first paragraph of the proof of the preceding theorem.

Lemma 1. If a nonempty convex set S is bounded from below, then $\lambda(S)$ is not empty.

This lemma is useful in Theorem 2.6.1 and in Exercises 2 and 3, which follow.

Exercises

1. Show that the closure of a convex set is convex.
2. Suppose that $\Theta = \{\theta_1, \cdots, \theta_k\}$ and that the risk set S is closed from below and bounded. Show that for every prior distribution $(p_1, \cdots, p_k)$ a Bayes rule with respect to $(p_1, \cdots, p_k)$ exists.
3. If S is bounded from below and closed from below and δ_0 is admissible, then $(R(\theta_1, \delta_0), \cdots, R(\theta_k, \delta_0)) \in \lambda(S)$.
4. Give a counterexample to Exercise (3), (a) if S is not required to be closed from below, (b) if S is not required to be bounded from below.

2.6 Existence of a Minimal Complete Class

The important question of the existence of a minimal complete class of decision rules is now answered in the affirmative when the set S is bounded below and closed below. The following theorem states that the class D_0 of decision rules, corresponding to points in the lower boundary $\lambda(S)$, is a minimal complete class.

Theorem 1. Suppose that $\Theta = \{\theta_1, \cdots, \theta_k\}$ and that the risk set S is bounded from below and closed from below. The class of decision rules

$$D_0 = \{\delta \in D^*: (R(\theta_1, \delta), \cdots, R(\theta_k, \delta)) \in \lambda(S)\} \tag{2.6}$$

is then a minimal complete class.

Proof. First we shall show that D_0 is a complete class. Let δ be any rule not in D_0 and let

$$\mathbf{x} = (R(\theta_1, \delta), \cdots, R(\theta_k, \delta)).$$

Then $\mathbf{x} \in S$, but $\mathbf{x} \notin \lambda(S)$. Let $S_1 = Q_\mathbf{x} \cap \bar{S}$; S_1 is nonempty, convex [since the closure of a convex set is convex (Exercise 2.5.1), and the intersection of two convex sets is convex (Exercise 1.7.2)], and bounded from below. Thus, from Lemma 2.5.1, $\lambda(S_1)$ is not empty. Let $\mathbf{y} \in \lambda(S_1)$; then $\{\mathbf{y}\} = Q_\mathbf{y} \cap \bar{S}_1$. Furthermore, $\mathbf{y} \in Q_\mathbf{x}$ because $\mathbf{y} \in \bar{S}_1 = \overline{Q_\mathbf{x} \cap \bar{S}} \subset \bar{Q}_\mathbf{x} = Q_\mathbf{x}$. Finally, $\mathbf{y} \in \lambda(S)$ because

$$\{\mathbf{y}\} = Q_\mathbf{y} \cap \bar{S}_1 = Q_\mathbf{y} \cap \overline{Q_\mathbf{x} \cap \bar{S}} = Q_\mathbf{y} \cap Q_\mathbf{x} \cap \bar{S} = Q_\mathbf{y} \cap \bar{S}.$$

Thus, because S is closed from below, there exists a rule $\delta_0 \in D_0$ for which

$$\mathbf{y} = (R(\theta_1, \delta_0), \cdots, R(\theta_k, \delta_0))$$

and which is better than δ, since $\mathbf{y} \in Q_\mathbf{x} \sim \{\mathbf{x}\}$. This proves D_0 complete.

From Theorem 2.4.1 every rule in D_0 is admissible. Hence no proper subclass of D_0 could be complete, because (see Lemma 2.1.1) every complete class must contain all admissible rules. This proves D_0 minimal complete and proves the theorem.

The following corollary of Theorem 1 is an immediate consequence of Theorem 2.1.1.

Corollary 1. The class D_0 consists exactly of the admissible rules.

Exercises

1. Find the minimal complete class of decision rules for the last example in Section 1.7.
2. Find the minimal complete class of decision rules for Exercise 1.7.3.

2.7 The Separating Hyperplane Theorem

Further progress in this chapter is impossible without the help of a famous theorem—one of the great theorems of mathematics—the separating hyperplane theorem, which states roughly that any two disjoint convex sets can be separated by a plane. Versions of this theorem are valid in quite general linear spaces, where it is equivalent to the Hahn-Banach theorem. An exposition containing much interesting related material may be found in Valentine (1964). A version valid for Euclidean spaces, entailing the convenience of the use of some vector notation, is presented in this section. We represent the elements of E_k by k-dimensional column vectors, written in boldface type, $\mathbf{x} \in E_k$. Its transpose, a row vector, is denoted by $\mathbf{x}^T = (x_1, x_2, \cdots, x_k)$. The inner product of two vectors $\mathbf{x}$ and $\mathbf{y}$ is written as $\mathbf{x}^T\mathbf{y} = \sum_1^k x_i y_i$. The origin, or zero vector, in E_k is denoted by $\mathbf{0} = (0, 0, \cdots, 0)^T$. We prove the separating hyperplane theorem by using some preliminary lemmas.

Lemma 1. If S is a closed convex subset of E_k and $\mathbf{0} \notin S$, there exists a vector $\mathbf{p} \in E_k$ such that $\mathbf{p}^T\mathbf{x} > 0$ for all $\mathbf{x} \in S$.

Remark. This is a version of the separating hyperplane theorem when one of the sets consists of one point, the origin $\mathbf{0}$. The origin lies in the plane $\mathbf{p}^T\mathbf{x} = 0$, and the inequality $\mathbf{p}^T\mathbf{x} > 0$ for all $\mathbf{x} \in S$ means that all points in the set S lie on one side of this plane.

Proof. For every real number $\alpha > 0$ let B_α be the sphere of radius α centered at the origin, $B_\alpha = \{\mathbf{x} \in E_k : \mathbf{x}^T\mathbf{x} \leq \alpha^2\}$. Let A be the set of real numbers $\alpha > 0$ for which B_α intersects S, $A = \{\alpha : B_\alpha \cap S \neq \emptyset\}$ ($\emptyset$ = the empty set). Because the lemma is trivial if S is empty, we consider only the case in which S, hence A, is nonempty. Then the greatest lower bound of A, $a = \text{glb } A$, is finite because A is nonempty and positive because S is closed and $\mathbf{0} \notin S$.

1. *$B_a \cap S$ is not empty.* As $\alpha \to a$ from above, $B_\alpha \cap S$ is a decreasing collection of nonempty compact sets whose limit $B_a \cap S$ is therefore nonempty.

Let $\mathbf{p}$ denote any point of $B_a \cap S$. (In fact, there is only one point in $B_a \cap S$, the point of S closest to the origin, but this fact is immaterial to our proof.)

2. *For all* $\mathbf{x} \in S$, $\mathbf{p}^T(\mathbf{x} - \mathbf{p}) \geq 0$. Let $f(\beta)$ denote the square of the distance from the origin to the point $\beta\mathbf{x} + (1 - \beta)\mathbf{p}$ for a fixed $\mathbf{x} \in S$, $\mathbf{x} \neq \mathbf{p}$.

$$\begin{aligned} f(\beta) &= (\beta\mathbf{x} + (1 - \beta)\mathbf{p})^T(\beta\mathbf{x} + (1 - \beta)\mathbf{p}) \\ &= \beta^2(\mathbf{x} - \mathbf{p})^T(\mathbf{x} - \mathbf{p}) + 2\beta\mathbf{p}^T(\mathbf{x} - \mathbf{p}) + \mathbf{p}^T\mathbf{p}. \end{aligned} \qquad (2.7)$$

This is a quadratic function of β whose square term has a positive coefficient. Thus there is a unique minimum value of $f(\beta)$, assumed when $\beta = \beta_0$;

$$\beta_0 = -\frac{\mathbf{p}^T(\mathbf{x} - \mathbf{p})}{(\mathbf{x} - \mathbf{p})^T(\mathbf{x} - \mathbf{p})}. \tag{2.8}$$

Because $f(1) = \mathbf{x}^T\mathbf{x} \geq \mathbf{p}^T\mathbf{p} = f(0)$, it is clear that $\beta_0 < 1$. Furthermore, since $\beta\mathbf{x} + (1 - \beta)\mathbf{p} \in S$ when $0 < \beta < 1$ from the convexity of S, it is clear that β_0 cannot be between zero and one without contradicting the fact that no point of S is closer to the origin than $\mathbf{p}$. Hence $\beta_0 \leq 0$ or, equivalently, $\mathbf{p}^T(\mathbf{x} - \mathbf{p}) \geq 0$.

This immediately completes the proof, for it entails $\mathbf{p}^T\mathbf{x} \geq \mathbf{p}^T\mathbf{p} > 0$ for all $\mathbf{x} \in S$.

The next lemma implies that a convex set and its closure have the same interior.

Lemma 2. If S is a convex subset of E_k, A is an open subset of E_k, and $A \subset \bar{S}$, then $A \subset S$.

Proof. Let $\mathbf{x}_0 \in A$; we are to show that $\mathbf{x}_0 \in S$. We translate x_0 to the origin: let

$$S' = \{\mathbf{x} = \mathbf{z} - \mathbf{x}_0, \mathbf{z} \in S\} \quad \text{and} \quad A' = \{\mathbf{x} = \mathbf{z} - \mathbf{x}_0, \mathbf{z} \in A\}.$$

Now, S' is a convex set, A' is open, and $A' \subset \bar{S}'$. Moreover, $\mathbf{0} \in A'$, and we will be finished when we show that $\mathbf{0} \in S'$.

Let $\mathbf{e}_i$ denote the unit vector with 1 in the ith component and 0 in the other components and let $\mathbf{1}$ denote the vector with 1 in each component. Because A' is open and $\mathbf{0} \in A'$, there exists an $\epsilon > 0$ such that $-\epsilon\mathbf{1}$ and $\epsilon\mathbf{e}_i$ for $i = 1, \cdots, k$ are elements of A'. Because $A' \subset \bar{S}'$, we may find $(k + 1)$ sequences $\mathbf{x}_i^{(n)}$, $i = 1, \cdots, k + 1$, of elements of S' for which $\mathbf{x}_i^{(n)} \to \epsilon\mathbf{e}_i$ for $i = 1, \cdots, k$ and $\mathbf{x}_{k+1}^{(n)} \to -\epsilon\mathbf{1}$ as $n \to \infty$. Let $\mathbf{X}^{(n)}$ represent the $k \times k$ matrix whose columns are the vectors $\mathbf{x}_1^{(n)}, \cdots, \mathbf{x}_k^{(n)}$, that is, $\mathbf{X}^{(n)} = (\mathbf{x}_1^{(n)}\mathbf{x}_2^{(n)} \cdots \mathbf{x}_k^{(n)})$. Clearly, $\mathbf{X}^{(n)} \to \epsilon\mathbf{I}$, where $\mathbf{I}$ is the identity matrix. Hence, from some n on, $\mathbf{X}^{(n)}$ will be nonsingular. Let $\boldsymbol{\mu}^{(n)} = [\mathbf{X}^{(n)}]^{-1}\mathbf{x}_{k+1}^{(n)}$, so that $\boldsymbol{\mu}^{(n)} \to -\mathbf{1}$. Hence, from some n on, all components of $\boldsymbol{\mu}^{(n)}$ will be negative. Let N be such a value of n. Then

$$\mathbf{x}_{k+1}^{(N)} = \mathbf{X}^{(N)}\boldsymbol{\mu}^{(N)} = \sum_1^k \mu_i^{(N)}\,\mathbf{x}_i^{(N)}.$$

In short, $\mathbf{x}_{k+1}^{(N)} + \sum_1^k(-\mu_i^{(N)})\mathbf{x}_i^{(N)} = \mathbf{0}$, so that $\mathbf{0}$ may be written as a linear combination with positive coefficients of elements of S'. Norming the coefficients so that their sum is one shows that $\mathbf{0} \in S'$, for S' is convex. This completes the proof.

The following theorem is similar to Lemma 1 but does not assume that S is closed and does not conclude that the inequality is strict. It is called *the supporting hyperplane theorem* when $\mathbf{x}_0$ is in the boundary of S because the hyperplane $\mathbf{p}^T\mathbf{x} = \mathbf{p}^T\mathbf{x}_0$ is then tangent to S and keeps S on one side.

Theorem 1. If S is a convex subset of E_k and $\mathbf{x}_0$ is not an interior point of S (that is, either $\mathbf{x}_0 \notin S$ or $\mathbf{x}_0$ is a boundary point of S), then there exists a vector $\mathbf{p} \in E_k$, $\mathbf{p} \neq \mathbf{0}$, such that $\mathbf{p}^T\mathbf{x} \geq \mathbf{p}^T\mathbf{x}_0$ for all $\mathbf{x} \in S$.

Proof. Because $\mathbf{x}_0$ is not an interior point of S, $\mathbf{x}_0$ is not an interior point of $\bar{S}$ from Lemma 2. Hence there is a sequence $\mathbf{y}_n \notin \bar{S}$ for which $\mathbf{y}_n \to \mathbf{x}_0$. We shall translate the origin to $\mathbf{y}_n$, successively, and apply Lemma 1. To this end, let

$$S_n = \{\mathbf{z}:\mathbf{z} = \mathbf{x} - \mathbf{y}_n,\ \mathbf{x} \in S\}.$$

Then $\bar{S}_n$ is a closed convex set (using Exercise 2.5.1), and $\mathbf{0} \notin \bar{S}_n$. From Lemma 1 there exist vectors $\mathbf{p}_n \in E_k$ such that $\mathbf{p}_n{}^T\mathbf{z} > 0$ for all $\mathbf{z} \in \bar{S}_n$, or, equivalently, such that $\mathbf{p}_n{}^T(\mathbf{x} - \mathbf{y}_n) > 0$ for all $\mathbf{x} \in \bar{S}$. Let $\mathbf{q}_n = \mathbf{p}_n/\sqrt{\mathbf{p}_n{}^T\mathbf{p}_n}$. Then $\mathbf{q}_n{}^T\mathbf{q}_n = 1$. Because the unit sphere in E_k is compact, there exists a limit point $\mathbf{p}$ of the $\mathbf{q}_n$ and a subsequence $\mathbf{q}_{n'} \to \mathbf{p}$. Hence $\mathbf{q}_{n'}{}^T(\mathbf{x} - \mathbf{y}_{n'}) \to \mathbf{p}^T(\mathbf{x} - \mathbf{x}_0)$; but $\mathbf{q}_{n'}{}^T(\mathbf{x} - \mathbf{y}_{n'}) > 0$ for all $\mathbf{x} \in S$ implies $\mathbf{p}^T(\mathbf{x} - \mathbf{x}_0) \geq 0$ for all $\mathbf{x} \in S$, as was to be proved.

Next, the famous separating hyperplane theorem. Its conclusion implies that the hyperplane $\mathbf{p}^T\mathbf{x} = c$, where $c = \inf_{\mathbf{x} \in S_1} \mathbf{p}^T\mathbf{x}$ divides E_k into two parts, one of which contains S_1 and the other, S_2.

Theorem 2. *The separating hyperplane theorem.* Let S_1 and S_2 be disjoint convex subsets of E_k. Then there exists a vector $\mathbf{p} \neq \mathbf{0}$ such that $\mathbf{p}^T\mathbf{y} \leq \mathbf{p}^T\mathbf{x}$ for all $\mathbf{x} \in S_1$ and all $\mathbf{y} \in S_2$.

Proof. Let $S = \{\mathbf{z}:\mathbf{z} = \mathbf{x} - \mathbf{y}$ for some $\mathbf{x} \in S_1$ and $\mathbf{y} \in S_2\}$.

1. *S is convex.* Let $\mathbf{z}_1$ and $\mathbf{z}_2$ be elements of S and let $0 < \beta < 1$. We

are to show $\beta\mathbf{z}_1 + (1 - \beta)\mathbf{z}_2 \in S$. Find $\mathbf{x}_1 \in S_1$, $\mathbf{x}_2 \in S_1$, $\mathbf{y}_1 \in S_2$, $\mathbf{y}_2 \in S_2$ such that $\mathbf{z}_1 = \mathbf{x}_1 - \mathbf{y}_1$ and $\mathbf{z}_2 = \mathbf{x}_2 - \mathbf{y}_2$. Then

$$\beta\mathbf{z}_1 + (1 - \beta)\mathbf{z}_2 = (\beta\mathbf{x}_1 + (1 - \beta)\mathbf{x}_2) - (\beta\mathbf{y}_1 + (1 - \beta)\mathbf{y}_2) \in S,$$

for $(\beta\mathbf{x}_1 + (1 - \beta)\mathbf{x}_2) \in S_1$ and $(\beta\mathbf{y}_1 + (1 - \beta)\mathbf{y}_2) \in S_2$ from convexity.

2. $\mathbf{0} \notin S$, for if $\mathbf{0} \in S$ there would be points $\mathbf{x} \in S_1$ and $\mathbf{y} \in S_2$ such that $\mathbf{x} - \mathbf{y} = \mathbf{0}$; that is, $\mathbf{x} = \mathbf{y}$ would contradict the assumption that S_1 and S_2 are disjoint.

3. From Theorem 1 there exists a vector $\mathbf{p} \neq \mathbf{0}$ such that $\mathbf{p}^T\mathbf{z} \geq 0$ for all $\mathbf{z}$ in S. Thus $\mathbf{p}^T(\mathbf{x} - \mathbf{y}) \geq 0$ for all $\mathbf{x} \in S_1$ and $\mathbf{y} \in S_2$, completing the proof.

The rest of this chapter may be considered as diverse applications of the separating hyperplane theorem (or Theorem 1). As a first application, we settle some questions raised in Section 1.7. In the following lemma $\mathbf{Z}$ represents a column vector whose transpose $(Z_1, \cdots, Z_k)$ is a k-tuple of random variables. The expectation of $\mathbf{Z}$ is denoted by $E\mathbf{Z} = (EZ_1, \cdots, EZ_k)^T$.

Lemma 3. If S is a convex subset of E_k and $\mathbf{Z}$ is a k-dimensional random vector for which $P(\mathbf{Z} \in S) = 1$ and for which $E\mathbf{Z}$ exists and is finite, then $E\mathbf{Z} \in S$.

Proof. Let $\mathbf{Y} = \mathbf{Z} - E\mathbf{Z}$ and let S' be the translation of S by $E\mathbf{Z}$, $S' = \{\mathbf{y}:\mathbf{y} = \mathbf{z} - E\mathbf{Z} \text{ for some } \mathbf{z} \in S\}$. Then S' is convex, $P\{\mathbf{Y} \in S'\} = 1$ and $E\mathbf{Y} = \mathbf{0}$. We will be finished when we show that $\mathbf{0} \in S'$.

We proceed by induction on k. The lemma is obviously true if $k = 0$, because then $\mathbf{Y}$ is degenerate at $\mathbf{0}$. Now suppose that the lemma is true for all nonnegative integers up to and including $k - 1$. We are to show the lemma is true for $k (k \geq 1)$.

Suppose that $\mathbf{0} \notin S'$. Then by Theorem 1 there exists a vector $\mathbf{p} \neq \mathbf{0}$ such that $\mathbf{p}^T\mathbf{y} \geq 0$ for all $\mathbf{y} \in S'$. Hence the random variable $U = \mathbf{p}^T\mathbf{Y}$ has expectation zero, yet $P\{U \geq 0\} = 1$. This implies that $P\{U = 0\} = 1$. Thus, with probability one, $\mathbf{Y}$ lies in the hyperplane $\mathbf{p}^T\mathbf{y} = 0$. Let $S'' = S' \cap \{\mathbf{y}:\mathbf{p}^T\mathbf{y} = 0\}$. Then S'' is a convex subset of $(k - 1)$-dimensional Euclidean space for which $P\{Y \in S''\} = 1$ and $E\mathbf{Y} = \mathbf{0}$. By the induction hypothesis $\mathbf{0} \in S''$. Because $S'' \subset S'$, this contradicts $\mathbf{0} \notin S'$ and completes the proof.

As a corollary, we prove that the risk set S is the convex hull of the nonrandomized risk set S_0 [see (1.24)]. In line with our representation of points in E_k as column vectors, we also take points of S and S_0 as column vectors. This should cause no confusion.

Corollary 1. S is the convex hull of S_0 [see (1.24)].

Proof. That S contains the convex hull of S_0 follows from Lemma 1.7.1. Now let C be any convex set containing S_0. We must show that C contains S. Let $\mathbf{z} \in S$. Then there is a $\delta \in D^*$ such that

$$\mathbf{z}^T = (R(\theta_1, \delta), \cdots, R(\theta_k, \delta));$$

but

$$(R(\theta_1, \delta), \cdots, R(\theta_k, \delta)) = E[(R(\theta_1, Z), \cdots, R(\theta_k, Z))],$$

where Z is a random variable taking values in D whose distribution is given by δ. Because $\mathbf{z}$ is thus the expectation of a random vector taking values in S_0, it follows from Lemma 3 that $\mathbf{z} \in C$. This proves $S \subset C$ as required.

Exercises

1. Let S be an arbitrary convex subset of E_k and consider the game $(\Theta, \mathcal{A}, L)$ for which $\Theta = \{\theta_1, \cdots, \theta_k\}$, $\mathcal{A} = S$, and $L(\theta_j, \mathbf{a}) = a_j =$ the jth component of vector $\mathbf{a}$. Suppose, further, that the random variable X, available to the statistician, is degenerate at zero for all values of $\theta \in \Theta$. The game (Θ, D^*, R) has risk set S.
2. Let Θ be finite and suppose that the set D is compact and that $R(\theta, d)$ is continuous in d for each $\theta \in \Theta$. The risk set S is bounded and closed.
3. Using Lemma 1, prove that if S_1 and S_2 are closed disjoint convex subsets of E_k and at least one of them is bounded there will be a vector $\mathbf{p} \in E_k$ such that $\sup_{\mathbf{y} \in S_2} \mathbf{p}^T\mathbf{y} < \inf_{\mathbf{x} \in S_1} \mathbf{p}^T\mathbf{x}$.
4. Find a counterexample to Exercise 3 if both sets are unbounded.
5. A convex set $S \subset E_k$ is said to be *strictly convex*, if, whenever $\mathbf{x}$ and $\mathbf{y}$ are in the boundary of S, $\beta\mathbf{x} + (1 - \beta)\mathbf{y}$ is in the interior of S for all $0 < \beta < 1$. Using Theorem 1, prove the following extension. If S is a strictly convex subset of E_k and $\mathbf{x}_0$ is not an interior point of S, then there exists a vector $\mathbf{p} \in E_k$ such that $\mathbf{p}^T\mathbf{x} > \mathbf{p}^T\mathbf{x}_0$ for all $\mathbf{x} \in S$, $\mathbf{x} \neq \mathbf{x}_0$.

2.8 Essential Completeness of the Class of Nonrandomized Decision Rules

In this and the following two sections we present three important applications of the separating hyperplane theorem. The first of these is to the proof of a general inequality due to Jensen (1906), which involves the notion of a convex function. A real-valued function, $f(\mathbf{x})$, defined on a convex subset S of E_k, is said to be *convex* if, whenever $\mathbf{x} \in S$, $\mathbf{y} \in S$ and $0 < \beta < 1$, the function satisfies the inequality

$$f(\beta\mathbf{x} + (1 - \beta)\mathbf{y}) \le \beta f(\mathbf{x}) + (1 - \beta) f(\mathbf{y}). \tag{2.9}$$

Thus a function is convex if the line segment joining any two points on its graph lies above or on the graph. The word "convex", when applied to a function, apparently refers to the fact that the graph of the function outlines a convex set; that is, the set

$$S_1 = \{(z_1, \cdots, z_{k+1})^T : \text{for some } \mathbf{x} \in S,\ \mathbf{x}^T = (z_1, \cdots, z_k) \text{ and } f(\mathbf{x}) \le z_{k+1}\} \tag{2.10}$$

is a convex set in E_{k+1}, as may easily be seen by applying the convexity of S and the inequality (2.9) (the proof is an exercise). Examples of convex functions on (any convex subset of) the real line are $f(x) = x^2$, $f(x) = |x|$, $f(x) = e^x$, and so on.

Lemma 1. (*Jensen's inequality.*) Let $f(\mathbf{x})$ be a convex real-valued function defined on a nonempty convex subset S of E_k, and let $\mathbf{Z}$ be a k-dimensional random vector with finite expectation $E\mathbf{Z}$ for which $P(\mathbf{Z} \in S) = 1$. Then $E\mathbf{Z} \in S$ and

$$f(E\mathbf{Z}) \le Ef(\mathbf{Z}). \tag{2.11}$$

The main idea of the proof may be viewed geometrically when $k = 1$ and $S = E_1$. This idea, which the student should retain, is as follows. The point $(EZ, f(EZ))$ is on the boundary of the convex set S_1 (see Fig. 2.3). Hence there exists a supporting hyperplane at $(EZ, f(EZ))$. Call this line $y = mx + b$. Because $(EZ, f(EZ))$ is on this line, it may be written $y = f(EZ) + m(x - EZ)$, and because this line is never above the curve $y = f(x)$ we have $f(x) \ge f(EZ) + m(x - EZ)$ for all x. Replacing x with Z and taking expectations, we find that

$$Ef(Z) \ge f(EZ) + m(EZ - EZ) = f(EZ).$$

We now give the formal proof of Jensen's inequality.

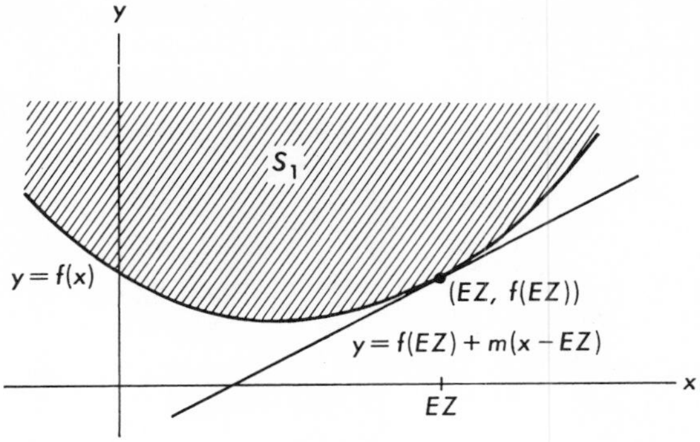

Fig. 2.3

Proof. We use an induction on k similar to that of the proof of Lemma 2.7.3. When $k = 0$, (2.11) is trivially satisfied, for $\mathbf{Z}$ is identically zero. Now suppose that the lemma is true for all nonnegative integers up to and including $k - 1$. We are to show that the lemma is true for k (where $k \geq 1$).

Lemma 2.7.3 implies that $E\mathbf{Z} \in S$. The point $(E\mathbf{Z}, f(E\mathbf{Z}))$ is a boundary point of the convex set S_1 of (2.10). Hence, by the supporting hyperplane theorem (Theorem 2.7.1), there exists a $(k + 1)$-dimensional vector $\mathbf{p} \neq \mathbf{0}$ such that for all $(z_1, \cdots, z_{k+1})^T \in S_1$

$$\sum_{j=1}^{k+1} p_j z_j \geq \sum_{j=1}^{k} p_j EZ_j + p_{k+1} f(E\mathbf{Z}). \tag{2.12}$$

First note that p_{k+1} cannot be negative, for letting $z_{k+1} \to \infty$ would eventually contradict inequality (2.12). Now by replacing z_{k+1} with $f(\mathbf{z})$ $(\mathbf{z}^T = (z_1, \cdots, z_k))$ and $\mathbf{z}$ with the random vector $\mathbf{Z}$, we rewrite (2.12) as

$$p_{k+1} f(E\mathbf{Z}) \leq p_{k+1} f(\mathbf{Z}) + \sum_{j=1}^{k} p_j (Z_j - EZ_j). \tag{2.13}$$

We distinguish two cases. If $p_{k+1} > 0$, we may take expectations on both sides of (2.13), divide by p_{k+1}, and obtain statement (2.11). If $p_{k+1} = 0$, (2.13) implies that the random variable $U = \sum_1^k p_j(Z_j - EZ_j)$ is nonnegative. Yet the expectation of U is zero. This implies that $P\{U = 0\} = 1$ or that $\mathbf{Z}$ gives all its mass to the $(k - 1)$-dimensional convex set $S^1 = S \cap \{z: \sum_1^k p_j(z_j - EZ_j) = 0\}$. The induction hypothesis then completes the proof.

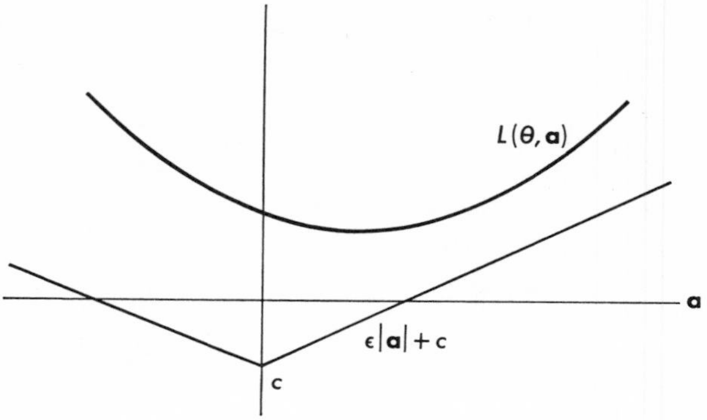

Fig. 2.4

Jensen's inequality is now used to prove the main objective of this section—a theorem on the completeness of the nonrandomized decision rules when the loss is convex.

Theorem 1. Let $\mathcal{A}$ be a convex subset of E_k and let $L(\theta, \mathbf{a})$ be a convex function of $\mathbf{a} \in \mathcal{A}$ for all $\theta \in \Theta$. If for some $\theta' \in \Theta$ there exists an $\epsilon > 0$ and a c such that $L(\theta', \mathbf{a}) \geq \epsilon \mid \mathbf{a} \mid + c$, then for every $P \in \mathcal{A}^*$, there exists an $\mathbf{a}_0 \in \mathcal{A}$ such that $L(\theta, \mathbf{a}_0) \leq L(\theta, P)$ for all $\theta \in \Theta$.

Proof. Let $P \in \mathcal{A}^*$ and let $\mathbf{Z}$ denote a random vector with values in $\mathcal{A}$ whose distribution is given by P. Then $E\mathbf{Z}$ is finite, for $\epsilon E \mid \mathbf{Z} \mid + c \leq EL(\theta', \mathbf{Z}) = L(\theta', P)$, which is finite by the definition of $\mathcal{A}^*$ (Definition 1.5.1). Thus we may apply Jensen's inequality to obtain

$$L(\theta, P) = EL(\theta, \mathbf{Z}) \geq L(\theta, E\mathbf{Z}) = L(\theta, \mathbf{a}_0), \tag{2.14}$$

where $\mathbf{a}_0 = E\mathbf{Z} \in \mathcal{A}$, thus completing the proof.

Remark. The strange-looking condition that the loss function be bounded below for some fixed θ by a function of the form $\epsilon \mid \mathbf{a} \mid + c$ (Fig. 2.4), where $\epsilon > 0$ ($\mid \mathbf{a} \mid$ represents the length of the vector $\mathbf{a}$), is equivalent in the context of the theorem to the condition that for some θ, $L(\theta, \mathbf{a}) \to \infty$ as $\mid \mathbf{a} \mid \to \infty$ (Exercise 3). Such a condition is automatically satisfied if $\mathcal{A}$ is a bounded set. This condition is used only to ensure that every element of $\mathcal{A}^*$ has a finite expected value. The theorem is not necessarily valid if this condition is removed, as the example in Exercise 4 shows.

This theorem has certain general implications on the completeness of the nonrandomized decision rules in the statistical decision problem $(\Theta, \mathfrak{D}, \hat{R})$, where the statistician chooses a behavioral decision rule in $\mathfrak{D}$. Elements of $\mathfrak{D}$ are maps from $\mathfrak{X}$ into $\mathcal{A}^*$—but because, under the conditions of the theorem for every element of $\mathcal{A}^*$ there is an element of $\mathcal{A}$ with no larger risk, we should be able to restrict attention to maps from $\mathfrak{X}$ into $\mathcal{A}$. In other words, *if $\mathcal{A}$ is a convex subset of E_k, and if, for all $\theta \in \Theta$, $L(\theta, \mathbf{a})$ is a convex function of $\mathbf{a} \in \mathcal{A}$, such that $L(\theta, \mathbf{a}) \to \infty$ as $|\mathbf{a}| \to \infty$, then the class of nonrandomized decision rules D is essentially complete for the statistical decision problem* $(\Theta, \mathfrak{D}, \hat{R})$. This statement, however, has a slight flaw; namely, given a mapping $\delta(x)$ from $\mathfrak{X}$ into $\mathcal{A}^*$, we may replace each $\delta(x) \in \mathcal{A}^*$ with an element $d(x) \in \mathcal{A}$ for which $L(\theta, d(x)) \leq L(\theta, \delta(x))$ for all θ[$d(x)$ may be taken as the mean of the distribution $\delta(x)$, as in Theorem 1]. But $\delta \in \mathfrak{D}$ does not imply that $d \in D$; in other words, $E_\theta L(\theta, d(X))$ may not exist because of lack of measurability, even though $E_\theta L(\theta, \delta(X))$ does exist. A proper answer to this difficulty involves discussion of the definition of the expectation E_θ and of the functions for which expectation exists. Such a discussion is beyond the scope of this book. We assume that these difficulties have been overcome.

For various other theorems which assert the essential completeness of the class of nonrandomized decision rules the reader is referred to Dvoretsky, Wald, and Wolfowitz (1951); as an example of the type of theorem to be found there, we mention the following. *If Θ is finite, if $\mathcal{A}$ is finite, and if $P_\theta(x)$ for each $\theta \in \Theta$ has no point masses, the class of nonrandomized rules is essentially complete.*

Exercises

1. Given a convex function $f(\mathbf{x})$ defined on a convex subset S of E_k, show that the set S_1 of (2.10) is convex.

2. Let $f(x)$ be a convex function defined on an interval of the real line. (a) Show that $f(x)$ is continuous at all interior points of the interval. (b) Show by a counterexample that $f(x)$ is not necessarily continuous at boundary points of the interval.

3. Let $\mathcal{A}$ be a convex subset of E_k and let $L(\mathbf{a})$ be a convex function of $\mathbf{a} \in \mathcal{A}$. Then there exists an $\epsilon > 0$ and a constant c such that $L(\mathbf{a}) \geq \epsilon |\mathbf{a}| + c$ if, and only if, $L(\mathbf{a}) \to \infty$ as $|\mathbf{a}| \to \infty$.

4. Let $\Theta = \mathcal{A} = (0, \infty)$ and let $L(\theta, a) = e^{-\theta a}$. (Regardless of the value of θ, the larger the statistician chooses a, the better.) Let

F be the distribution of any random variable Z on $(0, \infty)$ with infinite first moment. We may take $F \in \mathcal{A}^*$, for $L(\theta, F) = EL(\theta, Z)$ is finite for all θ. Show that there is no $a \in \mathcal{A}$ for which $L(\theta, a) \leq L(\theta, F)$ for all θ. (Check θ close to zero.)

5. (a) Let $f(x)$ be a function defined on the interval (a, b) with finite second derivative at every point of (a, b). Show that $f(x)$ is convex if and only if $f''(x) \geq 0$ for all $x \in (a, b)$.
(b) Let $f(\mathbf{x})$ be a function defined on an open convex subset S of E_k for which all second partial derivatives exist and are finite. Show that $f(\mathbf{x})$ is convex if, and only if, the matrix $\ddot{f}(\mathbf{x})$ of second partial derivatives

$$\ddot{f}(\mathbf{x}) = \left(\frac{\partial^2}{\partial x_i \partial x_j} f(\mathbf{x})\right)$$

is nonnegative definite for all $\mathbf{x} \in S$ (that is, $\mathbf{a}^T \ddot{f}(\mathbf{x})\mathbf{a} \geq 0$ for all $\mathbf{a} \in E_k$ and all $\mathbf{x} \in S$).
(c) Let $f(\mathbf{x})$ be a function defined on a convex subset S of E_k. Show that $f(\mathbf{x})$ is convex if, and only if, for all $\mathbf{x} \in S$ and $\mathbf{y} \in S$, $h(\alpha) = f(\alpha\mathbf{x} + (1 - \alpha)\mathbf{y})$ is a convex function of α for $0 \leq \alpha \leq 1$.

6. Show that $f(x, y) = -x^p y^{1-p}$ where $0 < p < 1$ is a convex function in the first quadrant $\{x > 0, y > 0\}$.

7. Use Exercise 6 to prove Hölder's inequality: for positive random variables X and Y with finite means, $EX^pY^{1-p} \leq (EX)^p(EY)^{1-p}$ for $0 \leq p \leq 1$.

8. Let $\Theta = \mathcal{A} = [0, 1]$ and let $L(\theta, a) = (\theta - a)^2$. For given $\theta \in \Theta$ let the observable random variable X have the binomial distribution, $f_X(x) = \binom{2}{x}\theta^x(1 - \theta)^{2-x}$, $x = 0, 1, 2$. Find a nonrandomized decision rule d which has smaller risk than the randomized rule $\delta \in D^*$ that chooses d_1 and d_2 each with probability 1/2, where $d_1(x) = x/2$ and $d_2(x) \equiv 1/2$. Find the risk functions $R(\theta, d)$ and $R(\theta, \delta)$.

9. A real-valued function, $f(\mathbf{x})$, defined on a convex subset S of E_k is said to be *strictly convex* if whenever $\mathbf{x} \in S$, $\mathbf{y} \in S$, and $0 < \beta < 1$, the function satisfies the strict inequality $f(\beta\mathbf{x} + (1 - \beta)\mathbf{y}) < \beta f(\mathbf{x}) + (1 - \beta) f(\mathbf{y})$. The function $f(x) = x^2$ on the real line is strictly convex, whereas $f(x) = |x|$ is convex but not strictly so. Prove that in Jensen's inequality if $f(\mathbf{x})$ is assumed strictly convex, and if $\mathbf{Z}$ is assumed to be nondegenerate it can be concluded that $f(E\mathbf{Z}) < Ef(\mathbf{Z})$, the inequality being strict.

10. If the nonrandomized risk set S_0 of (1.24) is closed and D is essentially complete, the risk set S is closed from below. (More generally, if $D' \subset D^*$ and D' is essentially complete and $S' = \{\mathbf{z}:\mathbf{z} = (R(\theta_1, \delta), \cdots, R(\theta_k, \delta)), \delta \in D'\}$ is closed, then S is closed from below.)

2.9 The Minimax Theorem

The second of our three main applications of the separating hyperplane theorem is to the fundamental theorem of game-theory. In this section we state and prove this theorem; in Section 2.11 we discuss methods of finding minimax rules.

Recall from Sections 1.6 and 2.2 that a decision rule δ_0 is said to be minimax if $\sup_\tau r(\tau, \delta_0) = \bar{V}$, where $\bar{V}$ is the upper value defined by

$$\bar{V} = \inf_{\delta \epsilon D^*} \sup_{\tau \epsilon \Theta^*} r(\tau, \delta). \tag{2.15}$$

Similarly, a prior distribution τ_0 is said to be least favorable if

$$\inf_{\delta} r(\tau_0, \delta) = \underline{V},$$

where $\underline{V}$ is the lower value defined by

$$\underline{V} = \sup_{\tau \epsilon \Theta^*} \inf_{\delta \epsilon D^*} r(\tau, \delta). \tag{2.16}$$

It is always true that

$$\underline{V} \leq \bar{V}, \tag{2.17}$$

because for all τ and all δ

$$\inf_{\delta' \epsilon D^*} r(\tau, \delta') \leq \sup_{\tau' \epsilon \Theta^*} r(\tau', \delta) \tag{2.18}$$

and because, after taking $\sup_\tau$ of the left side and $\inf_\delta$ of the right side, the inequality still holds. The minimax theorem asserts that under certain conditions equality actually holds in (2.17). When equality does hold in (2.17), we say the game has a value and the value is denoted by $V = \bar{V} = \underline{V}$. That not all games have a value is shown by an example following the theorem. Secondary assertions of the minimax theorem are that under certain conditions a minimax decision rule δ_0 and a least favorable distribution τ_0 actually exist.

Theorem 1. *The Minimax Theorem.* If for a given decision problem (Θ, D, R) with finite $\Theta = \{\theta_1, \cdots, \theta_k\}$ the risk set S is bounded below, then

$$\inf_{\delta \in D^*} \sup_{\tau \in \Theta^*} r(\tau, \delta) = \sup_{\tau \in \Theta^*} \inf_{\delta \in D^*} r(\tau, \delta) = V, \tag{2.19}$$

and there exists a least favorable distribution τ_0. Moreover, if S is closed from below, then there exists an admissible minimax decision rule δ_0, and δ_0 is Bayes with respect to τ_0.

Proof. It is always true that $\underline{V} \leq \bar{V}$; we must show that $\bar{V} \leq \underline{V}$. Let $V = \text{lub}\,\{\alpha : Q_{\alpha\mathbf{1}} \cap S = \emptyset\}$, where $\mathbf{1}$ (Fig. 2.5) is the vector with a one in each coordinate, $\mathbf{1} = (1, 1, \cdots, 1)^T$. Then for every n there exists a rule δ_n such that

$$R(\theta_j, \delta_n) \leq V + \frac{1}{n} \quad \text{for all } j. \tag{2.20}$$

Hence $r(\tau, \delta_n) \leq V + 1/n$ for all τ and $\sup_\tau r(\tau, \delta_n) \leq V + 1/n$ for all n, implying that $\bar{V} \leq V$. We will finish the proof of (2.19) when we show that $V \leq \underline{V}$.

Denote the interior of the set $Q_{V\mathbf{1}}$ by $Q_{V\mathbf{1}}{}^0$ and note that $Q_{V\mathbf{1}}{}^0$ and S are disjoint convex sets, so that there must be a hyperplane $\mathbf{p}^T\mathbf{x} = c$, which separates $Q_{V\mathbf{1}}{}^0$ and S, say $\mathbf{p}^T\mathbf{x} \geq c$ for $\mathbf{x} \in S$ and $\mathbf{p}^T\mathbf{x} \leq c$ for $\mathbf{x} \in Q_{V\mathbf{1}}{}^0$. Each coordinate p_j must be nonnegative, for if $p_j < 0$ for some

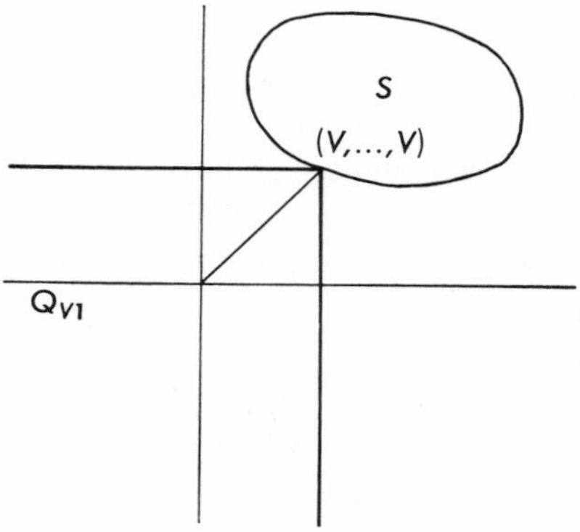

Fig. 2.5

j we may take $x_j \to -\infty$; the other coordinates of $\mathbf{x}$ are fixed, $\mathbf{x} \in Q_{V\mathbf{1}}{}^0$, so that $\mathbf{p}^T\mathbf{x} \to +\infty$, contradicting $\mathbf{p}^T\mathbf{x} \leq c$ for $\mathbf{x} \in Q_{V\mathbf{1}}{}^0$. Also, we may take $\sum p_j = 1$ (by dividing both sides of $\mathbf{p}^T\mathbf{x} = c$ by $\sum p_j > 0$). Thus $\mathbf{p}$ may be taken as a prior distribution, τ_0, for nature. Because $\mathbf{p}^T\mathbf{x} \leq c$ for all $\mathbf{x} \in Q_{V\mathbf{1}}{}^0$, letting $\mathbf{x} \to V\mathbf{1}$ implies that $V \leq c$. Thus for all δ

$$r(\tau_0, \delta) = \sum p_i R(\theta_i, \delta) \geq c \geq V, \tag{2.21}$$

so that $\underline{V} = \sup_\tau \inf_\delta r(\tau, \delta) \geq \inf_\delta r(\tau_0, \delta) \geq V$, thus completing the proof of (2.19). That τ_0 is least favorable follows from (2.21).

Now suppose, moreover, that S is closed from below. Consider the δ_n of (2.20) and let $\mathbf{y}_n = (R(\theta_1, \delta_n), \cdots, R(\theta_k, \delta_n))$. Because the $\mathbf{y}_n$ are bounded, there exists a limit point, $\mathbf{y}$, of $\mathbf{y}_n$. Obviously, $\mathbf{y} \in \bar{S}$. Then $Q_\mathbf{y} \cap \bar{S} \neq \emptyset$, and, from the lemma in Section 2.5, $\lambda(Q_\mathbf{y} \cap \bar{S}) \neq \emptyset$. Let $\mathbf{z} \in \lambda(Q_\mathbf{y} \cap \bar{S})$. Then, because $Q_\mathbf{y} \cap \bar{S} \cap Q_\mathbf{z} = \{\mathbf{z}\}$, we have $\mathbf{z} \in Q_\mathbf{y}$ and $Q_\mathbf{z} \cap \bar{S} = \{\mathbf{z}\}$ so that $\mathbf{z} \in \lambda(S)$. Because S is closed from below, $\mathbf{z} \in S$. Any δ_0 for which $\mathbf{z} = (R(\theta_1, \delta_0), \cdots, R(\theta_k, \delta_0))$ is admissible (Theorem 2.4.1) and satisfies $r(\tau, \delta_0) \leq V$, for $R(\theta_j, \delta_0) \leq V$ for all j. Furthermore, $r(\tau_0, \delta_0) = V$ from (2.21), which shows that δ_0 is Bayes with respect to τ_0.

Here is an example that shows that if Θ is not necessarily finite (2.19) of the minimax theorem does not necessarily hold.

Let $\Theta = \mathcal{A} = \{1, 2, 3, \cdots\}$, the set of all positive integers. Let

$$L(\theta, a) = \begin{cases} +1 & \text{if } a < \theta, \\ 0 & \text{if } a = \theta, \\ -1 & \text{if } a > \theta. \end{cases} \tag{2.22}$$

He who chooses the larger integer wins. Assume that the random variable X is degenerate at zero for all $\theta \in \Theta$. Then

$$\sup_\tau r(\tau, \delta) = 1 \qquad \text{for all } \delta, \tag{2.23}$$

and

$$\inf_\delta r(\tau, \delta) = -1 \quad \text{for all } \tau, \tag{2.24}$$

so that $\underline{V} = -1 \neq \bar{V} = +1$. In such a case we say that the game does not have a value.

In the extension of the minimax theorem to infinite Θ we employ the notion of a lower semicontinuous function. A real-valued function f is continuous if, and only if, the sets $\{x:f(x) < c\}$ and $\{x:f(x) > c\}$ are open sets for all real numbers c. A real-valued function f is said to be *lower semicontinuous* if for all real numbers c the set $\{x:f(x) > c\}$ is open, or, equivalently, if $\{x:f(x) \leq c\}$ is closed for all c. Recall that a set S is defined as *compact* if every open covering of S contains a finite subcovering (that is, if $\{O_\lambda\}$ is a family of open sets for which $S \subset \bigcup_\lambda O_\lambda$, there exists a finite subcollection $O_{\lambda_1}, \cdots, O_{\lambda_k}$ for which

$$S \subset \bigcup_{i=1}^{k} O_{\lambda_i}).$$

It is immediate from this definition that a decreasing sequence of nonempty closed sets contained in a compact set has a nonempty intersection.

The following two lemmas state some elementary facts about lower semicontinuous functions.

Lemma 1. If $f(x)$ is a lower semicontinuous function defined on a compact set, then $f(x)$ achieves its infimum. (In particular, f is bounded below.)

Proof. Let $K = \inf_x f(x)$ and let $K_1 > K_2 > \cdots$ be a decreasing sequence of numbers converging to K from above, $K_n \to K$. Then $\{x:f(x) \leq K_n\}$ is closed (hence compact—because a closed subset of a compact set is compact) and nonempty. Hence the intersection of these sets is also nonempty. For any x in this intersection, $f(x) = K$, as was to be proved.

The proof of the following lemma is an exercise. Note that $g(x)$ may take on infinite values.

Lemma 2. If $\{f_\theta(x), \theta \in \Theta\}$ is a family of lower semicontinuous functions, then $g(x) = \sup f_\theta(x)$ is lower semicontinuous (that is, the supremum of any family of lower semicontinuous functions is lower semicontinuous.)

In the following theorem it is assumed that we can find a topology on a set C for which C is compact and for which a large number of functions are lower semicontinuous. Any set having a topology with a finite number of open sets is compact; but for such topologies there are few continuous or lower semicontinuous functions. On the other hand, all functions are continuous on a set, given the discrete topology (in which all sets are open); but a set with a discrete topology is not compact unless it is finite. Thus the assumption on the topology of C in the following theorem is that a compromise is available: enough open sets so that the functions $R(\theta, \delta)$ are lower semicontinuous in $\delta \in C$ but not so many open sets that C is not compact.

Theorem 2. Let $C \subset D^*$ be an essentially complete class for the game (Θ, D, R). Assume that there is a topology on C such that (a) C is compact and (b) $R(\theta, \delta)$ is lower semicontinuous in $\delta \in C$ for all $\theta \in \Theta$. Then the game has a value and the statistician has a minimax strategy.

Proof. Let $\bar{V} = \inf_\delta \sup_\theta R(\theta, \delta)$ be the upper value of the game (Θ, D, R). If $\bar{V} = +\infty$, any rule is a minimax rule for the statistician. If $\bar{V} \neq \infty$, then, since $\sup_\theta R(\theta, \delta)$ is a lower semicontinuous function defined on a compact set, it achieves its infimum at some point $\delta_0 \in C$; $\sup_\theta R(\theta, \delta_0) = \bar{V}$, which implies that δ_0 is a minimax strategy.

To show that the game has a value, let M be an arbitrary fixed number, $M < \bar{V}$, and let $S_\theta = \{\delta \in C: R(\theta, \delta) > M\}$. From the lower semicontinuity of R, each set S_θ is an open subset of C. Furthermore, for each $\delta \in C$ there exists a $\theta \in \Theta$ such that $\delta \in S_\theta$, so that $\{S_\theta\}$ forms an open covering of C. Since C is compact, there exists a finite subcovering $\{S_{\theta_1}, S_{\theta_2}, \cdots, S_{\theta_k}\}$. Hence

$$\inf_{\delta \in C} \sup_i R(\theta_i, \delta) \geq M.$$

For the game (Θ_M, C, R), where $\Theta_M = \{\theta_1, \cdots, \theta_k\}$, the first part of Theorem 1 is applicable, since Lemma 1 implies that the risk set is bounded below. Therefore this game has a value $V_M \geq M$ and there exists a least favorable distribution $\{p_1, \cdots, p_k\}$ on $\{\theta_1, \cdots, \theta_k\}$; that is,

$$\inf_{\delta \in C} \sum_1^k p_i R(\theta_i, \delta) = V_M .$$

Because C is essentially complete, we have

$$\inf_{\delta \in D^*} \sum_1^k p_i R(\theta_i, \delta) = V_M \geq M,$$

and because it is valid for all $M < \bar{V}$ we have

$$\underline{V} = \sup_{\tau\epsilon\Theta^*} \inf_{\delta\epsilon D^*} r(\tau, \delta) \geq \bar{V},$$

which shows that the game has a value and completes the proof.

In the main applications of this theorem C is taken to be either the set D^* of all randomized decision rules or the set D of nonrandomized decision rules when we know that such a set is an essentially complete class. It should be noted that Θ^* may be taken to be the set of all finite distributions over Θ (since each $\Theta_M{}^*$ is such) and that the theorem is in its strongest form in this case.

Exercises

1. Prove (2.23) and (2.24). Show that in this example any rule $\delta \in D^*$ is minimax for the statistician.

2. Give an example of a game with a value (take Θ finite) for which there exists a minimax rule δ_0 that is Bayes with respect to some prior distribution τ_0, yet τ_0 is not least favorable. Show that if, in addition, $r(\tau_0, \delta_0) = V$, then τ_0 is least favorable.

3. Show that a real-valued function $f(x)$ is lower semicontinuous if and only if for every x_0 in its domain

$$\liminf_{x \to x_0} f(x) \geq f(x_0).$$

4. Prove Lemma 2.

5. Show, using Theorem 2, that if D is finite then the game has a value and the statistician has a minimax strategy.

6. Using Lemma 2.4.1, show that with finite $\Theta = \{\theta_1, \cdots, \theta_k\}$, if the risk set is bounded below and closed below, there exists a minimax rule for the statistician that is a mixture of at most $k + 1$ nonrandomized decision rules.

2.10 The Complete Class Theorem

The third main application of the separating hyperplane theorem is to the following converse to Theorem 2.3.2. It should be noted that no hypotheses are required of the risk set S for this theorem.

Theorem 1. If δ is admissible and Θ is finite, then δ is Bayes (with respect to some prior distribution).

Proof. If δ is admissible, then $Q_{\mathbf{x}} \cap S = \{\mathbf{x}\}$, where $\mathbf{x} = (R(\theta_1, \delta), \cdots, R(\theta_k, \delta))$. Thus, because $Q_{\mathbf{x}} \sim \{\mathbf{x}\}$ and S are disjoint convex sets, there exists a vector $\mathbf{p} \neq \mathbf{0}$ such that $\mathbf{p}^T\mathbf{y} \leq \mathbf{p}^T\mathbf{z}$ for all $\mathbf{y} \in Q_{\mathbf{x}} \sim \{\mathbf{x}\}$, and $\mathbf{z} \in S$. If some coordinate p_j of the vector $\mathbf{p}$ were negative, then, by taking $\mathbf{y}$ so that y_j is sufficiently negative, we would have $\sum p_j y_j > \sum p_j x_j$. Hence $p_j \geq 0$ for all j. We may normalize $\mathbf{p}$ so that $\sum p_j = 1$. Because $\mathbf{p}$ is now a probability distribution over Θ and $\sum p_j R(\theta_j, \delta) \leq \mathbf{p}^T\mathbf{z}$ for all $\mathbf{z} \in S$, δ is a Bayes rule with respect to $\mathbf{p}$.

Theorem 1 contains the essential part of Theorem 2, which is a version of the complete class theorem for finite θ.

Theorem 2. *The Complete Class Theorem.* If, for a given decision problem (Θ, D, R) with finite Θ, the risk set S is bounded from below and closed from below, then the class of all Bayes rules is complete and the admissible Bayes rules form a minimal complete class.

Proof. The complete class theorem follows immediately from Corollary 2.6.1 and Theorem 1 above.

This theorem may be extended to the case in which Θ is arbitrary in a manner analogous to the extension of the minimax theorem in the preceding section. As before, the risk functions $R(\theta, \delta)$ are assumed only to be lower semicontinuous, although no applications to other than continuous risk functions are presented in this book. However, such applications exist. Moreover, an exceedingly general complete class theorem due to Le Cam (1955) is based in part on the extended theorem (Theorem 3) and uses lower semicontinuity in an essential way. For these reasons, and because the proofs are not simplified by restricting the loss functions to be continuous, it seems worthwhile to require lower semicontinuity only.

Theorem 3. Under the assumptions of Theorem 2.9.2, the class of extended Bayes rules in C is essentially complete.

Proof. Let δ_0 be any decision rule and consider the game (Θ, D, W), in which $W(\theta, d) = R(\theta, d) - R(\theta, \delta_0)$. Then $W(\theta, \delta)$ is lower semicontinuous in $\delta \in C$, with C compact. Hence by Theorem 2.9.2 there exists a minimax decision rule δ_W in C which is extended Bayes with respect to the finite distributions over Θ (as in the proof of Theorem 2.9.2

or, using Exercise 2.2.3 with Θ^* as the set of finite distributions over Θ). It is therefore extended Bayes in the game (Θ, D, R); for

$$\sum_1^k p_i W(\theta_i, \delta_W) \leq \inf_\delta \sum_1^k p_i W(\theta_i, \delta) + \epsilon$$

implies

$$\sum_1^k p_i R(\theta_i, \delta_W) \leq \inf_\delta \sum_1^k p_i R(\theta_i, \delta) + \epsilon.$$

Furthermore, δ_W is as good as δ_0 in the game (Θ, D, R); for

$$\sup_\theta W(\theta, \delta_W) \leq \sup_\theta W(\theta, \delta_0) = 0$$

implies that

$$R(\theta, \delta_W) \leq R(\theta, \delta_0) \qquad \text{for all } \theta.$$

Although Theorem 3 seems quite general, its application to specific statistical problems is limited. To obtain a more general theorem is beyond the scope of this book. However, the student armed with the knowledge of Theorem 3 may attack the more general complete class theorem of Le Cam with applications to many statistical problems.

It should be noted that the class of extended Bayes rules referred to in Theorem 3 may be taken as the class of rules which for every $\epsilon > 0$ are ϵ-Bayes with respect to the finite distributions over Θ. For this problem we may take Θ^* to be the set of finite distributions over Θ. Theorem 3 is stated in its strongest form for this choice of Θ^*.

Exercises

1. Let Θ consist of two points, $\Theta = \{1/3, 2/3\}$, let $\mathcal{A}$ be the real line, and let $L(\theta, a) = (\theta - a)^2$. A coin is tossed once and the probability of heads is θ. Note that because the loss is convex attention may be restricted to nonrandomized decision rules, which are functions from the set $\{H, T\}$ (H = heads and T = tails) into $\mathcal{A}$. Thus the set D of nonrandomized decision rules $(d(H), d(T))$ may be considered as the Euclidean plane (x, y), $x = d(H)$, $y = d(T)$.
   (a) Find a Bayes rule with respect to the prior distribution, giving probability π to $\theta = 1/3$ and probability $1 - \pi$ to $\theta = 2/3$.
   (b) Give a rough plot of the set of all nonrandomized Bayes rules as a subset of the plane.

(c) Show that the set of nonrandomized Bayes rules is minimal essentially complete.

(d) Note that the problem of finding a minimal essentially complete class by the method of Section 2.6 is quite difficult, for the risk set S, or the nonrandomized risk set S_0 of (1.24), is quite difficult to obtain explicitly.

2. A coin has unknown probability $\theta \in \Theta = [0, 1]$ of coming up heads. It is required to estimate θ, ($\mathcal{A} = [0, 1]$), on the basis of one toss of the coin. The loss function is given as $L(\theta, a) = (\theta - a)^2$. The set of nonrandomized decision rules may be taken to be the unit square (x, y), $0 \leq x \leq 1$, $0 \leq y \leq 1$, $x = d(H)$, and $y = d(T)$.

(a) Find a Bayes rule with respect to a given distribution τ. Show that any other prior distribution τ', with the same first two moments as τ, has the same Bayes rule.

(b) Plot in the unit square the set of all nonrandomized Bayes rules.

(c) Do these rules form an essentially complete class?

3. Let Θ be finite and suppose that the risk set is bounded from below and closed from below. Let B be the class of all nonrandomized Bayes rules. Prove that the class B^* of elements of D^*, which give all their mass to B, is a complete class of decision rules. (Show that B^* is the class of Bayes rules.)

4. Let $\Theta = \{\theta_1, \theta_2\}$, $\mathcal{A} = \{a_1, a_2\}$, and $L(\theta_i, a_j) = 0$ if $i = j$, $L(\theta_i, a_j) = 1$ if $i \neq j$. Let X be a random variable whose distribution, if θ_i is the true state of nature, is binomial with sample size n and probability θ_i.

$$f(x \mid \theta_i) = \binom{n}{x} \theta_i^x (1 - \theta_i)^{n-x}, \qquad x = 0, 1, \cdots, n,$$

where $\theta_1 = 3/4$ and $\theta_2 = 1/2$.

(a) Describe the set of behavioral decision rules $\mathfrak{D}$ as a subset of $(n + 1)$-dimensional Euclidean space.

(b) Find $\hat{R}(\theta_i, \mathbf{p})$ where $\theta_i \in \Theta$ and $\mathbf{p} = (p_0, p_1, \cdots, p_n) \in \mathfrak{D}$.

(c) Find all Bayes rules with respect to the prior distributions where nature chooses θ_1 with probability π and θ_2 with probability $1 - \pi$, $0 \leq \pi \leq 1$.

(d) Is the class of all Bayes rules complete for this problem?

5. Let $\mathcal{B}$ denote the class of all Bayes rules (extended Bayes rules). Show that if $\mathcal{B}$ is essentially complete $\mathcal{B}$ is complete.

2.11 Solving for Minimax Rules

We present two methods that may be used to find minimax rules for the statistician.

In the first method we guess at a least favorable distribution. In a decision problem (Θ, D, R) the space D^* of decision rules for the statistician may be complex because of the gathering of information by way of the random quantity X, whereas the space Θ^* of prior distributions may be simple. Often we can guess which prior distribution we would least like to be told that nature is using. After guessing a distribution, call it τ_0, as least favorable, we must find a δ_0, which is Bayes with respect to τ_0, and check somehow that δ_0 is minimax. There are various ways of determining whether a given rule is minimax. The method most useful in the present context is based on the following theorem.

Theorem 1. If δ_0 is Bayes with respect to τ_0 and, for all $\theta \in \Theta$,

$$R(\theta, \delta_0) \leq r(\tau_0, \delta_0), \tag{2.25}$$

then the game has a value, δ_0 is a minimax rule, and τ_0 is least favorable.

Proof. The conclusions of the theorem follow immediately from the inequalities

$$\bar{V} \leq \sup_\theta R(\theta, \delta_0) \leq r(\tau_0, \delta_0) \leq \inf_\delta r(\tau_0, \delta) \leq \underline{V}.$$

It may sometimes be suspected that the least favorable distribution is not a proper distribution or, in other words, that the minimax rule is an extended Bayes rule. In such cases the following extension of Theorem 1 is useful.

Theorem 2. If δ_n is Bayes with respect to τ_n, if $r(\tau_n, \delta_n) \to C$, and if $R(\theta, \delta_0) \leq C$ for all θ, then the game has a value and δ_0 is a minimax rule.

The proof is an exercise.

In the second method of finding minimax rules we hunt for an equalizer rule. *An equalizer decision rule* is a rule δ_0 such that $R(\theta, \delta_0) = C$, some constant, for all $\theta \in \Theta$. The reason that we suspect that minimax rules are equalizer rules is based on the following lemma. (For an extension that allows Θ to be the real line see Exercise 2.)

Lemma 1. Suppose that the game (Θ, D, R) with Θ finite, has a value V and that a minimax rule δ_0 exists. Then, for any $\theta \in \Theta$ that receives positive weight from any least favorable distribution,

$$R(\theta, \delta_0) = V. \tag{2.26}$$

Proof. Because δ_0 is minimax, we must have $R(\theta, \delta_0) \leq V$ for all $\theta \in \Theta$. Suppose that for some $\theta' \in \Theta$, $R(\theta', \delta_0) < V$ and that θ' receives a positive probability from some least favorable distribution τ_0. Then,

$$V = r(\tau_0, \delta_0) = \sum_{\theta} \tau_0(\theta)\, R(\theta, \delta_0) < V,$$

a contradiction that proves the lemma.

This lemma implies that if there exists a least favorable distribution τ_0 that gives positive weight to each state of nature, then every minimax rule is an equalizer rule. Many problems in game theory, in which a minimax rule is to be obtained, are solved by finding an equalizer rule; or, because (2.26) does not necessarily hold for all θ, by finding a rule that is almost an equalizer rule in that the risk is a constant for all θ deemed important.

To determine whether an equalizer rule is a minimax rule it is perhaps simplest to use the following theorem. Because this theorem is a slightly weakened version of Theorem 2, the proof is omitted.

Theorem 3. If an equalizer rule is extended Bayes, it is a minimax rule.

The second method may now be summarized: find an equalizer rule if possible and check to see if it is a Bayes rule or an extended Bayes rule.

We now give examples of these two methods. Consider first the problem of finding the minimax rule for Exercise 2.10.1. In that exercise we know that θ is either 1/3 or 2/3 and we are to estimate θ by a point on the real line on the basis of one toss of a coin with probability θ of heads; we use squared error loss. Because the loss function is convex, a minimax rule, if any exists, may be found among the nonrandomized decision rules. A reasonable guess at a least favorable distribution τ_0 is that τ_0 gives probability 1/2 to both states of nature. If we denote a nonrandomized decision rule as a point (x, y) in the plane, with the interpretation that we estimate θ to be x if heads is observed and y if tails is observed, the Bayes risk with respect to τ_0 is

$$\begin{aligned} r(\tau_0, (x, y)) &= (1/2)\, R(2/3, (x, y)) + (1/2)\, R(1/3, (x, y)) \\ &= (1/2)[(2/3)(2/3 - x)^2 + (1/3)(2/3 - y)^2] \\ &\quad + (1/2)[(1/3)(1/3 - x)^2 + (2/3)(1/3 - y)^2]. \end{aligned} \tag{2.27}$$

The values of x and y that minimize this expression are easily found by differentiation to be

$$\delta_0 : x = 5/9, \qquad y = 4/9. \tag{2.28}$$

This is a Bayes rule with respect to τ_0, which is admissible by Theorem 2.3.2. To show that it is minimax it is sufficient to show that it satisfies inequality (2.25). It is easily checked from (2.27) that

$$r(\tau_0, \delta_0) = R(2/3, \delta_0) = R(1/3, \delta_0) = 2/81.$$

Thus δ_0 is minimax, τ_0 is least favorable, and the value of the game is $V = 2/81$. We note in passing that δ_0 is an equalizer rule and that Theorem 3, as well as Theorem 1, implies that δ_0 is minimax.

As a second example, consider finding the minimax rule in the problem of Exercise 2.10.2, in which a probability p, $0 \leq p \leq 1$, is to be estimated on the basis of a single toss of a coin having probability p of coming up heads. The loss is taken to be squared error $L(p, \hat{p}) = (p - \hat{p})^2$ so that attention may be restricted to nonrandomized decision rules. In this problem it appears to be hard to guess at a least favorable distribution, so we search for an equalizer decision rule. A nonrandomized decision rule may be represented by a point (x, y) in the unit square $\{(x, y\} : 0 \leq x \leq 1, 0 \leq y \leq 1\}$, with the interpretation that we estimate p to be x if heads is observed and y if tails is observed. The risk function is

$$\begin{aligned} R(p, (x, y)) &= p(p - x)^2 + (1 - p)(p - y)^2 \\ &= p^2(1 - 2x + 2y) + p(x^2 - 2y - y^2) + y^2. \end{aligned} \tag{2.29}$$

If (x, y) is to be an equalizer rule, we must have the coefficients of p^2 and p equal to zero. This occurs only for the decision rule

$$\delta_0 : x = 3/4, \qquad y = 1/4. \tag{2.30}$$

Thus the only equalizer rule is to estimate p as $3/4$ if heads is observed and $1/4$ if tails is observed, the constant risk being $y^2 = 1/16$ from (2.29). By Theorem 3 this rule will be minimax if it is Bayes. To search for a distribution with respect to which it is Bayes, we rewrite the risk function

$$R(p, (x, y)) = px^2 - 2p^2x + (1 - p)y^2 - 2p(1 - p)y + p^2.$$

For any prior distribution τ on $[0, 1]$ with $Ep = m_1$ and $Ep^2 = m_2$ we have as the Bayes risk

$$r(\tau, (x, y)) = m_1x^2 - 2m_2x + (1 - m_1)y^2 - 2(m_1 - m_2)y + m_2 .$$

The Bayes rule with respect to τ is that (x, y), which minimizes this Bayes risk and is easily found by taking derivatives to be

$$x = \frac{m_2}{m_1}, \qquad y = \frac{m_1 - m_2}{1 - m_1}.$$

If this is to yield the rule (2.30), it is necessary and sufficient that

$$m_1 = 1/2, \qquad m_2 = 3/8. \tag{2.31}$$

It is obvious that many such distributions exist; for example, the Beta distribution with $\alpha = 1/2$ and $\beta = 1/2$ has these first two moments. Thus δ_0 is minimax, the value is $V = 1/16$, and any distribution with the first two moments as in (2.31) is least favorable.

This may be extended by the use of Exercise 1.8.8 to the case in which the observable X has the Binomial distribution with sample size n and probability of success p. It was seen in Exercise 1.8.8 that the Bayes rule with respect to the Beta distribution with parameters α and β is $d(X) = (X + \alpha)/(n + \alpha + \beta)$. We can *hope* that among these rules there is an equalizer rule; by Theorem 3 such a rule would automatically be minimax. The risk function of the rule d is

$$\begin{aligned} R(p, d) &= E_p(d(X) - p)^2 \\ &= \frac{p^2[(\alpha + \beta)^2 - n] + p[n - 2\alpha(\alpha + \beta)] + \alpha^2}{(n + \alpha + \beta)^2}. \end{aligned}$$

This risk is independent of p if the coefficients of p^2 and p are equal to zero or, equivalently, for $\alpha > 0$ and $\beta > 0$ if $\alpha = \beta = \sqrt{n}/2$. Hence our hope is realized and the minimax rule is

$$d_0(X) = \frac{X + \sqrt{n}/2}{n + \sqrt{n}}. \tag{2.32}$$

See Steinhaus (1957) for an elementary exposition.

As a third and final example we consider the estimation of the mean of a normal distribution with variance 1, based on a sample of size 1, using quadratic loss. Thus we have $\Theta = \mathcal{A} = E_1$ and $L(\theta, a) = (\theta - a)^2$, and the distribution of the observable X is normal with mean θ and variance 1. There is one obvious equalizer strategy, namely, $d(X) = X$. The risk function for this rule is simply the variance of X, $R(\theta, d) = 1$. This estimate was investigated in Sec. 1.8, in which it was shown that it is an extended Bayes rule. Therefore, from Theorem 3, this estimate is minimax.

Exercises

1. Prove Theorem 2.
2. Let Θ be the real line and assume that $R(\theta, \delta)$ is a continuous function of θ for all $\delta \in D^*$ and that the game has a value V. If δ_0 is minimax, then for any θ in the support of any least favorable distribution $R(\theta, \delta_0) = V$.
3. Prove Theorem 3.
4. Prove the following alternative to Theorem 3: if an equalizer rule is admissible, it is minimax.
5. Show by a counterexample that in Exercise 4 it cannot be concluded that the game has a value or that the minimax admissible equalizer rule is Bayes. (*Hint*. In the example of a game without a value in Section 2.9, add a point 0 to $\mathcal{A}$ and define $L(\theta, 0) \equiv 0$.)
6. Let $\Theta = \{\theta_1, \theta_2\}$, $\mathcal{A} = [0, \pi/2]$ = the closed interval on the real line from 0 to $\pi/2$ and let $L(\theta_1, a) = -\cos a$ and $L(\theta_2, a) = -\sin a$. A coin is tossed once with probability P_θ of heads, where $P_{\theta_1} = 1/3$ and $P_{\theta_2} = 2/3$. Find $R(\theta_1, d)$ and $R(\theta_2, d)$ for all $d \in D$. Find the Bayes rule δ_0 with respect to the prior distribution τ_0 which gives probability 1/2 to each state of nature. Show that δ_0 is minimax and that τ_0 is least favorable.

7. Let Θ consist of two points $\{\theta_1, \theta_2\}$, let $\mathcal{A}$ be the closed unit interval $[0, 1]$, and let the loss function be

$$L(\theta_1, a) = a^2$$

$$L(\theta_2, a) = 1 - a.$$

Note that this loss is convex in a for each $\theta \in \Theta$. A coin is tossed once. The probability of heads is 1/3 if θ_1 is the true state of nature and 2/3 if θ_2 is the true state of nature.
(a) Represent the class D of decision rules as a subset of the plane.
(b) Find $R(\theta_1, (x, y))$ and $R(\theta_2, (x, y))$ for $(x, y) \in D$.
(c) Find the class of all nonrandomized Bayes rules. Plot this class as a subset of D.
(d) Find a minimax rule among the class of all Bayes rules.

8. Let X have a binomial distribution with n trials and probability θ of success, $0 < \theta < 1$. Using the loss function

$$L(\theta, a) = (\theta - a)^2/(\theta(1 - \theta)),$$

show that $d(X) = X/n$ is a minimax estimate of θ, with constant risk $1/n$ (see Exercise 1.8.9).

9. Let Θ be the half-open interval $[0, 1)$ (that is, $0 \leq \theta < 1$) let $\mathcal{A}$ be the closed interval $[0, 1]$, and let $L(\theta, a) = (\theta - a)^2/(1 - \theta)$. Let the observable X have the geometric distribution

$$f(x \mid \theta) = (1 - \theta)\theta^x, \qquad x = 0, 1, 2, \cdots.$$

(a) Write the risk function $R(\theta, d)$ for $d \in D$ as a power series in θ.
(b) Show that the only nonrandomized equalizer rule is $d(0) = 1/2$, $d(1) = d(2) = \cdots = 1$.
(c) Show that a rule in D is Bayes with respect to a distribution τ if and only if $d(i) = \mu_{i+1}/\mu_i$, $i = 0, 1, 2, \cdots$, where the μ_i are the moments of the distribution of τ.
(d) Show that the rule in (b) is extended Bayes, hence minimax.

10. Let Θ be the set of all distributions over $[0, 1]$, let $\mathcal{A} = [0, 1]$ and let $L(\theta, a) = (\mu_1 - a)^2$, where μ_1 is the mean of the distribution θ. Let $X_1, \cdots, X_n$ be a sample size n from the distribution θ and let $X = X_1 + X_2 + \cdots + X_n$.
(a) Show that if d_0 is the decision rule of formula (2.32) then

$$R(\theta, d_0) = n(\mu_2 - \mu_1 + 1/4)/(n + \sqrt{n})^2$$

(where μ_2 is the second moment about zero)and $R(\theta, d_0)$ is maximized if, and only if, θ gives all its mass to zero and one.

(b) Hence conclude from Theorem 1 that d_0 is minimax. (Note that Theorem 3 does not apply, for d_0 is not an equalizer rule.)

11. Let $\Theta = (0, \infty)$, $\mathcal{A} = [0, \infty)$, let X have the Poisson distribution, $\mathcal{P}(\theta)$, (see Sec. 3.1), and let $L(\theta, a) = (\theta - a)^2/\theta$.

(a) Show that the usual estimator $d_0(x) = x$ is an equalizer rule.

(b) Show that the usual estimator d_0 is generalized Bayes with respect to Lebesgue measure on $(0, \infty)$.

(c) Find Bayes decision rules with respect to the prior distributions $\tau_{\alpha,\beta}$, where $\tau_{\alpha,\beta}$ represents the gamma distribution $\mathcal{G}(\alpha, \beta)$.

(d) Show that d_0 is extended Bayes, hence minimax.

12. Let $\Theta = (0, \infty)$, $\mathcal{A} = [0, \infty)$, let X have the discrete distribution with probability mass function

$$f(x \mid \theta) = \binom{r + x - 1}{x} \theta^x (\theta + 1)^{-(r+x)}, \qquad x = 0, 1, 2, \cdots$$

(the negative binomial distribution reparametrized so that

$$E_\theta X = r\theta),$$

where r is some known positive integer, and let

$$L(\theta, a) = (\theta - a)^2/(\theta(\theta + 1)).$$

(a) Show that the usual estimator, $d_0(x) = x/r$, is an equalizer rule. ("Usual" here means "maximum likelihood" or "best unbiased.")

(b) Show that the usual estimator, d_0, is generalized Bayes with respect to Lebesgue measure on $(0, \infty)$, provided $r > 1$. What happens when $r = 1$?

(c) Find Bayes decision rules with respect to the prior distributions $\tau_{\alpha,\beta}$ with densities

$$f(\theta \mid \alpha, \beta) = \frac{\Gamma(\alpha + \beta)}{\Gamma(\alpha)\,\Gamma(\beta)} \theta^{\alpha-1} (\theta + 1)^{-(\alpha+\beta)} I_{(0,\infty)}(\theta)$$

(the distribution of $\theta = Z/(1 - Z)$, where Z has the Beta distribution $\mathcal{B}e(\alpha, \beta)$).

(*Ans*: $d_{\alpha,\beta}(x) = (\alpha + x - 1)/(\beta + r + 1)$.)

(d) Show that $d(x) = x/(r + 1)$ is minimax. (Note that d_0 is *not* minimax, hence not admissible.)

13. Show that the estimate d_0 of formula (2.32) is an admissible estimate for the problem considered there.

14. Let X have the hypergeometric distribution $\mathcal{H}(n, \theta, M)$, let θ have the beta-binomial distribution $\mathcal{BB}(\alpha, \beta, M)$ (see Section 3.1), and let $L(\theta, a) = (\theta - a)^2$.
(a) Show that the unconditional distribution of X is $\mathcal{BB}(\alpha, \beta, n)$.
(b) Show that the conditional distribution of $\theta - x$, given $X = x$, is $\mathcal{BB}(x + \alpha, n - x + \beta, M - n)$.
(c) Show that the Bayes estimate of θ is

$$d(x) = \frac{(M + \alpha + \beta)x + \alpha(M - n)}{(n + \alpha + \beta)}.$$

(d) Find the risk function of the estimates $\hat{\theta} = aX + b$.
(e) Show that the estimate $\hat{\theta} = a_0X + b_0$ has a constant risk equal to b_0^2 when

$$a_0 = M/(n(1 + \delta)), \qquad b_0 = (M - a_0n)/2,$$

where $\delta = \sqrt{(M - n)/(n(M - 1))}$.
(f) Show that the estimate $\hat{\theta} = a_0X + b_0$ is minimax (Bayes with respect to $\mathcal{BB}(\alpha, \beta, n)$, when $\alpha = \beta = b_0/(a_0 - 1)$). Take as a numerical example $N = 9$ and $n = 3$.

15. Let $\Theta = [0, 1] = \mathcal{A}$ and let the loss be $L(\theta, a) = (1 - \theta)a + \theta(1 - a)$. Let the random observable X have any distribution depending on θ. Show that the decision rule $d(x) \equiv \frac{1}{2}$ is a minimax rule. (Clearly in some problems this rule is not very good. See, for example, Exercise 4.3.5.)

The following exercise, pointed out to me by Herman Rubin, puts the minimax principle of choosing among decision rules in an even more unfavorable light, because the minimax rule is unique.

16. Let $\Theta = (0, 1]$, $\mathcal{A} = [0, 1]$, and $L(\theta, a) = \min((\theta - a)^2/\theta^2, 2)$. Let the random observable X have the binomial distribution $\mathcal{B}(n, \theta)$. Show that the rule $d(x) \equiv 0$ is the unique minimax rule.

CHAPTER 3

Distributions and Sufficient Statistics

3.1 Useful Univariate Distributions

Chapter 2 dealt with the development of theorems in decision theory which are valid in very general situations. Restrictions were placed only on the form of the parameter space, the space of actions of the statistician, and the loss or risk function. In this and subsequent chapters we attempt to take advantage of the form of the distribution of the random variable X which the statistician observes and whose distributions depend on the true unknown state of nature. It seemed worthwhile to collect in one place the main distributions used frequently as examples and counterexamples in the sequel. This section, therefore, may be considered a review of some of the material in a probability and statistics course. See Wilks (1962), Raiffa and Schlaifer (1961), or Parzen (1960).

Discrete Distributions

THE BINOMIAL DISTRIBUTION. The frequency function or probability mass function of the binomial distribution is

$$f(x) = \binom{n}{x} p^x (1 - p)^{n-x} \qquad x = 0, 1, 2, \cdots, n, \tag{3.1}$$

where n is a positive integer and $0 \le p \le 1$. This distribution is denoted by $\mathcal{B}(n, p)$. The notation $X \in \mathcal{B}(n, p)$ means that the random variable X has a binomial distribution with parameters n and p. The binomial distribution $\mathcal{B}(n, p)$ represents the distribution of the total number of successes in n independent trials when p is the probability of success at each individual trial. The mean of this distribution is np, and the variance is $np(1 - p)$. The characteristic function of this distribution [that is, $\phi(t) = E \exp(itX)$, where $X \in \mathcal{B}(n, p)$] is $\phi(t) = (1 + p(e^{it} - 1))^n$.

THE NEGATIVE BINOMIAL DISTRIBUTION. If independent trials each with probability p of success are performed until the rth failure occurs, the number of observed successes follows the negative binomial distribution with probability mass function

$$f(x) = \binom{r + x - 1}{x} (1 - p)^r p^x, \qquad x = 0, 1, 2, \cdots, \tag{3.2}$$

where r is a positive integer and $0 \le p < 1$. This distribution is denoted by $\mathcal{NB}(r, p)$. It has mean $rp/(1 - p)$, variance $rp/(1 - p)^2$, and characteristic function $\phi(t) = (1 - p)^r/(1 - pe^{it})^r$. An important special case is the *geometric distribution*, $\mathcal{NB}(1, p)$.

THE POISSON DISTRIBUTION. The probability mass function of the Poisson distribution is

$$f(x) = \frac{e^{-\lambda}\lambda^x}{x!} \qquad x = 0, 1, 2, \cdots, \tag{3.3}$$

where $\lambda > 0$. This distribution arises naturally as the distribution of the number of events that occur in the time interval $(0, 1)$, when the events are occurring in a Poisson process at rate λ per unit time. We denote this distribution by $\mathcal{P}(\lambda)$. It has mean λ, variance λ, and characteristic function $\phi(t) = \exp[\lambda(e^{it} - 1)]$. This distribution is also the limit of $\mathcal{B}(n, \lambda/n)$ as $n \to \infty$, which may easily be seen by taking the limit of the characteristic function of $\mathcal{B}(n, \lambda/n)$ as $n \to \infty$.

THE HYPERGEOMETRIC DISTRIBUTION. The hypergeometric distribution, denoted by $\mathcal{H}(n, m, M)$, has probability mass function

$$f(x) = \frac{\binom{n}{x}\binom{M - n}{m - x}}{\binom{M}{m}} = \frac{\binom{m}{x}\binom{M - m}{n - x}}{\binom{M}{n}},$$

$$x = max\ (0, m + n - M), \cdots, \min\ (m, n), \tag{3.4}$$

where m, n, and M are nonnegative integers, $m \leq M$, $n \leq M$. The mean of this distribution is mnM^{-1}, and the variance is

$$nm(M - n)(M - m)M^{-2}(M - 1)^{-1}.$$

In drawing a sample of size n without replacement from a population of M items, m of which have a property A, the distribution of the number of items of the sample with property A is $\mathcal{H}(n, m, M)$. Note that $\mathcal{H}(n, m, M) = \mathcal{H}(m, n, M)$.

THE BETA-BINOMIAL DISTRIBUTION. The distribution with probability mass function

$$f(x) = \binom{n}{x} \frac{\Gamma(\alpha + \beta)\,\Gamma(x + \alpha)\,\Gamma(n + \beta - x)}{\Gamma(\alpha)\,\Gamma(\beta)\,\Gamma(n + \alpha + \beta)}, \qquad x = 0, 1, \cdots, n, \tag{3.5}$$

where n is a nonnegative integer and where $\alpha > 0$ and $\beta > 0$, is known as the beta-binomial distribution, denoted by $\mathcal{BB}(\alpha, \beta, n)$ (see Exercise 1). The mean of this distribution is $n\alpha/(\alpha + \beta)$ and the variance is $n\alpha\beta(n + \alpha + \beta)(\alpha + \beta)^{-2}(\alpha + \beta + 1)^{-1}$. The special case $\mathcal{BB}(1, 1, n)$ is the uniform distribution over the integers $0, 1, \cdots, n$,

$$f(x) = \frac{1}{n + 1}, \qquad x = 0, 1, \cdots, n.$$

Absolutely Continuous Distributions

To simplify the writing of the densities of these distributions we introduce the use of the indicator function of a set. If A is a subset of the real line, the indicator function of A is written $I_A(x)$ which by definition is that function of a real variable x which is equal to one if $x \in A$ and equal to zero if $x \notin A$. Similarly, if B is a subset of a Euclidean space, then $I_B(\mathbf{x})$ is one if $\mathbf{x} \in B$ and zero if $\mathbf{x} \notin B$.

THE UNIFORM DISTRIBUTION. The distribution over the real line with density

$$f(x) = \frac{1}{\beta - \alpha} I_{(\alpha,\beta)}(x) \tag{3.6}$$

is known as the uniform distribution over the interval (α, β) and is denoted by $\mathcal{U}(\alpha, \beta)$. The mean of this distribution is $(\alpha + \beta)/2$ and the variance is $(\beta - \alpha)^2/12$.

THE NORMAL DISTRIBUTION. The density of the normal distribution with mean μ and variance σ^2 is

$$f(x) = \frac{1}{\sqrt{2\pi}\,\sigma} \exp\left(\frac{-(x - \mu)^2}{2\sigma^2}\right), \tag{3.7}$$

where $\sigma > 0$. Its importance is due mainly to the central limit theorem, although many other properties of the normal distribution also distinguish it. The next section is devoted to some of these other distinguishing properties. This distribution is denoted by $\mathcal{N}(\mu, \sigma^2)$. Its characteristic function is

$$\phi(t) = \exp\,(i\mu t - \tfrac{1}{2}\sigma^2 t^2). \tag{3.8}$$

THE GAMMA DISTRIBUTION. The distribution with density

$$f(x) = \frac{1}{\Gamma(\alpha)\beta^\alpha}\, e^{-x/\beta}\, x^{\alpha-1}\, I_{(0,\infty)}(x), \tag{3.9}$$

where $\alpha > 0$ and $\beta > 0$ and where $\Gamma(\alpha)$ represents the well-known gamma function

$$\Gamma(\alpha) = \int_0^\infty e^{-x}\, x^{\alpha-1}\, dx \tag{3.10}$$

is called the gamma distribution and is denoted by $\mathcal{G}(\alpha, \beta)$. This distribution arises as the distribution of the time of the occurrence of the αth event after time zero of a Poisson process in which events occur at a rate $\lambda = \beta^{-1}$. The mean of $\mathcal{G}(\alpha, \beta)$ is $\alpha\beta$, the variance is $\alpha\beta^2$, and the characteristic function is $\phi(t) = (1 - i\beta t)^{-\alpha}$. Two special cases are particularly important: $\mathcal{G}(1, \beta)$, *the exponential distribution*, and $\mathcal{G}(n/2, 2)$, *the χ^2-distribution with n degrees of freedom*, sometimes denoted by χ_n^2.

THE BETA DISTRIBUTION. An important distribution in nonparametric statistics is the beta distribution with density

$$f(x) = \frac{\Gamma(\alpha + \beta)}{\Gamma(\alpha)\,\Gamma(\beta)}\, x^{\alpha-1}(1 - x)^{\beta-1}\, I_{[0,1]}(x), \tag{3.11}$$

where $\alpha > 0$ and $\beta > 0$. It arises as the distribution of the αth-order statistic (from the bottom) of a sample of size $\alpha + \beta - 1$ from a $\mathcal{U}(0, 1)$

distribution. This distribution is denoted by $\mathcal{B}e(\alpha, \beta)$. The mean of this distribution is $\alpha/(\alpha + \beta)$ and the variance is $\alpha\beta/(\alpha + \beta)^2(\alpha + \beta + 1)$. The beta distribution with $\alpha = 1$ and $\beta = 1$ is obviously identical to the $\mathcal{U}(0, 1)$ distribution.

THE CAUCHY DISTRIBUTION. A distribution agreeable in many respects but disagreeable in others is the Cauchy distribution with density

$$f(x) = \frac{\beta}{\pi} \frac{1}{\beta^2 + (x - \alpha)^2}, \tag{3.12}$$

where $\beta > 0$, and cumulative distribution function

$$F(x) = \frac{1}{\pi}\left[\arctan \frac{(x - \alpha)}{\beta} + \frac{\pi}{2}\right]. \tag{3.13}$$

The mean and variance of this distribution do not exist, yet it is symmetric about its median α, and the semi-interquartile range (that is, half the distance between the first and third quartiles) is β. Its characteristic function is

$$\phi(t) = \exp(it\alpha - \beta \mid t \mid). \tag{3.14}$$

This distribution plays a role in the central limit theorem for infinite variances. A notorious feature of the Cauchy distribution is that if $X_1, \cdots, X_n$ are independent random variables, all having the distribution (3.12), then their average $(1/n) \sum X_i$ has the distribution (3.12) also. This distribution is denoted by $\mathcal{C}(\alpha, \beta)$.

THE LOGISTIC DISTRIBUTION. The distribution with cumulative distribution function

$$F(x) = \{1 + \exp[-(x - \alpha)/\beta]\}^{-1} \tag{3.15}$$

and density

$$f(x) = \frac{\exp[-(x - \alpha)/\beta]}{\beta\{1 + \exp[-(x - \alpha)/\beta]\}^2} = \frac{1}{2(1 + \cosh((x - \alpha)/\beta))\beta}, \tag{3.16}$$

where $\beta > 0$ is known as the logistic distribution and denoted by $\mathcal{L}(\alpha, \beta)$. This distribution is symmetric about α.

THE NONCENTRAL t-DISTRIBUTION. The noncentral t-distribution with ν degrees of freedom and noncentrality parameter μ is *defined* as the distribution of $T = Y/\sqrt{Z/\nu}$, where Y and Z are independent random variables, $Y \in \mathfrak{N}(\mu, 1)$ and $Z \in \chi_\nu^2(\chi_\nu^2 = \mathcal{G}(\nu/2, 2))$. This distribution, denoted by $t_\nu(\mu)$, has density

$$f(t) = c_\nu(t^2 + \nu)^{-(\nu+1)/2} \exp\left[-\frac{\nu\mu^2}{2(t^2+\nu)}\right]$$

$$\times \int_0^\infty \exp\left[-\tfrac{1}{2}\left(x - \frac{\mu t}{\sqrt{t^2+\nu}}\right)^2\right] x^\nu \, dx, \qquad (3.17)$$

where

$$c_\nu = \frac{\nu^{\nu/2}}{\sqrt{\pi}\,\Gamma(\nu/2) 2^{(\nu-1)/2}}.$$

For tables, see Resnikoff and Lieberman (1957), Locks, Alexander, and Byars (1963), and D. B. Owen (1963). The distribution $t_\nu(0)$ is the (central) t-distribution with ν degrees of freedom and is more often denoted by t_ν. Its density is

$$f(t) = \frac{\Gamma((\nu+1)/2)}{\sqrt{\nu\pi}\,\Gamma(\nu/2)} (t^2/\nu + 1)^{-(\nu+1)/2},$$

which when $\nu = 1$ reduces to the Cauchy distribution, $\mathcal{C}(0, 1)$.

THE NONCENTRAL χ^2-DISTRIBUTION. The distribution with density

$$f(x) = \sum_{j=0}^{\infty} p_{\gamma^2/2}(j) f_{n+2j}(x), \qquad (3.18)$$

where $p_{\gamma^2/2}(j)$ is the probability mass function of the Poisson distribution $\mathcal{P}(\gamma^2/2)$, and where $f_{n+2j}(x)$ is the density of the gamma distribution $\mathcal{G}((n+2j)/2, 2) = \chi_{n+2j}^2$, is known as the noncentral χ^2-distribution with n degrees of freedom and noncentrality parameter γ^2. This distribution is denoted by $\chi_n^2(\gamma^2)$. The distribution $\chi_n^2(0)$ is the (central) χ^2-distribution, χ_n^2. It is clear from (3.18) that $\chi_n^2(\gamma^2)$ is the marginal distribution of a variable X, whose distribution given θ is $\chi_{n+2\theta}^2$, when θ has the Poisson distribution $\mathcal{P}(\gamma^2/2)$. The main property of $\chi_n^2(\gamma^2)$ is that it is the distribution of $Y_1^2 + \cdots + Y_n^2$, where $Y_1, \cdots, Y_n$ are independent random variables, $Y_i \in \mathfrak{N}(\mu_i, 1)$ and $\gamma^2 = \sum_1^n \mu_i^2$. For tables, see Fix (1949) and Fix, Hodges, and Lehmann (1959).

THE NONCENTRAL $\mathcal{F}$-DISTRIBUTION. The noncentral $\mathcal{F}$-distribution with r and n degrees of freedom and noncentrality parameter γ^2, denoted by $\mathfrak{F}_{r,n}(\gamma^2)$, is *defined* as the distribution of $X = (Y/r)/(Z/n)$, where Y and Z are independent random variables with $Y \in \chi_r^2(\gamma^2)$ and $Z \in \chi_n^2$. The distribution $\mathfrak{F}_{r,n}(0)$ is known as the (central) $\mathfrak{F}$-distribution with r and n degrees of freedom, usually denoted simply as $\mathfrak{F}_{r,n}$. The distribution $\mathfrak{F}_{r,n}(\gamma^2)$ has density

$$f(x/\gamma^2) = \sum_{j=0}^{\infty} c_j p_{\gamma^2/2}(j) x^{j-1+(r/2)} (rx + n)^{-j-[(r+n)/2]}, \qquad (3.19)$$

where

$$c_j = r^{j+(r/2)} n^{n/2} \frac{\Gamma(j + [(r+n)/2])}{\Gamma(j + (r/2))\Gamma(n/2)}$$

and where $p_{\gamma^2/2}(j)$ is the probability mass function of the Poisson distribution $\mathcal{P}(\gamma^2/2)$. A set of charts for the noncentral $\mathfrak{F}$-distribution may be found in Pearson and Hartley (1951) and in Fox (1956).

Exercises

1. Let Y have a beta distribution $\mathcal{B}e(\alpha, \beta)$ and let the conditional distribution of X given $Y = y$ be binomial, $\mathcal{B}(n, y)$.
(a) Show that the marginal distribution of X is beta-binomial, $\mathcal{BB}(\alpha, \beta, n)$.
(b) Show that $EX = n\alpha/(\alpha + \beta)$. [Use $EX = E(E(X \mid Y))$.]
(c) Show that the variance of X is

$$n\alpha\beta(n + \alpha + \beta)(\alpha + \beta)^{-2}(\alpha + \beta + 1)^{-1}.$$

2. Suppose that $Y \in \mathcal{B}(M, p)$ and that the conditional distribution of X given $Y = y$ is $\mathcal{H}(n, y, M)$
(a) Show that the unconditional distribution of X is $\mathcal{B}(n, p)$.
(b) Show that the conditional distribution of $Y - x$, given $X = x$, is $\mathcal{B}(M - n, p)$.

3. Derive the density of the noncentral t-distribution and show that it satisfies (3.17).

4. Show that if $Y_1, \cdots, Y_n$ are independent random variables with $Y_i \in \mathfrak{N}(\mu_i, 1)$, then $X = Y_1^2 + \cdots + Y_n^2 \in \chi_n^2(\gamma^2)$, where

$$\gamma^2 = \mu_1^2 + \cdots + \mu_n^2.$$

5. Derive the density of the noncentral $\mathfrak{F}$-distribution and show that it satisfies (3.19).

6. Show that if X has the central $\mathcal{F}$-distribution $\mathcal{F}_{r,n}$, then

$$Y = rX/(rX + n)$$

has the Beta distribution $\mathcal{B}e(r/2, n/2)$.

7. An urn contains n black balls and $(\alpha + \beta - 1)$ white balls. Balls are drawn out one at a time at random without replacement until α white balls are in the sample. Let X be the number of black balls in the sample. Show that $X \in \mathcal{BB}(\alpha, \beta, n)$.

3.2 The Multivariate Normal Distribution

It has already been mentioned that the expectation of a random vector $\mathbf{X} = (X_1, \cdots, X_n)^T$ is the vector of expectations,

$$E\mathbf{X} = (EX_1, \cdots, EX_n)^T.$$

More generally, the expectation of a random matrix (Y_{ij}) $i = 1, \cdots, m, j = 1, \cdots, n$ is defined as the matrix of expectations (EY_{ij}) $i = 1, \cdots, m, j = 1, \cdots, n$. With this definition, the reader should check if $\mathbf{Y}$ is an $m \times n$ random matrix and $\mathbf{A}$ is an $n \times k$ matrix of constants, then $E(\mathbf{YA}) = (E\mathbf{Y})\mathbf{A}$ and $E(\mathbf{A}^T\mathbf{Y}^T) = \mathbf{A}^T(E\mathbf{Y}^T)$ (Exercise 1).

The covariance matrix of a random vector $\mathbf{X} = (X_1, \cdots, X_n)^T$ is defined as the matrix of covariances

$$(\text{Cov}\ (X_i, X_j))\ i = 1, \cdots, n, j = 1, \cdots, n.$$

Using matrix notation, this covariance matrix may be written

$$\text{Cov}\,\mathbf{X} = E(\mathbf{X} - E\mathbf{X})\ (\mathbf{X} - E\mathbf{X})^T, \tag{3.20}$$

where the right side is the expectation of the $n \times n$ matrix obtained by multiplying an $n \times 1$ matrix by a $1 \times n$ matrix (Exercise 2). Not all $n \times n$ matrices may be covariance matrices; for example, all elements on the main diagonal must be nonnegative, for they are variances, $\text{Cov}\ (X_i, X_i) = \text{Var}\ X_i$. The following lemma shows that the covariance matrices are exactly the symmetric, nonnegative definite matrices.

Recall from your undergraduate linear algebra course (or Paige and Swift (1961), 287, or Tucker (1962), 143) that every real symmetric matrix $\mathbf{A}$ can be diagonalized. In other words, *for every real symmetric matrix* $\mathbf{A}$ *there exists a real orthogonal matrix* $\mathbf{P}$ (that is, a real matrix such that $\mathbf{P}^T\mathbf{P} = \mathbf{I} = \mathbf{PP}^T$) *such that* $\mathbf{PAP}^T$ *is a diagonal matrix* (a matrix with zeros in all spots off the main diagonal). From this noteworthy theorem—a version of the spectral theorem—will follow most of the results in matrix theory that we shall need. As an important example, we show how to extract a square root of a symmetric nonnegative definite matrix.

An $n \times n$ matrix $\mathbf{A}$ is said to be *positive definite* if $\mathbf{b}^T\mathbf{A}\mathbf{b} > 0$ for all n-dimensional vectors $\mathbf{b} \neq \mathbf{0}$. $\mathbf{A}$ is said to be *nonnegative definite* if $\mathbf{b}^T\mathbf{A}\mathbf{b} \geq 0$ for all $\mathbf{b}$.

Given a symmetric nonnegative definite matrix $\mathbf{A}$, we show how to find a symmetric nonnegative definite matrix $\mathbf{B}$ such that $\mathbf{BB} = \mathbf{A}$. This matrix is called a nonnegative square root of $\mathbf{A}$, and we write $\mathbf{B} = \mathbf{A}^{1/2}$. Given a symmetric nonnegative definite matrix $\mathbf{A}$, find a matrix $\mathbf{P}$ such that $\mathbf{PP}^T = \mathbf{I} = \mathbf{P}^T\mathbf{P}$ and $\mathbf{PAP}^T = \mathbf{D}$, where $\mathbf{D}$ is a diagonal matrix. Because $\mathbf{A}$ is nonnegative definite, all the diagonal elements of $\mathbf{D}$ are nonnegative (Exercises 3 and 4). Let $\mathbf{D}^{1/2}$ denote the matrix of the nonnegative square roots of the elements of $\mathbf{D}$ so that $\mathbf{D}^{1/2}\mathbf{D}^{1/2} = \mathbf{D}$. Finally, let $\mathbf{B} = \mathbf{P}^T\mathbf{D}^{1/2}\mathbf{P}$. Then $\mathbf{B}$ is symmetric and nonnegative definite (Exercise 3). Furthermore,

$$\mathbf{BB} = \mathbf{P}^T\mathbf{D}^{1/2}\mathbf{PP}^T\mathbf{D}^{1/2}\mathbf{P} = \mathbf{P}^T\mathbf{D}^{1/2}\mathbf{D}^{1/2}\mathbf{P} = \mathbf{P}^T\mathbf{DP} = \mathbf{P}^T(\mathbf{PAP}^T)\mathbf{P} = \mathbf{A},$$

so that $\mathbf{B} = \mathbf{A}^{1/2}$. (Note that $\mathbf{B}$ is positive definite if $\mathbf{A}$ is.)

Lemma 1. Every covariance matrix is symmetric and nonnegative definite. Every symmetric and nonnegative definite matrix is a covariance matrix. If Cov $\mathbf{X}$ is not positive definite, then with probability one, $\mathbf{X}$ lies in some hyperplane $\mathbf{b}^T\mathbf{X} = c$ with $\mathbf{b} \neq \mathbf{0}$.

Proof. Cov $\mathbf{X}$ is symmetric because Cov (x_i, x_j) = Cov (x_j, x_i). Furthermore,

$$\begin{aligned} \mathbf{b}^T(\mathrm{Cov}\,\mathbf{X})\mathbf{b} &= \mathbf{b}^T(E(\mathbf{X} - E\mathbf{X})(\mathbf{X} - E\mathbf{X})^T)\mathbf{b} \\ &= E\mathbf{b}^T(\mathbf{X} - E\mathbf{X})(\mathbf{X} - E\mathbf{X})^T\mathbf{b} \\ &= E\{(\mathbf{b}^T(\mathbf{X} - E\mathbf{X}))^2\} \geq 0, \end{aligned} \tag{3.21}$$

which proves that Cov $\mathbf{X}$ is nonnegative definite. If, for some $\mathbf{b} \neq \mathbf{0}$, $E\{[\mathbf{b}^T(\mathbf{X} - E\mathbf{X})]^2\} = 0$, then $P(\mathbf{b}^T\mathbf{X} = \mathbf{b}^TE\mathbf{X}) = 1$, so that with probability one $\mathbf{X}$ lies in the hyperplane $\mathbf{b}^T\mathbf{X} = c$, where $c = \mathbf{b}^TE\mathbf{X}$.

Now let $\boldsymbol{\Sigma}$ be an arbitrary symmetric nonnegative definite matrix. Let $\mathbf{A}$ be the nonnegative square root of $\boldsymbol{\Sigma}$. Let $\mathbf{X}$ be a vector of independent random variables with zero means and unit variances. Then cov $\mathbf{X} = \mathbf{I}$. Now let $\mathbf{Y} = \mathbf{AX}$. Then, $E\mathbf{Y} = \mathbf{A}(E\mathbf{X}) = \mathbf{0}$ and

$$\begin{aligned} \text{Cov}\,\mathbf{Y} &= E(\mathbf{YY}^T) = E(\mathbf{AX})(\mathbf{AX})^T \\ &= \mathbf{A}E\mathbf{XX}^T\mathbf{A}^T = \mathbf{AA}^T = \mathbf{\Sigma}. \end{aligned} \tag{3.22}$$

Thus $\mathbf{\Sigma}$ is a covariance matrix, completing the proof.

Having completed the preliminaries, we now define the multivariate normal distributions. Defining these distributions by means of the density is not satisfactory because some of these distributions lack a proper density. Instead, we use a general definition which is appropriate for extension to infinite-dimensional linear spaces as well.

Definition 1. An n-dimensional random vector $\mathbf{X}$ is said to have a *multivariate or n-dimensional normal distribution*, if for every n-dimensional vector $\mathbf{u}$ the random variable $\mathbf{u}^T\mathbf{X}$ has a normal distribution (possibly degenerate) on the real line.

The normal distribution on the real line is a univariate or one-dimensional normal distribution by this definition, for if a random variable X has a normal distribution then so has the random variable uX for any real number u.

One advantage of defining the multivariate normal distribution in this way is that the following lemma is an immediate consequence. This lemma states that linear transformations of multivariate normal distributions are multivariate normal.

Lemma 2. If $\mathbf{X}$ has an n-dimensional normal distribution, then for any k-dimensional vector $\mathbf{n}$ of constants and any $k \times n$ matrix $\mathbf{A}$ of constants, the random vector $\mathbf{Y} = \mathbf{AX} + \mathbf{n}$ has a k-dimensional normal distribution.

Proof. Let $\mathbf{u}$ be an arbitrary k-dimensional vector. We are to show that $\mathbf{u}^T\mathbf{Y}$ is normally distributed; but because $\mathbf{X}$ has a multivariate normal distribution $(\mathbf{u}^T\mathbf{A})\mathbf{X}$ is normally distributed and so is $\mathbf{u}^T\mathbf{AX} + \mathbf{u}^T\mathbf{n} = \mathbf{u}^T\mathbf{Y}$, completing the proof.

To compute the characteristic function of the multivariate normal distribution recall that the joint characteristic function of random variables $X_1, X_2, \cdots, X_n$ is defined as

$$\varphi_X(\mathbf{u}) = E \exp [i(u_1X_1 + \cdots + u_nX_n)] = E \exp (i\mathbf{u}^T\mathbf{X}). \quad (3.23)$$

If $\mathbf{X}$ has a multivariate normal distribution, then, since the right side of (3.23) represents $\varphi_{\mathbf{u}^T\mathbf{X}}(1)$ (the characteristic function of the random variable $\mathbf{u}^T\mathbf{X}$ evaluated at 1), the characteristic function of $\mathbf{X}$ may be derived by using (3.8):

$$\varphi_X(\mathbf{u}) = \varphi_{\mathbf{u}^T\mathbf{X}}(1) = \exp [iE\mathbf{u}^T\mathbf{X} - \tfrac{1}{2} \operatorname{Var} (\mathbf{u}^T\mathbf{X})]. \quad (3.24)$$

With the notation $\boldsymbol{\mu} = E\mathbf{X}$ and $\boldsymbol{\Sigma} = \operatorname{Cov} \mathbf{X}$, it follows that

$$E\mathbf{u}^T\mathbf{X} = \mathbf{u}^T\boldsymbol{\mu} \quad \text{and} \quad \operatorname{Var} (\mathbf{u}^T\mathbf{X}) = E\mathbf{u}^T(\mathbf{X} - \boldsymbol{\mu})(\mathbf{X} - \boldsymbol{\mu})^T\mathbf{u} = \mathbf{u}^T\boldsymbol{\Sigma}\mathbf{u}.$$

Hence

$$\varphi_X(\mathbf{u}) = \exp (i\mathbf{u}^T\boldsymbol{\mu} - \tfrac{1}{2}\mathbf{u}^T\boldsymbol{\Sigma}\mathbf{u}). \quad (3.25)$$

Conversely, if (3.25) represents a characteristic function, it is the characteristic function of a multivariate normal distribution, as is easily seen by computing

$$\varphi_{\mathbf{u}^T\mathbf{X}}(t) = \varphi_X(t\mathbf{u}) = \exp [i(\mathbf{u}^T\boldsymbol{\mu})t - \tfrac{1}{2}(\mathbf{u}^T\boldsymbol{\Sigma}\mathbf{u})t^2]$$

which is of the form (3.8). Because the characteristic function determines the distribution uniquely, the multivariate normal distribution is determined once its mean vector $\boldsymbol{\mu}$ and covariance matrix $\boldsymbol{\Sigma}$ are given.

To see that (3.25) actually represents a characteristic function if $\boldsymbol{\Sigma}$ is a covariance matrix, let $\mathbf{Z} = (Z_1, \cdots, Z_n)^T$ be a vector of independent random variables, each having a normal distribution with mean zero and variance one. The characteristic function of $\mathbf{Z}$ is

$$\begin{aligned}\varphi_Z(\mathbf{u}) &= E \exp \Big(i \sum_{j=1}^{n} u_jZ_j\Big) = \prod_{j=1}^{n} E \exp (iu_jZ_j) \\ &= \prod_{j=1}^{n} \exp (-u_j^2/2) = \exp (-\tfrac{1}{2}\mathbf{u}^T\mathbf{u}).\end{aligned} \quad (3.26)$$

Now, let $\mathbf{A}$ be the symmetric nonnegative definite square root of $\boldsymbol{\Sigma}$ and let $\mathbf{Y} = \mathbf{AZ} + \boldsymbol{\mu}$. Then

$$\begin{aligned}\varphi_Y(\mathbf{u}) &= E \exp (i\mathbf{u}^T\mathbf{AZ} + i\mathbf{u}^T\boldsymbol{\mu}) = \exp (i\mathbf{u}^T\boldsymbol{\mu})\varphi_Z(\mathbf{Au}) \\ &= \exp (i\mathbf{u}^T\boldsymbol{\mu} - \tfrac{1}{2}\mathbf{u}^T\mathbf{AAu}) = \exp (i\mathbf{u}^T\boldsymbol{\mu} - \tfrac{1}{2}\mathbf{u}^T\boldsymbol{\Sigma}\mathbf{u})\end{aligned} \quad (3.27)$$

which shows that (3.25) is indeed a characteristic function if $\boldsymbol{\Sigma}$ is a covariance matrix. This proves the following lemma.

Lemma 3. Functions of the form (3.25) where $\boldsymbol{\Sigma}$ is a symmetric nonnegative definite matrix are characteristic functions of multivariate normal distributions. Every multivariate normal distribution has a characteristic function of the form (3.25), where $\boldsymbol{\mu}$ is the mean vector and $\boldsymbol{\Sigma}$ is the covariance matrix of the distribution. We denote this distribution by $\mathfrak{N}(\boldsymbol{\mu}, \boldsymbol{\Sigma})$.

The distribution with characteristic function (3.26) is thus $\mathfrak{N}(\mathbf{0}, \mathbf{I})$.

Now we seek the probability density of the multivariate normal distribution. Note that the variable degenerate at zero is normal by our definition. A probability density for this normal variable does not exist. Similarly, when the covariance matrix $\boldsymbol{\Sigma}$ is not positive definite, all the probability mass lies in some hyperplane (Lemma 1) and again the probability density does not exist. In such a case we say that the multivariate normal distribution is *singular*, and in this case the covariance matrix $\boldsymbol{\Sigma}$ is singular (Exercise 5). However, when the multivariate normal distribution is *nonsingular*, that is when $\boldsymbol{\Sigma}$ is positive definite, hence nonsingular, the multivariate probability density does exist and is easily computed.

Lemma 4. When the covariance matrix $\boldsymbol{\Sigma}$ is nonsingular, the density of the multivariate normal distribution with characteristic function (3.25) exists and is given by

$$f(\mathbf{y}) = (2\pi)^{-n/2}(\det \boldsymbol{\Sigma})^{-1/2} \exp\left[-\tfrac{1}{2}(\mathbf{y} - \boldsymbol{\mu})^T \boldsymbol{\Sigma}^{-1}(\mathbf{y} - \boldsymbol{\mu})\right]. \quad (3.28)$$

Proof. The distribution with characteristic function (3.25) is the distribution of $\mathbf{Y} = \mathbf{AZ} + \boldsymbol{\mu}$, where $\mathbf{A}$ is the symmetric positive definite square root of $\boldsymbol{\Sigma}$ and where $\mathbf{Z} \in \mathfrak{N}(\mathbf{0}, \mathbf{I})$. The density of $\mathbf{Z}$ is the product of the marginal densities

$$\begin{aligned} f_{\mathbf{Z}}(\mathbf{z}) &= f_{Z_1}(z_1) \cdots f_{Z_n}(z_n) \\ &= (2\pi)^{-n/2} \exp\left(-\tfrac{1}{2}\mathbf{z}^T\mathbf{z}\right). \end{aligned}$$

The inverse transformation is $\mathbf{Z} = \mathbf{A}^{-1}(\mathbf{Y} - \boldsymbol{\mu})$, whose Jacobian is

$$|J| = \det \mathbf{A}^{-1} = \det \boldsymbol{\Sigma}^{-1/2} = (\det \boldsymbol{\Sigma})^{-1/2}$$

(Exercises 6 and 7). Hence the density of $\mathbf{Y}$ is

$$\begin{aligned} f_Y(\mathbf{y}) &= f_Z(\mathbf{A}^{-1}(\mathbf{y} - \boldsymbol{\mu}))\,|J| \\ &= (2\pi)^{-n/2}(\det \boldsymbol{\Sigma})^{-1/2} \exp\left[-\tfrac{1}{2}(\mathbf{y} - \boldsymbol{\mu})^T \mathbf{A}^{-1}\mathbf{A}^{-1}(\mathbf{y} - \boldsymbol{\mu})\right], \end{aligned}$$

which reduces immediately to (3.28), completing the proof.

As a final lemma, we demonstrate another important property of the multivariate normal distribution. When two random variables are independent, their covariance is zero (provided it exists), but a zero covariance does not imply that the variables are independent. However, for the multivariate normal distribution a vanishing covariance does imply independence.

Lemma 5. If $\mathbf{Y} \in \mathfrak{N}(\boldsymbol{\mu}, \boldsymbol{\Sigma})$, then $Y_1, Y_2, \cdots, Y_n$ are mutually independent if and only if $\boldsymbol{\Sigma}$ is a diagonal matrix.

Proof. If $\boldsymbol{\Sigma}$ is not a diagonal matrix, there is a nonzero off-diagonal element. This gives a nonzero covariance between two of the Y_j; therefore they cannot be independent. Conversely, if $\boldsymbol{\Sigma}$ is a diagonal matrix,

$$\boldsymbol{\Sigma} = \begin{pmatrix} \sigma_1^2 & 0 & \cdots & 0 \\ 0 & \sigma_2^2 & & \\ \vdots & & \ddots & \\ 0 & & & \sigma_n^2 \end{pmatrix},$$

the characteristic function (3.25) factors

$$\begin{aligned} \varphi_Y(\mathbf{u}) &= \exp\left(i \sum_{j=1}^{n} u_j \mu_j - \tfrac{1}{2} \sum_{j=1}^{n} u_j^2 \sigma_j^2\right) \\ &= \prod_{j=1}^{n} \exp\left(i u_j \mu_j - \tfrac{1}{2} u_j^2 \sigma_j^2\right), \end{aligned}$$

which proves the independence.

This lemma can be extended easily to show that if $(Y_1, \cdots, Y_n)$ is multivariate normal and no variable of $(Y_1, \cdots, Y_r)$ is correlated with any variable of $(Y_{r+1}, \cdots, Y_n)$, then $(Y_1, \cdots, Y_r)$ is completely independent of $(Y_{r+1}, \cdots, Y_n)$ (Exercise 10).

We conclude this section with an example of two normal random variables X and Y, whose joint distribution is not two-dimensional normal. Let the joint density of X and Y be

$$f_{X,Y}(x, y) = \frac{2}{2\pi} \exp\left[-\tfrac{1}{2}(x^2 + y^2)\right] I_A(x, y),$$

where $A = [(x, y): xy > 0]$. The marginal distribution of X (hence Y) is easily seen to be normal. Yet (X, Y) does not have a bivariate normal distribution. This provides an example of two normal random variables X and Y, whose sum $X + Y$ does not have a normal distribution. Another example is given in Exercise 9, which exhibits two normal random variables that are uncorrelated but not independent.

Exercises

1. Let $\mathbf{Y}$ be an $m \times n$ matrix of random variables with finite expectations. Let $\mathbf{A}$ be an $n \times k$ matrix of constants. Then $E(\mathbf{YA}) = (E\mathbf{Y})\mathbf{A}$ and $E(\mathbf{A}^T\mathbf{Y}^T) = \mathbf{A}^T E\mathbf{Y}^T$.
2. Let $\mathbf{X} = (X_1, \cdots, X_n)^T$ be a random vector with a finite covariance matrix

$$\operatorname{Cov} \mathbf{X} = \begin{pmatrix} \operatorname{Var} X_1 & \operatorname{Cov}(X_1, X_2) & \cdots & \operatorname{Cov}(X_1, X_n) \\ \operatorname{Cov}(X_1, X_2) & \operatorname{Var} X_2 & & \cdot \\ \cdot & & & \cdot \\ \cdot & & & \cdot \\ \cdot & & & \cdot \\ \operatorname{Cov}(X_1, X_n) & & \cdots & \operatorname{Var} X_n \end{pmatrix}.$$

   Prove formula(3.20).
3. Show that if $\mathbf{A}$ is symmetric nonnegative (positive) definite and $\mathbf{B} = \mathbf{QAQ}^T$ for some square (nonsingular) matrix $\mathbf{Q}$, then $\mathbf{B}$ is symmetric nonnegative (positive) definite.
4. The diagonal elements of a nonnegative (positive) definite matrix $\mathbf{A}$ are nonnegative (positive).
5. Let $\mathbf{A}$ be a symmetric nonnegative definite matrix. Then $\mathbf{A}$ is nonsingular if and only if $\mathbf{A}$ is positive definite.
6. Let $\mathbf{Y} = \mathbf{AX} + \boldsymbol{\mu}$ be a linear transformation from $\mathbf{X}$ to $\mathbf{Y}$, where $\mathbf{A}$ is a nonsingular square matrix. Show that the Jacobian $\det(\partial y_i/\partial x_j)$ of this transformation is the determinant of $\mathbf{A}$.

7. Using the fact that for two square matrices det $\mathbf{AB}$ = det $\mathbf{A}\cdot$det $\mathbf{B}$, show (a) for a nonsingular matrix $\mathbf{A}$, det $\mathbf{A}^{-1} = (\det \mathbf{A})^{-1}$; (b) for a nonnegative definite matrix $\mathbf{A}$, det $\mathbf{A}^{1/2} = (\det \mathbf{A})^{1/2}$.

8. Show that the bivariate normal distribution with mean (μ_1, μ_2) and covariance matrix

$$\begin{pmatrix} \sigma_1^2 & \sigma_{12} \\ \sigma_{12} & \sigma_2^2 \end{pmatrix},$$

where $\sigma_1^2 \sigma_2^2 > \sigma_{12}^2$, has a probability density

$$f(x, y) = (2\pi\sigma_1\sigma_2\sqrt{1 - \rho^2})^{-1}$$

$$\exp\left\{-\frac{1}{2(1 - \rho^2)}\left[\frac{(x - \mu_1)^2}{\sigma_1^2} - 2\rho\frac{(x - \mu_1)(y - \mu_2)}{\sigma_1 \sigma_2} + \frac{(y - \mu_2)^2}{\sigma_2^2}\right]\right\},$$

where ρ is the correlation coefficient $\rho = \sigma_{12}/\sigma_1\sigma_2$.

9. Let X have a normal distribution with mean zero and variance one. Let c be a nonnegative number and let $Y = -X$ if $|X| \le c$ and $Y = X$ if $|X| > c$. Then Y also has a normal distribution with zero mean and variance one. The covariance of X and Y is a continuous function of c, going from $+1$, when $c = 0$, to -1 when c tends to ∞. Therefore for some value of c X and Y are uncorrelated, yet X and Y are as far from being independent as possible, each being a function of the other. ($c = 1.538\cdots$)

10. Suppose that $(Y_1, \cdots, Y_n)$ has a multivariate normal distribution and that Cov $(Y_i, Y_j) = 0$ for all $i = 1, \cdots, r$ and $j = r + 1, \cdots, n$. Then $(Y_1, \cdots, Y_r)$ and $(Y_{r+1}, \cdots, Y_n)$ are completely independent.

3.3 Sufficient Statistics

Often a significant simplification of the decision problem may be obtained by a reduction in the complexity of the observable quantity X by means of sufficient statistics. To paraphrase R. A. Fisher, who introduced this concept, a sufficient statistic summarizes the whole of the relevant information supplied by the sample. We illustrate this notion by an example.

A coin with unknown probability p, $0 \le p \le 1$, of heads is tossed independently n times. If we let X_i be zero if the outcome of the ith toss is tails and one if the outcome of the ith toss is heads, the random variables $X_1, X_2, \cdots, X_n$ are independent and identically distributed, the common distribution being

$$P(X_j = x) = p^x(1-p)^{1-x} \quad \text{for} \quad x = 0, 1. \tag{3.29}$$

If we are looking at the outcome of this sequence of tosses in order to make a guess of the value of p, it is clear that the important thing to consider is the total number of heads and tails. It is hard to see how the information concerning the order of heads and tails can help us once we know the total number of heads. In fact, if we let T denote the total number of heads, $T = \sum_{i=1}^{n} X_i$, then intuitively the conditional distribution of $(X_1, X_2, \cdots, X_n)$, given $T = j$, is uniform over the $\binom{n}{j}$ n-tuples which have j ones and $n - j$ zeros; that is, given that $T = j$, the distribution of $(X_1, X_2, \cdots, X_n)$ may be obtained by choosing completely at random the j places in which ones go and putting zeros in the other spots. *This may be done not knowing* p. Thus, once we know the total number of heads, being given the rest of the information about $(X_1, X_2, \cdots, X_n)$ is like being told the value of a random variable whose distribution does not depend on p at all. In other words, the total number of heads carries all the information the sample has to give about the unknown parameter p. The total number of heads is a sufficient statistic for p.

A rigorous definition of a sufficient statistic involves the notion of a conditional distribution. A rigorous and general definition of conditional probability demands a larger background than is required for this course. However, we assume that the reader is familiar with conditional probability distributions for discrete variables and for absolutely continuous variables with sufficiently smooth probability densities, as contained in Parzen's book (1960). We give a definition of a sufficient statistic valid for the discrete variables and absolutely continuous variables with sufficiently smooth densities to which our applications are restricted.

Definition 1. Let X denote a random variable (or vector) whose distribution depends on a parameter $\theta \in \Theta$. A real-valued (or vector-valued) function T of X is said to be *sufficient* for θ if the conditional distribution of X, given $T = t$, is independent of θ (except perhaps for t in a set A for which $P_\theta(T \in A) = 0$ for all θ).

We illustrate by proving that the total $T = \sum_{i=1}^{n} X_i$ is a sufficient statistic for p. This means *proving* that the conditional distribution of $(X_1, X_2, \cdots, X_n)$, given $T = t$, is independent of p. This conditional distribution is

$$f_{X_1,\cdots,X_n|T=t}(x_1, \cdots, x_n \mid p) = \frac{P(X_1 = x_1, \cdots, X_n = x_n, T = t \mid p)}{P(T = t \mid p)}. \tag{3.30}$$

The denominator of this expression is the binomial probability

$$P(T = t \mid p) = \binom{n}{t} p^t(1 - p)^{n-t}. \tag{3.31}$$

The numerator is zero except when $x_1 + \cdots + x_n = t$ and each $x_i = 0$ or 1, and then

$$\begin{aligned} P(X_1 = x_1, \cdots, X_n = x_n, T = t \mid p) &= P(X_1 = x_1, \cdots, X_n = x_n \mid p) \\ &= p^{x_1}(1 - p)^{1-x_1} \cdots p^{x_n}(1 - p)^{1-x_n} \\ &= p^{\Sigma x_i}(1 - p)^{n-\Sigma x_i}. \end{aligned} \tag{3.32}$$

Thus

$$f_{X_1,\cdots,X_n|T=t}(x_1, \cdots, x_n \mid p) = \binom{n}{t}^{-1} \tag{3.33}$$

when $\sum x_i = t$ and each $x_i = 0$ or 1 and is equal to zero otherwise. This distribution is for all $t = 0, 1, 2, \cdots, n$ independent of p, which proves the sufficiency of T.

We illustrate by another example the problems involved with absolutely continuous distributions. Let $X_1, X_2, \cdots, X_n$ be a sample of size n from a normal distribution with mean θ and variance one. It might be suspected that $\bar{X} = (1/n) \sum X_i$ is a sufficient statistic for θ. This is correct and is proved below. More simply, we prove that $T = \sum X_i$ is sufficient for θ. We transform the variables $X_1, \cdots, X_n$ to $T, Y_2, \cdots, Y_n$, where

$$T = \sum X_i,\ Y_2 = X_2 - X_1,\ Y_3 = X_3 - X_1, \cdots, Y_n = X_n - X_1,$$

a one-to-one transformation. If the distribution of $Y_2, \cdots, Y_n$, given $T = t$, is independent of θ, then the distribution of $T, Y_2, \cdots, Y_n$, given $T = t$, is independent of θ, and hence the distribution of $X_1, X_2, \cdots, X_n$, given $T = t$, is independent of θ.

But each variable in the set $\{Y_2, \cdots, Y_n\}$ is uncorrelated with T [$\text{Cov}(T, Y_j) = \text{Cov}(T, X_j) - \text{Cov}(T, X_1) = 0$ by symmetry], hence $(Y_2, \cdots, Y_n)$ is independent of T. Therefore the conditional distribution of $(Y_2, \cdots, Y_n)$, given $T = t$, is exactly the unconditional distribution of $(Y_2, \cdots, Y_n)$, which is multivariate normal with mean $(0, \cdots, 0)$ [$EY_j = EX_j - EX_1 = 0$] and covariance matrix

$$\text{Cov}(Y_2, \cdots, Y_n) = \begin{pmatrix} 2 & 1 & \cdots & 1 \\ 1 & 2 & \cdots & 1 \\ \vdots & \vdots & & \vdots \\ 1 & 1 & \cdots & 2 \end{pmatrix} \tag{3.34}$$

Because this distribution is independent of θ, T is a sufficient statistic for θ.

Incidently, this analysis shows that $\bar{X}$ is independent of $(X_2 - X_1, \cdots, X_n - X_1)$, and therefore of any function of the differences of the X_i. For example, it follows immediately that $\bar{X}$ and $s^2 = (1/n) \sum (X_i - \bar{X})^2$ are independent.

Our definition of a sufficient statistic is not very convenient. Not only must we be able to guess what statistic is sufficient, but we must also go through a tedious process of determining whether our guess was correct. The following theorem remedies this situation. Although stated and proved for discrete variables, it is valid for absolutely continuous variables as well.

Theorem 1. *The Factorization Theorem.* Let X be a discrete random quantity whose probability mass function $f(x \mid \theta)$ depends on a parameter $\theta \in \Theta$. A function $T = t(X)$ is sufficient for θ if, and only if, the frequency function factors into a product of a function of $t(x)$ and θ and a function of x alone; that is,

$$f(x \mid \theta) = g(t(x), \theta)\, h(x). \tag{3.35}$$

Proof. Here $f(x \mid \theta)$ represents the probability mass function $f(x \mid \theta) = P_\theta(X = x)$. Suppose that T is sufficient for θ. Because the conditional distribution of X given T is independent of θ, we may write

$$\begin{aligned} P_\theta(X = x) &= P_\theta(X = x,\ T = t(x)) \\ &= P_\theta(T = t(x))\, P(X = x \mid T = t(x)), \end{aligned} \tag{3.36}$$

provided the conditional probability is well defined. Hence for those x for which $P_\theta(X = x) = 0$ for all θ we define $h(x) = 0$, and for those x for which $P_\theta(X = x) > 0$ for some θ we define

$$h(x) = P_\theta(X = x \mid T = t(x)),$$

which is independent of the particular θ we choose. With $g(t(x), \theta) = P_\theta(T = t(x))$, the factorization (3.35) is immediate. Conversely, suppose a factorization of the form (3.35) holds and fix t_0 for which $P_\theta(T = t_0) > 0$ for some $\theta \in \Theta$. Then

$$P_\theta(X = x \mid T = t_0) = \frac{P_\theta(X = x,\, T = t_0)}{P_\theta(T = t_0)}. \tag{3.37}$$

The numerator is zero for all θ whenever $t(x) \neq t_0$, and when $t(x) = t_0$ the numerator is simply $P_\theta(X = x)$. The denominator may be written

$$\begin{aligned} P_\theta(T = t_0) &= \sum_{x \in A(t_0)} P_\theta(X = x) \\ &= \sum_{x \in A(t_0)} g(t(x), \theta)\, h(x), \end{aligned}$$

where $A(t_0) = [x : t(x) = t_0]$. Hence

$$P_\theta(X = x \mid T = t_0) = \begin{cases} 0 & \text{if } t(x) \neq t_0, \\ \dfrac{g(t_0, \theta)\, h(x)}{g(t_0, \theta) \sum_{x' \in A(t_0)} h(x')} & \text{if } t(x) = t_0. \end{cases} \tag{3.38}$$

Thus $P_\theta(X = x \mid T = t_0)$ is independent of θ for all t_0 and θ for which it is defined. This completes the proof.

This theorem is valid not only for discrete variables but for quite arbitrary families of distributions. It is suggested that the reader familiar with measure theory consult Halmos and Savage (1948) for details. Surprisingly, the general formulation given there does not contain the theorem just proved, for in the general formulation it is assumed that the family of probability measures P_θ is dominated by a single σ-finite measure. In the discrete case that is equivalent to assuming that there is a denumerable set S for which $P_\theta(S) = 1$ for all $\theta \in \Theta$ (the set S not depending on θ). However, the general formulation would be more than enough to cover the absolutely continuous case with which we must now deal. The statement of the theorem for which we shall sketch the proof is that under certain regularity conditions a vector $\mathbf{T} = (T_1, \cdots, T_r)$ of functions of $\mathbf{X} = (X_1, \cdots, X_n)$ [that is, $\mathbf{T} = \mathbf{t}(\mathbf{X})$] is sufficient for θ if, and only if,

$$f_{\mathbf{X}}(\mathbf{x} \mid \theta) = g(\mathbf{t}(\mathbf{x}), \theta)\, h(\mathbf{x})$$

(except perhaps for x in a set A for which $P_\theta(A) = 0$ for all θ). The exact statements of the regularity conditions under which such a factorization theorem is valid seem hardly worth mentioning in detail, for in the general formulation we get by with making only the usual requirements of measurability. The reader interested in the exact statements of the regularity conditions under which the following proof of the factorization theorem is valid for absolutely continuous variables may consult the original paper of Neyman (1935), wherein the first rigorous proof of a factorization theorem is given. However, in the sequel we use the factorization theorem in the absolutely continuous case without bothering about regularity conditions.

Sketch of the proof in the absolutely continuous case. First we add new functions of x, $u_{r+1}(x), \cdots, u_n(x)$, so that the transformation $x_1, \cdots, x_n \to t_1, \cdots, t_r, u_{r+1}, \cdots, u_n$ is one-to-one and smooth enough for the Jacobian to exist. The densities transform thus:

$$\begin{aligned} f_{\mathbf{X}}(\mathbf{x} \mid \theta) &= f_{\mathbf{T},\mathbf{U}}(\mathbf{t}, \mathbf{u} \mid \theta) \left| \frac{\partial(\mathbf{t}, \mathbf{u})}{\partial \mathbf{x}} \right| \\ &= f_{\mathbf{T}}(\mathbf{t} \mid \theta) f_{\mathbf{U}\mid\mathbf{T}=\mathbf{t}}(\mathbf{u} \mid \theta) \left| \frac{\partial(\mathbf{t}, \mathbf{u})}{\partial \mathbf{x}} \right|. \end{aligned} \tag{3.39}$$

If $\mathbf{T}$ is sufficient, then $f_{\mathbf{U}\mid\mathbf{T}=\mathbf{t}}(\mathbf{u} \mid \theta)$ is independent of θ, giving the required factorization. Conversely, if a factorization exists, then

$$\begin{aligned} f_{\mathbf{T},\mathbf{U}}(\mathbf{t}, \mathbf{u} \mid \theta) &= f_{\mathbf{X}}(\mathbf{x}(\mathbf{t}, \mathbf{u}) \mid \theta) \left| \frac{\partial \mathbf{x}}{\partial(\mathbf{t}, \mathbf{u})} \right| \\ &= g(\mathbf{t} \mid \theta)\, h(\mathbf{x}(\mathbf{t}, \mathbf{u})) \left| \frac{\partial \mathbf{x}}{\partial(\mathbf{t}, \mathbf{u})} \right|, \end{aligned} \tag{3.40}$$

so that

$$f_{\mathbf{T}}(\mathbf{t} \mid \theta) = g(\mathbf{t} \mid \theta) \int h(\mathbf{x}(\mathbf{t}, \mathbf{u})) \left| \frac{\partial \mathbf{x}}{\partial(\mathbf{t}, \mathbf{u})} \right| du, \tag{3.41}$$

and we have

$$f_{\mathbf{U}|\mathbf{T}=\mathbf{t}}(\mathbf{u} \mid \theta) = \frac{f_{\mathbf{T},\mathbf{U}}(\mathbf{t}, \mathbf{u} \mid \theta)}{f_{\mathbf{T}}(\mathbf{t} \mid \theta)}$$

$$= h(\mathbf{x}(\mathbf{t}, \mathbf{u})) \left| \frac{\partial \mathbf{x}}{\partial (\mathbf{t}, \mathbf{u})} \right| \bigg/ \int h(\mathbf{x}(\mathbf{t}, \mathbf{u})) \left| \frac{\partial \mathbf{x}}{\partial (\mathbf{t}, \mathbf{u})} \right| d\mathbf{u}, \tag{3.42}$$

independent of θ. Thus the distribution of $\mathbf{U}$, given $\mathbf{T}$, is independent of θ; hence the distribution of $(\mathbf{T}, \mathbf{U})$, given $\mathbf{T}$, is independent of θ, and the distribution of $\mathbf{X}$, given $\mathbf{T}$, is independent of θ.

We turn now to some more examples. Consider first a sample $X_1, \cdots, X_n$ from $\mathfrak{N}(\mu, \sigma^2)$. The joint density of $X_1, \cdots, X_n$ is

$$f_{X_1,\cdots,X_n}(x_1, \cdots, x_n \mid \mu, \sigma) = (2\pi\sigma^2)^{-n/2} \times \exp\left[-(2\sigma^2)^{-1} \sum_{i=1}^{n} (x_i - \mu)^2\right] \tag{3.43}$$

If μ is a known quantity, then from the factorization theorem $\sum(X_i - \mu)^2$ is a sufficient statistic for σ^2. [In this case the function $h(\mathbf{x})$ may be taken identically equal to one.] Let $\bar{x} = (1/n) \sum x_i$ and $s^2 = (1/n) \sum (x_i - \bar{x})^2$, so that the density (3.43) may be written

$$f_{X_1,\cdots,X_n}(x_1, \cdots, x_n \mid \mu, \sigma) = (2\pi\sigma^2)^{-n/2} \times \exp(-ns^2/2\sigma^2) \cdot \exp[-n(\bar{x} - \mu)^2/2\sigma^2]. \tag{3.44}$$

If σ^2 is a known quantity, then from the factorization theorem $\bar{X}$ is a sufficient statistic for μ. (This was essentially proved earlier.) If both μ and σ^2 are unknown, the pair $(\bar{x}, s^2)$ is a sufficient statistic for (μ, σ^2).

As a second example, consider a sample $X_1, \cdots, X_n$ from the uniform distribution $\mathfrak{U}(\alpha, \beta)$. The joint density is

$$f_{X_1,\cdots,X_n}(x_1, \cdots, x_n) = (\beta - \alpha)^{-n} \prod_{i=1}^{n} I_{(\alpha,\beta)}(x_i). \tag{3.45}$$

This may be written as

$$f_{X_1,\cdots,X_n}(x_1, \cdots, x_n) = (\beta - \alpha)^{-n} I_{(\alpha,\infty)}(\min x_i)\, I_{(-\infty,\beta)}(\max x_i) \tag{3.46}$$

to show (a) if α is known, max x_i is a sufficient statistic for β; (b) if β is known, min x_i is a sufficient statistic for α; (c) if both α and β are unknown, (min X_i, max X_i) is a sufficient statistic for (α, β).

Exercises

1. Assume that $X_1, X_2, \cdots, X_n$ are independent random variables.
(a) If $X_j \in \mathcal{B}(n_j, p)$ for $j = 1, \cdots, n(n_j$ known), then $\sum X_j$ is sufficient for p. The distribution of $\sum X_j$ is $\mathcal{B}(\sum n_j, p)$.
(b) If $X_j \in \mathcal{NB}(r_j, p)$ for $j = 1, \cdots, n(r_j$ known), then $\sum X_j$ is sufficient for p. The distribution of $\sum X_j$ is $\mathcal{NB}(\sum r_j, p)$.
(c) If $X_j \in \mathcal{P}(\lambda)$ for $j = 1, \cdots, n$, then $\sum X_j$ is sufficient for λ. The distribution of $\sum X_j$ is $\mathcal{P}(n\lambda)$.
(d) If $X_j \in \mathcal{G}(\alpha, \beta)$ for $j = 1, \cdots, n$, then $(\prod X_j, \sum X_j)$ is sufficient for (α, β). If α is known, $\sum X_j$ is sufficient for β. If β is known, $\prod X_j$ is sufficient for α. The distribution of $\sum X_j$ is $\mathcal{G}(n\alpha, \beta)$.
(e) If $X_j \in \mathcal{B}e(\alpha, \beta)$ for $j = 1, 2, \cdots, n$, then $(\prod X_j, \prod(1 - X_j))$ is sufficient for (α, β). If α is known, then $\prod(1 - X_j)$ is sufficient for β. If β is known, then $\prod X_j$ is sufficient for α.
2. *Undominated case.* Suppose that $X_1, \cdots, X_n$ are independent, identically distributed, random variables with probability mass function
$$f(x \mid \theta, p) = (1 - p)p^{x-\theta}, \qquad x = \theta, \theta + 1, \theta + 2, \cdots,$$
where θ and p are unknown parameters, $0 < p < 1$. Then $(\min X_j, \sum X_j)$ is sufficient for (θ, p). If p is known, then $\min X_j$ is sufficient for θ. If θ is known, then $\sum X_j$ is sufficient for p.
3. Show that if $X_1, \cdots, X_n$ is a sample from any family of distributions, $F(x \mid \theta)$, then the vector of order statistics $X_{(1)}, X_{(2)}, \cdots, X_{(n)}$, where $X_{(1)}$ = smallest X_i, $X_{(2)}$ = next smallest $X_i, \cdots, X_{(n)}$ = largest X_i is sufficient for θ.
4. If $\mathbf{X}_1, \mathbf{X}_2, \cdots, \mathbf{X}_n$ is a sample from a k-dimensional normal distribution $\mathcal{N}(\boldsymbol{\mu}, \boldsymbol{\Sigma})$, the sample mean and covariance,
$$\bar{\mathbf{X}} = \frac{1}{n}\sum_{i=1}^{n} \mathbf{X}_i, \quad \mathbf{S}^2 = \frac{1}{n}\sum_{i=1}^{n} (\mathbf{X}_i - \bar{\mathbf{X}})(\mathbf{X}_i - \bar{\mathbf{X}})^T$$
is sufficient for $(\boldsymbol{\mu}, \boldsymbol{\Sigma})$. (Treat the case of nonsingular $\boldsymbol{\Sigma}$ first.)

3.4 Essentially Complete Classes of Rules Based on Sufficient Statistics

We now formalize, in decision theoretic terminology, the notion that a sufficient statistic, according to the definition, carries all the information the sample has to give about the true value of the parameter. This formalization states that the class of decision rules based on a sufficient statistic forms an essentially complete class. The proof is carried out for the game $(\Theta, \mathfrak{D}, \hat{R})$, where behavioral randomized strategies are used by the statistician. A decision function $\delta(x)$ from the sample space to $\mathcal{A}^*$ is said to be based on a statistic $T(x)$, if $\delta(x)$ is a function of $T(x)$, that is, if $\delta(x) = \delta(x')$ whenever $T(x) = T(x')$.

Theorem 1. Consider the game $(\Theta, \mathcal{A}, L)$ where the statistician observes a random vector $\mathbf{X}$ whose distribution depends on θ. If T is a sufficient statistic for θ, then the set $\mathfrak{D}_0$ of decision rules in $\mathfrak{D}$, which are based on T, forms an essentially complete class in the game $(\Theta, \mathfrak{D}, \hat{R})$.

Proof. Let $\delta \in \mathfrak{D}$. We shall find an element δ_0 of $\mathfrak{D}_0$ which is as good as δ; in fact, we shall have $\hat{R}(\theta, \delta) = \hat{R}(\theta, \delta_0)$ for all $\theta \in \Theta$. The idea of the proof is as follows. After observing $T = t$, we choose X' at random from the distribution of X, given $T = t$. We may do so without knowing θ, for T is sufficient for θ. Then we choose an action in $\mathcal{A}$, according to the distribution $\delta(x')$, where x' is the observed value of X'. For each θ, X and X' have the same (unconditional) distribution; hence the risk functions must be identical. Because the second procedure involves no other knowledge of X than that given by T (plus a little extra randomization), this procedure must be in $\mathfrak{D}_0$.

More precisely, for a given $\delta \in \mathfrak{D}$, define $\delta_0(x)$ as that distribution over $\mathcal{A}$ obtained by first choosing X' at random from the distribution of X, given $T = t(x)$, and then, after observing $X' = x'$, choosing $a \in \mathcal{A}$ according to the distribution $\delta(x')$. The distribution $\delta_0(x)$ is the same for all x that give the same value of $t(x)$, hence $\delta_0 \in \mathfrak{D}_0$. Furthermore, if $t(x) = t$, then $L(\theta, \delta_0(x)) = E(L(\theta, \delta(X')) \mid T = t)$ [that is, the average loss using $\delta_0(x)$ is the average of the average losses using $\delta(x')$]. Hence

$$\begin{aligned}\hat{R}(\theta, \delta_0) &= E_\theta L(\theta, \delta_0(X)) = E_\theta E(L(\theta, \delta(X')) \mid T) \\ &= E_\theta E(L(\theta, \delta(X)) \mid T) = E_\theta(L(\theta, \delta(X))) = \hat{R}(\theta, \delta),\end{aligned}$$

completing the proof.

The rule δ_0 in this proof could be written symbolically as

$$\delta_0(t) = E(\delta(X) \mid T = t), \tag{3.47}$$

for $\delta_0(t)$ is the mixture of the distributions $\delta(x)$ for which $T(x) = t$. Note that even if δ were a nonrandomized decision rule δ_0 could be a randomized decision rule.

In those situations in which (a) the class of decision rules based on a sufficient statistic is essentially complete and (b) the nonrandomized decision rules form an essentially complete class (as in Theorem 2.8.1) the question is whether the nonrandomized rules based on a sufficient statistic form an essentially complete class. This does not follow directly, for the intersection of two essentially complete classes need not be essentially complete (Exercise 2.1.7). However, under the assumptions of Theorem 2.8.1, we can reason as follows. For any rule $\delta \in \mathfrak{D}$ there is a rule $\delta' \in \mathfrak{D}$ which is based on a sufficient statistic and is as good as δ. For such a rule δ' the proof of Theorem 2.8.1 yields a nonrandomized rule d, which is based on δ', hence on the sufficient statistic, and is as good as δ'. Thus d is as good as δ. The details of the proof of Theorem 2 are left to the reader.

Theorem 2. Let $\mathfrak{A}$ be a convex subset of E_k, let $L(\theta, \mathbf{a})$ be a convex function of $\mathbf{a} \in \mathfrak{A}$ for each $\theta \in \Theta$ and suppose that T is a sufficient statistic for θ. If, for some $\theta \in \Theta$, there exists an $\epsilon > 0$ and a c such that $L(\theta, \mathbf{a}) \geq \epsilon \mid \mathbf{a} \mid + c$, then the class D_0 of nonrandomized decision rules based on T is essentially complete in the game $(\Theta, \mathfrak{D}, \hat{R})$.

Closely related to this topic is the famous Rao-Blackwell theorem, which in conjunction with Theorem 2.8.1 may be used in place of Theorem 1 to give a proof of Theorem 2. Especially useful in the Rao-Blackwell theorem is the explicit formula by which a nonrandomized rule may be improved by a nonrandomized rule based on a sufficient statistic.

Theorem 3. (*Rao-Blackwell*). Let $\mathfrak{A}$ be a convex subset of E_k, let $L(\theta, \mathbf{a})$ be a convex function of $\mathbf{a} \in \mathfrak{A}$ for each $\theta \in \Theta$, and suppose that T is a sufficient statistic for θ. If $d(x)$ is a nonrandomized decision rule, then the nonrandomized decision rule based on T

$$\hat{\theta}(t) = E(d(X) \mid T = t), \tag{3.48}$$

provided this expectation exists (except perhaps for t in a set of probability zero), is as good as d.

Proof. Because T is sufficient for θ, we are correct in writing the expectation in (3.48) without a subscript θ (the distribution of X given

T does not depend on θ). If $\hat{\theta}$ defined by (3.48) exists, then from Jensen's inequality

$$E(L(\theta, d(X)) \mid T = t) \geq L(\theta, E(d(X) \mid T = t)) = L(\theta, \hat{\theta}(t)) \tag{3.49}$$

for all θ, so that

$$\begin{aligned} R(\theta, d) &= E_\theta L(\theta, d(X)) = E_\theta[E(L(\theta, d(X)) \mid T)] \\ &\geq E_\theta[L(\theta, \hat{\theta}(T))] = R(\theta, \hat{\theta}), \end{aligned} \tag{3.50}$$

thus completing the proof.

Frequently the Rao-Blackwell theorem is stated with the additional hypothesis that $d(X)$ is an unbiased estimate of θ (that is, $E_\theta d(X) \equiv \theta$), and an additional conclusion that $\hat{\theta}(T)$ is also unbiased. This follows immediately from (3.48), for

$$E_\theta \hat{\theta}(T) = E_\theta(E(d(X) \mid T)) = E_\theta d(X) \equiv \theta.$$

[See Rao (1945) or Blackwell (1947).]

We illustrate the use of (3.48) with some examples. Consider the problem of estimating the mean θ of a normal distribution with variance one from a sample of size n, $X_1, X_2, \cdots, X_n$, using squared error loss, $L(\theta, a) = (\theta - a)^2$. In Section 3.3 it was found that $T = \sum_1^n X_i$ is a sufficient statistic for θ. A reasonable estimate of θ, which many statisticians would prefer when the assumption of normality is doubtful but when symmetry seems reasonable, is the median of the X_i (defined here as the central value if n is odd and the mean of the two central values if n is even). If the assumption of normality is exactly satisfied, an improved estimate is $\hat{\theta}(T) = E(\text{median } X_i \mid T)$. An elementary argument (Exercise 1) involving symmetry shows that $\hat{\theta}(T) = (1/n)T = \bar{X}$. In fact, both estimates are unbiased and the variance of $\bar{X}$ is $1/n$, whereas the variance of the median is for large n approximately $\pi/2n$. Note that $\bar{X}$ is as good as median X_i no matter what the loss function happens to be, provided it is convex in a for each θ; for example, $L(\theta, a) = |\theta - a|$ or $L(\theta, a) = (\theta^2 - a)^2$.

With respect to the loss function $L(\theta, a) = (\theta^2 - a)^2$, neither of these estimates is very good. Because $EX_j^2 = 1 + \theta^2$, we might expect the estimate $U = (1/n)\sum X_j^2 - 1$ to be reasonable, but we know that $E(U \mid T)$ is bound to be as good if not better (again irrespective of the loss function, provided it is convex). The transformation of page 114 may be applied to $X_1, \cdots, X_n$, so that $X_1 = (1/n)(T - Y_2 - \cdots - Y_n)$. Then, using the independence of T and $(Y_2, \cdots, Y_n)$ and the fact that

Cov $(Y_i, Y_j) = 1, i \neq j$, Cov $(Y_i, Y_i) = 2$, we find that $E(X_1^2 \mid T) = \bar{X}^2 + (n - 1)/n$. By symmetry $E(X_j^2 \mid T) = E(X_1^2 \mid T)$ so that $E(U \mid T) = \bar{X}^2 - 1/n$.

An Example in Bio-assay. To avoid the impression that improving estimates by means of (3.48) is a rather academic exercise we mention a problem of a more practical nature described in an article by Berkson (1955). The bio-assay problem with which we are concerned is as follows. The experimenter may test the potency of a certain drug by giving groups of animals injections of the drug at certain levels. Each animal is assumed to show either of two possible responses, positive or negative. The probability of a positive response to a dosage at level x is denoted by $P(x)$. We consider the special problem in which the response curve is given by the distribution function $P(x) = [1 + e^{-(\alpha+\beta x)}]^{-1}$ (the logistic reparameterized) where $-\infty < \alpha < \infty$ and $0 < \beta < \infty$ are unknown parameters to be estimated. The experimenter chooses N dose levels, $x_1, \cdots, x_N$, and assigns $n_1, n_2, \cdots, n_N$ animals, respectively, to those levels. If $Y_1, Y_2, \cdots, Y_N$ represent the number of positive responses at respective levels $x_1, x_2, \cdots, x_N$, then $Y_i \in \mathcal{B}(n_i, P(x_i)), i = 1, 2, \cdots, N$, so that their joint probability mass function is

$$f_{\mathbf{Y}}(\mathbf{y}) = \prod_{i=1}^{N} \binom{n_i}{y_i} P(x_i)^{y_i}(1 - P(x_i))^{n_i - y_i}$$

$$= \left(\prod_{i=1}^{N} \binom{n_i}{y_i}(1 - P(x_i))^{n_i}\right) \exp\left(\sum_{i=1}^{N} (\alpha + \beta x_i) y_i\right). \quad (3.51)$$

Two time-honored estimates of the unknown parameters α and β are *the maximum likelihood estimates* (the values of α and β that maximize the likelihood function (3.51) in which the y_i are replaced by the observed values of the Y_i) and *the minimum* χ^2 *estimates* (the values of α and β that minimize the χ^2 statistic

$$\chi^2 = \sum_{i=1}^{N} \frac{n_i(P(x_i) - \hat{P}_i)^2}{P(x_i)(1 - P(x_i))}, \quad (3.52)$$

where $\hat{P}_i = Y_i/n_i$). To avoid the formidable task involved in the numerical computation of either of these estimates, Berkson in 1944 suggested alternative estimates which he called the *minimum logit-*χ^2 *estimates*. If p is a number between zero and one, then logit(p) is defined as $\log(p/(1 - p))$; thus logit $(P(x_i)) = \alpha + \beta x_i$. The logit-$\chi^2$ of

Berkson is

$$\text{logit-}\chi^2 = \sum_{i=1}^{N} n_i\hat{P}_i(1 - \hat{P}_i)(\alpha + \beta x_i - \text{logit}(\hat{P}_i))^2. \quad (3.53)$$

The minimum logit-χ^2 estimates, that is, the values of α and β which minimize this expression, are found by differentiation to be the solutions of two simultaneous linear equations and are thus easy to write out explicitly (Exercise 4).

Berkson originally suggested the minimum logit-χ^2 estimates on the basis of their simple evaluation. However, it was later found that these estimates have all the large sample optimal properties of the maximum likelihood and minimum χ^2 estimates. Furthermore, Berkson was prompted to compare the *small sample* behavior of these estimates with $N = 3$, $n_1 = n_2 = n_3 = 10$, and various choices of x_1, x_2, x_3, α and β. This led to the surprising result that in the cases Berkson investigated the minimum logit-χ^2 estimates are better in the sense of smaller mean squared error than either the maximum likelihood estimates or the minimum χ^2 estimates.

Now comes the topper to this story and the reason that this problem is being described in this section. It was then noticed that there are sufficient statistics for this problem, namely, $\sum_1^N Y_i$ and $\sum_1^N x_iY_i$ [as is easily seen from (3.51)], and that the minimum logit-χ^2 estimates of α and β are not functions of the sufficient statistics. Consequently, Berkson was able to improve on the minimum logit-χ^2 estimates by an application of (3.48), a process that he dubbed Rao-Blackwellization. The maximum likelihood estimates, on the other hand, are already functions of the sufficient statistics as is usually the case (see Exercise 3), so that these estimates are not subject to improvement by Rao-Blackwellization. As a result, the Rao-Blackwellized minimum logit-χ^2 estimates are a substantial improvement over the maximum likelihood estimates in the cases investigated by Berkson. Of course, the Rao-Blackwellized minimum logit-χ^2 estimates have lost one appealing feature that the ordinary minimum logit-χ^2 estimates had—their ease of evaluation.

Exercises

1. For the example in the text show that $E(\text{median } X_i \mid T) = T/n$.
2. Let $X_1, X_2, \cdots, X_n$ be a sample of size n from the uniform distribution $\mathcal{U}(\alpha, \beta)$ with α and β unknown. It is required to estimate the mean $\theta = (\alpha + \beta)/2$, with squared error loss $L((\alpha, \beta), a) = (\theta - a)^2$.

The sample mean $\bar{X}$ is a reasonable estimate. Show that $E(\bar{X} \mid \max X_i, \min X_i)$ is an improvement over $\bar{X}$ and that

$$E(\bar{X} \mid \max X_i, \min X_i) = \frac{\max X_i + \min X_i}{2}.$$

3. Show that if T is a sufficient statistic for a parameter θ and the factorization theorem holds, then the maximum likelihood estimate is a function of T.

4. (a) Find the minimum logit-χ^2 estimates of α and β when they can be determined for the bio-assay problem in the text.
(b) Take $N = 3$, $n_1 = n_2 = n_3 = 10$, $x_1 = -1$, $x_2 = 0$, $x_3 = 1$. Evaluate the minimum logit-χ^2 estimates of α and β when $Y_1 = 0$, $Y_2 = 4$, $Y_3 = 9$. Evaluate the Rao-Blackwellized minimum logit-χ^2 estimates in this case. ($\log 2 = 0.69315$, $\log 3 = 1.09861$) ($p(1-p)$ logit (p) is defined as zero if $p = 0$ or 1).

5. Suppose that T is a sufficient statistic for a parameter θ and that the factorization theorem holds. Show that (nonrandomized) Bayes rules are functions of T.

6. If, in the Rao-Blackwell theorem, $L(\theta, a)$ is strictly convex in a for each $\theta \in \Theta$ and the distribution of $d(X)$, given $T = t$, is nondegenerate for a set of t with positive probability under some θ, the rule of (3.48) is better than d.

3.5 Exponential Families of Distributions

Let $X_1, X_2, \cdots, X_n$ be a sample of size n from a distribution with distribution function $F(x \mid \theta)$, $\theta \in \Theta$. There is a trivial set of sufficient statistics for θ; the entire set of observations is always sufficient. The conditional distribution of $X_1, \cdots, X_n$, given $X_1, \cdots, X_n$, is degenerate at a point independent of θ. It is of interest to discover families of distributions, $F(x \mid \theta)$, for which there exists a sufficient statistic of fixed dimension k, $(T_1, \cdots, T_k)$, irrespective of the sample size n. The family of normal distributions $\mathfrak{N}(\mu, \sigma^2)$ is an example in which for any sample size the two-dimensional statistic $(\sum X_i, \sum X_i^2)$ (or, equivalently, the sample mean and the sample variance) is sufficient for (μ, σ^2). The main families of distributions with this property are the exponential families of distributions.

We consider only the cases in which the underlying distribution is either discrete or absolutely continuous. For a treatment of the general case the reader is referred to Lehmann's *Testing Statistical Hypotheses* (1959). The probability mass function in the discrete case, or the density in the absolutely continuous case is denoted by $f(x \mid \theta)$.

Definition 1. A family of distributions on the real line with probability mass function or density $f(x \mid \theta)$, $\theta \in \Theta$, is said to be an *exponential family of distributions* if $f(x \mid \theta)$ is of the form

$$f(x \mid \theta) = c(\theta)\, h(x) \exp\Big[\sum_{i=1}^{k} \pi_i(\theta)\, t_i(x)\Big]. \tag{3.54}$$

Because $f(x \mid \theta)$ is a probability mass function or density of a distribution, the function $c(\theta)$ is determined by the functions $h(x)$, $\pi_i(\theta)$, and $t_i(x)$ by means of the formulas

$$c(\theta)^{-1} = \sum_{x} h(x) \exp\Big[\sum_{i=1}^{k} \pi_i(\theta)\, t_i(x)\Big] \tag{3.55}$$

in the discrete case and

$$c(\theta)^{-1} = \int h(x) \exp\Big[\sum_{i=1}^{k} \pi_i(\theta)\, t_i(x)\Big]\, dx \tag{3.56}$$

in the absolutely continuous case.

The exponential families of distributions are frequently called the "Koopman-Pitman-Darmois" families of distributions from the fact that these three authors independently and almost simultaneously (1937–1938) studied their main properties.

If $X_1, \cdots, X_n$ is a sample of size n from an exponential family of distributions with mass function or density (3.54), then

$$\mathbf{T} = (T_1, \cdots, T_k) = \Big(\sum_{1}^{n} t_1(X_j), \cdots, \sum_{1}^{n} t_k(X_j)\Big)$$

is a sufficient statistic, as may be seen from the factorization theorem applied to the joint probability mass function or density

$$f(x_1, \cdots, x_n \mid \theta) = c(\theta)^n \Big(\prod_{j=1}^{n} h(x_j)\Big) \exp\Big[\sum_{i=1}^{k} \pi_i(\theta) \sum_{j=1}^{n} t_i(x_j)\Big]. \tag{3.57}$$

As an example, note that for the binomial distribution $\mathcal{B}(m, \theta)$ with m known and $0 < \theta < 1$ the probability mass function may be written

$$f(x \mid \theta) = (1 - \theta)^m \binom{m}{x} \exp\left[x(\log\theta - \log(1 - \theta))\right],$$

so that this family of distributions is an exponential family with $c(\theta) = (1 - \theta)^m$, $h(x) = \binom{m}{x}$ for $x = 0, 1, 2, \cdots, m$, $k = 1$, $\pi_1(\theta) = \log\theta - \log(1 - \theta)$, and $t_1(x) = x$. Hence for a sample of size n, $\sum_1^n X_j$ is sufficient for θ. Other examples of exponential families of distributions are the Poisson distributions $\mathcal{P}(\lambda)$, the negative binomial distributions $\mathcal{NB}(r, \theta)$ with r known, the normal distributions $\mathcal{N}(\mu, \sigma^2)$ with $\theta = (\mu, \sigma^2)$, the gamma distributions $\mathcal{G}(\alpha, \beta)$ with $\theta = (\alpha, \beta)$, and the beta distributions $\mathcal{B}e(\alpha, \beta)$ with $\theta = (\alpha, \beta)$. Examples of important families of distributions that are not exponential families are the uniform distributions $\mathcal{U}(\alpha, \beta)$ and the Cauchy distributions $\mathcal{C}(\alpha, \beta)$.

If $X_1, \cdots, X_n$ is a sample of size n from the exponential family (3.54), the marginal distributions of the sufficient statistic $\mathbf{T} = [\sum t_1(X_j), \cdots, \sum t_k(X_j)]$ also form an exponential family as Lemma 1 indicates.

Lemma 1. Let $X_1, \cdots, X_n$ be a sample from the exponential family (3.54). If (3.54) is a probability mass function, the distribution of the sufficient statistic $\mathbf{T}$ has probability mass function of the form

$$f_{\mathbf{T}}(\mathbf{t} \mid \theta) = c_0(\theta)\, h_0(\mathbf{t}) \exp\left[\sum_{i=1}^{k} \pi_i(\theta) t_i\right]. \tag{3.58}$$

If (3.54) is a density and a k-dimensional density of $\mathbf{T}$ exists, this density has the form of (3.58).

Proof in the discrete case. Because the joint mass function of $X_1, \cdots, X_n$ is given by (3.57), the joint mass function of $T_1, \cdots, T_k$ is

$$\begin{aligned} f_{\mathbf{T}}(\mathbf{t} \mid \theta) &= P_\theta(T_1 = t_1, \cdots, T_k = t_k) \\ &= c(\theta)^n\, h_0(\mathbf{t}) \exp\left[\sum_{i=1}^{k} \pi_i(\theta) t_i\right], \end{aligned}$$

where $h_0(\mathbf{t})$ is the sum of $\prod_{j=1}^n h(x_j)$ over all $(x_1, \cdots, x_n)$ for which $\sum_{j=1}^n t_1(x_j) = t_1, \cdots, \sum_{j=1}^n t_k(x_j) = t_k$, completing the proof.

A proof of Lemma 1 for densities that satisfy sufficient regularity conditions may be carried through in a manner similar to the sketch of the proof of the factorization theorem in the absolutely continuous case. However, this lemma is valid without stringent regularity conditions (as, for example, in Lehmann (1959), 52). For this reason we omit the proof in the absolutely continuous case.

It is worthwhile to investigate a little of the structure of exponential families of distributions. If each π_i is taken to be a parameter in (3.54), so that

$$f(x \mid \boldsymbol{\pi}) = c(\boldsymbol{\pi})\, h(x) \exp\left[\sum_{i=1}^{k} \pi_i t_i(x)\right], \tag{3.59}$$

where $\boldsymbol{\pi} = (\pi_1, \cdots, \pi_k)$, we say that the exponential family has been given its *natural parametrization*. The *natural parameter space* Π is that subset of E_k for which the formula (3.59) represents a density, that is, for which $c(\boldsymbol{\pi})$ is finite and positive:

$$c(\boldsymbol{\pi})^{-1} = \int h(x) \exp\left[\sum_{i=1}^{k} \pi_i t_i(x)\right] dx < \infty \tag{3.60}$$

(with the integral replaced by a sum in the discrete case). Obviously, for any $\theta \in \Theta$, the point $(\pi_1(\theta), \cdots, \pi_k(\theta))$ must lie in Π.

Lemma 2. The natural parameter space Π is convex.

Proof. Let $\boldsymbol{\pi} = (\pi_1, \cdots, \pi_k)$ and $\boldsymbol{\pi}' = (\pi_1', \cdots, \pi_k')$ be points of Π and let $0 < \alpha < 1$. Then by Hölder's inequality (see Exercise 2.8.7)

$$\int h(x) \exp\left[\sum_{i=1}^{k} (\alpha\pi_i' + (1-\alpha)\pi_i) t_i(x)\right] dx$$

$$= c(\boldsymbol{\pi})^{-1} \int \left\{\exp\left[\sum_{i=1}^{k} (\pi_i' - \pi_i)\, t_i(x)\right]\right\}^{\alpha} 1^{1-\alpha}$$

$$\times\, c(\boldsymbol{\pi})\, h(x) \exp\left[\sum_{i=1}^{k} \pi_i t_i(x)\right] dx$$

$$\leq c(\boldsymbol{\pi})^{-1} \left\{\int c(\boldsymbol{\pi})\, h(x) \exp\left[\sum_{i=1}^{k} \pi_i' t_i(x)\right] dx\right\}^{\alpha}.$$

This shows that $\alpha\boldsymbol{\pi}' + (1-\alpha)\boldsymbol{\pi} \in \Pi$, thus completing the proof. (If the distribution is discrete, the integral sign should be replaced everywhere by a summation sign.)

Although the natural parameter space Π is a convex subset of E_k, it may be of effective dimension less than k, for it may be contained in some hyperplane of E_k. If this is so, the π_i satisfy some linear equation, and one of the π_i may be eliminated by solving for it and substituting the result into (3.59), thus reducing the number of parameters by one. We may assume that any such reduction in the number of parameters has already been made. In other words, we may assume without loss of generality that Π contains an open set in E_k.

An important regularity property of exponential families of distributions is contained in Lemma 3, whose proof we omit because it is contained in Lehmann (1959), 52–53, and because it does not lend much insight. [See also Widder (1946), 240–241.]

Lemma 3. If $\phi(x_1, \cdots, x_n)$ is a function for which the integral

$$\int \cdots \int \phi(x_1, \cdots, x_n) \exp\left[\sum_{i=1}^{k} \pi_i \sum_{j=1}^{n} t_i(x_j)\right] \prod_{j=1}^{n} h(x_j)\, dx_j$$

exists for all $\boldsymbol{\pi}$ in the natural parameter space Π, this integral is an analytic function of $\boldsymbol{\pi}$ at all interior points of Π, and derivatives of all orders with respect to $\boldsymbol{\pi} \in \Pi$ may be passed beneath the integral sign. (For discrete exponential families the integral is replaced by a sum.)

In particular, the function $c(\boldsymbol{\pi})$ in (3.59) is an analytic function of $\boldsymbol{\pi}$ at all interior points of Π, hence also the function $E_{\pi}\phi(X_1, \cdots, X_n)$ for bounded measurable ϕ. If in (3.54) the functions $\boldsymbol{\pi}(\theta)$ are analytic functions of θ on the real line, then $E_\theta\phi(X_1, \cdots, X_n)$ is also analytic in θ for bounded measurable ϕ.

One important characteristic of exponential families of distributions is that the set on which the density is positive does not depend on the parameter θ. The set on which the density (mass function) (3.54) is zero is exactly the set on which $h(x) = 0$. For this reason the uniform distributions $\mathfrak{U}(\alpha, \beta)$ are clearly not an exponential family.

The importance of exponential families of distributions stems from the following result. If a distribution with density (mass function) $f(x \mid \theta)$ has the property that there exists a sufficient statistic $(T_1, \cdots, T_k)$ of fixed dimension, whatever the size of the sample drawn from this distribution, and if the set on which $f(x \mid \theta)$ is zero does not depend on θ and certain mild regularity conditions are satisfied, the distributions form an exponential family. If the set on which $f(x \mid \theta)$ is zero is allowed to depend on θ, there are other families of distributions—the family of uniform distributions is an example—for which there exists a fixed dimensional sufficient statistic, whatever the sample size. These are sometimes referred to as "nonregular" families of distributions admitting a fixed dimensional sufficient statistic, whereas the "regular" families of distributions are the exponential families. A general exposition of these problems at an elementary level is contained in an article by Dynkin (1951). We consider first some "nonregular" families of distributions which, like the family of uniform distributions, have sufficient statistics of fixed dimension due precisely to the fact that the set on which $f(x \mid \theta)$ is zero depends on θ.

Let $h(x)$ be a positive function of x. Let $\pi_1(\theta)$ and $\pi_2(\theta)$ be extended real-valued functions of θ for which $\pi_1(\theta) < \pi_2(\theta)$ ($\pi_1(\theta)$ is allowed to be $-\infty$ and $\pi_2(\theta)$ is allowed to be $+\infty$). Then, provided

$$c(\theta)^{-1} = \int_{\pi_1(\theta)}^{\pi_2(\theta)} h(x)\, dx$$

is finite,

$$f(x \mid \theta) = c(\theta)\, h(x) I_{(\pi_1(\theta), \pi_2(\theta))}(x), \qquad (3.61)$$

represents a density on the real line. The joint density of a sample of size n, $X_1, \cdots, X_n$, from this distribution is

$$f_{X_1,\cdots,X_n}(x_1, \cdots, x_n \mid \theta)$$

$$= c(\theta)^n \left(\prod_{j=1}^{n} h(x_j)\right) I_{(\pi_1(\theta),\infty)}(\min x_j) I_{(-\infty,\pi_2(\theta))}(\max x_j), \qquad (3.62)$$

so that from the factorization theorem ($\min X_j$, $\max X_j$) is a sufficient statistic for θ. In the natural parametrization for families of this type π_1 and π_2 are numbers rather than functions and the natural parameter space Π is the half-plane

$$\Pi = \left\{(\pi_1, \pi_2): -\infty \leq \pi_1 < \pi_2 \leq +\infty, \int_{\pi_1}^{\pi_2} h(x)\, dx < \infty\right\},$$

the density being

$$f(x \mid \boldsymbol{\pi}) = c(\boldsymbol{\pi})\, h(x) I_{(\pi_1, \pi_2)}(x). \qquad (3.63)$$

The uniform distributions are a special case in which $h(x) \equiv 1$.

We can generalize the families given by (3.54) and (3.61) by the families with densities of the form

$$f(x \mid \theta) = c(\theta)\, h(x) I_{(\pi_1(\theta), \pi_2(\theta))}(x) \exp\left[\sum_{i=3}^{k} \pi_i(\theta)\, t_i(x)\right], \qquad (3.64)$$

where $\pi_1(\theta) < \pi_2(\theta)$ for all θ. By the same methods used earlier we see that for a sample $X_1, \cdots, X_n$ the set of statistics $(\min X_j, \max X_j, \sum t_3(X_j), \cdots, \sum t_k(X_j))$ is sufficient for θ. As an example, the family of negative exponential distributions

$$f(x \mid \theta) = \beta^{-1} I_{(\alpha, \infty)}(x) \exp\left[-\frac{(x - \alpha)}{\beta}\right], \qquad \beta > 0, \qquad (3.65)$$

is of the form (3.64) with $\theta = (\alpha, \beta)$, $c(\theta) = \beta^{-1} \exp(\alpha/\beta)$, $h(x) \equiv 1$, $\pi_1(\theta) = \alpha$, $\pi_2(\theta) = \infty$, $\pi_3(\theta) = -1/\beta$, and $t_3(x) = x$. For a sample $X_1, \cdots, X_n$ the statistic $(\min X_j, \sum X_j)$ is sufficient for θ. If α is known, $\sum X_j$ is sufficient for β; if β is known, $\min X_j$ is sufficient for α.

Exercises

1. Show that the following families of distributions are exponential families.

(a) $\mathcal{P}(\lambda)$, $\lambda > 0$.
(b) $\mathfrak{NB}(r, \theta)$, r known, $0 < \theta < 1$.
(c) $\mathfrak{N}(\mu, \sigma^2)$, $\theta = (\mu, \sigma^2)$, $\sigma^2 > 0$.
(d) $\mathcal{G}(\alpha, \beta)$, $\theta = (\alpha, \beta)$, $\alpha > 0$ $\beta > 0$.
(e) $\mathcal{B}e(\alpha, \beta)$, $\theta = (\alpha, \beta)$, $\alpha > 0$ $\beta > 0$.

In each case make an explicit choice of the functions $c(\theta)$, $h(x)$, $\pi_i(\theta)$, and $t_i(x)$, exhibit a form of the sufficient statistic $\mathbf{T}$, and describe the natural parameter space.

2. Let $f(x \mid \boldsymbol{\pi})$ be given by (3.59). Show that the expectations and covariances of $t_1(X), \cdots, t_k(X)$ may be derived from the function $c(\boldsymbol{\pi})$ via the formulas

$$E_{\boldsymbol{\pi}} t_j(X) = -\frac{\partial}{\partial \pi_j} \log c(\boldsymbol{\pi}),$$

$$\operatorname{Cov}_{\boldsymbol{\pi}}(t_i(X), t_j(X)) = E_{\boldsymbol{\pi}}[(t_i(X) - E_{\boldsymbol{\pi}} t_i(X))(t_j(X) - E_{\boldsymbol{\pi}} t_j(X))]$$

$$= -\frac{\partial^2}{\partial \pi_i\, \partial \pi_j} \log c(\boldsymbol{\pi}),$$

valid at interior points of the natural parameter space.

3. Let $X_1, \cdots, X_n$, $n \geq 2$, be a sample from a distribution with density (3.63). Show that the distribution of the sufficient statistic $\mathbf{T} = (T_1, T_2) = (\min X_i, \max X_i)$ has the form

$$f(\mathbf{t} \mid \boldsymbol{\pi}) = c(\boldsymbol{\pi})^n h_0(\mathbf{t}) I_{(\pi_1,\infty)}(t_1) I_{(-\infty,\pi_2)}(t_2).$$

3.6 Complete Sufficient Statistics

Because a great reduction in the complexity of the data may occasionally be achieved by means of sufficient statistics, it is worthwhile to know how far such a reduction can be carried for a given problem. The smallest amount of data that is still sufficient for the parameter is called a *minimal sufficient statistic* or a *necessary and sufficient statistic.* The reader is referred to Dynkin (1951) in which it is shown that the likelihood ratio function (that is, $f(\mathbf{x} \mid \theta)/f(\mathbf{x} \mid \theta_0)$ as a function of θ for fixed $\theta_0 \in \Theta$) is, in the regular case, minimal sufficient. We restrict our discussion to a property that is somewhat stronger than minimality, namely completeness.

Definition 1. A sufficient statistic, T, for a parameter $\theta \in \Theta$ is said to be *complete* if for every real-valued function g,

$$E_\theta g(T) \underset{\theta}{\equiv} 0$$

implies

$$P_\theta\{g(T) = 0\} \underset{\theta}{\equiv} 1;$$

it is said to be *boundedly complete* if, for every bounded real-valued function g,

$$E_\theta g(T) \underset{\theta}{\equiv} 0 \quad \text{implies} \quad P_\theta\{g(T) = 0\} \underset{\theta}{\equiv} 1.$$

In other words, T is a (boundedly) complete sufficient statistic if it is sufficient and if, whenever g is a (bounded) function for which $E_\theta g(T)$ exists and is equal to zero for all $\theta \in \Theta$, we have $g(t) = 0$ for all t, except perhaps for t in a set N for which $P_\theta\{T \in N\} = 0$ for all $\theta \in \Theta$. Completeness implies bounded completeness but not conversely (see Exercise 6). We prefer bounded completeness in the hypotheses of theorems and completeness in the conclusions.

As an example, let $X_1, \cdots, X_n$ be a sample from the binomial distributions $\mathcal{B}(m, \theta)$. Then $T = \sum_1^n X_j$ is sufficient for θ and $T \in \mathcal{B}(N, \theta)$ where $N = nm$. If

$$E_\theta g(T) = \sum_{t=0}^{N} g(t) \binom{N}{t} \theta^t (1 - \theta)^{N-t}$$

is identically zero for $0 < \theta < 1$, then on canceling $(1 - \theta)^N$ we obtain

$$\sum_{t=0}^{N} g(t) \binom{N}{t} \left(\frac{\theta}{1 - \theta}\right)^t \underset{\theta}{\equiv} 0.$$

If a polynomial of degree N or, more generally, if a convergent power series $\sum a_n z^n$ is zero for z in some open interval, each of the coefficients a_n must be zero. This expression is a polynomial of degree N in $\theta/(1 - \theta)$, identically equal to zero, so that $g(t)\binom{N}{t}$ hence $g(t)$, is zero for $t = 0, 1, \cdots, N$. Therefore $P_\theta\{g(T) = 0\} = 1$ for all θ, proving that T is a complete sufficient statistic.

As another example, let $X_1, \cdots, X_n$ be a sample from the uniform distribution $\mathcal{U}(0, \theta)$, $\theta > 0$. Then $T = \max X_j$ is sufficient for θ, and the distribution of T has density (Exercise 1)

$$f_T(t \mid \theta) = n\theta^{-n} t^{n-1} I_{(0,\theta)}(t). \tag{3.66}$$

Hence, if

$$E_\theta g(T) = n\theta^{-n} \int_0^\theta g(t) t^{n-1}\, dt$$

is identically equal to zero for $\theta > 0$,

$$\int_0^\theta g(t) t^{n-1}\, dt \underset{\theta}{\equiv} 0.$$

This implies that $g(t) = 0$ for all $t > 0$, except perhaps for a set of t having Lebesgue measure zero. [If the Riemann integral is used, then $g(t)$ must be continuous at all points $t > 0$ except for a set of Lebesgue measure zero; at such points of continuity the fundamental theorem of calculus shows that $g(t)$ is zero.] Hence $P_\theta\{g(T) = 0\} = 1$ for all $\theta > 0$, so that T is a complete sufficient statistic.

If $X_1, \cdots, X_n$ is a sample from $\mathcal{U}(\theta - \frac{1}{2}, \theta + \frac{1}{2})$, then $(\min X_j, \max X_j)$ is sufficient for θ (in fact, minimal sufficient) but

$$E_\theta(\max X_j - \min X_j - ((n - 1)/(n + 1))) = 0$$

(Exercise 2), so that (min X_j, max X_j) is not a complete sufficient statistic. If $X_1, \cdots, X_n$ is a sample from $\mathfrak{N}(\theta, \theta^2)$, then $(\sum X_j, \sum X_j^2)$ is a sufficient statistic (in fact, minimal sufficient), yet

$$E_\theta(2(\sum X_j)^2 - (n+1)(\sum X_j^2)) \equiv 0$$

(Exercise 3). In these examples a complete sufficient statistic does not exist. In general, a two-dimensional sufficient statistic for a one-dimensional parameter is not complete.

For exponential families with the natural parameterization, however, there does exist a complete sufficient statistic.

Lemma 1. If $X_1, \cdots, X_n$ is a sample from the distribution with probability mass function or density

$$f(x \mid \boldsymbol{\pi}) = c(\boldsymbol{\pi})\, h(x) \exp\left[\sum_{i=1}^{k} \pi_i t_i(x)\right],$$

if $n \geq k$, and the natural parameter space Π contains an open set in E_k, then

$$\mathbf{T} = \left(\sum_1^n t_i(X_j), \cdots, \sum_1^n t_k(X_j)\right)$$

is a complete sufficient statistic for $\boldsymbol{\pi}$.

For a proof of this lemma in a more general setting the reader is referred to Lehmann (1959), 132. The idea is as follows. If $g(\mathbf{t})$ is a function for which $E_{\boldsymbol{\pi}} g(\mathbf{T}) \equiv 0$, then

$$\int g(\mathbf{t})\, h_0(\mathbf{t}) \exp\left[\sum_{i=1}^{k} \pi_i t_i\right] d\mathbf{t} \equiv 0.$$

From the unicity of the Laplace transform $g(\mathbf{t})\, h_0(\mathbf{t})$ is equal to zero for almost all $\mathbf{t}$. Thus for almost all $\mathbf{t}$ for which $h_0(\mathbf{t}) \neq 0$ we have $g(\mathbf{t}) = 0$, so that $P_{\boldsymbol{\pi}}\{g(\mathbf{T}) = 0\} = 1$.

As examples of this lemma, $\mathbf{T}$ is a complete sufficient statistic for any of the families of distributions of Exercise 3.5.1.

Minimum Variance Unbiased Estimates. One of the uses of the notion of completeness of sufficient statistics is in the problem of finding minimum variance unbiased estimates. In the underlying decision problem for estimating a real parameter γ, Θ is a subset of E_k, $\mathcal{A}$ is the real line, and $L(\theta, a) = (\gamma(\theta) - a)^2$, where $\gamma(\theta)$ is a real-valued function of θ. Let $\mathbf{T} = (T_1, \cdots, T_k)$ be a sufficient statistic for θ. The sufficiency of $\mathbf{T}$ and the convexity of L in $a \in \mathcal{A}$ for each $\theta \in \Theta$ imply that the statistician may restrict attention to the nonrandomized decision rules which are functions of $\mathbf{T}$.

Suppose that for some reason the statistician is interested only in unbiased estimates of $\gamma(\theta)$, that is, estimates, $d(\mathbf{T})$, for which $E_\theta\, d(\mathbf{T}) = \gamma(\theta)$ for all $\theta \in \Theta$. If $\mathbf{T}$ is a complete sufficient statistic, there is essentially *at most one* unbiased estimate based on $\mathbf{T}$ for any function of θ. Thus, if the statistician has already found an unbiased estimate $\hat{\lambda}$ of $\gamma(\theta)$, the unbiased estimate $E(\hat{\lambda} \mid \mathbf{T})$, based on $\mathbf{T}$, must be the best unbiased estimate of $\gamma(\theta)$. Note that with squared error loss the risk function of an unbiased estimate is the variance of the estimate, so that the best unbiased estimate in this problem may be referred to as the minimum variance unbiased estimate: among all unbiased estimates of $g(\theta)$ that based on $\mathbf{T}$, if it exists, has smallest variance for each value of θ.

As an example, let $X_1, \cdots, X_n$ be a sample from the distribution $\mathfrak{N}(\theta, 1)$. From the lemma $T = \sum_1^n X_j$ is a complete sufficient statistic for θ. Furthermore, $\bar{X} = T/n$ is an unbiased estimate of θ. Hence $\bar{X}$ is a minimum variance unbiased estimate of θ. On the other hand, $E_\theta \bar{X}^2 = (1/n) + \theta^2$, so that $\bar{X}^2 - (1/n)$ is a minimum variance unbiased estimate of θ^2.

It should be pointed out that being minimum variance unbiased is an extremely dubious optimum property for an estimate when there is a complete sufficient statistic, because we automatically restrict attention to estimates based on such statistics; if, in order to choose an estimate among the many available, we further restrict attention to unbiased estimates and the sufficient statistic is complete, there is only one such estimate. Being best in a class consisting of one element is no optimum property at all. Unbiasedness itself is not an optimum property. Indeed, the minimum variance unbiased estimate, $\bar{X}^2 - (1/n)$, of θ^2 in the preceding paragraph is not admissible, for it occasionally estimates negative values for a parameter known to be positive; the estimate $\max(0, \bar{X}^2 - (1/n))$ is strictly better (that is, has strictly smaller mean squared error for all values of θ).

A few more examples will emphasize this point. Let $X_1, \cdots, X_n$ be a sample from the distribution $\mathfrak{N}(0, \sigma^2)$. Then $T = \sum_1^n X_j^2$ is a complete sufficient statistic for σ^2. The maximum likelihood estimate, $s^2 = T/n$, is an unbiased estimate of σ^2, hence is minimum variance unbiased. However, it is not admissible. Consider the class of estimates $\hat{\sigma}_c^2 = cT$. Because T/σ^2 has a χ_n^2-distribution [that is, $\mathcal{G}(n/2, 2)$], $E_{\sigma^2}T = n\sigma^2$ and $E_{\sigma^2}T^2 = n(n+2)\sigma^4$. Therefore the risk function of the estimate $\hat{\sigma}_c^2$ is its mean squared error,

$$E_{\sigma^2}(\hat{\sigma}_c^2 - \sigma^2)^2 = (c^2n(n+2) - 2cn + 1)\sigma^4.$$

This is minimized by taking $c = 1/(n+2)$. Thus the estimate $T/(n+2)$ has strictly smaller risk than the usual estimate T/n. At least $T/(n+2)$ has the optimum property of being best out of the class of estimates $\{\hat{\sigma}_c^2\}$. A similar example may be found in Exercise 2.11.12.

As a final example, consider the case of the telephone operator who is given a new switchboard to care for and who after working 10 minutes wonders if he would be missed if he took a 20-minute coffee break. He assumes that calls are coming in to this switchboard as a Poisson process at the unknown rate of λ calls per 10 minutes. On the basis of the number X of calls received within the first 10 minutes [X, a sufficient statistic for this problem, has the Poisson distribution $\mathcal{P}(\lambda)$], he wants to estimate the probability that no calls will be received within the next 20 minutes, this probability being $\theta = e^{-2\lambda}$. If he is enamored with unbiased estimates, he will look for an estimate $\hat{\theta}(X)$ for which

$$E_\lambda\hat{\theta}(X) = \sum_{x=0}^{\infty} \hat{\theta}(x) \frac{e^{-\lambda}\lambda^x}{x!} \equiv e^{-2\lambda}.$$

After multiplying both sides by e^λ and expanding $e^{-\lambda}$ in a power series, he would obtain

$$\sum_{x=0}^{\infty} \hat{\theta}(x) \frac{\lambda^x}{x!} \equiv \sum_{x=0}^{\infty} (-1)^x \frac{\lambda^x}{x!}.$$

Two convergent power series can be equal only if corresponding coefficients are equal. The only unbiased estimate of $\theta = e^{-2\lambda}$ is $\hat{\theta}(x) = (-1)^x$. Thus he would estimate the probability of receiving no calls within the next 20 minutes as $+1$ if he received an even number of calls in the last 10 minutes and as -1 if he received an odd number of calls in the last 10 minutes. This ridiculous estimate nonetheless is a minimum variance unbiased estimate.

Exercises

1. Let $X_1, \cdots, X_n$ be a sample of size n from the uniform distribution $\mathcal{U}(0, \theta)$. Show that the density of $T = \max X_j$ exists and is given by (3.66).

2. Let $X_1, \cdots, X_n$ be a sample of size n from the uniform distribution $\mathcal{U}(0, 1)$. Show that $E \max X_j = n/(n+1)$. Hence deduce

$$E \min X_j = \frac{1}{n+1},$$

and, if $X_1, \cdots, X_n$ is a sample of size n from $\mathcal{U}(\theta - \frac{1}{2}, \theta + \frac{1}{2})$, deduce

$$E_\theta(\max X_j - \min X_j) = \frac{n-1}{n+1}.$$

3. Let $X_1, \cdots, X_n$ be a sample of size n from the normal distribution, $\mathcal{N}(\mu, \sigma^2)$. Show that

$$E(\sum_1^n X_j)^2 = n\sigma^2 + n^2\mu^2 \quad \text{and} \quad E\sum_1^n X_j^2 = n\sigma^2 + n\mu^2.$$

Hence deduce

$$E(2(\sum X_j)^2 - (n+1)(\sum X_j^2)) = n(n-1)(\mu^2 - \sigma^2).$$

4. Let $X_1, \cdots, X_n$ be a sample of size n from the "nonregular" distribution admitting a fixed dimensional sufficient statistic of (3.63). Consider the distribution of the sufficient statistic $\mathbf{T}$ given in Exercise 3.5.3 and suppose that we are using the Riemann integral, so that if $Eg(\mathbf{T})$ exists then $g(\mathbf{t})\, h_0(\mathbf{t})$ is continuous except on a set N of Lebesgue measure zero. Show that $\mathbf{T}$ is a complete sufficient statistic for $\boldsymbol{\pi}$.

5. Let X be a sample of size 1 from $\mathcal{N}(0, \sigma^2)$. Show that X is not a complete sufficient statistic for σ^2. Show that X^2 is a complete sufficient statistic for σ^2.

6. Let X have a discrete distribution with probability mass function

$$f(x \mid \theta) = \begin{cases} \theta & \text{if } x = -1, \\ (1-\theta)^2\theta^x & \text{if } x = 0, 1, 2, \cdots, \end{cases}$$

where $0 < \theta < 1$. Show that X is a boundedly complete sufficient statistic for θ but that X is not a complete sufficient statistic for θ.

3.7 Continuity of the Risk Function

In Theorem 2.3.3 and in Exercise 2.11.2 we made the assumption that the risk function $R(\theta, \delta)$ is a continuous function of θ for all $\delta \in D^*$.

In this section we show that this assumption is satisfied in many important cases and give an application to the proof of admissibility of the minimax estimate $\bar{X}$ of the mean of a normal distribution using squared error loss.

We present two theorems on the continuity of the risk function. In the first, strong conditions are placed on the loss function and weak conditions, on the family of distributions. The second theorem complements the first in that weak conditions are placed on the loss function, whereas strong conditions are placed on the family of distributions. These theorems deal only with the nonrandomized decision rules. The reason for this is that conditions under which an integral with respect to d of a function $R(\theta, d)$, continuous in θ for each d, is continuous in θ are easily found in the literature. To add such conditions to Theorems 1 and 2 would confuse rather than clarify. In addition, when the nonrandomized decision rules form an essentially complete class, continuity of the risk for nonrandomized rules is all that matters.

In Theorems 1 and 2 we use the Lebesgue integral. In particular, we assume that there is a σ-finite measure μ defined on a fixed σ-field, $\mathcal{B}$, of subsets of the sample space $\mathfrak{X}$ and that, for each $\theta \in \Theta$, P_θ is absolutely continuous with respect to μ. The density of P_θ with respect to μ is denoted by $f(x \mid \theta)$. We assume further that the space D of nonrandomized decision rules contains only rules d for which $L(\theta, d(x))$ is a $\mathcal{B}$-measurable function of x for all $\theta \in \Theta$, and

$$R(\theta, d) = E_\theta L(\theta, d(X)) = \int L(\theta, d(x))\, f(x \mid \theta)\, d\mu(x)$$

exists and is finite for all $\theta \in \Theta$.

Theorem 1. If

(a) $L(\theta, a)$ is bounded.

(b) $L(\theta, a)$ is continuous in θ uniformly for $a \in \mathcal{A}$ [that is, $\sup_a | L(\theta, a) - L(\theta_0, a) | \to 0$ as $\theta \to \theta_0$ for each $\theta_0 \in \Theta$].

(c) For all bounded $\mathcal{B}$-measurable functions φ, the integral

$$\int \varphi(x)\, f(x \mid \theta)\, d\mu(x)$$

is a continuous function of θ.

Then, for all $d \in D$, the risk function $R(\theta, d)$ is continuous in θ.

Proof. Let $\epsilon > 0$ and fix $\theta_0 \in \Theta$. Note that $\int L(\theta_0, d(x)) f(x \mid \theta)\, d\mu$ exists and is finite for all θ_0 and $\theta \in \Theta$, for $L(\theta_0, d(x))$ is bounded and measurable. Hence

$$| R(\theta, d) - R(\theta_0, d) | \leq \int | L(\theta, d(x)) - L(\theta_0, d(x)) | f(x \mid \theta)\, d\mu$$

$$+ \left| \int L(\theta_0, d(x)) (f(x \mid \theta) - f(x \mid \theta_0))\, d\mu \right|. \quad (3.67)$$

From (b) there exists a $\delta_1 > 0$ such that $| \theta - \theta_0 | < \delta_1$ implies that $| L(\theta, d(x)) - L(\theta_0, d(x)) | < \epsilon$ for all $x \in \mathfrak{X}$. From (a) and (c) there exists a $\delta_2 > 0$ such that $| \theta - \theta_0 | < \delta_2$ implies that the second integral on the right side of (3.67) is less that ϵ. Hence, if $| \theta - \theta_0 | < \min(\delta_1, \delta_2)$, then $| R(\theta, d) - R(\theta_0, d) | < 2\epsilon$, thus completing the proof.

Condition (c) of Theorem 1 is relatively weak. Lemma 3.5.3 shows that it is satisfied for exponential families of distributions. It is also satisfied for absolutely continuous distributions of which θ is a one-dimensional location or scale parameter. As an example in which (a) and (b) are satisfied but (c) is not and the risk function is not continuous, see Blackwell's example in Section 4.5. The main drawback of Theorem 1 is that the loss function is required to be bounded and thus cannot, for example, be squared error loss. The next theorem allows squared error loss but requires that the distributions be a one-parameter exponential family.

Theorem 2. Let Θ be the real line. Suppose that

(a) For all θ_1 and $\theta_2 \in \Theta$ there exist functions $B_1(\theta_1, \theta_2)$ and $B_2(\theta_1, \theta_2)$, bounded on the compact sets of $\Theta \times \Theta$, such that for all $a \in \mathfrak{A}$

$$| L(\theta_2, a) | \leq B_1(\theta_1, \theta_2) | L(\theta_1, a) | + B_2(\theta_1, \theta_2). \quad (3.68)$$

(b) $L(\theta, a)$ is continuous in θ for each $a \in \mathfrak{A}$.

(c) $f(x \mid \theta) = c(\theta) h(x) \exp [Q(\theta) T(x)]$, where $Q(\theta)$ is a continuous increasing function.

Then, for all $d \in D$, $R(\theta, d)$ is a continuous function of θ.

Proof. Let $\epsilon > 0$ and fix $\theta_0 \in \Theta$. Note that (a) implies that whenever

$$\int L(\theta_1, d(x)) f(x \mid \theta_1)\, d\mu(x) < \infty,$$

then

$$\int L(\theta_2, d(x))\, f(x \mid \theta_1)\, d\mu(x) < \infty.$$

Hence

$$| R(\theta_0, d) - R(\theta, d) | \leq \left| \int (L(\theta_0, d(x)) - L(\theta, d(x)))\, f(x \mid \theta_0)\, d\mu \right|$$

$$+ \int | L(\theta, d(x)) || f(x \mid \theta_0) - f(x \mid \theta) |\, d\mu. \qquad (3.69)$$

From (a) this last integrand is bounded for $| \theta - \theta_0 | < \delta$ by

$$(B_1 | L(\theta_0, d(x)) | + B_2)\, h(x)\, | c(\theta_0) \exp [Q(\theta_0) T(x)] - c(\theta) \exp [Q(\theta)\, T(x)] |,$$

where

$$B_i = \sup_{\theta_0 - \delta < \theta < \theta_0 + \delta} B_i(\theta_0, \theta).$$

This, in turn, is bounded by

$$(B_1 | L(\theta_0, d(x)) | + B_2)\, h(x)\, (c(\theta_0) \exp [Q(\theta_0)\, T(x)] + c \exp [Q_0(x)\, T(x)], \qquad (3.70)$$

where

$$c = \sup_{\theta_0 - \delta < \theta < \theta_0 + \delta} c(\theta)$$

and

$$Q_0(x) = \begin{cases} Q(\theta_0 + \delta) & \text{if} \quad T(x) > 0, \\ Q(\theta_0 - \delta) & \text{if} \quad T(x) < 0. \end{cases}$$

Because (3.70) is an integrable function, the Lebesgue bounded convergence theorem may be applied to show that the second integral of (3.69) tends to zero as $\theta \to \theta_0$.

The first integrand of (3.69) is bounded for $| \theta - \theta_0 | < \delta$ by

$$((B_1 + 1) | L(\theta_0, d(x)) | + B_2)\, f(x \mid \theta_0),$$

an integrable function, so that, again by the Lebesgue bounded convergence theorem, the first term on the right side of (3.69) tends to zero as $\theta \to \theta_0$.

Admissibility of $\overline{X}$ as an Estimate of the Mean of a Normal Distribution Using Squared Error Loss. Let $X_1, \cdots, X_n$ be a sample of size n from a normal distribution with unknown mean θ and known variance. The statistician is to estimate θ when the loss is squared error $L(\theta, a) = (\theta - a)^2$. A sufficient statistic for the problem is $T = \bar{X}$, whose distribution is also normal with mean θ and known variance. For simplicity, we take var $T = 1$ so that $T \in \mathfrak{N}(\theta, 1)$. The usual (minimax) estimate of θ is T, which we now show to be admissible.

Theorem 2 implies that, for every nonrandomized decision rule d, $R(\theta, d)$ is continuous in θ: conditions (b) and (c) are obviously satisfied, whereas condition (a) is satisfied with $B_1(\theta_1, \theta_2) = 2$ and $B_2(\theta_1, \theta_2) = 2(\theta_1 - \theta_2)^2$.

Suppose that $d'(t) = t$ is not an admissible decision rule. Then there exists a nonrandomized rule d'' such that $R(\theta, d'') \leq R(\theta, d')$ for all θ and $R(\theta_0, d'') < R(\theta_0, d')$ for some θ_0. Because $R(\theta, d')$ and $R(\theta, d'')$ are continuous functions of θ, there exists an $\epsilon > 0$ such that, for $|\theta - \theta_0| < \epsilon$, $R(\theta, d'') < R(\theta, d') - \epsilon$.

Now consider the prior distribution $\tau_\sigma = \mathfrak{N}(0, \sigma^2)$ and the Bayes rule d_σ with respect to τ_σ. Then $d_\sigma(t) = t\sigma^2/(1 + \sigma^2)$ and the Bayes risk is $r(\tau_\sigma, d_\sigma) = \sigma^2/(1 + \sigma^2)$ (see Section 1.8). Thus

$$r(\tau_\sigma, d') - r(\tau_\sigma, d_\sigma) = 1 - \frac{\sigma^2}{1 + \sigma^2} = \frac{1}{1 + \sigma^2}.$$

However,

$$\begin{aligned}\sigma[r(\tau_\sigma, d'') &- r(\tau_\sigma, d')] \\ &= \sigma \int [R(\theta, d'') - R(\theta, d')] \frac{1}{\sqrt{2\pi}\sigma} \exp\left(-\frac{\theta^2}{2\sigma^2}\right) d\theta \\ &< -\frac{\epsilon}{\sqrt{2\pi}} \int_{\theta_0-\epsilon}^{\theta_0+\epsilon} \exp\left(-\frac{\theta^2}{2\sigma^2}\right) d\theta\end{aligned}$$

so that

$$\begin{aligned}0 &\leq \sigma[r(\tau_\sigma, d'') - r(\tau_\sigma, d_\sigma)] \\ &= \sigma[r(\tau_0, d'') - r(\tau_\sigma, d')] + \sigma[r(\tau_\sigma, d') - r(\tau_\sigma, d_\sigma)] \\ &< -\frac{\epsilon}{\sqrt{2\pi}} \int_{\theta_0-\epsilon}^{\theta_0+\epsilon} \exp\left(-\frac{\theta^2}{2\sigma^2}\right) d\theta + \frac{\sigma}{1 + \sigma^2},\end{aligned}$$

which converges as $\sigma \to \infty$ to $-(2\epsilon^2/\sqrt{2\pi})$ and yields a contradiction that completes the proof.

Exercises

1. (Dvoretsky, Wald, and Wolfowitz). Let

$$\Theta = [-2, -1] \cup [1, 2], \qquad \mathcal{A} = \{a_0, a_1\}$$

and

$$L(\theta, a_0) = I_{[-2,-1]}(\theta), \; L(\theta, a_1) = I_{[1,2]}(\theta).$$

Let $Y \in \mathfrak{N}(\theta, 1)$ and let the observable random variable X be defined by

$$X = \begin{cases} |Y| & \text{if} \quad |Y| > 1, \\ Y & \text{if} \quad |Y| \le 1. \end{cases}$$

Consider the following behavioral decision rule δ^0, in which

$$\delta_x{}^0(a_1) = \begin{cases} 1 & \text{if} \quad -1 < y < 0, \\ 0 & \text{if} \quad 0 < y < 1, \\ \frac{1}{2} & \text{if} \quad |y| > 1. \end{cases}$$

Show that δ^0 is admissible by demonstrating that it is Bayes with respect to the uniform distribution on Θ and applying Theorem 2.3.3. Show that the distribution of X forms an exponential family; hence X is a complete sufficient statistic. Using this, show that any decision rule that is as good as δ^0 is essentially the same as δ^0, so that the nonrandomized rules do not form an essentially complete class. (This example shows that the Dvoretsky-Wald-Wolfowitz theorem (p. 79) is not necessarily valid if Θ is not finite.)

2. Show that $\bar{X}$ is admissible as an estimate of the mean of a normal distribution using absolute error loss.

CHAPTER 4

Invariant Statistical Decision Problems

4.1 Invariant Decision Problems

The invariance principle was mentioned and illustrated in Section 1.6. We proceed immediately to the definition.

The invariance principle involves groups of transformations over the three spaces around which decision theory is built: the parameter space Θ, the action space $\mathcal{A}$, and the sample space denoted by $\mathfrak{X}$ and assumed to be a subset of n-dimensional Euclidean space E_n. The most basic is the group of transformations over $\mathfrak{X}$. A transformation g from $\mathfrak{X}$ into itself is said to be *onto* $\mathfrak{X}$ if the range of g is the whole of $\mathfrak{X}$, that is, if for every $x_1 \in \mathfrak{X}$ there exists an $x_2 \in \mathfrak{X}$ such that $g(x_2) = x_1$. A transformation g from $\mathfrak{X}$ into itself is said to be *one-to-one* if $g(x_1) = g(x_2)$ implies $x_1 = x_2$.

Let $\mathcal{G}$ denote a group of measurable transformations from $\mathfrak{X}$ into itself. The group operation is *composition*: if g_1 and g_2 are transformations from $\mathfrak{X}$ into itself, g_2g_1 is defined as the transformation $x \to g_2(g_1(x))$. The identity transformation is denoted by e. A set of transformations on a space is a group if it is closed under composition and inverse. The inverse of a transformation g (that is, a transformation g^{-1} such that $g^{-1}g = e$) exists if and only if g is one-to-one and onto. Hence all transformations in $\mathcal{G}$ are automatically one-to-one and onto. The assumption that $g \in \mathcal{G}$ be measurable is made to ensure that whenever X is a random variable with values in $\mathfrak{X}$, then $g(X)$ is also a random variable. We first define what it means for a family of distributions P_θ, $\theta \in \Theta$, on $\mathfrak{X}$ to be invariant under $\mathcal{G}$.

Definition 1. The family of distributions P_θ, $\theta \in \Theta$, is said to be *invariant under the group* $\mathcal{G}$, if for every $g \in \mathcal{G}$ and every $\theta \in \Theta$ there exists a unique $\theta' \in \Theta$ such that the distribution of $g(X)$ is given by $P_{\theta'}$ whenever the distribution of X is given by P_θ.

The θ' uniquely determined by g and θ is denoted by $\bar{g}(\theta)$.

Important Formula. The condition that the family of distributions P_θ be invariant under $\mathcal{G}$ is that for every (measurable) set $A \subset \mathfrak{X}$

$$P_\theta\{g(X) \in A\} = P_{\bar{g}(\theta)}\{X \in A\}. \tag{4.1}$$

In terms of expectations, this is equivalent to saying that for every (integrable) real-valued function ϕ

$$E_\theta\, \phi(g(X)) = E_{\bar{g}(\theta)}\, \phi(X), \tag{4.2}$$

where E_θ refers to the expectation when the distribution of the random variable X is given by P_θ.

Identifiability. A parameter θ for a family of distributions P_θ, $\theta \in \Theta$, is said to be *identifiable* if distinct values of θ correspond to distinct distributions; that is, θ is identifiable if $\theta \neq \theta'$ implies $P_\theta \neq P_{\theta'}$. If the parameter θ of the family of distributions of Definition 1 is identifiable, the unicity of θ' in that definition will be automatically satisfied. Conversely, if a family of distributions P_θ, $\theta \in \Theta$, is invariant under $\mathcal{G}$, the unicity of θ' implies that θ is identifiable.

When g is fixed, $\bar{g}(\theta)$ as a function of θ is a transformation of Θ into itself. The following lemma shows that the set $\bar{\mathcal{G}}$ of such transformations is a group.

Lemma 1. If a family of distributions P_θ, $\theta \in \Theta$, is invariant under $\mathcal{G}$, then $\bar{\mathcal{G}} = \{\bar{g}: g \in \mathcal{G}\}$ is a group of transformations of Θ into itself.

Proof. If the distribution of X is given by P_θ, the distribution of $g_1(X)$ is given by $P_{\bar{g}_1(\theta)}$; hence, the distribution of $g_2(g_1(X)) = g_2g_1(X)$

is given by both $P_{\bar{g}_2(\bar{g}_1(\theta))}$ and $P_{\overline{g_2 g_1}(\theta)}$. From unicity, it follows that

$$\overline{g_2 g_1} = \bar{g}_2 \bar{g}_1. \tag{4.3}$$

This shows that $\bar{\mathcal{G}}$ is closed under composition. It also shows that $\bar{\mathcal{G}}$ is closed under inverses if we let $g_2 = g_1^{-1}$ and note that $\bar{e}$ is the identity in $\bar{\mathcal{G}}$. This completes the proof.

All transformations $\bar{g} \in \bar{\mathcal{G}}$ are automatically one-to-one and onto.

Equation (4.3) shows that $\bar{\mathcal{G}}$ is a homomorphic image of $\mathcal{G}$. That $\bar{\mathcal{G}}$ and $\mathcal{G}$ are not necessarily isomorphic is seen by letting $\mathfrak{X}$ be the real line and P_θ the normal distribution with mean 0 and variance $\theta \in (0, \infty) = \Theta$. The group $\mathcal{G} = \{e, g\}$, where g is the transformation $g(x) = -x$, obviously leaves the family P_θ, $\theta \in \Theta$ invariant. Yet, $\bar{g} = \bar{e}$ so that $\bar{\mathcal{G}}$ consists of one element.

We now define what it means for a given decision problem $(\Theta, \mathcal{A}, L)$ with observable quantity X taking values in $\mathfrak{X}$ with distribution P_θ to be invariant under a group $\mathcal{G}$ of transformations of $\mathfrak{X}$ onto itself.

Definition 2. A decision problem, consisting of the game $(\Theta, \mathcal{A}, L)$ and distributions P_θ over $\mathfrak{X}$, is said to be *invariant under the group* $\mathcal{G}$ if the family of distributions P_θ, $\theta \in \Theta$, is invariant under $\mathcal{G}$ and if the loss function is invariant under $\mathcal{G}$ in the sense that for every $g \in \mathcal{G}$ and $a \in \mathcal{A}$ there exist a unique $a' \in \mathcal{A}$ such that

$$L(\theta, a) = L(\bar{g}(\theta), a') \qquad \text{for all } \theta \in \Theta. \tag{4.4}$$

The requirement that a' be unique is not restrictive, for if it were not unique there would exist an $a'' \in \mathcal{A}$ such that

$$L(\bar{g}(\theta), a') = L(\bar{g}(\theta), a'') \qquad \text{for all } \theta.$$

Because $\bar{g}$ is onto, this implies

$$L(\theta, a') = L(\theta, a'') \qquad \text{for all } \theta.$$

Thus we could remove a'' from the space of actions of the statistician without changing the problem at all. The a' uniquely determined by g and a is denoted by $\tilde{g}(a)$.

Lemma 2. If a decision problem is invariant under a group $\mathcal{G}$, then $\tilde{\mathcal{G}} = \{\tilde{g} : g \in \mathcal{G}\}$ is a group of transformations of $\mathcal{A}$ into itself.

Proof. We show that $\tilde{\mathcal{G}}$ is closed under composition by showing

$$\tilde{g}_2\tilde{g}_1 = (\widetilde{g_2g_1}). \tag{4.5}$$

Indeed, we have

$$L(\theta, a) = L(\bar{g}_1(\theta), \tilde{g}_1(a)) = L(\bar{g}_2\bar{g}_1(\theta), \tilde{g}_2\tilde{g}_1(a)) = L(\overline{g_2g_1}(\theta), \tilde{g}_2\tilde{g}_1(a)),$$

which with unicity proves (4.5). The rest of the proof follows immediately.

Equation (4.5) reflects the fact that $\tilde{\mathcal{G}}$ is a homomorphic image of $\mathcal{G}$. In fact, it is easy to see more: that $\tilde{\mathcal{G}}$ is a homomorphic image of $\bar{\mathcal{G}}$.

EXAMPLE 1. Suppose that $X \in \mathfrak{N}(\theta, 1)$, $\Theta = \mathcal{A} =$ the real line and that $L(\theta, a) = (\theta - a)^2$. Consider the group $\mathcal{G}$ of translations $g_c(x) = x + c$. The distribution of $g_c(X)$ is $\mathfrak{N}(\theta + c, 1)$. Hence the distributions are invariant under $\mathcal{G}$ and $\bar{g}_c(\theta) = \theta + c$. Furthermore,

$$L(\theta, a) = L(\bar{g}_c(\theta), \tilde{g}_c(a))$$

is satisfied for all θ when $\tilde{g}_c(a) = a + c$ so that the loss is invariant; hence the decision problem is invariant under $\mathcal{G}$. The groups $\bar{\mathcal{G}}$ and $\tilde{\mathcal{G}}$ are groups of translations also.

EXAMPLE 2. Let X have the binomial distribution $\mathcal{B}(n, \theta)$, where n is known and $\Theta = [0, 1]$. Suppose, in addition, that $\mathcal{A} = [0, 1]$ and $L(\theta, a) = W(\theta - a)$, some even function of $\theta - a$. Let $\mathcal{G}$ be the group consisting of e (the identity transformation) and g, where $g(x) = n - x$. The distribution of $g(X)$ is clearly $\mathcal{B}(n, 1 - \theta)$ so that the distributions are invariant under $\mathcal{G}$ and $\bar{g}(\theta) = 1 - \theta$. With $\tilde{g}(a) = 1 - a$, the loss is clearly invariant also, and the decision problem is invariant.

EXAMPLE 3. Suppose $\mathbf{X} \in \mathfrak{N}(\theta_1\mathbf{1}, \theta_2^2\mathbf{I})$, where $\mathbf{1} = (1, 1, \cdots, 1)$ and $\mathbf{I}$ is the identity matrix. Thus the components of $\mathbf{X}$ are independent normal variables, and $\Theta =$ the half plane $[(\theta_1, \theta_2) : \theta_2 > 0]$. Let $\mathcal{A}$ be the real line and let $L(\boldsymbol{\theta}, a) = (\theta_1 - a)^2/\theta_2^2$. Consider the group $\mathcal{G}$ of transformations $g_{b,c}(\mathbf{x}) = b\mathbf{x} + c\mathbf{1}$, where $b \neq 0$. The distribution of $g_{b,c}(\mathbf{X})$ is $\mathfrak{N}((b\theta_1 + c)\mathbf{1}, b^2\theta_2^2\mathbf{I})$. Hence the distributions are invariant under $\mathcal{G}$ and $\bar{g}_{b,c}(\theta_1, \theta_2) = (b\theta_1 + c, |b| \theta_2)$. Furthermore, the loss is invariant if we take $\tilde{g}_{b,c}(a) = ba + c$, for

$$L(\bar{g}_{b,c}(\boldsymbol{\theta}), \tilde{g}_{b,c}(a)) = \frac{(b\theta_1 + c - ba - c)^2}{b^2\theta_2^2} = L(\boldsymbol{\theta}, a).$$

Thus the decision problem is invariant under $\mathcal{G}$.

EXAMPLE 4. Let $\mathbf{X} \in \mathfrak{N}(\theta_1\mathbf{1}, \theta_2^2\mathbf{I})$, where $\Theta = [(\theta_1, \theta_2):\theta_2 > 0]$. Let $\mathcal{A}$ consist of two points $\mathcal{A} = \{0, 1\}$ and let the loss function be defined as

$$L(\boldsymbol{\theta}, 0) = \begin{cases} 1 & \text{if} \quad \theta_1 > 0, \\ 0 & \text{if} \quad \theta_1 \le 0, \end{cases}$$

$$L(\boldsymbol{\theta}, 1) = \begin{cases} 1 & \text{if} \quad \theta_1 \le 0, \\ 0 & \text{if} \quad \theta_1 > 0. \end{cases}$$

Consider the group $\mathcal{G}$ of transformations $g_b(\mathbf{x}) = b\mathbf{x}$ where $b > 0$. The distribution of $g_b(\mathbf{X})$ is $\mathfrak{N}(b\theta_1\mathbf{1}, b^2\theta_2^2\mathbf{I})$ so that the distributions are invariant under $\mathcal{G}$ and $\bar{g}_b(\boldsymbol{\theta}) = (b\theta_1, b\theta_2)$. Furthermore, if $\tilde{g}_b(a) = a$ for all $a \in \mathcal{A}$, it is clear that the loss is invariant. Hence the decision problem is invariant. Note that $\tilde{\mathcal{G}}$ is not isomorphic to $\bar{\mathcal{G}}$ here.

Exercises

1. If a decision problem is invariant under a group $\mathcal{G}$ and $\mathcal{G}_1$ is a subgroup of $\mathcal{G}$, the decision problem is invariant under $\mathcal{G}_1$.
2. If transformations g_1 and g_2 leave the family of distributions invariant in the sense that the distribution of $g_i(X)$ is given by $P_{\bar{g}_i(\theta)}$ whenever the distribution of X is given by P_θ, the transformation g_2g_1 also leaves the family of distributions invariant. If g_1 is bimeasurable, one-to-one and onto, and $\bar{g}_1$ is one-to-one and onto, then g_1^{-1} leaves the family of distributions invariant. (Thus *groups* of transformations that leave the family of distributions invariant are the natural things to study.)
3. Let Θ be the set of all continuous distribution functions on the real line, let $\mathcal{A}$ be the real line, and let $L(F, a)$, where $F \in \Theta$ and $a \in \mathcal{A}$, be some function of $F(a)$, $L(F, a) = W(F(a))$. Let $\mathbf{X} = (X_1, \cdots, X_n)$ be a vector of independent random variables each with distribution function F, and consider the set $\mathcal{G}$ of all transformations of the form $g_\varphi(x_1, \cdots, x_n) = (\varphi(x_1), \cdots, \varphi(x_n))$, where φ is a continuous increasing one-to-one function of the real line onto itself. Show that this decision problem is invariant under the group $\mathcal{G}$ and find the groups $\bar{\mathcal{G}}$ and $\tilde{\mathcal{G}}$. [If $W(F(a)) = (F(a) - \frac{1}{2})^2$, this is the problem of estimating the median of a continuous distribution.]
4. Let $\Theta = E_n$, $\mathcal{A} = (0, \infty)$, and suppose the loss depends on $\boldsymbol{\theta}$ only through its length $L(\boldsymbol{\theta}, a) = W(|\boldsymbol{\theta}|, a)$. Let the observable vector $\mathbf{X} \in E_n$ have distribution $\mathfrak{N}(\boldsymbol{\theta}, \mathbf{I})$, where $\boldsymbol{\theta} \in \Theta$, and consider the

group $\mathcal{G}$ of orthogonal transformations of E_n. Show that the decision problem is invariant under $\mathcal{G}$ and find $\bar{\mathcal{G}}$ and $\tilde{\mathcal{G}}$.

5. Let Θ be the set of $n \times n$ diagonal covariance matrices $\mathbf{\Sigma}$, let $\mathcal{A}$ be arbitrary, and let the loss be a function only of det $\mathbf{\Sigma}$ and a, $L(\mathbf{\Sigma}, a) = W(\det \mathbf{\Sigma}, a)$. Let the observable vector $\mathbf{X} \in E_n$ have the distribution $\mathfrak{N}(\mathbf{0}, \mathbf{\Sigma})$ and consider the group $\mathcal{G}$ of diagonal matrices of determinant ± 1. Show that the decision problem is invariant under $\mathcal{G}$ and find $\bar{\mathcal{G}}$ and $\tilde{\mathcal{G}}$.

4.2 Invariant Decision Rules

In Section 1.6 we gave an argument supporting the contention that the statistician should be willing to use an invariant decision rule in a particular estimation problem. We now give the argument in the general case.

Consider a decision problem $(\Theta, \mathcal{A}, L)$ with an observable random quantity X whose distribution P_θ depends on θ and assume that the decision problem is invariant under a group $\mathcal{G}$ of transformations on the sample space $\mathfrak{X}$. The decision problem in which the statistician observes a random quantity $Y = g(X)$ whose distribution is P_ϕ, $\phi \in \Theta$ [where $\phi = \bar{g}(\theta)$], and must choose a point $b \in \mathcal{A}$, with loss function $L(\phi, b)$, is exactly the same problem as the one above, since for any $a \in \mathcal{A}$ that he may choose in the original problem the choice $b = \tilde{g}(a)$ yields the same loss in the new problem [$L(\theta, a) = L(\bar{g}(\theta), \tilde{g}(a)) = L(\phi, b)$]. Thus the decision to take action a_0 if $X = x_0$ in the first problem is equivalent to the decision to take action $\tilde{g}(a_0)$ if $Y = g(x_0)$ in the second. Because the two problems are identical, he should be willing to take action $\tilde{g}(a_0)$ if $X = g(x_0)$ in the original problem. Thus, we should have $d(g(x_0)) = \tilde{g}(d(x_0))$ for all $g \in \mathcal{G}$. Rules that satisfy this equality are called invariant nonrandomized decision rules.

Definition. 1 Given an invariant decision problem, a nonrandomized decision rule $d \in D$ is said to be *invariant under* $\mathcal{G}$ if for all $x \in \mathfrak{X}$ and all $g \in \mathcal{G}$.

$$d(g(x)) = \tilde{g}(d(x)). \tag{4.6}$$

A randomized decision rule $\delta \in D^*$ is said to be *invariant* if δ, as a probability distribution over D, gives all its mass to the subset of invariant nonrandomized rules. A behavioral decision rule $\delta \in \mathfrak{D}$ is said to be *invariant* if for all $x \in \mathfrak{X}$ and all $g \in \mathcal{G}$, $\delta(g(x)) = \tilde{g}\delta(x)$, where by $\tilde{g}\delta$ we mean the distribution of $\tilde{g}Z$ when Z has distribution δ.

To understand the notion of a behavioral decision rule it is useful to denote the distribution chosen by a rule $\delta \in \mathfrak{D}$ when $X = x$ is observed by δ_x rather than $\delta(x)$. The probability of a given (measurable) set $A \subset \mathcal{A}$ by the distribution δ_x may then be denoted by $\delta_x(A)$. With this notation, a behavioral decision rule $\delta \in \mathfrak{D}$ is invariant if

$$\delta_{g(x)}(A) = \delta_x(\tilde{g}^{-1}(A)) \tag{4.7}$$

for all $x \in \mathfrak{X}$, all $g \in \mathcal{G}$, and all (measurable) sets $A \subset \mathcal{A}$. On the right side of (4.7) we have the probability that $Z \in \tilde{g}^{-1}(A)$ or equivalently $\tilde{g}Z \in A$, when Z has the distribution given by δ_x. By Definition 1 this must be $\delta_{g(x)}(A)$, the left side of (4.7). It is perhaps simpler to write this equation in the form

$$\delta_x(A) = \delta_{g(x)}(\tilde{g}(A)) \tag{4.8}$$

for all g, x, and A.

(Equations 4.7 and 4.8 entail the supposition that for all $\tilde{g} \in \tilde{\mathcal{G}}$, $\tilde{g}(A)$ is measurable whenever A is measurable; in other words, that the σ-field of subsets of $\mathcal{A}$ on which the probabilities δ are defined is invariant under $\tilde{\mathcal{G}}$ or, equivalently, that all $\tilde{g}$ are measurable functions.)

Two points $\theta_1, \theta_2 \in \Theta$ are said to be *equivalent* if there exists a $\bar{g} \in \bar{\mathcal{G}}$ such that $\theta_1 = \bar{g}(\theta_2)$. This is an equivalence relation (Exercise 1), which breaks Θ up into equivalence classes, or *orbits*. A very important property of an invariant decision rule is that *the risk function of an invariant decision rule δ is constant on orbits, or, more precisely,*

$$R(\theta, \delta) = R(\bar{g}(\theta), \delta) \tag{4.9}$$

for all $\theta \in \Theta$ and all $\bar{g} \in \bar{\mathcal{G}}$. For a nonrandomized invariant rule d, (4.9) may be proved as follows:

$$\begin{aligned}
R(\theta, d) &= E_\theta L(\theta, d(X)) \\
&= E_\theta L(\bar{g}(\theta), \tilde{g}(d(X))) && \text{(invariance of loss)} \\
&= E_\theta L(\bar{g}(\theta), d(g(X))) && \text{(invariance of } d\text{)} \\
&= E_{\bar{g}(\theta)}\, L(\bar{g}(\theta), d(X))) && \text{(invariance of distributions)} \\
& && \text{[see (4.2)]} \\
&= R(\bar{g}(\theta), d). && (4.10)
\end{aligned}$$

From the definition of a randomized invariant rule it is immediate that (4.9) holds for all $\delta \in D^*$, because the equality (4.10) is just averaged over some distribution on D. Equation 4.9 holds also for behavioral invariant rules $\delta \in \mathfrak{D}$. This result requires a new proof, for behavioral invariant rules are somewhat more general than randomized invariant rules, as indicated later. First, we show something slightly more general than (4.9) for behavioral invariant rules for use in the next section.

Let δ be any rule in $\mathfrak{D}$ and define δ^g as the rule such that $\delta_x{}^g$ is the distribution $\tilde{g}^{-1}\delta_{g(x)}$; that is, $\delta_x{}^g$ is the distribution of $\tilde{g}^{-1}Z$ when Z has distribution $\delta_{g(x)}$ or, equivalently, for every $x \in \mathfrak{X}$ and every measurable set $A \subset \mathcal{A}$,

$$\delta_x{}^g(A) = \delta_{g(x)}(\tilde{g}A). \tag{4.11}$$

From this equation and (4.8) it is clear that $\delta \in \mathfrak{D}$ is invariant if, and only if, $\delta^g = \delta$ for all $g \in \mathcal{G}$.

Lemma 1. If a decision problem is invariant under $\mathcal{G}$, then for any decision rule $\delta \in \mathfrak{D}$ and any $g \in \mathcal{G}$

$$\hat{R}(\theta, \delta^g) = \hat{R}(\bar{g}(\theta), \delta). \tag{4.12}$$

Proof. By definition $\hat{R}(\theta, \delta^g) = E_\theta L(\theta, \delta_X{}^g)$. The function $L(\theta, \delta)$ is defined for $\theta \in \Theta$ and $\delta \in \mathcal{A}^*$ as in (1.8). The invariance of the loss function implies that $L(\theta, \delta) = L(\bar{g}\theta, \tilde{g}\delta)$ for all $\theta \in \Theta$, $\delta \in \mathcal{A}^*$, and $g \in \mathcal{G}$. Hence

$$L(\theta, \delta_x{}^g) = L(\theta, \tilde{g}^{-1}\delta_{g(x)}) = L(\bar{g}\theta, \delta_{g(x)}), \tag{4.13}$$

so that

$$\begin{aligned} \hat{R}(\theta, \delta^g) &= E_\theta L(\bar{g}\theta, \delta_{g(X)}) \\ &= E_{\bar{g}\theta} L(\bar{g}\theta, \delta_X) \qquad \text{(invariance of distributions)} \\ &= \hat{R}(\bar{g}\theta, \delta), \end{aligned} \tag{4.14}$$

thus completing the proof.

Theorem 1. If a decision problem is invariant under $\mathcal{G}$, then for any invariant decision rule $\delta \in \mathfrak{D}$, $\hat{R}(\theta, \delta) = \hat{R}(\bar{g}(\theta), \delta)$ for all $\theta \in \Theta$ and all $\bar{g} \in \bar{\mathcal{G}}$.

Proof. If $\delta \in \mathfrak{D}$ is invariant, then $\delta^g = \delta$ for all $g \in \mathcal{G}$, so that this theorem follows immediately from Lemma 1.

The Relation Between Randomized Invariant and Behavioral Invariant Decision Rules. Consider whether randomized invariant rules and behavioral invariant rules are equivalent in the sense (see Section 1.5) that for every randomized invariant rule there is a behavioral invariant rule with the same risk function and conversely. Unfortunately, this is not the case, and behavioral invariant rules are more general than randomized invariant rules. To see why this is so, we make an analysis similar to that in Section 1.5.

A randomized invariant rule chooses at random a function Y_x from the sample space into $\mathfrak{A}$ with the property that for all $g \in \mathcal{G}$

$$Y_{g(x)} = \tilde{g}Y_x . \tag{4.15}$$

Such a rule specifies the joint distribution of all the Y_x. A behavioral invariant rule chooses for each x in the sample space a point Y_x at random in $\mathfrak{A}$ with the property that for all $x \in \mathfrak{X}$ and $g \in \mathcal{G}$ the distribution of $Y_{g(x)}$ is identical to the distribution of $\tilde{g}Y_x$:

$$\text{dist } Y_{g(x)} \equiv \text{dist } \tilde{g}Y_x . \tag{4.16}$$

No joint distributions of the Y_x are specified for behavioral rules. It is immediately clear that for every randomized invariant rule there is an equivalent behavioral invariant rule (just forget about the joint distributions). To show the converse we would have to show that for any set of marginal distributions of the Y_x for which (4.16) holds there exists a choice of a distribution for the random function Y_x with these given marginals for which (4.15) holds.

Equation (4.15) implies that once some Y_{x_0} is determined then all Y_x for x in the same orbit as x_0 are determined also. Hence we might try to define the distribution of the random function Y_x as follows. Choose a set $\mathfrak{X}_e \subset \mathfrak{X}$ that consists of one point in each orbit and let the Y_x for $x \in \mathfrak{X}_e$ be independent, each with the distribution given by the behavioral rule. Then define Y_x for $x \notin \mathfrak{X}_e$ by (4.15). The difficulty lies in the fact that for some orbits Y_x may not be uniquely defined: there may exist an x and a $g \neq e$ such that $g(x) = x$. However, if the validity of (4.16) for all g for which $g(x) = x$ implies that

$$Y_x = \tilde{g}Y_x \quad \text{with probability one} \tag{4.17}$$

for all such g, then (4.15) can be satisfied. In two different circumstances it is easy to see that randomized and behavioral invariant rules are equivalent. First, if $\tilde{\mathcal{G}}$ consists only of the identity transformation (as is usually the case in hypothesis testing problems), clearly (4.17) is satisfied. Second, define an orbit to be of multiplicity one if for any x and x' in this orbit there exists only one $g \in \mathcal{G}$ such that $g(x) = x'$. If all orbits are of multiplicity one (or, more generally, if the set of orbits of multiplicity one has probability one under P_θ for all $\theta \in \Theta$), then again (4.17) is satisfied.

As an example of these considerations, take a decision problem with $\mathcal{A} = \mathfrak{X} = E_2$ which is invariant under the group $\mathcal{G}$ of rotations on $\mathfrak{X}$ and suppose $\tilde{\mathcal{G}} = \mathcal{G}$. For any x other than the origin the orbit containing x is of multiplicity one. Any nonrandomized invariant rule must map the origin of $\mathfrak{X}$ into the origin of $\mathcal{A}$. Any randomized invariant rule must choose a function that does the same. But a behavioral invariant rule may map the origin into any distribution on $\mathcal{A}$ that is invariant under $\tilde{\mathcal{G}}$, that is, any spherically symmetric distribution. This does not matter so long as $P_\theta\{\text{origin}\} = 0$ for all $\theta \in \Theta$, but if $P_\theta\{\text{origin}\} > 0$ for some θ the behavioral invariant rules are more general than the randomized invariant rules. If $\mathfrak{X} = \mathcal{A} = E_3$ and $\mathcal{G} = \tilde{\mathcal{G}} =$ the rotations, there are no orbits of multiplicity one, and the behavioral invariant rules are more general than the randomized invariant rules, regardless of the distributions P_θ. (See also Exercises 6 and 8.)

Invariant Rules and Prior Information. We have said that a statistician "should" be willing to use an invariant decision rule. This remark, which belongs to the realm of statistical philosophy, requires some modification here. The trouble is that usually—nearly always—the statistician has prior information about the position of the true value of the parameter that he would like to use in choosing a decision rule. If this information is not invariant with respect to the group $\bar{\mathcal{G}}$, he cannot be expected to use an invariant rule. It is only when he is unwilling to use this information or when he feels that ignoring the information will leave the chosen decision rule approximately the same that an invariant decision rule may be expected.

We must also mention that in spite of the argument at the beginning of this section we cannot hold that the statistician is acting irrationally by using a noninvariant decision rule, even when he is not acting on any prior information. In fact, for some invariant decision problems, there are no invariant decision rules (see Exercise 7). As further evidence, some examples are given later in which a best invariant rule exists but is not admissible.

Exercises

1. Let $\bar{\mathcal{G}}$ be a group of transformations on Θ. Define θ_1 as equivalent to $\theta_2 (\theta_1 \equiv \theta_2)$ if there exists a $\bar{g} \in \bar{\mathcal{G}}$ such that $\bar{g}(\theta_1) = \theta_2$. Show that $\equiv$ is an equivalence relation [i.e., show $\theta_1 \equiv \theta_1$ (reflexivity), $\theta_1 \equiv \theta_2 \Rightarrow \theta_2 \equiv \theta_1$ (symmetry), and $\theta_1 \equiv \theta_2$, $\theta_2 \equiv \theta_3 \Rightarrow \theta_1 \equiv \theta_3$ (transitivity)].
2. In Examples 1 through 4 of Section 4.1 find the form of the invariant nonrandomized decision rules (that is, write out (4.6)). In Example 1 show that the invariant nonrandomized decision rules are exactly the functions $g(x) = x + b$. In Example 4 find the form of the invariant behavioral rules.
3. In Exercise 4.1.3 the order statistics $(Y_1, \cdots, Y_n)$, $Y_1 < Y_2 < \cdots < Y_n$, are sufficient for θ. Consider Exercise 4.1.3 when the observable variables are $Y_1, \cdots, Y_n$ and show that the problem is still invariant under the group $\mathcal{G}$ considered there. Show that the nonrandomized invariant decision rules are of the form $d(Y_1, \cdots, Y_n) = Y_j$ for some j. Show that the risk function of such rules is independent of $F \in \Theta$. (Note that there is more than one orbit so that Theorem 1 does not apply.)
4. In Exercise 4.1.4 show that the class of invariant nonrandomized rules are the nonrandomized rules based on $|\mathbf{X}|$ and that the risk function for such rules depends only on $|\boldsymbol{\theta}|$.
5. In Exercise 4.1.5 show that the class of invariant nonrandomized rules are the nonrandomized rules based on $|\prod_{i=1}^{n} X_i|$ (provided this product is not zero) and that the risk function of such rules depends only on det $\boldsymbol{\Sigma}$.
6. Let $\Theta = \{-1, +1\}$, let $\mathcal{A} = \mathfrak{X} = \{-1, 0, +1\}$, let $P_\theta\{X = 0\} = P_\theta\{X = \theta\} = \frac{1}{2}$, and let $L(\theta, a) = 1$ if $\theta \neq a$ and 0 if $\theta = a$. This problem is invariant under the group $\mathcal{G} = \{e, g\}$, where $g(x) = -x$. Find $\bar{\mathcal{G}}$ and $\tilde{\mathcal{G}}$. Show that the behavioral invariant rule δ, where δ_x is degenerate at x if $x = \pm 1$ and where δ_0 gives probability $\frac{1}{2}$ to $+1$ and -1, is minimax; yet there is no equivalent randomized invariant rule. In fact, no randomized invariant rule is minimax.
7. (Stein). Let $\mathbf{X}$ and $\mathbf{Y}$ be independent k-dimensional random vectors, $k \geq 2$, $\mathbf{X} \in \mathfrak{N}(\mathbf{0}, \boldsymbol{\Sigma})$ and $\mathbf{Y} \in \mathfrak{N}(\mathbf{0}, \Delta\boldsymbol{\Sigma})$. Let $\Theta = \{(\Delta, \boldsymbol{\Sigma})\}$, where $\Delta > 0$ and where $\boldsymbol{\Sigma}$ is a nonsingular covariance matrix, and let the loss function depend only on Δ and a, $L(\Delta, a)$, where $\mathcal{A}$ is an arbitrary set. Let $\mathcal{G}$ be the group of transformations $g_{\mathbf{B}}(\mathbf{X}, \mathbf{Y}) = (\mathbf{BX}, \mathbf{BY})$, where $\mathbf{B}$ is an arbitrary nonsingular matrix.
   (a) Show that the decision problem is invariant under $\mathcal{G}$ and find the groups $\bar{\mathcal{G}}$ and $\tilde{\mathcal{G}}$.

(b) The probability that **X** and **Y** are linearly dependent is zero for all $\theta \in \Theta$, so that such points may be removed from the sample space without changing the problem. Show that on the rest of the sample space invariant decision rules are independent of the observations.
(c) Suppose $\mathcal{A} = (0, \infty)$ and $L(\Delta, a) = W(a/\Delta)$, some function of a/Δ. Then the problem is invariant under the group $\mathcal{G}$ of transformations $g_{\mathbf{B}, c}(\mathbf{X}, \mathbf{Y}) = (\mathbf{BX}, c\mathbf{BY})$, $\mathbf{B}$ nonsingular, $c > 0$. Find $\bar{\mathcal{G}}$ and $\tilde{\mathcal{G}}$. Show that there are no invariant decision rules for this problem.

8. Suppose that X and Y are independent Poisson variables, $X \in \mathcal{P}(\theta_1)$, $Y \in \mathcal{P}(\theta_2)$, where $\Theta = \{(\theta_1, \theta_2) : \theta_1 > 0, \theta_2 > 0\}$. Let $\mathcal{A} = \{0, 1\}$, and let

$$L(\boldsymbol{\theta}, 0) = \begin{cases} 1 & \text{if } \theta_1 < \theta_2 \\ 0 & \text{if } \theta_1 \geq \theta_2 \end{cases}$$

and

$$L(\boldsymbol{\theta}, 1) = \begin{cases} 1 & \text{if } \theta_1 > \theta_2 \\ 0 & \text{if } \theta_1 \leq \theta_2. \end{cases}$$

(a) Show that this problem is invariant under the group $\mathcal{G}$ generated by $g(x, y) = (y, x)$, and find the groups $\bar{\mathcal{G}}$ and $\tilde{\mathcal{G}}$.
(b) Show that there do not exist any nonrandomized invariant rules.
(c) Describe the class of behavioral invariant rules.
(d) Find an invariant rule that is as good as any other invariant rule. (Such a rule is called a best invariant rule.)

4.3 Admissible and Minimax Invariant Rules

In subsequent sections we shall find the best decision rule in the class of invariant rules for many specific decision problems. We should like to know if such a rule is still a good one when considered as a member of the class of all rules. For example, if a rule is minimax in the class of all invariant rules, will it be minimax in the class of all rules? If so, then in a search for a minimax rule we may restrict attention to the invariant rules. The following theorem states that when $\mathcal{G}$ is finite there is an affirmative answer to this question.

Theorem 1. Suppose that a given decision problem is invariant under a finite group $\mathcal{G}$. Then, if there exists a minimax rule, there exists a minimax rule which is (behavioral) invariant. If a rule is minimax within the class of behavioral invariant rules, it is minimax.

Proof. We will show that for any $\delta \in \mathfrak{D}$ there exists a behavioral invariant rule δ^I such that

$$\sup_\theta \hat{R}(\theta, \delta^I) \leq \sup_\theta \hat{R}(\theta, \delta).$$

Both conclusions of the theorem follow immediately. Let

$$\mathcal{G} = \{g_1, \cdots, g_N\}$$

and define a rule δ^I as follows. For each $x \in \mathfrak{X}$ and (measurable) $A \subset \mathcal{A}$ let

$$\delta_x{}^I(A) = \frac{1}{N} \sum_{i=1}^N \delta_{g_i(x)}(\tilde{g}_i(A)). \tag{4.18}$$

[δ^I represents the distribution that chooses a $g_i \in \mathcal{G}$ at random, each g_i with probability $1/N$, and then uses the distribution δ^{g_i} as defined by (4.11). Thus, even though δ may be a nonrandomized rule, δ^I is a randomized rule.]

Then δ^I is an invariant rule, since for every $g \in \mathcal{G}$

$$\delta^I_{g(x)}(\tilde{g}(A)) = \frac{1}{N} \sum_{i=1}^N \delta_{g_i(g(x))}(\tilde{g}_i(\tilde{g}(A)))$$

$$= \frac{1}{N} \sum_{i=1}^N \delta_{g_i(x)}(\tilde{g}_i(A)) = \delta_x{}^I(A) \tag{4.19}$$

for all $x \in \mathfrak{X}$, so that (4.8) holds. Moreover,

$$\sup_\theta \hat{R}(\theta, \delta^I) = \sup_\theta \frac{1}{N} \sum_{i=1}^N \hat{R}(\theta, \delta^{g_i})$$

$$= \sup_\theta \frac{1}{N} \sum_{i=1}^N \hat{R}(\bar{g}_i(\theta), \delta) \qquad \text{(Lemma 4.2.1)}$$

$$\leq \frac{1}{N} \sum_{i=1}^N \sup_\theta \hat{R}(\bar{g}_i(\theta), \delta)$$

$$= \frac{1}{N} \sum_{i=1}^N \sup_\theta \hat{R}(\theta, \delta) = \sup_\theta \hat{R}(\theta, \delta), \tag{4.20}$$

thus completing the proof.

This theorem is also true if $\mathcal{G}$ is a compact topological group. The proof works if $1/N\sum_{i=1}^{N}$ is replaced everywhere by $\int d\mu$, where μ is a Haar (invariant) probability measure. For noncompact groups the Haar measure is not finite and cannot be made into a probability measure. Without additional restrictions, the Theorem 1 does not hold for noncompact groups (see Exercise 4). A problem in which $\mathcal{G}$ is the group of translations on the real line is treated in Theorem 4.5.2. The general problem is treated in Kiefer (1957). See also H. Kudo (1955).

Theorem 2. Suppose that a given decision problem is invariant under a finite group $\mathcal{G}$. If an invariant rule $\delta_0 \in \mathfrak{D}$ is admissible within the class of all invariant rules, it is admissible.

Proof. Suppose δ_0 is not admissible. Then there is a rule $\delta \in \mathfrak{D}$ such that $\hat{R}(\theta, \delta) \leq \hat{R}(\theta, \delta_0)$ for all $\theta \in \Theta$, and $\hat{R}(\theta_0, \delta) < \hat{R}(\theta_0, \delta_0)$ for some $\theta_0 \in \Theta$. Let δ^I be the rule of (4.18). That δ^I is invariant is proved in Theorem 1. Also

$$\begin{aligned}
\hat{R}(\theta, \delta^I) &= \frac{1}{N}\sum_{i=1}^{N} \hat{R}(\theta, \delta^{g_i}) \\
&= \frac{1}{N}\sum_{i=1}^{N} \hat{R}(\bar{g}_i(\theta), \delta) && \text{(Lemma 4.2.1)} \\
&\leq \frac{1}{N}\sum_{i=1}^{N} \hat{R}(\bar{g}_i(\theta), \delta_0) \\
&= \frac{1}{N}\sum_{i=1}^{N} \hat{R}(\theta, \delta_0) && \text{(Theorem 4.2.1)} \\
&= \hat{R}(\theta, \delta_0). && (4.21)
\end{aligned}$$

Thus δ^I is as good as δ_0. However,

$$\begin{aligned}
\hat{R}(\theta_0, \delta^I) &= \frac{1}{N}\sum_{i=1}^{N} \hat{R}(\bar{g}_i(\theta_0), \delta) \\
&< \frac{1}{N}\sum_{i=1}^{N} \hat{R}(\bar{g}_i(\theta_0), \delta_0)
\end{aligned}$$

(since one of the $\bar{g}_i$ is the identity)

$$= \hat{R}(\theta_0, \delta_0), \tag{4.22}$$

so that δ^I is, in fact, better than δ_0. This contradicts the assumption that δ_0 is admissible out of the class of all invariant rules and proves the theorem.

An example is given in Section 4.5 of a decision problem invariant under a noncompact group $\mathcal{G}$ for which Theorem 2 does not hold.

Sufficiency and Invariance. Often a statistical decision problem will allow a reduction of the data by both sufficiency and invariance. Several questions arise.

1. Is it possible to apply both methods? For example, if a given decision problem admits a sufficient statistic T for the parameter and is invariant under some finite group, we know that if there is a minimax rule then there is a minimax rule based on T and there is an invariant minimax rule. Does there exist in this situation an invariant minimax rule based on T?

2. Given that it is possible to apply both methods in either order, when will the two orders lead to the same reduction of the data? In other words, when is sufficiency followed by invariance equivalent to invariance followed by sufficiency? These and related questions are treated in a paper by Hall, Wijsman, and Ghosh (1965).

We shall try to avoid these difficulties by reducing the data by means of sufficiency before searching for a group of transformations under which the problem is invariant. If T is sufficient for the parameter, we pretend that T is the observable random quantity and consider only groups of transformations on the space of values of T that leave the problem invariant.

This procedure of reducing the decision problem by sufficiency as much as possible before trying to reduce it by means of the invariance principle has one advantage. In some decision problems the principles of sufficiency and invariance lead to the same reduction of the data. For example, if $X \in \mathfrak{N}(0; \sigma^2)$, then the problem of estimating σ^2 with squared error loss is invariant under the transformation $g(x) = -x, (\bar{g}(\sigma^2) = \sigma^2$, $\tilde{g}(a) = a$, whatever the loss function), so that the invariant decision rules are such that $\delta(x) = \delta(-x)$ (that is, δ is a function of x^2). Because X^2 is a sufficient statistic, the principles of sufficiency and invariance both lead to restricting attention to the class of rules which are functions of X^2. Of course, it is preferable to use the principle of sufficiency in such situations, for the rules based on a sufficient statistic form an essentially complete class, whereas the invariant rules, in general, do not. Thus, in reducing the data as far as possible by means of sufficient statistics before searching for groups of transformations which leave the problem invariant, we avoid this pitfall.

Invariance and Convex Loss. Suppose that a given decision problem (after reduction by sufficiency, if possible) is invariant under a finite group and that the loss is convex in $a \in \mathcal{A}$ (a convex subset of a Euclidean space), as in Theorem 2.8.1. If there is a minimax rule, then there is an invariant minimax rule and a nonrandomized minimax rule. Naturally the question is—is there a nonrandomized invariant minimax rule? I do not know the complete answer to this question. However, if the group $\tilde{\mathcal{G}}$ of transformations on $\mathcal{A}$ contains only linear transformations (that is, transformations of the form $\tilde{g}(\mathbf{a}) = \mathbf{Ba} + \mathbf{c}$), the answer is yes. To see this, we cannot take a nonrandomized minimax rule and construct the invariant minimax rule by (4.18), because this new rule may be randomized. Instead, we may take an invariant minimax rule δ and try to form a nonrandomized rule as good by the method in Section 2.8. This entails replacing each distribution $\delta(x)$ over $\mathcal{A}$ by its expected value. It remains to be shown that this nonrandomized minimax rule will still be invariant. Let Z represent a random variable over $\mathcal{A}$ with distribution $\delta(x)$, so that $d(x) = EZ$ is the first moment of the distribution $\delta(x)$. If δ is invariant, the distribution of $\tilde{g}Z$ is given by $\delta(g(x))$; hence

$$d(g(x)) = E\tilde{g}Z.$$

If $\tilde{g}$ is a linear transformation on $\mathcal{A}$, then

$$d(g(x)) = E\tilde{g}Z = \tilde{g}EZ = \tilde{g}d(x),$$

showing that d is an invariant nonrandomized rule as good as δ, because the loss is convex. If δ is minimax, so is d.

The following examples may be considered as applications of Theorems 1 and 2.

EXAMPLE 1. Let $X_1, X_2, \cdots, X_n$ represent the outcomes of n independent tosses of a coin; $X_i = 1$ if the ith toss is heads and $X_i = 0$ otherwise where the probability of heads is p, $p \in [0, 1]$. We wish to estimate p

with loss $L(p, a) = L(p - a)$, where L is an even function [that is, $L(x) = L(-x)$]. This problem is invariant under the group of permutations of the subscripts of $(X_1, \cdots, X_n)$, for if $g(x_1, \cdots, x_n) = (x_{\nu_1}, \cdots, x_{\nu_n})$, where $(\nu_1, \cdots, \nu_n)$ is a permutation of $(1, \cdots, n)$, then $\bar{g}(p) = p$ and $\tilde{g}(a) = a$. The invariant nonrandomized rules are those for which $d(g(x_1, \cdots, x_n)) = d(x_1, \cdots, x_n)$, which, since each x_i is zero or one, means that $d(x_1, \cdots, x_n)$ must be a function of the number of ones or, equivalently, a function of $T = \sum_{i=1}^{n} X_i$. However, T is sufficient for this problem. As previously mentioned we use the principle of sufficiency instead of the principle of invariance to reduce the problem to considering only functions of T, for such functions constitute an essentially complete class. Now T has a binomial distribution $\mathcal{B}(n, p)$. This problem is invariant under the transformation $g(t) = n - t$, for the distribution of $g(T)$ is $\mathcal{B}(n, 1 - p)$, so that $\bar{g}(p) = 1 - p$ and $\tilde{g}(a) = 1 - a$. Theorem 1 implies that in the search for a minimax rule attention may be restricted to the invariant rules. An invariant rule $\delta \in D^*$ gives all its mass to the set of nonrandomized rules for which $d(g(t)) = \tilde{g}(d(t))$, or equivalently $d(n - t) = 1 - d(t)$. Thus, instead of having to choose at random a rule d that specifies $d(0), d(1), \cdots, d(n)$ as unrelated numbers, we need to consider only the rules for which $d(0) = 1 - d(n)$, $d(1) = 1 - d(n - 1), \cdots$, and so on, and if n is even $d(n/2) = 1/2$. Thus the principle of invariance has reduced the complexity of the space of nonrandomized decision rules by reducing its dimension from $n + 1$ to $n/2$ if n is even and to $(n + 1)/2$ if n is odd.

Example 2. Consider the problem of estimating p on the basis of one toss of a coin with probability p of heads, using absolute error loss, $L(p, a) = |p - a|$. The second example of Section 2.11 is this problem using squared error loss, and it was seen that the minimax rule estimates p to be 1/4 if tails appears and p to be 3/4 if heads appears. We are interested in seeing what difference will appear in the minimax rule if absolute error loss is used. Here, however, we avail ourselves of the results of Example 1 and immediately restrict attention to rules for which $d(0) = 1 - d(1)$; for in this problem the loss is convex and the problem is invariant under a group $\mathcal{G}$ for which $\tilde{\mathcal{G}}$ contains only linear transformations. This reduces the dimensionality of the space of decision rules under consideration from two to one.

An invariant nonrandomized rule d is denoted by x where $x = d(0) = 1 - d(1)$. The risk function is then

$$R(p, x) = (1 - p)\,|p - x| + p\,|p - (1 - x)|. \tag{4.23}$$

By Theorem 4.2.1, the risk function is constant on orbits when an invariant rule is used, so that

$$R(p, x) = R(1 - p, x). \tag{4.24}$$

Thus for invariant rules the risk is symmetric about $p = \frac{1}{2}$. This is easily checked by (4.23).

Suppose first that $x \leq \frac{1}{2}$. Then for $0 \leq p \leq x$

$$\begin{aligned} R(p, x) &= (1 - p)(x - p) + p(1 - p - x) \\ &= -2px + x, \end{aligned} \tag{4.25}$$

a linear function of p with negative slope. For $x \leq p \leq 1/2$

$$\begin{aligned} R(p, x) &= (1 - p)(p - x) + p(1 - p - x) \\ &= -2p^2 + 2p - x, \end{aligned} \tag{4.26}$$

a quadratic function of p with maximum at $p = 1/2$. Thus when $x \leq 1/2$,

$$\begin{aligned} \sup_p R(p, x) &= \max\left(R(0, x), R(\tfrac{1}{2}, x)\right) \\ &= \max\left(x, \tfrac{1}{2} - x\right). \end{aligned} \tag{4.27}$$

For $x \leq 1/2$ (Fig. 4.1) this maximum is minimized when $x = 1/4$, the maximum risk there being 1/4. For $x > 1/2$, obviously $\sup_p R(p, x) \geq R(0, x) = x > 1/2$, so that out of all invariant rules the one that minimizes the maximum risk is $x = 1/4$: estimate 1/4 if tails appears and 3/4 if heads appears. From Theorem 1 this rule is minimax out of the class of all rules. It is easy to see from this analysis that this rule is admissible out of the class of all invariant rules, so that by Theorem 2 it is admissible.

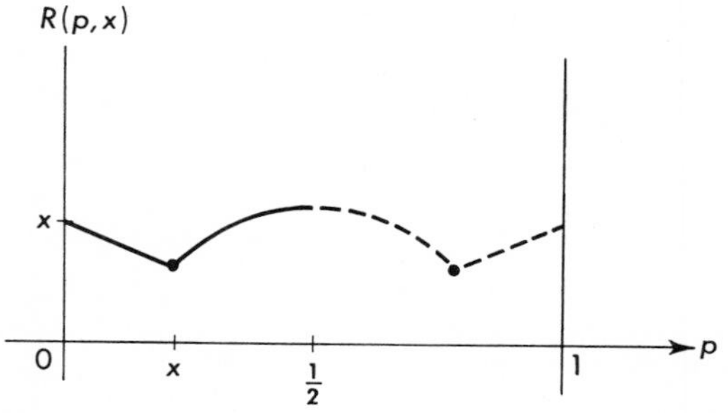

Fig. 4.1

EXAMPLE 3. Consider two coins, one fair (with probability 1/2 of coming up heads) and the other biased (with probability θ of coming up heads, $\theta \in [0, 1]$). Each coin is tossed once, and we are to estimate θ with quadratic loss $(\theta - a)^2$. However, we do not know which coin (if either) is biased.

Let X be zero or one, depending on whether the first coin comes up tails or heads, and let Y be zero or one, depending on whether the second coin comes up tails or heads. The distribution of (X, Y) is either $[\mathcal{B}(1, 1/2), \mathcal{B}(1, \theta)]$, to be denoted by $P_{(\theta,1)}$, or $[\mathcal{B}(1, \theta), \mathcal{B}(1, 1/2)]$, to be denoted by $P_{(\theta,2)}$. Thus the parameter space is $\Theta = \{(\theta, j) : j = 1$ or 2 and $0 \le \theta \le 1\}$. The loss function is $L((\theta, j), a) = (\theta - a)^2$, where $a \in \mathcal{A} = [0, 1]$. The points $(1/2, 1)$ and $(1/2, 2)$ of Θ are to be considered as identical.

This problem is invariant under the transformation

$$g_1(x, y) = (y, x), \quad \bar{g}_1(\theta, j) = (\theta, 3 - j), \quad \tilde{g}_1(a) = a.$$

It is also invariant under the transformation

$$g_2(x, y) = (1 - x, 1 - y), \quad \bar{g}_2(\theta, j) = (1 - \theta, j), \quad \tilde{g}_2(a) = 1 - a.$$

Thus the problem is invariant under the group (of four elements) generated by g_1 and g_2. Because the loss function is convex and all elements of $\tilde{\mathcal{G}}$ are linear transformations, the problem of finding a minimax rule may be restricted to searching through the class of invariant nonrandomized rules. A general nonrandomized decision rule for this problem may estimate four different numbers $d(0, 0)$, $d(1, 0)$, $d(0, 1)$, and $d(1, 1)$. For such a rule to be invariant under g_1 we must have $d(1, 0) = d(0, 1)$. For such a rule to be invariant under g_2 we must have $d(0, 0) = 1 - d(1, 1)$ and $d(1, 0) = 1 - d(0, 1)$. Hence $d(1, 0) = d(0, 1) = 1/2$ for all invariant rules and, if $z = d(0, 0)$, then $d(1, 1) = 1 - z$. We denote such an invariant rule by z; the risk function is

$$R((\theta, i), z) = (1/2)(1 - \theta)(\theta - z)^2 + (1/2)(1 - \theta)(\theta - 1/2)^2 \\ + (1/2)\theta(\theta - 1/2)^2 + (1/2)\theta(\theta - (1 - z))^2.$$

This risk is independent of i, for

$$R((\theta, 1), z) = R(\bar{g}_1(\theta, 1), z) = R((\theta, 2), z).$$

It is symmetric about $\theta = 1/2$, for

$$R((\theta, i), z) = R(\bar{g}_2(\theta, i), z) = R((1 - \theta, i), z).$$

Thus

$$R((\theta, i), z) = 2z\theta^2 - 2z\theta + z^2/2 + 1/8$$

is a quadratic function of θ with a maximum in the interval $[0, 1]$ at $\theta = 0$ and $\theta = 1$. Thus

$$\sup_{(\theta, i)} R((\theta, i), z) = R((0, i), z) = z^2/2 + 1/8.$$

This supremum is minimized by taking $z = 0$. The minimax invariant rule, hence minimax rule by Theorem 1, is to estimate $(X + Y)/2$ as the value of θ.

Invariant Prior Distributions. A prior distribution τ on Θ is said to be invariant under $\bar{\mathcal{G}}$ if for all $\bar{g} \in \bar{\mathcal{G}}$, $\bar{g}\tau = \tau$ where by $\bar{g}\tau$ we mean the distribution of $\bar{g}T$ when T has distribution τ.

Because the probability that $\bar{g}T \in A$ is the same as the probability that $T \in \bar{g}^{-1}A$, a distribution τ is invariant if for all measurable sets $A \subset \Theta$ and all $\bar{g} \in \bar{\mathcal{G}}$, $\tau(\bar{g}^{-1}A) = \tau(A)$.

An alternative method of finding minimax rules is to search among the invariant rules that are Bayes with respect to invariant prior distributions. That this may succeed is indicated by the following theorem.

Theorem 3. Suppose that a given decision problem is invariant under a finite group $\mathcal{G}$.

(a) If δ_0 is Bayes with respect to τ and δ_0 is invariant, there exists an invariant prior distribution τ_0 with respect to which δ_0 is Bayes.

(b) If δ is Bayes with respect to τ_0 and τ_0 is invariant, there exists an invariant rule δ_0 which is Bayes with respect to τ_0.

(c) If there exists a least favorable distribution τ, there exists an invariant least favorable distribution τ_0.

Proof of (a). Define

$$\tau_0 = N^{-1} \sum_{g \in \mathcal{G}} \bar{g}\tau,$$

where N is the number of elements in $\mathcal{G}$. Then, for all rules δ,

$$r(\tau_0, \delta) = \frac{1}{N} \sum_{g \in \mathcal{G}} r(\bar{g}\tau, \delta) = \frac{1}{N} \sum_{g \in \mathcal{G}} E\hat{R}(\bar{g}T, \delta)$$

$$= \frac{1}{N} \sum_{g \in \mathcal{G}} E\hat{R}(T, \delta^g) = \frac{1}{N} \sum_{g \in \mathcal{G}} r(\tau, \delta^g)$$

by Lemma 4.2.1, where T has distribution τ. We are given that $r(\tau, \delta_0) \leq r(\tau, \delta)$ for all rules δ. Hence

$$r(\tau_0, \delta) - r(\tau_0, \delta_0) = \frac{1}{N} \sum_{g \in \mathcal{G}} [r(\tau, \delta^g) - r(\tau, \delta_0{}^g)]$$

$$= \frac{1}{N} \sum_{g \in \mathcal{G}} [r(\tau, \delta^g) - r(\tau, \delta_0)] \geq 0,$$

because δ_0 is invariant, thus completing the proof.

The proofs of (b) and (c) are left as exercises.

Exercises

1. Extend Example 3 to n coins, $n - 1$ of which are known to be fair, the other having probability θ of coming up heads. Again the problem is to estimate θ with squared error loss on the basis of one toss of each coin. Show that the minimax rule estimates θ as the proportion of heads in n coins.
2. Find the least favorable distribution in Example 2. (*Hint.* From Lemma 2.11.1 the least favorable distribution must give all its weight to points p for which $R(p, 1/4) = v = 1/4$. Use Theorem 3(c)).
(*Ans.* $\tau(0) = \tau(1) = 1/4, \tau(1/2) = 1/2$)
3. Find the least favorable distribution in Example 3.
4. In Stein's example in Exercise 4.2.7(a)(b), suppose that $\mathcal{A} = (0, \infty)$, $L(\Delta, a) = 0$ if $|\Delta - a| \leq \Delta/2$, and $L(\Delta, a) = 1$ if $|\Delta - a| > \Delta/2$. Let $\mathcal{G}$ be the group of transformations $g_B(\mathbf{X}, \mathbf{Y}) = (\mathbf{BX}, \mathbf{BY})$ for nonsingular $\mathbf{B}$. Show that for any rule δ invariant under $\bar{\mathcal{G}}$, $\sup_\theta R(\theta, \delta) = 1$ so that any invariant rule is minimax within the class of invariant rules. Yet, if $d(\mathbf{X}, \mathbf{Y}) = |Y_1/X_1|$, then $R(\theta, d)$ is a constant less than one.
5. Let $\Theta = [0, 1] = \mathcal{A}$, let the loss function be

$$L(\theta, a) = (1 - \theta)a + \theta(1 - a),$$

and let X have the binomial distribution $\mathcal{B}(n, \theta)$. Because the loss is convex, we may restrict attention to nonrandomized decision rules.
(a) Show that this problem is invariant under the group of two elements generated by $g(x) = n - x$ and find $\bar{g}(\theta)$ and $\tilde{g}(a)$.
(b) Find the form of the nonrandomized invariant decision rules.
(c) Show that the invariant rule $d(x) = 0$ for $x < n/2$, $d(n/2) = 1/2$, $d(x) = 1$ for $x > n/2$ is a best invariant decision rule. Compare with Exercise 2.11.15.

6. Prove Theorem 3(b).

7. Prove Theorem 3(c).

4.4 Location and Scale Parameters

The principle of invariance is particularly useful in problems dealing with location and scale parameters.

We consider a real-valued random variable X whose distribution P_θ depends on a real parameter $\theta \in \Theta \subset E_1$, the real line. $F_X(x \mid \theta)$ is used to denote the distribution function of the random variable X when θ is the true value of the parameter.

Definition 1. A real parameter $\theta \in \Theta$ is said to be a *location parameter* for the distribution of a random variable X if $F_X(x \mid \theta)$ is a function only of $x - \theta$ [then $F_X(x \mid \theta) = F(x - \theta)$, where F is a distribution function.]

The following lemma gives alternative definitions of the notion of a location parameter.

Lemma 1. (a) θ is a location parameter for the distribution of X if, and only if, the distribution of $X - \theta$, when θ is the true value of the parameter, is independent of θ. (b) If the distributions of X are absolutely continuous with density $f_X(x \mid \theta)$, then θ is a location parameter for the distribution of X if, and only if, $f_X(x \mid \theta) = f(x - \theta)$ for some density $f(x)$.

The proof of this lemma is left to the student.

Example 1. If $X \in \mathfrak{N}(\theta, \sigma^2)$, σ^2 known, then θ is a location parameter for X. If X has the Cauchy distribution of (3.12) or (3.13), then α is a location parameter for the distribution of X for each fixed β. If $X \in \mathfrak{U}(\theta, \theta + 1)$, then θ is a location parameter for the distribution of X.

Note. If θ is a location parameter for the distribution of X, so is $\mu = \theta + c$ for any constant c.

Definition 2. A positive real parameter $\theta \in \Theta$ is said to be a *scale parameter* for the distribution of the random variable X if $F_X(x \mid \theta)$ is a function only of x/θ [then $F_X(x \mid \theta) = F(x/\theta)$, where F is a distribution function].

Lemma 2. (a) θ is a scale parameter for the distribution of X if, and only if, the distribution of X/θ when θ is the true value of the parameter is independent of θ. (b) If the distributions of X are absolutely continuous with density $f_X(x \mid \theta)$, then θ is a scale parameter for the distribution of X if, and only if, $f_X(x \mid \theta) = (1/\theta)f(x/\theta)$ for some density $f(x)$.

Again, the proof is left to the student.

EXAMPLE 2. If $X \in \mathfrak{N}(\mu\theta, \theta^2)$, μ known, then θ is a scale parameter for the distribution of X. If X has a Cauchy distribution of (3.12) or (3.13), then β is a scale parameter for the distribution of X when $\alpha = 0$. If $X \in \mathfrak{U}(0, \theta)$, then θ is a scale parameter for the distribution of X. If $X \in \mathcal{G}(\alpha, \beta)$, α known, then β is a scale parameter for the distribution of X.

Note. If θ is a scale parameter for the distribution of X, so is $\sigma = \theta/c$ for any constant $c > 0$.

If θ is a location parameter for the distribution of X, then $\sigma = e^\theta$ is a scale parameter for the distribution of $Y = e^X$. This follows immediately from parts (a) of Lemmas 1 and 2.

In fact, the general scale parameter family of distributions may be constructed as follows. Choose random variables X and Y whose distributions both have θ as a location parameter and choose three nonnegative numbers p_1, p_2, p_3 whose sum is one. Then, e^θ is a scale parameter of the distribution of the random variable Z, defined as equal to e^X with probability p_1, $-e^Y$ with probability p_2, and zero with probability p_3.

Conversely, if θ is a scale parameter of the distribution of a random variable Z, then $\log\theta$ is a location parameter of the conditional distribution of $\log Z$, given $Z > 0$, and of the conditional distribution of $\log(-Z)$, given $Z < 0$.

These remarks will be useful in passing from problems involving location parameters to those involving scale parameters.

Location-Scale Parameters. We complete this section with mention of distributions that have both locations and scale parameters in a certain sense.

Definition 3. A two-dimensional parameter (μ, σ), with $\sigma > 0$, is said to be a *location-scale parameter* for the distribution of a random variable X if $F_X(x \mid \mu, \sigma)$ is a function only of $(x - \mu)/\sigma$ [then $F_X(x \mid \mu, \sigma) = F((x - \mu)/\sigma)$ where F is a distribution function].

Lemma 3. (a) (μ, σ) is a location-scale parameter for the distribution of X if, and only if, the distribution of $(X - \mu)/\sigma$, when (μ, σ) is the true value of the parameter, is independent of (μ, σ). (b) If the distributions of X are absolutely continuous with density $f_X(x \mid \mu, \sigma)$, then (μ, σ) is a location-scale parameter for the distribution of X if, and only if, $f_X(x \mid \mu, \sigma) = (1/\sigma)f((x - \mu)/\sigma)$ for some density $f(x)$.

If $X \in \mathfrak{N}(\mu, \sigma^2)$, then (μ, σ) is a location-scale parameter for X. If $X \in \mathfrak{U}(\mu - \sqrt{3}\sigma, \mu + \sqrt{3}\sigma)$, then (μ, σ) is a location-scale parameter for X (and σ^2 is the variance of X). If X has the Cauchy distribution of (3.12) or (3.13), then (α, β) is a location-scale parameter for X.

It is important to notice that if (μ, σ) is a location-scale parameter for the distribution of X then μ is a true location parameter (that is, the distribution of $X - \mu$ does not depend on μ for each fixed $\sigma > 0$). However, σ does not satisfy our definition of a scale parameter for the distribution of X unless $\mu = 0$.

4.5 Minimax Estimates of Location Parameters

Consider the decision problem of estimating a parameter $\theta \in \Theta = \mathfrak{R}$ the real line, in which θ is a location parameter of the distribution of the observable random variable X. We shall suppose the loss function is a function of $(a - \theta)$ alone, $L(\theta, a) = L(a - \theta)$. This problem is clearly invariant under the group $\mathcal{G}$ of translations $g_c(x) = x + c$, with $\bar{g}_c(\theta) = \theta + c$, and $\tilde{g}(a) = a + c$. An invariant nonrandomized estimate for this problem satisfies (4.6), which in this case reduces to

$$d(x + c) = d(x) + c \tag{4.28}$$

for all x and all c. Putting $x = 0$ in this equation gives $d(c) = c + d(0)$ for all c. Thus every invariant nonrandomized decision rule has the form

$$d_b(x) = x - b, \tag{4.29}$$

where $b = -d_b(0)$ is an arbitrary real number. From Theorem 4.2.1, the risk function of any invariant decision rule $\delta \in D^*$ satisfies

$$R(\theta, \delta) = R(\theta + c, \delta) \tag{4.30}$$

for all $\theta \in \Theta$ and all c. Thus the risk is independent of θ. (See Exercise 4.) However, we do not call the rule d_b of (4.29) a decision rule at all unless $R(\theta, d_b)$ is finite for all $\theta \in \Theta$. The risk of the nonrandomized rule d_b is

$$R(\theta, d_b) = E_\theta L(X - b - \theta) = E_0 L(X - b), \tag{4.31}$$

for it is independent of θ (where E_0 represents the expectation when $\theta = 0$). Hence if $E_0 L(X - b)$ exists and is finite for some b, the corresponding estimate of the form (4.29) is an invariant nonrandomized decision rule for this problem.

We seek now a best invariant rule (that is, an invariant rule that is as good as any other invariant rule). Among *nonrandomized* invariant rules, the rule that minimizes the quantity (4.31) is obviously best. The following lemma shows that in trying to find a best invariant rule attention may be restricted to the nonrandomized rules.

Lemma 1. If every nonrandomized invariant rule is an equalizer rule (that is, has constant risk), the nonrandomized invariant decision rules form an essentially complete class among the class of all randomized invariant rules.

Proof. Let δ be any invariant rule in D^*. By definition, δ gives all its probability mass to the class of invariant nonrandomized rules. Because all such rules have constant risk, the risk $R(\theta, \delta)$ is a constant also and is the average of the risks $R(\theta, d)$, averaged over the distribution δ. It follows that there is a nonrandomized invariant rule d such that $R(\theta, d) \leq R(\theta, \delta)$ (that is, the mean of a distribution is greater than or equal to its minimum value), thus completing the proof.

This lemma is not necessarily true for behavioral invariant rules. In other words, even if all invariant rules are equalizers there may be an invariant behavioral rule for which no nonrandomized invariant rule is at least as good. An example may be found in Exercise 4.2.6. It was indicated in Section 4.2 that behavioral and randomized rules are equivalent, provided that all orbits of $\mathfrak{X}$ have multiplicity one. This condition is satisfied for the problem of estimating a location parameter, so that as a corollary of Lemma 1 we have Theorem 1.

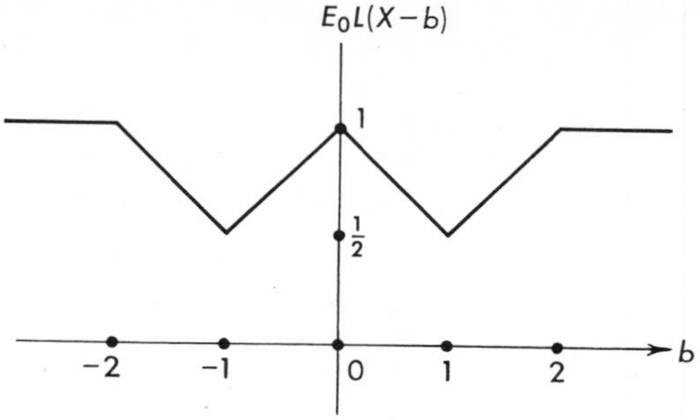

Fig. 4.2

Theorem 1. In the problem of estimating a location parameter with loss $L(\theta, a) = L(a - \theta)$, if $E_0L(X - b)$ exists and is finite for some b and if there exists a b_0 such that

$$E_0L(X - b_0) = \inf_b E_0L(X - b), \tag{4.32}$$

where the infimum is taken over all b for which $E_0L(X - b)$ exists, then $d(x) = x - b_0$ is a best invariant rule. It has constant risk equal to the left side of (4.32).

Example 1. If $L(\theta, a) = (a - \theta)^2$ and X has finite variance, then $b_0 = E_0X$, the expected value of X when $\theta = 0$. If $L(\theta, a) = |a - \theta|$ and X has finite first moment, then b_0 is a median of the distribution of X when $\theta = 0$.

The fact that the rule $d(x) = x - b_0$ has a constant risk and is best out of all invariant rules should suggest to the reader the possibility that $d(x)$ is a minimax rule. Exercise 2.11.4 says that if $d(x)$ is admissible (which it ought to be because it is a best invariant rule), it will be minimax, for it has constant risk. Unfortunately, it is not always true that a best invariant estimate of a location parameter is admissible, as an example due to Blackwell (1951) shows. Let X take the value $\theta + 1$ with probability 1/2 and $\theta - 1$ with probability 1/2 and let the loss function be

$$L(\theta, a) = L(a - \theta) = \begin{cases} |a - \theta| & \text{if } |a - \theta| \leq 1, \\ 1 & \text{if } |a - \theta| > 1. \end{cases} \tag{4.33}$$

This is a problem of estimating a location parameter, with loss a function of $a - \theta$. It is easily seen (Fig. 4.2) that

$$E_0 L(X - b) = \begin{cases} 1 - (1/2)\,|b| & \text{if } |b| \le 1, \\ (1/2)\,|b| & \text{if } 1 < |b| \le 2, \\ 1 & \text{if } |b| > 2. \end{cases} \tag{4.34}$$

This obviously takes its minimum when $b = +1$ or -1. Thus the rules $d_1(X) = X + 1$ and $d_2(X) = X - 1$ are both best invariant for this problem. The constant risk is $R(\theta, d_1) = 1/2$. However, neither of these rules is admissible. The rule

$$d_0(X) = \begin{cases} X + 1 & \text{if } X < 0, \\ X - 1 & \text{if } X \ge 0 \end{cases} \tag{4.35}$$

is not invariant but has risk function

$$R(\theta, d_0)$$

$$= 1/2\begin{cases} L(\theta, \theta + 2) & \text{if } \theta + 1 < 0 \\ L(\theta, \theta) & \text{if } \theta + 1 \ge 0 \end{cases} + 1/2\begin{cases} L(\theta, \theta) & \text{if } \theta - 1 < 0, \\ L(\theta, \theta - 2) & \text{if } \theta - 1 \ge 0, \end{cases}$$

$$= 1/2\begin{cases} 1 & \text{if } \theta < -1 \\ 0 & \text{if } \theta \ge -1 \end{cases} + 1/2\begin{cases} 0 & \text{if } \theta < 1, \\ 1 & \text{if } \theta \ge 1, \end{cases}$$

$$= \begin{cases} 0 & \text{if } -1 \le \theta < 1, \\ 1/2 & \text{otherwise.} \end{cases} \tag{4.36}$$

Because d_0 is better than d_1 or d_2, the latter are not admissible.

A more important example, due to Stein (1956) [see also James and Stein (1961)], involves estimating the mean of a multivariate normal distribution with squared error loss. The covariance matrix is assumed known and for simplicity is taken to be the identity matrix. Let $\mathbf{X}_1, \mathbf{X}_2, \cdots, \mathbf{X}_n$ be a sample from the k-dimensional normal distribution

$\mathfrak{N}(\boldsymbol{\theta}, \mathbf{I})$, let $\Theta = \mathcal{A} = E_k$, and let

$$L(\boldsymbol{\theta}, \mathbf{a}) = |\boldsymbol{\theta} - \mathbf{a}|^2 = \sum(\theta_i - a_i)^2.$$

For $k = 1$, it was seen in Section 2.11 that the sample mean is a minimax estimate and in Section 3.7 that it is an admissible estimate. It is true in the k-dimensional case that $\bar{\mathbf{X}} = (1/n) \sum \mathbf{X}_i$ is a minimax estimate (see Exercise 5), and we would naturally suspect that $\bar{\mathbf{X}}$ would be admissible also. Stein has shown that for $k = 2$ $\bar{\mathbf{X}}$ is admissible but that for $k > 2$ $\bar{\mathbf{X}}$ is *not* admissible. An estimate that improves on $\bar{\mathbf{X}}$ when $k \geq 3$ for all values of θ is

$$d_0(\mathbf{X}) = \bar{\mathbf{X}}\left(1 - \frac{(k-2)}{|\bar{\mathbf{X}}|^2}\right).$$

This improved estimate takes $\bar{\mathbf{X}}$ and shifts it toward the origin (or past the origin if $|\bar{\mathbf{X}}|^2 < k - 2$). The origin, of course, has nothing to do with the problem, and the corresponding estimate that shrinks $\bar{\mathbf{X}}$ toward an arbitrary point $\mathbf{b}$, $d(\mathbf{X}) = \mathbf{b} + (\bar{\mathbf{X}} - \mathbf{b})(1 - (k-2)|\bar{\mathbf{X}} - \mathbf{b}|^{-2})$ also improves on the estimate $\bar{\mathbf{X}}$ for all $\boldsymbol{\theta}$, the greatest improvement being for $\boldsymbol{\theta}$ near $\mathbf{b}$. There is a close analogy to Blackwell's example in which estimates that improve on the best invariant estimate may be obtained by shifting toward an arbitrary point b:

$$d_b(x) = (x + 1)I_{(-\infty,b)}(x) + (x - 1)I_{(b,\infty)}.$$

In spite of these examples, it is true in general that the best invariant estimate of a location parameter is minimax as Theorem 2 shows. In the proof it is worthwhile to use the notation of the Riemann-Stieltjes integral instead of the expectation sign. The integral $\int \phi(x)\, dF(x)$, for those not familiar with the Riemann-Stieltjes integral, may be viewed as a different notation for $E\,\phi(X)$, where X is a random variable with distribution function F and

$$\int_a^b \phi(x)\, dF(x)$$

may be viewed as a notation for $E\,\phi(X)I_{[a,b)}(X)$, where I is the indicator function (see Section 3.1). For example, if F is absolutely continuous with density function f, so that $dF(x) = f(x)\, dx$, then

$$\int_a^b \phi(x)\, dF(x) = \int_a^b \phi(x) f(x)\, dx,$$

and, if F is discrete with probability mass function $p(x)$, then

$$\int_a^b \phi(x)\, dF(x) = \sum_{a \leq x_i < b} \phi(x_i)\, p(x_i),$$

the summation being taken over all i for which $a \leq x_i < b$.

Let the distribution of the random variable X be denoted by

$$F_X(x \mid \theta) = F(x - \theta),$$

so that $F(x)$ represents the distribution of X when $\theta = 0$. We shall let R_0 denote the minimum risk $E_0L(X - b_0)$ attainable with invariant rules (4.32). We know under the assumptions of Theorem 1 that as $N \to \infty$

$$\int_{-N}^{N} L(x - b)\, dF(x) \to E_0L(X - b) \geq R_0 = \inf_b E_0L(X - b) \tag{4.37}$$

for all b.

Theorem 2. Under the assumptions of Theorem 1, if L is bounded below and if for every $\epsilon > 0$ there exists an N such that

$$\int_{-N}^{N} L(x - b)\, dF(x) \geq R_0 - \epsilon \tag{4.38}$$

for all b, then the best invariant rule is minimax.

Proof. We may and shall assume without loss of generality that $L(x) \geq 0$, for if $L(x) \geq -B$ the validity of the theorem for the loss function $L'(x) = L(x) + B$ implies the validity for the loss function $L(x)$.

For the proof we use Theorem 2.11.3. Thus we shall exhibit a sequence τ_n of prior distributions for nature and show that

$$\lim_{n\to\infty} \inf_\delta r(\tau_n, \delta) = R_0;$$

then, because the best invariant rule has constant risk R_0, it will be minimax. We might guess that the least favorable distribution for nature would be Lebesgue measure over the entire real line. Because this is not a probability distribution, we choose τ_M as an approximation of some sort, namely the uniform distribution over the interval $(-M, M)$.

Let $\epsilon > 0$ and find N to satisfy condition (4.38). Let $d(x)$ be any nonrandomized decision rule and let $M > N$. Then, since integrals may be interchanged because of the nonnegativity of the integrand,

$$\begin{aligned}
r(\tau_M, d) &= \frac{1}{2M}\int_{-M}^{M} R(\theta, d)\, d\theta = \frac{1}{2M}\int_{-M}^{M}\int L(d(y) - \theta)\, dF(y - \theta)\, d\theta \\
&= \frac{1}{2M}\int_{-M}^{M}\int L(d(x + \theta) - \theta)\, dF(x)\, d\theta \\
&= \frac{1}{2M}\int\int_{-M}^{M} L(d(x + \theta) - \theta)\, d\theta\, dF(x) \\
&= \frac{1}{2M}\int\int_{x-M}^{x+M} L(d(z) - z + x)\, dz\, dF(x) \\
&= \frac{1}{2M}\int\int_{z-M}^{z+M} L(x + d(z) - z)\, dF(x)\, dz \\
&\geq \frac{1}{2M}\int_{-(M-N)}^{M-N}\int_{z-M}^{z+M} L(x + d(z) - z)\, dF(x)\, dz \\
&\geq (R_0 - \epsilon)\frac{2M - 2N}{2M}
\end{aligned} \tag{4.39}$$

The last inequality follows, since for $z \in (-(M - N), (M - N))$ the inside integral is not less than the integral in (4.38). Because the inequality (4.39) is true for all nonrandomized $d \in D$, it is also true for randomized $\delta \in D^*$. Hence

$$\liminf_{M\to\infty}\ \inf_{\delta} r(\tau_M, \delta) \geq \lim_{M\to\infty} (R_0 - \epsilon)\frac{2M - 2N}{2M} = (R_0 - \epsilon) \tag{4.40}$$

for all $\epsilon > 0$. Hence the left side of (4.40) is not less than R_0, thus completing the proof.

If condition (4.38) does not hold, Theorem 2 may not be true (Exercise 7). Two lemmas show that in two important cases condition (4.38) of Theorem 2 is satisfied.

Lemma 2. If L is bounded, condition (4.38) is satisfied.

Proof. If $L(x) < B$, then find N such that $P_0\{|X| > N\} \leq \epsilon/B$. Then

$$\begin{aligned}\int_{-N}^{N} L(x - b)\, dF(x) &= E_0 L(X - b) - \int_{-\infty}^{-N} L(x - b)\, dF(x) - \int_{N}^{\infty} L(x - b)\, dF(x) \\ &\geq R_0 - BP_0\{|X| > N\} \\ &\geq R_0 - \epsilon, \end{aligned} \tag{4.41}$$

thus completing the proof.

This proof involves the demonstration that the convergence of the limit in (4.37) is uniform in b. In Lemma 3 the convergence in (4.37) may not be uniform in b, but condition (4.38) still holds.

Lemma 3. If L is continuous and $L(x) \to +\infty$ as $x \to \pm\infty$, then condition (4.38) is satisfied.

Proof. As in the proof of Theorem 2, we may assume without loss of generality that L is bounded below by zero, for the hypotheses of this lemma imply that L is bounded below. Then

$$R_N(b) = \int_{-N}^{N} L(x - b)\, dF(x) \tag{4.42}$$

is

(a) nondecreasing in N for each fixed b,
(b) continuous in b, and
(c) $R_N(b) \to +\infty$ as $b \to \pm\infty$ (for N large enough so that $P\{|X| < N\} > 0$).

From (c) there is an N_0 and an interval $[A, B]$ such that for b outside the interval $[A, B]$, $R_{N_0}(b) \geq R_0 + 1$. From (a) this inequality is valid for all $N \geq N_0$. Because $\inf_b R_N(b) \leq R_0$, (b) implies that the infimum of $R_N(b)$ is assumed at some point to be $b_N \in [A, B]$, when $N \geq N_0$. Hence there exists a limit point b' of b_N, and we shall assume that $b_n \to b'$. Since, for $n > N > N_0$

$$R_N(b_n) \leq R_n(b_n) \leq R_0, \tag{4.43}$$

we have $R_N(b') \leq R_0$ for all N; but, as $N \to \infty$,

$$R_N(b') \to E_0 L(X - b') \geq R_0,$$

so that from (4.43) $R_n(b_n) \to R_0$. Thus there exists an N such that $R_N(b_N) > R_0 - \epsilon$, thus completing the proof.

Exercises

1. Let X have a Cauchy distribution with unknown median θ and known semi-interquartile range β and let the loss function be

$$L(\theta, a) = \begin{cases} 0 & \text{if } |a - \theta| \leq c, \\ 1 & \text{if } |a - \theta| > c, \end{cases}$$

where c is a given positive number. Find the best invariant estimate of θ based on X and note that it does not depend on β or c. (See Exercise 1.8.5.)

2. Let θ be a location parameter for the distribution of a random variable X with finite variance and suppose that X is a complete sufficient statistic for θ. In the problem of estimating θ with quadratic loss $L(\theta, a) = (a - \theta)^2$, show that the best invariant estimate is the best unbiased estimate also.

3. Let θ be a scale parameter for the distribution of X and suppose that the loss function is a function only of a/θ, that is, $L(\theta, a) = L(a/\theta)$. State and prove a theorem for this problem analogous to Theorem 1.

4. A group $\bar{\mathcal{G}}$ of transformations on a set Θ is said to be *transitive* if for every θ_1 and $\theta_2 \in \Theta$, there exists a $\bar{g} \in \bar{\mathcal{G}}$ such that $\bar{g}(\theta_1) = \theta_2$. Show that if a decision problem is invariant under a group $\mathcal{G}$ and $\bar{\mathcal{G}}$ is transitive on Θ every invariant rule is an equalizer rule.

5. Let $\mathbf{X}_1, \cdots, \mathbf{X}_n$ be a sample from the k-dimensional normal distribution $\mathfrak{N}(\boldsymbol{\theta}, \mathbf{I})$. Show that $\bar{\mathbf{X}}$ is a minimax estimate of $\boldsymbol{\theta}$, using squared error loss, by generalizing to k dimensions the one-dimensional proof found in Section 2.11.

6. In Blackwell's example consider the randomized rule δ which chooses a point b according to a strictly increasing distribution function $F(b)$ and then uses the nonrandomized rule

$$d_b = (x + 1) I_{(-\infty, b)}(x) + (x - 1) I_{(b, \infty)}(x).$$

Show that the randomized rule δ improves on the best invariant rule at each value of θ.

7. (Blackwell and Girshick). Let the distribution of $X - \theta$ have probability mass function

$$f(x) = \frac{1}{x(x+1)}, \qquad x = 1, 2, \cdots$$

and let the loss function be $L(\theta, a) = \max(a - \theta, 0)$.
(a) Show that the risk function of every invariant rule is identically $+\infty$.
(b) Consider the noninvariant nonrandomized rule $d(x) = x - c\,|x|$, where $c > 1$. Show that

$$R(\theta, d) \le \sum_{a<k<b} (k+1)^{-1},$$

where

$$a = \begin{cases} \dfrac{c\,|\theta|}{c+1} & \text{for } \theta < 0 \\ 0 & \text{for } \theta \ge 0 \end{cases} \quad \text{and} \quad b = \begin{cases} \dfrac{c\,|\theta|}{c-1} & \text{for } \theta < 0, \\ 0 & \text{for } \theta \ge 0. \end{cases}$$

Hence show that

$$R(\theta, d) \le \log\left[2\,\frac{c+1}{c-1}\right],$$

so that no invariant estimate is minimax.
(c) Kiefer (1957) points out that this counterexample does not depend entirely on the fact that all invariant rules have infinite risk. If $L(x) = 1$ when x is an integer and $L(x) = \max(x, 0)$ otherwise, any invariant rule $d(x) = x - u$ for u an integer has constant risk equal to one; yet, if c is irrational and large, the decision rule of (b) has maximum risk less than one.

8. Let $X \in \mathfrak{N}(\mu + 2, 1)$. Note that μ is a location parameter. Find the best invariant estimate of μ, using squared error loss.

9. Under the assumptions of Theorem 1 show that the best invariant estimate is also generalized Bayes with respect to Lebesgue measure on the whole real line.

10. Let X and Y be random variables whose joint distribution, given parameters θ_1 and θ_2, has density of the form

$$f_{X,Y}(x, y \mid \theta_1, \theta_2) = f(x - \theta_1, y - \theta_2)$$

for some function f. The decision problem is to estimate $(\theta_1 + \theta_2)/2$ with squared error loss

$$L((\theta_1, \theta_2), a) = \left(\frac{\theta_1 + \theta_2}{2} - a\right)^2,$$

where $\mathcal{A}$ is the real line and Θ is the plane.
(a) Show that the problem is invariant under the group of location changes $\mathcal{G} = (g_{b,c})$, $g_{b,c}(x, y) = (x + b, y + c)$ and find the transformations $\bar{g}_{b,c}$ and $\tilde{g}_{b,c}$.
(b) Show that in the search for best invariant rules attention may be restricted to the nonrandomized invariant rules.
(c) Find the form of the nonrandomized invariant rules.
(d) Find the best invariant rule.

4.6 Minimax Estimates for the Parameters of a Normal Distribution

The results contained in Theorem 4.5.2 and Lemmas 4.5.2 and 4.5.3 are quite general. As a first example consider the problem of estimating the mean θ of a normal distribution with known variance σ^2, based on a sample of size n, $X_1, X_2, \cdots, X_n$. The sample mean $\bar{X}$ is a sufficient statistic for θ, and $\bar{X}$ has distribution $\mathfrak{N}(\theta, \sigma^2/n)$. For each fixed σ^2 and n, θ is a location parameter for the distribution of $\bar{X}$. If the loss function is quadratic $[L(\theta, a) = (a - \theta)^2]$, it follows that $\bar{X}$ is the best invariant estimate of θ from Theorem 4.5.1 ($b_0 = E_0\bar{X} = 0$) and that $\bar{X}$ is minimax from Theorem 4.5.2 and Lemma 4.5.3. More generally, $\bar{X}$ is best invariant and minimax for a large class of loss functions as the following corollary to Theorem 4.5.2 shows. [Wolfowitz (1950).]

Theorem 1. If $X_1, \cdots, X_n$ is a sample from a normal distribution with mean θ and known variance σ^2, then $\bar{X}$ is a best invariant estimate of θ and a minimax estimate of θ, provided that the loss function is a nondecreasing function of $|a - \theta|$ and that $E_0L(\bar{X})$ exists and is finite.

Proof. To show that $\bar{X}$ is best invariant, we must also show, according to Theorem 4.5.1, that

$$E_0L(\bar{X} - b) \geq E_0L(\bar{X}) \qquad \text{for every } b.$$

Since by symmetry

$$E_0 L(\bar{X} - b) = E_0 L(\bar{X} + b),$$

we may restrict attention to $b > 0$. Denoting the density of $\mathfrak{N}(0, \sigma^2/n)$ by $f(x)$, we have

$$\begin{aligned}
E_0[L(\bar{X} - b) - L(\bar{X})] &= \int [L(x - b) - L(x)] f(x)\, dx \\
&= \int_{-\infty}^{b/2} [L(x - b) - L(x)] f(x)\, dx \\
&\quad + \int_{b/2}^{\infty} [L(x - b) - L(x)] f(x)\, dx \\
&= \int_{b/2}^{\infty} [L(-y) - L(b - y)] f(b - y)\, dy \\
&\quad - \int_{b/2}^{\infty} [L(x) - L(x - b)] f(x)\, dx \\
&= \int_{b/2}^{\infty} [L(y) - L(y - b)] \\
&\quad \times [f(y - b) - f(y)]\, dy \geq 0, \qquad (4.44)
\end{aligned}$$

using the fact that both L and f are even functions (Fig. 4.3). Because both factors of the integrand are nonnegative, the integral itself is nonnegative.

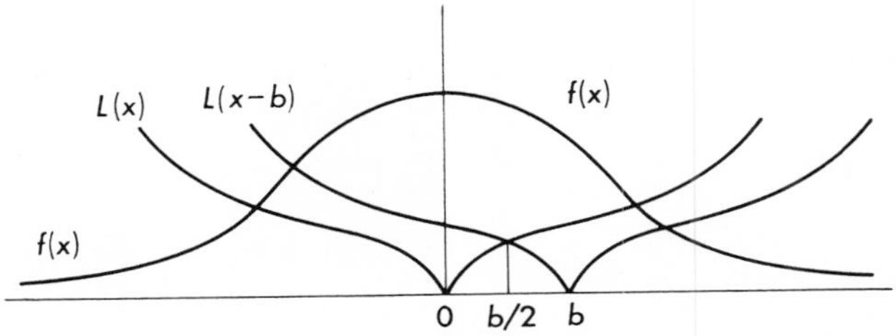

Fig. 4.3

To show that $\bar{X}$ is minimax, we shall also show that the conditions of Theorem 4.5.2 are satisfied for the distribution of $\bar{X}$; but

$$\int_{-N}^{N} L(x - b)\, f(x)\, dx \geq \int_{-N}^{N} L(x)\, f(x)\, dx \tag{4.45}$$

for all b by an argument similar to that used in (4.44). Because the right side of (4.45) converges to $E_0 L(\bar{X}) = R_0$, there exists an N such that

$$\int_{-N}^{N} L(x - b)\, f(x)\, dx \geq R_0 - \epsilon$$

for all b, thus completing the proof.

Scale Parameters. Suppose that θ is a scale parameter of the distribution of a random variable X and that the loss function is a function only of a/θ, say $L(\theta, a) = L(a/\theta)$. Such a problem is invariant under the group of changes of scale $g_c(X) \to cX$, where $c > 0$ [hence $\bar{g}_c(\theta) = c\theta$ and $\tilde{g}_c(a) = ca$]. For the sake of simplicity, we treat here only the case in which

$$P\{X > 0 \mid \theta\} = 1. \tag{4.46}$$

If (4.46) is satisfied for one value of θ, it is satisfied for all values of θ, because θ is a scale parameter. We shall find the best invariant rule for this problem by transforming it to a location parameter problem. The foregoing problem is identical to estimating a location parameter $\theta' = \log \theta$ for the distribution of $X' = \log X$ when the loss function

$$L'(\theta', a') = L(e^{\theta'}, e^{a'}) = L(e^{a' - \theta'}).$$

If we define $L'(x) = L(e^x)$, then $L'(\theta', a') = L'(a' - \theta')$. The best invariant rule for this new problem is to estimate θ' by $X' - b_0'$, where b_0' is that value of b' which minimizes $E(L'(X' - b') \mid \theta' = 0)$. Hence the best invariant estimate of $\log \theta$ is $\log X - \log b_0$, where $b_0 = e^{b_0'}$ minimizes

$$E\{L[\exp (X' - b')] \mid \theta = 1\} = E[L(X/b) \mid \theta = 1].$$

Thus the best invariant estimate of θ is X/b_0, where b_0 is that value of b which minimizes $E(L(X/b) \mid \theta = 1)$ (see Exercise 4.5.3, in which (4.46) is not used). The problem of showing that a best invariant rule is minimax is easily reduced to an application of Theorem 4.5.2.

In the problem of estimating a scale parameter a squared error loss does not seem so appropriate as it did in the location parameter problem. It might be more appropriate to let the loss be squared error in $\log \theta$, that is,

$$L(\theta, a) = (\log a - \log \theta)^2 = (\log (a/\theta))^2.$$

This is in line with the remarks of the preceding paragraph.

As an example, consider the problem of estimating the variance σ^2 of a normal distribution with mean zero on the basis of a sample of size n, $X_1, \cdots, X_n$, when the loss is squared error in $\log \sigma^2$; that is,

$$L(\sigma^2, a) = (\log a - \log \sigma^2)^2. \tag{4.47}$$

Then

$$Z = \sum_{i=1}^{n} X_i^2$$

is a sufficient statistic for σ^2, and σ^2 is a scale parameter for the distribution of Z, for Z/σ^2 has a χ^2-distribution with n degrees of freedom, namely $\mathcal{G}(n/2, 2)$, independent of σ^2. The best invariant estimate of $\log \sigma^2$ is then $\log Z - b_0$, where b_0 is that value of b which minimizes

$$E[(\log Z - b)^2 \mid \sigma^2 = 1].$$

Hence

$$\begin{aligned} b_0 = E(\log Z \mid \sigma^2 = 1) &= [\Gamma(n/2) 2^{n/2}]^{-1} \int_0^\infty e^{-z/2} z^{n/2-1} \log z \, dz \\ &= \Gamma(n/2)^{-1} \int_0^\infty e^{-y} y^{n/2-1} (\log y + \log 2) \, dy \\ &= \dot{\Psi}(n/2) + \log 2, \end{aligned}$$

where Ψ represents the digamma, or logarithmic derivative of gamma,

$$\Psi(\alpha) = \frac{d}{d\alpha} \log \Gamma(\alpha) = \frac{1}{\Gamma(\alpha)} \int_0^\infty e^{-x} x^{\alpha-1} \log x \, dx.$$

Transforming to obtain an estimate of σ^2, we find that $Z/(2 \exp[\Psi(n/2)])$ is the best invariant estimate of σ^2 using the loss (4.47). It is interesting to compare this estimate with those suggested in Section 3.6, using squared error loss $(\sigma^2 - a)^2$. In that section it was shown that Z/n was the best unbiased estimate using this loss and that $Z/(n + 2)$ was the best estimate using this loss out of the class of all estimates of the form cZ. This is equivalent to saying that $Z/(n + 2)$ is best invariant for the loss function $L(\sigma^2, a) = (a/\sigma^2 - 1)^2$ (see Exercise 3). It turns out (see Jahnke and Emde (1945)) that $2 \exp [\Psi(n/2)] \sim (n - 1)$, so that the best invariant estimate of σ^2 using the loss (4.47) is close to $Z/(n - 1)$. The actual values of $2 \exp [\Psi(n/2)]$ may easily be computed from the tables of Ψ in Jahnke and Emde (1945). Certain values are shown in the following table:

n	$2 \exp \Psi(n/2)$
1	0.281
2	1.123
3	2.074
4	3.053
5	4.041
6	5.033
7	6.028
8	7.023

Exercises

1. Let $X_1, \cdots, X_n$ be a sample of size n from the exponential distribution with location parameter θ, having the density

$$f(x \mid \theta) = e^{-(x-\theta)} I_{(\theta,\infty)}(x)$$

Find the best invariant estimate of θ when the loss function is

(a) $L(\theta, a) = (a - \theta)^2$;
(b) $L(\theta, a) = |a - \theta|$;

$$\text{(c)}\ L(\theta, a) = \begin{cases} 0 & \text{if } |a - \theta| \le c, \\ 1 & \text{if } |a - \theta| > c. \end{cases}$$

2. Let $X_1, \cdots, X_n$ be a sample from the geometric distribution with unknown location parameter θ and known geometric parameter $0 < p < 1$, having probability mass function

$$f(x \mid \theta) = (1 - p)p^{x-\theta}, \qquad x = \theta, \theta + 1, \theta + 2, \cdots.$$

Find the best invariant estimate of θ when the loss function is

$$\text{(a)}\ L(\theta, a) = (a - \theta)^2, \qquad \text{(b)}\ L(\theta, a) = \begin{cases} 0 & \text{if } a = \theta, \\ 1 & \text{if } a \neq \theta. \end{cases}$$

3. Let $X_1, \cdots, X_n$ be a sample from the gamma distribution $\mathcal{G}(\alpha, \beta)$ with α known and β unknown. Find the best invariant estimate of β when the loss function is $L(\beta, a) = (a - \beta)^2/\beta^2$.

Show that the best unbiased estimate of β is also the best invariant estimate of β when the loss function is $L(\beta, a) = (a/\beta) - 1 - \log(a/\beta)$.

4. Let $X_1, \cdots, X_n$ be a sample from the distribution with density for $\theta > 0$:

$$f(x \mid \theta) = \frac{\theta}{x^2} I_{(\theta,\infty)}(x).$$

Find the best invariant estimate of θ if the loss function is

(a) $L(\theta, a) = \left(\dfrac{a}{\theta} - 1\right)^2, \quad n \geq 3;$

(b) $L(\theta, a) = |\log a - \log \theta|;$

(c) $L(\theta, a) = \left|\dfrac{a}{\theta} - 1\right|, \quad n \geq 2.$

5. Let $X_1, \cdots, X_n$ be a sample from the uniform distribution $\mathcal{U}(0, \theta)$. Find the best invariant estimate of θ if the loss function is

(a) $L(\theta, a) = \left(\dfrac{a}{\theta} - 1\right)^2;$

(b) $L(\theta, a) = \begin{cases} 0 & \text{if } \dfrac{1}{c} \leq \dfrac{a}{\theta} \leq c, \quad c > 1 \\ 1 & \text{otherwise;} \end{cases}$

(c) $L(\theta, a) = \left|\dfrac{a}{\theta} - 1\right|.$

6. Let $X_1, \cdots, X_n$ be a sample from $\mathfrak{N}(0, \sigma^2)$ and suppose that the problem is to estimate σ with loss $L(\sigma, a) = (\sigma - a)^2/\sigma^2$. Find the best invariant estimate and compare with the corresponding estimates using losses $(a/\sigma^2 - 1)^2$ and $(\log a - \log \sigma^2)^2$.

7. Let $X_1, \cdots, X_n (n \geq 2)$ be a sample from $\mathfrak{N}(\mu, \sigma^2)$, let Θ be the half-plane $(\mu, \sigma)\sigma > 0$, let $\mathcal{A}$ be the real line, and let the loss function be

$$L((\mu, \sigma), a) = \frac{(v\mu + w\sigma - a)^2}{\sigma^2}$$

for some given numbers v and w. Restrict attention to the sufficient statistics $(\bar{x}, s)$, where

$$\bar{x} = n^{-1} \sum_1^n x_i, \; s^2 = n^{-1} \sum_1^n (x_i - \bar{x})^2.$$

(a) Show that the problem is invariant under change of location and scale,

$$g_{b,c}(\bar{x}, s) = (b\bar{x} + c, bs), b > 0,$$

and find the groups $\bar{\mathcal{G}}$ and $\tilde{\mathcal{G}}$.
(b) Show that all invariant nonrandomized rules have the form $d(\bar{x}, s) = v\bar{x} + ks$ for some constant k.
(c) Find the best invariant decision rule

$$\left(k = \frac{w\sqrt{n}\,\Gamma(n/2)}{\sqrt{2}\Gamma((n + 1)/2)}\right)$$

and show that its risk is

$$\frac{v^2}{n} + w^2\left(1 - \frac{\Gamma(n/2)^2}{\Gamma((n - 1)/2)\;\Gamma((n + 1)/2)}\right).$$

That this rule is also minimax follows from certain general theorems [see Kiefer (1957)] which tell when a best invariant rule is minimax. The natural Bayes approach of Section 2.11 does not provide a proof that this rule is minimax, as the following exercise shows. The trouble is that the posterior distribution seems to assign s^2 one too many degrees of freedom.

8. In Exercise 7 let the prior distribution of σ have density

$$f(\sigma) = \frac{2\lambda^\alpha}{\Gamma(\alpha)} \exp[-\lambda/\sigma^2]\sigma^{-2\alpha-1} I_{(0,\infty)}(\sigma)$$

(that is, $\sigma^{-2} \in \mathcal{G}(\alpha, 1/\lambda)$), and let the conditional prior distribution of μ, given σ, be normal with mean zero and variance $\gamma^2\sigma^2$ [that is, $\mu \in \mathfrak{N}(0, \gamma^2\sigma^2)$], where α, λ, and γ^2 are known.
(a) Show that the posterior distribution of (μ, σ), given $X_1, \cdots, X_n$, is $\sigma^{-2} \in \mathcal{G}(\alpha + n/2, 2/(ns^2 + 2\lambda + n\bar{x}^2/(1 + n\gamma^2)))$ and, given σ, $\mu \in \mathfrak{N}(n\gamma^2\bar{x}/(1 + n\gamma^2), \sigma^2\gamma^2/(1 + n\gamma^2))$.
(b) Show that the Bayes rule is $d(\mathbf{x}) = v\hat{\mu} + w\hat{\sigma}$, where

$$\hat{\mu} = \frac{n\gamma^2\bar{x}}{1 + n\gamma^2}$$

and

$$\hat{\sigma} = \frac{\Gamma((2\alpha + n + 1)/2)}{\Gamma((2\alpha + n + 2)/2)} \left(\frac{ns^2}{2} + \frac{n\bar{x}^2}{2(1 + n\gamma^2)} + \lambda\right)^{1/2}.$$

(c) Show that the minimum Bayes risk is

$$\frac{v^2\gamma^2}{1 + n\gamma^2} + w^2\left(1 - \frac{\Gamma((2\alpha + n + 1)/2)^2}{\Gamma((2\alpha + n)/2)\,\Gamma((2\alpha + n + 2)/2)}\right).$$

(d) This minimum Bayes risk approaches its maximum value as $\gamma \to \infty$ and $\alpha \to 0$, namely,

$$\frac{v^2}{n} + w^2\left(1 - \frac{\Gamma((n + 1)/2)^2}{\Gamma(n/2)\,\Gamma((n + 2)/2)}\right).$$

This is strictly less, provided $w \neq 0$, than the risk of the best invariant decision rule given in Exercise 7.
The rule,

$$d(\mathbf{x}) = v\bar{x} + w\frac{\Gamma((n + 1)/2)}{\Gamma((n + 2)/2)}\left(\frac{ns^2}{2}\right)^{1/2}$$

provides an example of a rule which is a limit of Bayes rules (and a generalized Bayes rule) but which is not an extended Bayes rule.

9. *Interval Estimates.* Let Θ be the real line, let $\mathcal{A}$ be the half-plane $\{(y, z): y \leq z\}$, and let the loss function be

$$L(\theta, (y, z)) = k(z - y) - I_{(y,z)}(\theta),$$

where k is some positive constant. Suppose that $X \in \mathfrak{N}(\theta, 1)$.
(a) Show that the problem is invariant under the group $\mathcal{G}$ of translations and find $\bar{\mathcal{G}}$ and $\tilde{\mathcal{G}}$.
(b) Find the form of the nonrandomized invariant decision rules.
(c) Find the best invariant decision rule.

10. Let Θ be the half plane $\{(\mu, \sigma): \sigma > 0\}$, let $\mathcal{A}$ be the half plane $\{(y, z): y \leq z\}$, and let the loss function be

$$L((\mu, \sigma), (y, z)) = k(z - y)\sigma^{-1} - I_{(y,z)}(\mu),$$

where $k > 0$. Let $X_1, \cdots, X_n$ be a sample from the normal distribution $\mathfrak{N}(\mu, \sigma^2)$. Restrict attention to the sufficient statistic $(\bar{X}, s)$, the sample mean and standard deviation.
(a) Show that the problem is invariant under change of location and scale, as in Exercise 7(a), and find the groups $\bar{\mathcal{G}}$, and $\tilde{\mathcal{G}}$.
(b) Show that all nonrandomized invariant decision rules have the form $(\bar{X} + a_1 s, \bar{X} + a_2 s)$, where $a_1 \leq a_2$.
(c) Show that the best invariant decision rule is the usual confidence interval $(\bar{X} - as, \bar{X} + as)$, where $a = ((n/(2\pi k^2))^{1/n} - 1)^{1/2}$, provided that $2\pi k^2 < n$.

11. Let X_1 and X_2 have a joint distribution with density on the sample space $\{(x_1, x_2): x_1 < x_2\}$

$$f_{X_1,X_2}(x_1, x_2 \mid \mu, \sigma) = \sigma^{-2} f\left(\frac{x_1 - \mu}{\sigma}, \frac{x_2 - \mu}{\sigma}\right),$$

so that $\Theta = \{(\mu, \sigma): \sigma > 0\}$. Suppose that $\mathcal{A}$ is the real line and that the loss is a function of $(v\mu + w\sigma - a)/\sigma$ for some fixed numbers v and w,

$$L((\mu, \sigma), a) = W\left(\frac{v\mu + w\sigma - a}{\sigma}\right).$$

(a) Show that the problem is invariant under the group $\mathcal{G}$ of transformations

$$g_{b,c}(x_1, x_2) = (bx_1 + c, bx_2 + c),$$

and find $\bar{g}_{b,c}(\mu, \sigma)$ and $\tilde{g}_{b,c}(a)$, $b > 0$.

(b) Show that every nonrandomized invariant rule is of the form

$$d_\alpha(x_1, x_2) = vx_1 + \alpha(x_2 - x_1)$$

for some number α.

(c) Show that the best invariant decision rule is d_α, where α is chosen to minimize $E_{(0,1)}\{W(w - vX_1 - \alpha(X_2 - X_1))\}$. ($E_{(0,1)}$ represents the expectation when $\mu = 0$ and $\sigma = 1$.)

(d) Show that for squared error

$$L((\mu, \sigma), a) = (v\mu + w\sigma - a)^2/\sigma^2$$

the best invariant decision rule is

$$d(x_1, x_2) = w \frac{(x_2 - x_1)E_{(0,1)}(X_2 - X_1)}{E_{(0,1)}(X_2 - X_1)^2} + v \frac{E_{(0,1)}\{(X_2 - X_1)(x_1X_2 - x_2X_1)\}}{E_{(0,1)}(X_2 - X_1)^2}$$

(assuming, of course, that X_1 and X_2 have finite second moments).

12. Let $Y_1, \cdots, Y_n$ be a sample from the uniform distribution $\mathcal{U}(\mu, \mu + \sigma)$. Then $(X_1, X_2) = (\min Y_i, \max Y_i)$ is sufficient for (μ, σ). Using the loss function of part (d) of Problem 11, show that the best invariant rule is

$$d(x_1, x_2) = w \frac{n + 2}{n} (x_2 - x_1) + v \left[x_1 - \frac{1}{n}(x_2 - x_1)\right].$$

The next two exercises involve the notion of an almost admissible decision rule.

Definition. When Θ is the real line, we define a decision rule δ_0 to be *almost admissible* if for every rule such that $R(\theta, \delta) \leq R(\theta, \delta_0)$ for all θ we have that $R(\theta, \delta) = R(\theta, \delta_0)$ for almost all θ (that is, except for θ in a set of Lebesgue measure zero).

13. Let $\Theta = \mathcal{A} =$ the real line, let $X \in \mathcal{N}(\theta, 1)$, and let $L(\theta, a)$ be a nondecreasing function of $|a - \theta|$, as in Theorem 1. Show that $d'(x) = x$ is an almost admissible rule. (Use a proof similar to that found in Example 3.7.1. Show that $d_\sigma(x) = \sigma^2 x/(\sigma^2 + 1)$ and

$$\sigma[r(\tau_\sigma, d') - r(\tau_\sigma, d_\sigma)] = \frac{1}{\sqrt{2\pi}} \int L(y)\sigma$$

$$\times \left[1 - \frac{\sqrt{\sigma^2 + 1}}{\sigma} \exp(-y^2/2\sigma^2)\right] \exp(-y^2/2)\, dy,$$

which tends to zero as $\sigma \to \infty$.)

14. Let $\Theta = \mathcal{A} =$ the real line and let θ be a location parameter for the distribution of the random variable X, which is assumed to be absolutely continuous, with density $f_X(x \mid \theta) = f(x - \theta)$. Suppose that the loss function $L(\theta, a)$ is strictly convex in a for each $\theta \in \Theta$. Show that any almost admissible nonrandomized rule is admissible. (*Hint.* If d_0 is almost admissible, and d_1 is better, then the set $S = \{x : d_0(x) \neq d_1(x)\}$ has positive Lebesgue measure. Show that

$$\int_S f(x - \theta)\, dx$$

is positive for θ in a set $\Theta_0 \subset \Theta$ of positive Lebesgue measure. Let $d'(x) = \frac{1}{2}[d_0(x) + d_1(x)]$. Then $R(\theta, d') \leq R(\theta, d_0)$ for all θ, with strict inequality for $\theta \in \Theta_0$.)

4.7 The Pitman Estimate

We now consider the problem of estimating a location parameter θ on the basis of a sample of size n observations ($n > 1$). More generally, we suppose that the observable variables $X_1, X_2, \cdots, X_n$ are not necessarily independent but have a joint distribution for which θ is a location parameter in the sense that the distribution of $X_1 - \theta, \cdots, X_n - \theta$ does not depend on θ. If a density exists, it means that

$$f_{X_1,\cdots,X_n}(x_1, \cdots, x_n \mid \theta) = f(x_1 - \theta, \cdots, x_n - \theta) \tag{4.48}$$

for some function f.

Decision problems involving such a location parameter may be reduced to one-dimensional problems by the following artifice. Consider the distribution of the variables $X_1, Y_2, Y_3, \cdots, Y_n$, where $Y_2 = X_2 - X_1$, $Y_3 = X_3 - X_1, \cdots, Y_n = X_n - X_1$. It is easy to see that the distribution of $\mathbf{Y} = (Y_2, \cdots, Y_n)$ does not depend on θ [since $Y_j =$

$(X_j - \theta) - (X_1 - \theta)$ and the distribution of $X_1 - \theta, \cdots, X_n - \theta$ does not depend on θ]. Thus we may pretend that $Y_2, \cdots, Y_n$ were observed first and that X_1 was then chosen from the conditional distribution of X_1, given $\mathbf{Y}$, a distribution with θ as a location parameter. This conditional problem (conditional on $\mathbf{Y}$) is intuitively equivalent to the original problem. For example, if the loss function is a function only of $a - \theta$, $L(\theta, a) = L(a - \theta)$, the preceding sections show that *the best invariant estimate of θ for the conditional problem is*

$$d_0(\mathbf{X}) = X_1 - b_0(\mathbf{Y}), \tag{4.49}$$

where $b_0(\mathbf{Y})$ is that number (provided it exists) for which

$$E_0(L(X_1 - b_0) \mid \mathbf{Y}) = \inf_b E_0(L(X_1 - b) \mid \mathbf{Y}). \tag{4.50}$$

In fact, the estimate (4.49) is best invariant for the original unconditional problem as well. To see this, we proceed as follows. The original problem is invariant under the group of translations $g_c(X_1, \cdots, X_n) = (X_1 + c, \cdots, X_n + c)$, with

$$\bar{g}_c(\theta) = \theta + c \quad \text{and} \quad \tilde{g}_c(a) = a + c.$$

An invariant nonrandomized decision rule must satisfy the equation

$$d(x_1 + c, \cdots, x_n + c) = d(x_1, \cdots, x_n) + c \tag{4.51}$$

for all $(x_1, \cdots, x_n)$ and all c. Thus every invariant nonrandomized estimate is of the form $d(\mathbf{X}) = X_1 - b(\mathbf{Y})$ [take $c = -X_1$ in (4.51)] for some function $b(\mathbf{Y})$. As in Lemma 4.5.1, since invariant rules have constant risk, there is for every randomized invariant rule a nonrandomized invariant rule which is just as good. Furthermore, for any nonrandomized invariant rule d

$$\begin{aligned} R(\theta, d) &= E_0 L(X_1 - b(\mathbf{Y})) = E\{E_0(L(X_1 - b(\mathbf{Y})) \mid \mathbf{Y})\} \\ &\geq E\{E_0(L(X_1 - b_0(\mathbf{Y})) \mid \mathbf{Y})\} \\ &= E_0 L(X_1 - b_0(\mathbf{Y})) = R(\theta, d_0), \end{aligned} \tag{4.52}$$

provided only that $X_1 - b_0(\mathbf{Y})$ is an estimate for the unconditional decision problem in the sense that $E_0 L(X_1 - b_0(\mathbf{Y}))$ is well defined. (It might not be, because $b_0(\mathbf{Y})$ is defined separately for each $\mathbf{Y}$ and, when considered as a function of $\mathbf{Y}$, might not be measurable or continuous anywhere). Equation 4.52 shows that d_0 of (4.49) is best invariant.

A generalization of this idea to a more general problem is mentioned in Section 5.6.

The special case of quadratic loss $L(x) = x^2$ is of particular interest. In this case $b_0(\mathbf{Y}) = E_0(X_1 \mid \mathbf{Y})$ and the corresponding estimate

$$d_0(\mathbf{X}) = X_1 - E_0(X_1 \mid \mathbf{Y}) \tag{4.53}$$

is known as the *Pitman estimate*. Before invariance as a general principle was propounded by Hunt and Stein in the mid-1940's, Pitman's paper (1939) utilized the main ideas and proposed the estimate (4.53). In this paragraph we assume that $X_1, \cdots, X_n$ has a density of the form (4.48) and derive the usual form of this estimate. The density of $X_1, Y_2, \cdots, Y_n$ when $\theta = 0$ is

$$f_{X_1,Y_2,\cdots,Y_n}(x_1, y_2, \cdots, y_n \mid 0) = f(x_1, y_2 + x_1, \cdots, y_n + x_1), \tag{4.54}$$

so that the marginal distribution of $\mathbf{Y}$ is

$$f_{Y_2,\cdots,Y_n}(y_2, \cdots, y_n) = \int f(x_1, y_2 + x_1, \cdots, y_n + x_1)\, dx_1\,. \tag{4.55}$$

The conditional distribution of X_1, given $\mathbf{Y}$, when $\theta = 0$, is therefore

$$f_{X_1 \mid Y_2=y_2,\cdots,Y_n=y_n}(x_1 \mid 0) = \frac{f(x_1, y_2 + x_1, \cdots, y_n + x_1)}{\int f(x_1, y_2 + x_1, \cdots, y_n + x_1)\, dx_1}. \tag{4.56}$$

Hence the Pitman estimate (4.53) is

$$d_0(\mathbf{X}) = X_1 - \frac{\int x_1 f(x_1, Y_2 + x_1, \cdots, Y_n + x_1)\, dx_1}{\int f(x_1, Y_2 + x_1, \cdots, Y_n + x_1)\, dx_1}. \tag{4.57}$$

Changing the dummy variable x_1 for a dummy variable θ by the formula $x_1 = X_1 - \theta$ or $\theta = X_1 - x_1$, we find the usual form of the estimate:

$$d_0(\mathbf{X}) = \frac{\int \theta f(X_1 - \theta, X_2 - \theta, \cdots, X_n - \theta)\, d\theta}{\int f(X_1 - \theta, X_2 - \theta, \cdots, X_n - \theta)\, d\theta}. \tag{4.58}$$

We conclude by remarking that a theorem analogous to Theorem 4.5.2 holds in the more general case treated in this section. Even though a decision rule is minimax for a conditional decision problem, it may be a very poor estimate for an unconditional decision problem (see Exercise 1). However, for the location parameter problem given, the estimate (4.49), if it is a decision rule for the unconditional problem, is minimax precisely in the situation in which Theorem 4.5.2 asserts that it is minimax in the conditional problem, namely, when condition (4.38) is satisfied for each conditional decision problem. More precisely, let $R_0(\mathbf{Y}) = E_0(L(X_1 - b_0(\mathbf{Y})) \mid \mathbf{Y})$ denote the conditional risk of the best invariant rule (4.49) for the conditional problem. We require that the loss function be a function only of $a - \theta$, $L(\theta, a) = L(a - \theta)$, that L be bounded from below and that for every $\epsilon > 0$ and every $\mathbf{y}$ there exist a finite number N, $(N_\epsilon(\mathbf{y}))$, such that

$$\int_{-N}^{N} L(x_1 - b)\, dF_{X_1|Y=\mathbf{y}}(x_1) \geq R_0(\mathbf{y}) - \epsilon \tag{4.59}$$

for all b. Under these conditions, we give a proof, valid for the discrete case, that the estimate (4.49) is minimax in the general problem.

Suppose without loss of generality that $L \geq 0$. Let $\epsilon > 0$ and find for each $\mathbf{y}$ a number $N(\mathbf{y})$ such that (4.59) holds for all b. Let $S_N = \{\mathbf{y}: N(\mathbf{y}) \leq N\}$, so that $P\{\mathbf{Y} \in S_N\} \to 1$ as $N \to \infty$. (Here we use the assumption that $\mathbf{Y}$ is discrete. If it were not, we would have no assurance that S_N is a measurable set for which $P\{\mathbf{Y} \in S_N\}$ is defined.) Then, for any nonrandomized rule d and $M > N$,

$$\begin{aligned}
r(\tau_M, d) &= \frac{1}{2M} \int_{-M}^{M} R(\theta, d)\, d\theta \\
&= \frac{1}{2M} \int_{-M}^{M} E\{E_\theta(L(\theta, d) \mid \mathbf{Y})\}\, d\theta \\
&= E\left\{\frac{1}{2M} \int_{-M}^{M} E_\theta(L(\theta, d) \mid \mathbf{Y})\, d\theta\right\} \\
&\geq E\left\{(R_0(\mathbf{Y}) - \epsilon)\left(\frac{2M - 2N}{2M}\right) I_{S_N}(\mathbf{Y})\right\}, \qquad (4.60)
\end{aligned}$$

using inequality (4.39). Because (4.60) is true for all $d \in D$, it is also true for all $\delta \in D^*$, and

$$\liminf_{M \to \infty} \inf_{\delta} r(\tau_M, \delta) \geq E\{(R_0(\mathbf{Y}) - \epsilon)\, I_{S_N}(\mathbf{Y})\}. \tag{4.61}$$

Because this is valid for all N and all ϵ, it is valid in the limit as $N \to \infty$ and $\epsilon \to 0$; this implies that

$$\liminf_{M\to\infty} \inf_{\delta} r(\tau_M, \delta) \geq E\, R_0(\mathbf{Y}) = R(\theta, d_0) \tag{4.62}$$

for all θ, thus completing the proof by Theorem 2.11.3.

Exercises

1. Consider a game in which nature chooses a point in $\Theta = [0, 1]$ and a coin is tossed with probability one half of heads. If heads comes up, the statistician chooses d_1 or d_2 with risk function $R(\theta, d_1) = \theta$ and $R(\theta, d_2) = 1 - \epsilon$, where $0 < \epsilon < \frac{1}{2}$. If the coin does not come up heads, the statistician chooses d_3 or d_4 with risk function $R(\theta, d_3) = 1 - \theta$ and $R(\theta, d_4) = 1 - \epsilon$. Show that the conditional minimax decision rule (conditional on the outcome of the toss of the coin) is not minimax or admissible in the general unconditional problem.
2. Obtain the best invariant estimate of θ based on a sample of n observations, $X_1, \cdots, X_n$, from the uniform distribution $\mathcal{U}(\theta - \frac{1}{2}, \theta + \frac{1}{2})$, using the loss functions
(a) $L(\theta, a) = (a - \theta)^2$ *Ans.* $(\min X_i + \max X_i)/2$.
(b) $L(\theta, a) = |a - \theta|$

$$\text{(c)}\quad L(\theta, a) = \begin{cases} 0 & \text{if } |a - \theta| \leq c \\ 1 & \text{if } |a - \theta| > c. \end{cases}$$

3. Let $X_1, X_2, \cdots, X_n$ be a sample from the half normal distribution with density

$$f(x \mid \theta) = \sqrt{\frac{2}{\pi}} \exp\left[-\tfrac{1}{2}(x - \theta)^2\right] I_{(\theta,\infty)}(x).$$

Show that the Pitman estimate of θ is

$$d(\mathbf{X}) = \bar{X} - \frac{\exp\left[-n(\min X_i - \bar{X})^2/2\right]}{\sqrt{2n\pi}\,\Phi(\sqrt{n}(\min X_i - \bar{X}))}$$

where Φ is the distribution function of the standard normal distribution, $\mathfrak{N}(0, 1)$.

4. Let $X_1, \cdots, X_n$ have a joint distribution given by the density

$$f(x_1, \cdots, x_n \mid \theta) = \theta^{-n} f(x_1/\theta, \cdots, x_n/\theta)$$

for some function f, which vanishes unless all coordinates are positive, where $\theta > 0$ is an unknown scale parameter. Suppose the loss is a function of a/θ, $L(\theta, a) = W(a/\theta)$. Find the best invariant estimate of θ, analogous to (4.49) and (4.50) [take $\mathbf{Y} = (X_1/X_n, \cdots, X_{n-1}/X_n)$]. When $W(a/\theta) = (a/\theta - 1)^2$, find the estimate in a form analogous to (4.58). *Ans:*

$$d(\mathbf{X}) = \frac{\int_0^\infty \theta^{-(n+2)} f(X_1/\theta, \cdots, X_n/\theta)\, d\theta}{\int_0^\infty \theta^{-(n+3)} f(X_1/\theta, \cdots, X_n/\theta)\, d\theta}.$$

5. Let $X_1, \cdots, X_n$ be a sample from the uniform distribution $\mathcal{U}(\theta, 2\theta)$, where $\Theta = (0, \infty) = \mathcal{A}$, and $L(\theta, a) = (\theta - a)^2/\theta^2$. Show that the best invariant decision rule is

$$d(\mathbf{X}) = \frac{(n+2)[(V/2)^{-(n+1)} - U^{-(n+1)}]}{(n+1)[(V/2)^{-(n+2)} - U^{-(n+2)}]},$$

where $U = \min X_i$ and $V = \max X_i$.

6. If $X_1, \cdots, X_n$ has density (4.49) and if $u(x_1, \cdots, x_n)$ is any invariant real-valued function (i.e., $u(x_1 - c, \cdots, x_n - c) = u(x_1, \cdots, x_n) - c$), show that $u(\mathbf{X}) = X_1 +$ some function of $\mathbf{Y}$, where $\mathbf{Y} = (X_2 - X_1, \cdots, X_n - X_1)$. Hence the best invariant estimate can be written in the form, analogous to (4.49) and (4.50),

$$d_0(\mathbf{X}) = u(\mathbf{X}) - b_0(\mathbf{Y}),$$

where $b_0(\mathbf{Y})$ is that number b which minimizes $E_0(L(u(\mathbf{X}) - b) \mid \mathbf{Y})$.

4.8 Estimation of a Distribution Function

The estimation of a continuous distribution function provides another example of the use of invariance in decision theory. Let $X_1, X_2, \cdots, X_n$ be a sample from a univariate distribution, whose distribution function $F(x)$ is known to be continuous but is otherwise unknown. The problem is to estimate F on the basis of $X_1, X_2, \cdots, X_n$ by an arbitrary, not necessarily continuous, distribution function $\hat{F}$, which we take to be continuous from the right. Thus Θ is the set of all continuous distribution functions on the real line, and $\mathcal{A}$ is the set of all distribution functions on the real line. We consider the following two loss functions.

$$L_1(F, \hat{F}) = \sup_x | F(x) - \hat{F}(x) |, \tag{4.63}$$

$$L_2(F, \hat{F}) = \int (F(x) - \hat{F}(x))^2 \, dF(x). \tag{4.64}$$

The famous Glivenko-Cantelli lemma [see Loève (1963)] states that if $\hat{F}$ is taken to be the sample distribution function

$$\hat{F}(x) = \frac{1}{n} \sum_{i=1}^{n} I_{[X_i, \infty)}(x), \tag{4.65}$$

then $L_1(F, \hat{F})$ converges to zero with probability one as n tends to infinity. The function L_2 is useful, for it is usually easier to manage analytically.

First we reduce the problem by means of sufficient statistics. For this problem the vector of order statistics $\mathbf{Y} = (Y_1, Y_2, \cdots, Y_n)$ is sufficient, where $Y_1 =$ the smallest of the X_i, $Y_2 =$ the next smallest, $\cdots$, and $Y_n =$ the largest (Exercise 3.3.3).

Second, we show that the decision problem is invariant under the group $\mathcal{G}$ of transformations,

$$g_\varphi(y_1, \cdots, y_n) = (\varphi(y_1), \cdots, \varphi(y_n)),$$

where φ is a continuous strictly increasing function from the real line onto the real line. If $\mathbf{Y}$ is a vector of order statistics from a distribution with distribution function $F(x)$, then $g_\varphi(\mathbf{Y})$ is a vector of order statistics from a distribution with distribution function $G(x) = P\{\varphi(X) < x\} = F(\varphi^{-1}(x))$. Hence the distributions are invariant under $\mathcal{G}$, and $\bar{g}_\varphi(F(x)) = F(\varphi^{-1}(x))$. The loss functions (4.63) and (4.64) are also invariant under $\mathcal{G}$, with $\tilde{g}_\varphi(\hat{F}(x)) = \hat{F}(\varphi^{-1}(x))$, as may easily be checked (Exercise 1).

Third, we find the form of the invariant decision rules. We will see that all nonrandomized invariant rules have constant risk, and that, from Lemma 4.5.1, the nonrandomized invariant rules are essentially complete within the class of randomized invariant rules. We still must show that the randomized invariant rules are equivalent to the behavioral invariant rules before we can restrict attention to the nonrandomized rules. This is done by checking (4.17). A behavioral rule chooses a distribution function $\hat{F}(x)$ at random after the observation $\mathbf{Y} = \mathbf{y}$ is made; this random function is denoted by $\hat{F}_{\mathbf{y}}(x)$. Such a behavioral rule is invariant if for every $g_\varphi \in \mathcal{G}$ the distribution of $\hat{F}_{g_\varphi(\mathbf{y})}(x)$ is identical to the distribution of $\hat{F}_{\mathbf{y}}(\varphi^{-1}(x))$. Let $\Phi_{\mathbf{y}} = \{\varphi : g_\varphi(\mathbf{y}) = \mathbf{y}\}$. Then, a behavioral invariant rule has the property that for all $\varphi \in \Phi_{\mathbf{y}}$ the distribution of $\hat{F}_{\mathbf{y}}(x)$ is the same as the distribution of $\hat{F}_{\mathbf{y}}(\varphi^{-1}(x))$. Now, let x_1, x_2, x_3 be three numbers that, for some $0 \le j \le n$, satisfy $y_j < x_1 < x_2 < x_3 < y_{j+1}$ ($y_0 = -\infty$, $y_{n+1} = +\infty$). Because there exists φ_1 and $\varphi_2 \in \Phi_{\mathbf{y}}$ such that $\varphi_1(x_1) = x_1$, $\varphi_1(x_2) = x_3$, and $\varphi_2(x_1) = x_2$, $\varphi_2(x_2) = x_3$, it follows that for any $\delta > 0$

$$P\{\hat{F}_{\mathbf{y}}(x_2) - \hat{F}_{\mathbf{y}}(x_1) > \delta\} = P\{\hat{F}_{\mathbf{y}}(x_3) - \hat{F}_{\mathbf{y}}(x_1) > \delta\}$$
$$= P\{\hat{F}_{\mathbf{y}}(x_3) - \hat{F}_{\mathbf{y}}(x_2) > \delta\}.$$

In other words, if

$$Z_1 = \hat{F}_{\mathbf{y}}(x_2) - \hat{F}_{\mathbf{y}}(x_1) \quad \text{and} \quad Z_2 = \hat{F}_{\mathbf{y}}(x_3) - \hat{F}_{\mathbf{y}}(x_2),$$

then

$$P\{Z_1 > \delta\} = P\{Z_1 + Z_2 > \delta\} = P\{Z_2 > \delta\},$$

yet

$$P\{Z_1 \ge 0\} = P\{Z_2 \ge 0\} = 1.$$

This implies that

$$P\{(Z_1 \le \delta, \text{ or } Z_2 \le \delta), \text{and } Z_1 + Z_2 > \delta\} = 0$$

(the shaded area in Fig. 4.4). Because it is true for all $\delta > 0$, it follows that $P\{Z_1 = 0\} = P\{Z_2 = 0\} = 1$. Thus a behavioral invariant rule

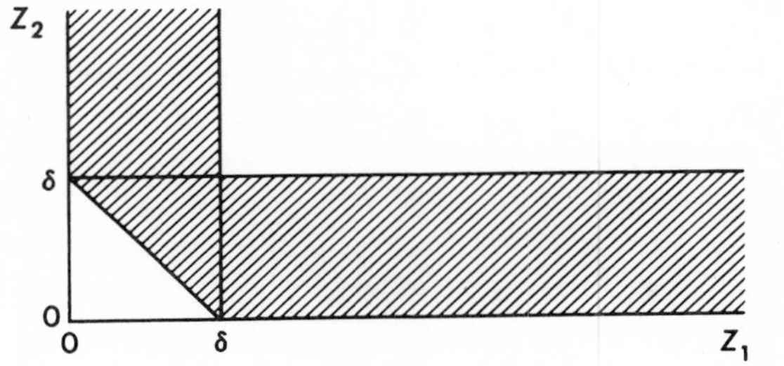

Fig. 4.4

has the property that if $y_j < x_1 < x_2 < y_{j+1}$ then $P\{\hat{F}_{\mathbf{y}}(x_1) = \hat{F}_{\mathbf{y}}(x_2)\} = 1$, or, equivalently, the probability is one that the distribution function $\hat{F}_{\mathbf{y}}(x)$ is discrete and has jumps only at points $y_1, y_2, \cdots, y_n$. This proves (4.17), so that behavioral and randomized invariant rules are equivalent for this problem. Furthermore, given any $\mathbf{y}$ and $\mathbf{y}'$, there exists a transformation $g_\varphi \in \mathcal{G}$ such that $g_\varphi(\mathbf{y}) = \mathbf{y}'$; hence the distribution of $\hat{F}_{\mathbf{y}'}(x)$ is determined from and equal to the distribution of $\hat{F}_{\mathbf{y}}(\varphi^{-1}(x))$. Thus a behavioral invariant rule has the form

$$\hat{F}_{\mathbf{y}}(x) = \sum_{i=1}^{n-1} U_i I_{[y_i, y_{i+1})}(x) + I_{[y_n, \infty)}(x), \tag{4.66}$$

where $\mathbf{U} = (U_1, \cdots, U_{n-1})$ is a random vector for which

$$P\{0 \le U_1 \le \cdots \le U_{n-1} \le 1\} = 1.$$

It may be deduced from this (or directly, as in Exercise 2) that a nonrandomized invariant decision rule has the form

$$\hat{F}_{\mathbf{Y}}(x) = \sum_{i=1}^{n-1} u_i I_{[Y_i, Y_{i+1})}(x) + I_{[Y_n, \infty)}(x), \tag{4.67}$$

where $0 \le u_1 \le \cdots \le u_{n-1} \le 1$.

To find a best invariant rule under loss function L_1 of (4.63) is indeed a difficult problem analytically. As far as I know, it is an unsolved problem. The loss function L_2 yields much more readily to analysis. In fact, we consider the following class of loss functions, which includes L_2, as a special case,

$$L(F, \hat{F}) = \int (F(x) - \hat{F}(x))^2 h(F(x))\, dF(x), \tag{4.68}$$

where h is some positive continuous function defined on $(0, 1)$. The decision problem, using the more general class of loss functions (4.68) (Exercise 1), is still invariant.

We evaluate the risk function of the nonrandomized invariant rules (4.67) using the loss function (4.68). For uniformity of notation we write (4.67) as

$$\hat{F}_{\mathbf{Y}}(x) = \sum_{i=0}^{n} u_i I_{[Y_i, Y_{i+1})}(x), \tag{4.69}$$

where $u_0 = 0$, $u_n = 1$, $Y_0 = -\infty$, and $Y_{n+1} = +\infty$. Then

$$L(F, \hat{F}) = \int \Big(F(x) - \sum_{i=0}^{n} u_i I_{[Y_i, Y_{i+1})}(x)\Big)^2 h(F(x))\, dF(x)$$

$$= \int_0^1 \Big(t - \sum_{i=0}^{n} u_i I_{[Y_i, Y_{i+1})}(F^{-1}(t))\Big)^2 h(t)\, dt$$

$$= \int_0^1 \Big(t - \sum_{i=0}^{n} u_i I_{[F(Y_i), F(Y_{i+1}))}(t)\Big)^2 h(t)\, dt$$

$$= \sum_{i=0}^{n} \int_{F(Y_i)}^{F(Y_{i+1})} (t - u_i)^2 h(t)\, dt. \tag{4.70}$$

Because $W_1 = F(Y_1), \cdots, W_n = F(Y_n)$ have the distribution of the order statistics from a sample of size n from the uniform distribution $\mathcal{U}(0, 1)$, we may write

$$R(F, \hat{F}) = \sum_{i=0}^{n} E \int_{W_i}^{W_{i+1}} (t - u_i)^2 h(t)\, dt,$$

where $W_0 = 0$ and $W_{n+1} = 1$; but

$$E \int_{W_i}^{W_{i+1}} (t - u_i)^2 h(t)\, dt$$

$$= \int_0^1 \int_0^{w_{i+1}} \int_{w_i}^{w_{i+1}} (t - u_i)^2 h(t)\, dt\, dF_{W_i, W_{i+1}}(w_i, w_{i+1})$$

$$= \int_0^1 (t - u_i)^2 h(t) \int_0^t \int_t^1 dF_{W_i, W_{i+1}}(w_i, w_{i+1})\, dt$$

$$= \int_0^1 (t - u_i)^2 h(t) \binom{n}{i} t^i (1 - t)^{n-i}\, dt,$$

so that

$$R(F, \hat{F}) = \sum_{i=0}^{n} \int_0^1 (t - u_i)^2 h(t) \binom{n}{i} t^i (1 - t)^{n-i}\, dt. \tag{4.71}$$

It is easily seen that the choice of u_i, which minimizes this expression is

$$u_i = \frac{\int_0^1 h(t)t^{i+1}(1 - t)^{n-i}\,dt}{\int_0^1 h(t)t^i(1 - t)^{n-i}\,dt} \tag{4.72}$$

for $i = 1, 2, \cdots, n - 1$.

In the specific case in which $h(t) = 1$, and in which we are using the loss L_2 of (4.64), the u_i of (4.72) become ratios of beta functions; therefore it is easily computed that $u_i = (i + 1)/(n + 2)$ for $i = 1, \cdots, n - 1$. Thus the best invariant estimate in this case is *not* the sample distribution function (4.65) but a distribution function that gives weight $2/(n + 2)$ to the largest and smallest observations and weight $1/(n + 2)$ to all others. The sample distribution function (4.65) is the best invariant estimate in the case $h(t) = (t(1 - t))^{-1}$, in which the loss function is

$$L(F, \hat{F}) = \int \frac{(F(x) - \hat{F}(x))^2}{F(x)(1 - F(x))}\,dF(x), \tag{4.73}$$

as may easily be checked from (4.72).

Sometimes it may not be required that the estimate $\hat{F}$ be a distribution function, in which case $\mathcal{A}$ may be enlarged to contain all nondecreasing functions on the real line or, more simply, all functions on the real line. Then the nonrandomized invariant estimates are of the form (4.69) without the restriction that $u_0 = 0$ and $u_n = 1$, and the best invariant estimate is given by (4.72) for $i = 0, 1, \cdots, n$. For example, in the case of the loss function L_2, the best invariant estimate is found in (4.69) with $u_i = (i + 1)/(n + 2)$ for $i = 0, 1, \cdots, n$. With the loss function (4.73), the sample distribution (4.65) is still the best invariant estimate, since, for the risk (4.71) to be finite, it is necessary that $u_0 = 0$ and $u_n = 1$.

Om P. Agarwal (1955) has studied these problems, using the loss functions

$$L(F, \hat{F}) = \int |F - \hat{F}|^r\,dF$$

and

$$L(F, \hat{F}) = \int |F - \hat{F}|^r/(F(1 - F))\, dF$$

for positive integers r.

We expect best invariant estimates to be minimax. However, it is an open question whether any of the best invariant estimates mentioned in this section are minimax in their respective problems.

Exercises

1. Show that in the problem of estimating a distribution function, the loss functions (4.63) and (4.68) are invariant under the groups mentioned.
2. Show directly, using (4.6), that the nonrandomized invariant decision rules for the problem in this section have form (4.67).
3. Let Θ be the set of all continuous distribution functions on the real line, let $\mathcal{A}$ be the real line, and let $L(F, a)$ be some function of $F(a)$, $L(F, a) = W(F(a))$. Let $X_1, \cdots, X_n$ be a sample of size n from the true distribution F, and let $\mathbf{Y} = (Y_1, \cdots, Y_n)$ be the vector of order statistics, a sufficient statistic for F. Consider the group $\mathcal{G}$ of all transformations of the form $g_\varphi(y_1, \cdots, y_n) = (\varphi(y_1), \cdots, \varphi(y_n))$, where φ is a continuous increasing one-to-one function from E_1 onto E_1.
(a) Show that this decision problem is invariant under the group $\mathcal{G}$, and find the groups $\bar{\mathcal{G}}$ and $\tilde{\mathcal{G}}$.
(b) Let $\Phi_{\mathbf{y}} = \{\varphi : \varphi(y) = y\}$, and let $Z_{\mathbf{y}}$ denote the random variable with values in $\mathcal{A}$ chosen by a given behavioral decision rule when $\mathbf{Y} = \mathbf{y}$ is observed. Show that if the distribution of $Z_{\mathbf{y}}$ is the same as the distribution of $\varphi(Z_{\mathbf{y}})$ for all $\varphi \in \Phi_{\mathbf{y}}$, then $Z_{\mathbf{y}} = \varphi(Z_{\mathbf{y}})$ with probability one. This proves (4.17), so that behavioral and randomized invariant rules are equivalent.
(c) Show that there are only n nonrandomized invariant rules, namely, $d_i(\mathbf{Y}) = Y_i$, $i = 1, \cdots, n$.
(d) If $W(F(a)) = (F(a) - \frac{1}{2})^2$ (the problem of estimating the median of a continuous distribution), find the best invariant estimate.
(e) If $W(F(a)) = F(a)$ for $F(a) > 0$ and $W(F(a)) = 1$ for $F(a) = 0$ (a problem in estimating the lower bound of a distribution), find the best invariant decision rule.

CHAPTER 5

Testing Hypotheses

5.1 The Neyman-Pearson Lemma

As mentioned in Section 1.3, statistical decision problems, in which the space of actions consists of two elements $\mathcal{A} = \{a_0, a_1\}$, are called problems of testing hypotheses. A nonrandomized decision rule for such problems has a characteristic form and may be considered as a measurable subset, ω, of the sample space $\mathfrak{X}$, $\omega \subset \mathfrak{X}$, with the understanding that if the observable variable X falls in ω action a_1 is taken and if X falls in $\omega^c = \mathfrak{X} \sim \omega$ action a_0 is taken. The risk function for such a nonrandomized rule is

$$\begin{aligned} R(\theta, \omega) &= (1 - P_\theta\{\omega\})\, L(\theta, a_0) + P_\theta\{\omega\}\, L(\theta, a_1) \\ &= L(\theta, a_0) + P_\theta\{\omega\}(L(\theta, a_1) - L(\theta, a_0)), \end{aligned} \tag{5.1}$$

where by $P_\theta\{\omega\}$ we mean the probability that X will fall in ω when θ is the true state of nature. The set ω is sometimes referred to as a *test* or a *critical region*. The function of θ, $P_\theta\{\omega\}$, is sometimes called *the power function corresponding to the test* ω. It is important to note that the risk function depends on the test ω only through the power function.

Randomized decision rules for this problem are rather messy. They involve choosing a set ω at random from the class of all measurable subsets of $\mathfrak{X}$. However, a behavioral decision rule is quite simple. A behavioral decision rule is a function ϕ defined on $\mathfrak{X}$ with values in the interval [0, 1], the interpretation being that if $X = x$ is observed the statistician takes action a_1 with probability $\phi(x)$, and action a_0 with probability $(1 - \phi(x))$. The nonrandomized rule ω in the preceding paragraph is in this notation:

$$\phi(x) = \begin{cases} 1 & \text{if} \quad x \in \omega, \\ 0 & \text{if} \quad x \notin \omega; \end{cases} \tag{5.2}$$

action a_1 is taken (with probability one) if X falls in ω, and action a_0 is taken (with probability one) if X falls in ω^c.

The risk function corresponding to an arbitrary behavioral rule ϕ is

$$R(\theta, \phi) = L(\theta, a_0) + E_\theta\, \phi(X)\,(L(\theta, a_1) - L(\theta, a_0)), \tag{5.3}$$

where $E_\theta\, \phi(X)$ is called *the power function corresponding to the decision rule (or test)* ϕ. Thus ϕ is not called a decision rule unless the power function $E_\theta\, \phi(X)$ exists for all θ. (In this chapter we deal with behavioral rather than randomized decision rules; but, for simplicity, we use R throughout rather than $\hat{R}$ to denote the risk.)

In the classical version of hypothesis testing problems the loss function has a special form; namely, for some disjoint sets Θ_0 and Θ_1, with $\Theta = \Theta_0 \cup \Theta_1$,

$$L(\theta, a_0) = I_{\Theta_1}(\theta) = \begin{cases} 1 & \text{if} \quad \theta \in \Theta_1, \\ 0 & \text{if} \quad \theta \in \Theta_0, \end{cases}$$

$$L(\theta, a_1) = I_{\Theta_0}(\theta). \tag{5.4}$$

It is in such decision problems that most of the terminology of the theory of hypothesis testing originates. Thus we speak of the "null hypothesis" H_0 that θ lies in Θ_0 and the "alternative hypothesis" H_1 that θ lies in Θ_1. One of these disjoint hypotheses is true, and the statistician must guess which one, the loss being zero if he guesses correctly and one if he guesses incorrectly; therefore a_0 may be considered as the action "accept H_0" and a_1, the action "accept H_1" or "reject H_0".

Simple Hypothesis versus Simple Alternative. We consider first in detail the very important special case in which both H_0 and H_1 are *simple*; that is, Θ_0 and Θ_1 each consist of exactly one element, $\Theta_0 = \{\theta_0\}$ and $\Theta_1 = \{\theta_1\}$. Because $\Theta = \{\theta_0, \theta_1\}$ is finite for this problem, we may consider the risk set $S = \{(R(\theta_0, \phi), R(\theta_1, \phi)) : \phi \in \mathfrak{D}\}$. In this case

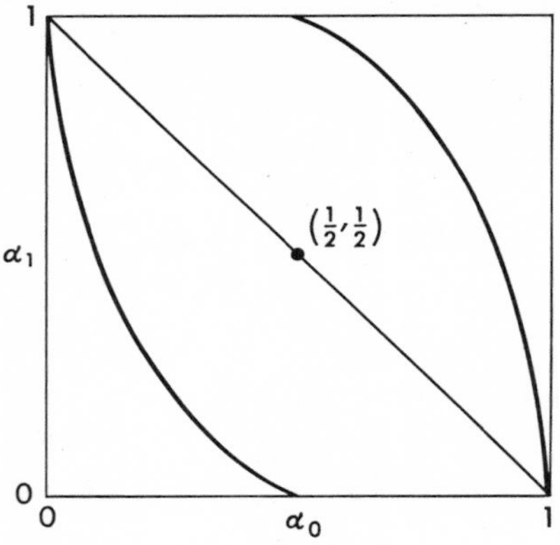

Fig. 5.1

the restriction that the loss function have the form of (5.4) is unnecessary. Given an arbitrary loss function, we may assume that $L(\theta_0, a_0) < L(\theta_0, a_1)$ and $L(\theta_1, a_0) > L(\theta_1, a_1)$ to avoid trivial cases. Then, if θ_0 is the true value of the parameter, we prefer to take action a_0, whereas if θ_1 is the true value of the parameter we prefer action a_1. The probability of rejecting hypothesis H_0 when true is called the *size* of the test ϕ and is denoted by α_0, whereas the probability of accepting H_0 when false is denoted by α_1. Thus

$$\begin{aligned} \alpha_0 &= E_{\theta_0} \phi(X) &&= \text{size of the test} \\ \alpha_1 &= E_{\theta_1}(1 - \phi(X)) &&= \text{probability of error of the second kind.} \end{aligned} \tag{5.5}$$

Instead of the risk set S, we shall consider the set $S' = \{(\alpha_0, \alpha_1) : \phi \in \mathfrak{D}\}$, which differs from S only in a change of location and scale along both axes. This set is convex (Fig. 5.1), contains the points (1, 0) and (0, 1) [put $\phi(X) \equiv 1$ and $\phi(X) \equiv 0$, respectively], and is symmetric about the point $(\frac{1}{2}, \frac{1}{2})$ [if a test ϕ gives (α_0, α_1), then the test $1 - \phi$ gives $(1 - \alpha_0, 1 - \alpha_1)$].

Definition. A test ϕ is said to be *best of size* α_0 for testing H_0 against H_1 if $E_{\theta_0} \phi(X) = \alpha_0$ and if for every test ϕ' for which $E_{\theta_0} \phi'(X) \le \alpha_0$ we have

$$E_{\theta_1}(1 - \phi(X)) \le E_{\theta_1}(1 - \phi'(X)). \tag{5.6}$$

that is, a test ϕ is best of size α_0 if, out of all tests of size not greater than α_0, ϕ has the smallest probability of error of the second kind. Note that if ϕ is an admissible test then ϕ is best of its size; that is, ϕ is best of size $\alpha_0 = E_{\theta_0}\,\phi(X)$. However, a best test of size α_0 is not necessarily admissible (see Exercise 1).

We now give a general method for finding the best tests of a simple hypothesis against a simple alternative by the fundamental lemma of Neyman and Pearson (1933). We give the proof of part (a) for the case in which the random observable X has an absolutely continuous distribution under both hypotheses. This theorem is true in general, the proof being identical to that given; but because we have no recourse to the Lebesgue integral the continuous and discrete cases are treated separately. The discrete case is left to the reader; the proof that follows may be used if the integral is everywhere replaced by a summation sign.

The Neyman-Pearson Lemma. Suppose that $\Theta = \{\theta_0, \theta_1\}$ and that the distributions of X have densities (or mass functions) $f(x \mid \theta_0) = f_0(x)$ and $f(x \mid \theta_1) = f_1(x)$.

(a) Any test $\phi(x)$ of the form

$$\phi(x) = \begin{cases} 1 & \text{if } f_1(x) > k\,f_0(x), \\ \gamma(x) & \text{if } f_1(x) = k\,f_0(x), \\ 0 & \text{if } f_1(x) < k\,f_0(x), \end{cases} \tag{5.7}$$

for some $k \geq 0$ and $0 \leq \gamma(x) \leq 1$, is best of its size for testing $H_0{:}\theta = \theta_0$ against $H_1{:}\theta = \theta_1$. Corresponding to $k = \infty$, the test

$$\phi(x) = \begin{cases} 1 & \text{if } f_0(x) = 0, \\ 0 & \text{if } f_0(x) > 0, \end{cases} \tag{5.8}$$

is best of size zero for testing H_0 against H_1.

(b) (*Existence*). For every α, $0 \leq \alpha \leq 1$ there exists a test of the form above with $\gamma(x) = \gamma$, a constant, for which $E_{\theta_0}\,\phi(X) = \alpha$.

(c) (*Uniqueness*). If ϕ' is a best test of size α for testing H_0 against H_1, then it has the form (5.7) or (5.8), except perhaps for a set of x with probability zero under H_0 and H_1.

Proof. (a) Choose any $\phi(x)$ of the form (5.7) and let $\phi'(x)$, $0 \leq \phi'(x) \leq 1$, be any test for which $E_{\theta_0}\,\phi'(X) \leq E_{\theta_0}\,\phi(X)$. We are to show that $E_{\theta_1}\,\phi'(x) \leq E_{\theta_1}\,\phi(X)$; but

$$\int (\phi(x) - \phi'(x))(f_1(x) - k f_0(x))\,dx \geq 0, \tag{5.9}$$

for the integrand is nonnegative. This implies that

$$E_{\theta_1}\,\phi(X) - E_{\theta_1}\,\phi'(X) \geq kE_{\theta_0}\,\phi(X) - kE_{\theta_0}\,\phi'(X) \geq 0, \tag{5.10}$$

as was to be shown. For the case $k = \infty$ any test ϕ' of size zero must be zero almost everywhere on the set $\{x:f_0(x) > 0\}$. Hence

$$E_{\theta_1}(\phi(X) - \phi'(X)) = \int_{\{x:f_0(x)=0\}} (1 - \phi'(x))\,f_1(x)\,dx \geq 0,$$

thus completing the proof of (a).

(b) A best test of size $\alpha = 0$ is found in (5.8) so that we may restrict attention to $0 < \alpha \leq 1$. The size of the test (5.7), when $\gamma(x) = \gamma$, is

$$\begin{aligned} E_{\theta_0}\,\phi(X) &= P_{\theta_0}\{f_1(X) > k f_0(X)\} + \gamma\,P_{\theta_0}\{f_1(X) = k f_0(X)\} \\ &= 1 - P_{\theta_0}\{Y \leq k\} + \gamma\,P_{\theta_0}\{Y = k\}, \end{aligned} \tag{5.11}$$

where $Y = f_1(X)/f_0(X)$. For fixed α, $0 < \alpha \leq 1$, we are to find k and γ so that $E_{\theta_0}\,\phi(X) = \alpha$, or

$$P_{\theta_0}\{Y \leq k\} - \gamma\,P_{\theta_0}\{Y = k\} = 1 - \alpha. \tag{5.12}$$

If there exists a k_0 for which $P_{\theta_0}\{Y \leq k_0\} = 1 - \alpha$, we take $\gamma = 0$, $k = k_0$, and are finished. If not, there exists a k_0 such that (Fig. 5.2)

$$P_{\theta_0}\{Y < k_0\} \leq 1 - \alpha < P_{\theta_0}\{Y \leq k_0\}.$$

Then we may take $k = k_0$ and

$$\gamma = \frac{P_{\theta_0}\{Y \leq k_0\} - (1 - \alpha)}{P_{\theta_0}\{Y = k_0\}}, \tag{5.13}$$

which satisfies (5.12) and $0 \leq \gamma \leq 1$.

(c) If $\alpha = 0$, the argument in (a) shows that $\phi(x) = 0$ almost everywhere on the set $\{x:f_0(x) > 0\}$. If ϕ' has minimum probability of the second kind of error, then $1 - \phi'(x) = 0$ almost everywhere on the set $\{x:f_1(x) > 0\} \sim \{x:f_0(x) > 0\}$. Thus ϕ' differs from the ϕ of (5.8) by a set of probability zero under either hypothesis.

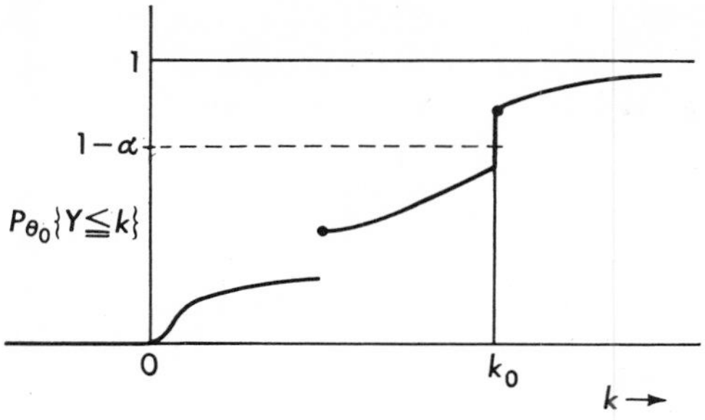

Fig. 5.2

If $\alpha > 0$, let ϕ be a best test of size α of the form (5.7). Then, because $E_{\theta_i}\,\phi(X) = E_{\theta_i}\,\phi'(X)$, $i = 0, 1$, the integral (5.9) must be equal to zero. But because the integrand is nonnegative it must be zero almost everywhere; that is to say, on the set for which $f_1(x) \neq kf_0(x)$ we have $\phi(x) = \phi'(x)$ almost everywhere. Thus, except for a set of probability zero, $\phi'(x)$ has the form (5.7) with the same value of k as $\phi(x)$, thus completing the proof.

As the proof of part (a) shows, the tests (5.7) are optimal even if the functions f_0 and f_1 assume negative values, that is, are not necessarily densities. The extended Neyman-Pearson lemma (Exercise 2), which generalizes this statement, is used in subsequent sections, and should be proved by the student. [For an extension of the test (5.8) which allows f_0 and f_1 to assume negative values, see Exercise 9.]

Corollary. The risk set S is closed from below (in fact, it is compact).

Proof. It is sufficient to show that $S' = \{(\alpha_0, \alpha_1) : \phi \in \mathfrak{D}\}$ is closed from below, for S is related to S' by a simple linear transformation with positive scale change. Let $\mathbf{z} \in \lambda(S')$, so that $\{\mathbf{z}\} = Q_{\mathbf{z}} \cap \bar{S}'$. If $\mathbf{z} = (\alpha_0, \alpha_1)$, there exists a best test, ϕ, of size α_0. The probability of error of the second kind cannot be less than α_1 without contradicting $\{\mathbf{z}\} = Q_{\mathbf{z}} \cap \bar{S}'$. If $E_{\theta_1}(1 - \phi(X)) > \alpha_1 + \epsilon$ and $\alpha_0 \neq 0$, then, from the convexity of S' and because $(0, 1) \in S'$ and $(1, 0) \in S'$, there must exist points $(\alpha_0, y) \in S'$ with $\alpha_1 \leq y < \alpha_1 + \epsilon$ for every ϵ. Hence, if $\alpha_0 \neq 0$, then $E_{\theta_1}(1 - \phi(X)) = \alpha_1$. If $\alpha_0 = 0$, this argument does not work, but, if ϕ' is the best test of size $\alpha_1 + \epsilon$ for testing H_1 against H_0, the probability of error of the second kind must be zero by the argument given, so that

$1 - \phi'$ is a better test of size zero than ϕ. This contradiction completes the proof.

This corollary in conjunction with Theorem 2.6.1 implies that the set of best tests forms a complete class. The statistician may choose a rule from this class by choosing a size, α_0, and then finding the best test of size α_0. From this point of view, the size of the test is merely an index by which to choose an admissible test. (That such a test is admissible must be checked separately if $\alpha_1 = 0$.) Another way of choosing a member of this complete class is, of course, by Bayes principle.

This corollary and the existence part of the Neyman-Pearson lemma follow immediately from the weak-compactness theorem for the set of test functions ϕ [see Lehmann (1959) for this theorem]. In fact, the weak-compactness theorem may be used to show a great deal more: namely, that *when both* Θ *and* $\mathcal{A}$ *are finite, the risk set* S *is compact,* hence bounded from below and closed from below. The reason we have not used this approach is that the weak-compactness theorem requires some elementary notions of measure theory with which we have not assumed the reader is familiar. As a compensation, though indeed a small one, it may be noted that the proof of the Neyman-Pearson lemma and the subsequent corollary are quite general in nature and apply to Riemann-integrable, as well as Lebesgue-measurable, functions, ϕ, whereas the weak-compactness theorem involves Lebesgue-measurable functions and is not applicable to Riemann-integrable functions.

Exercises

1. If ϕ is a best test of size α_0 and ϕ is not admissible, then

$$\alpha_1 = E_{\theta_1}(1 - \phi(X)) = 0.$$

2. *Generalized Neyman-Pearson Lemma.* Let $f_0(x), f_1(x), \cdots, f_n(x)$ be integrable functions on the real line. Let ϕ_0 be any integrable function of the form

$$\phi_0(x) = \begin{cases} 1 & \text{if } f_0(x) > k_1 f_1(x) + \cdots + k_n f_n(x), \\ \gamma(x) & \text{if } \qquad = \\ 0 & \text{if } f_0(x) < \sum_{j=1}^{n} k_j f_j(x), \end{cases}$$

where $0 \le \gamma(x) \le 1$. Then ϕ_0 maximizes the integral $\int \phi(x) f_0(x)\, dx$ out of all functions ϕ, $0 \le \phi \le 1$, such that

$$\int \phi(x) f_j(x)\, dx = \int \phi_0(x) f_j(x)\, dx$$

for $j = 1, \cdots, n$. If $k_j \ge 0$ for $j = 1, \cdots, n$, then ϕ_0 maximizes

$$\int \phi(x) f_0(x)\, dx$$

out of all functions ϕ, $0 \le \phi \le 1$, such that

$$\int \phi(x) f_j(x)\, dx \le \int \phi_0(x) f_j(x)\, dx, \qquad j = 1, \cdots, n.$$

3. In testing $H_0: X \in \mathcal{U}(0, 1)$ against $H_1: X \in \mathcal{U}(\frac{1}{2}, \frac{3}{2})$, plot the risk set S'.

4. In testing $H_0: X \in \mathcal{B}(2, \frac{1}{2})$ against $H_1: X \in \mathcal{B}(2, \frac{2}{3})$, plot the risk set S'.

5. In testing $H_0: X \in \mathcal{G}(1, 1)$ against $H_1: X \in \mathcal{G}(1, 2)$, plot the risk set S'.

6. Let $X \in \mathcal{C}(\theta, 1)$. Show that the test

$$\phi(x) = \begin{cases} 1 & \text{if} \quad 1 < x < 3, \\ 0 & \text{otherwise,} \end{cases}$$

is best of its size for testing $H_0: \theta = 0$ against $H_1: \theta = 1$. Give a rough plot of the power function as a function of θ.

7. Let ϕ be a best test of size α, $0 < \alpha < 1$, for testing $H_0: \theta = \theta_0$ against $H_1: \theta = \theta_1$. Show that $E_{\theta_1} \phi(x) > \alpha$ unless $P_{\theta_0} = P_{\theta_1}$.

8. Let $Z_1, \cdots, Z_k$ have the multinomial distribution

$$f_{Z_1,\cdots,Z_k}(z_1, \cdots, z_k \mid \theta_1, \cdots, \theta_k) = n! \prod_{i=1}^{k} \left(\frac{\theta_i^{z_i}}{z_i!}\right), \qquad 0 \le z_i, \ \sum_1^k z_i = n,$$

where $\theta_i > 0$ and $\sum_1^k \theta_i = 1$. Find the form of the best tests (5.7) and (5.8) for testing $H_0: \theta_i = \theta_i^0$, $i = 1, \cdots, k$, against $H_1: \theta_i = \theta_i'$, $i = 1, \cdots, k$. Consider the special case $k = 4$, $\theta_1^0 = .55$, $\theta_2^0 = .2$, $\theta_3^0 = .15$, $\theta_4^0 = .1$, $\theta_1' = .1$, $\theta_2' = .4$, $\theta_3' = .3$, $\theta_4' = .2$, and show how the computation of the probabilities of error may be calculated from the binomial distribution probabilities.

9. Let $f_0(x)$ and $f_1(x)$ be integrable functions on the real line and let

$$\phi_0(x) = \begin{cases} 1 & \text{if} \quad f_0(x) < 0 \quad \text{or} \quad (f_0(x) = 0 \quad \text{and} \quad f_1(x) \geq 0), \\ 0 & \text{if} \quad f_0(x) > 0 \quad \text{or} \quad (f_0(x) = 0 \quad \text{and} \quad f_1(x) < 0). \end{cases}$$

Show that ϕ_0 maximizes the integral $\int \phi(x)\, f_1(x)\, dx$ out of all functions ϕ, $0 \leq \phi \leq 1$, such that

$$\int \phi(x)\, f_0(x)\, dx = \int \phi_0(x)\, f_0(x)\, dx.$$

5.2 Uniformly Most Powerful Tests

We now focus our attention on the problem of testing a composite hypothesis $H_0:\theta \in \Theta_0$ against a composite alternative $H_1:\theta \in \Theta_1$. We shall say a hypothesis $H':\theta \in \Theta'$ is *composite* if Θ' consists of at least two elements. Here there is some loss of generality in assuming that the loss function is in the simple form of equation (5.4). Furthermore, the set of admissible rules is not so easily described as in Section 5.1. However, the notion of a best test of size α generalizes to this situation.

Definition 1. A test ϕ of $H_0:\theta \in \Theta_0$ against $H_1:\theta \in \Theta_1$ is said to have *size* α, if

$$\sup_{\theta \in \Theta_0} E_\theta\, \phi(X) = \alpha. \tag{5.14}$$

Definition 2. A test ϕ_0 is said to be *uniformly most powerful* (UMP) *of size* α for testing $H_0:\theta \in \Theta_0$ against $H_1:\theta \in \Theta_1$ if ϕ_0 is of size α and if for any other test ϕ of size at most α

$$E_\theta\, \phi_0(X) \geq E_\theta\, \phi(X) \tag{5.15}$$

for each $\theta \in \Theta_1$.

We would expect to be able to find in the class of tests of size α one that maximizes the power $E_{\theta_1}\,\phi(X)$ at a fixed element $\theta_1 \in \Theta_1$. However, there is no reason why this test should also maximize the power $E_{\theta_2}\,\phi(X)$ at another point $\theta_2 \in \Theta_1$, and in order to be *uniformly* most powerful a test must maximize the power $E_\theta\,\phi(X)$ for each $\theta \in \Theta_1$. It is not surprising, then, that UMP tests exist only in special circumstances.

To illustrate these ideas we first consider an important example. Let X be a random variable with a normal distribution of variance one and unknown mean θ. Let $\Theta_0 = (-\infty, \theta_0]$ and let $\Theta_1 = (\theta_0, +\infty)$. We seek a UMP test of size α for testing $H_0{:}\theta \in \Theta_0$, against $H_1{:}\theta \in \Theta_1$. (Note that the problem could easily be generalized to the problem in which a sample $X_1, \cdots, X_n$ is drawn from such a normal distribution, for, by sufficiency, the statistician can restrict his attention to $\bar{X}$, which has the same distribution as X except for the variance.) Thus we seek a test ϕ_0 that is uniformly best out of the class of all tests ϕ for which

$$E_\theta\,\phi(X) \leq \alpha \qquad \text{for all } \theta \leq \theta_0. \tag{5.16}$$

To solve this problem we first solve an easy related problem; namely, we find the best test ϕ_0 of size α for testing the simple hypothesis $H_0'{:}\theta = \theta_0$ against the simple alternative $H_1'{:}\theta = \theta_1$, where $\theta_1 > \theta_0$. By the Neyman-Pearson lemma

$$\phi_0(x) = \begin{cases} 1 & \text{if } \dfrac{1}{\sqrt{2\pi}} \exp[-\tfrac{1}{2}(x-\theta_1)^2] > \dfrac{k}{\sqrt{2\pi}} \exp[-\tfrac{1}{2}(x-\theta_0)^2], \\ \gamma & \text{if } \qquad\qquad = \\ 0 & \text{if } \qquad\qquad < \end{cases} \tag{5.17}$$

or, equivalently,

$$\phi_0(x) = \begin{cases} 1 & \text{if } x > k', \\ 0 & \text{otherwise,} \end{cases} \tag{5.18}$$

where $k' = (\theta_1^2/2 - \theta_0^2/2 + \log k)/(\theta_1 - \theta_0)$. [The γ in (5.17) may be chosen to be zero, for the probability that $X = k'$ is zero anyway.] Choosing k in $[0, \infty]$ is equivalent to choosing k' in $[-\infty, \infty]$. The best test of size α of H_0' against H_1' is obtained in (5.18) by choosing k' so that

$$E_{\theta_0}\,\phi_0(X) = \alpha. \tag{5.19}$$

We see that k' depends only on α and θ_0 and not otherwise on θ_1. Thus ϕ_0 of (5.18) is a UMP test of size α for testing $H_0':\theta = \theta_0$ against $H_1:\theta > \theta_0$; that is, ϕ_0 is UMP out of the class of all tests for which

$$E_{\theta_0}\,\phi(X) \le \alpha. \tag{5.20}$$

To show that ϕ_0 is UMP for testing $H_0:\theta \le \theta_0$ against $H_1:\theta > \theta_0$ involves a simple observation. First note that the power function $E_\theta\phi_0$ is an increasing function of θ with the value α at $\theta = \theta_0$. Hence ϕ_0 satisfies inequality (5.16); but because ϕ_0 is uniformly best out of the class of all tests satisfying (5.20) it is also uniformly best out of the class of all tests satisfying (5.16), for the latter class is a *subclass* of the former.

This is a rather remarkable result: that there exists a uniformly most powerful test of size α for testing the one-sided hypothesis $H_0:\theta \le \theta_0$ against the alternatives $H_1:\theta > \theta_0$, for any θ_0, when θ is the mean of the normal random variable X with known variance. Furthermore, this uniformly most powerful test has a very simple form; namely, reject H_0 if $X > k'$ and accept H_0 if $X \le k'$, where k' is chosen to make the size of the test equal to α. That is to say, the critical region is one-sided also. It is therefore of great interest to know to what one-parameter families of distributions, other than the normal distributions with unknown mean, this analysis extends. The answer is to families of distributions with monotone likelihood ratio.

Definition 3. A real parameter family of distributions is said to have *monotone likelihood ratio* if densities (probability mass functions) $f(x \mid \theta)$ exist such that whenever $\theta_1 < \theta_2$ the likelihood ratio

$$\frac{f(x \mid \theta_2)}{f(x \mid \theta_1)} \tag{5.21}$$

is a nondecreasing function of x in the set of its existence; that is, for x in the set of points for which at least one of $f(x \mid \theta_1)$ and $f(x \mid \theta_2)$ is positive. [If $f(x \mid \theta_1) = 0$ and $f(x \mid \theta_2) > 0$, the likelihood ratio (5.21) is defined as $+\infty$.]

Strictly speaking, this defines a family of distributions with nondecreasing likelihood ratio. Distributions with nonincreasing likelihood ratio may be treated by symmetry, that is, by a change of variable $Y = -X$ or by a reparametrization $\vartheta = -\theta$.

EXAMPLE 1. The one-parameter exponential family of distributions with density (probability mass function)

$$f(x \mid \theta) = c(\theta)\, h(x) \exp [Q(\theta)\, T(x)]$$

has likelihood ratio with $\theta_1 < \theta_2$

$$\frac{f(x \mid \theta_2)}{f(x \mid \theta_1)} = \frac{c(\theta_2)}{c(\theta_1)} \exp \{[Q(\theta_2) - Q(\theta_1)]\, T(x)\},$$

which is nondecreasing in x, provided that both $Q(\theta)$ and $T(x)$ are nondecreasing. [Given an arbitrary one-parameter exponential family of distributions, the nondecreasingness of $Q(\theta)$ and $T(x)$ can be obtained by a change of variable, $Y = T(x)$, and a reparametrization, $\vartheta = Q(\theta)$, if necessary.]

This example includes normal, binomial, negative binomial, Poisson, gamma, and beta distributions. Furthermore, if a sample of size n is taken from any of these distributions, the principle of sufficiency will immediately reduce the problem to a one-parameter exponential family of distributions on the real line.

EXAMPLE 2. The hypergeometric distribution $\mathcal{H}(n, m, M)$ with probability mass function (3.4) considering m as the only unknown parameter, has monotone likelihood ratio, because

$$\frac{P\{X = x \mid m + 1\}}{P\{X = x \mid m\}} = \frac{m + 1}{M - m} \frac{M - m - n + x}{m + 1 - x}$$

is an increasing function of x and the ratio $P\{X = x \mid m'\}/P\{X = x \mid m\}$ for $m' > m$ is a product of such functions.

EXAMPLE 3. The uniform distribution $\mathcal{U}(\theta, \theta + 1)$ has monotone likelihood ratio (Exercise 2).

EXAMPLE 4. Other examples include the uniform distributions $\mathcal{U}(0, \theta)$, $\theta > 0$, the noncentral t and noncentral F distributions, the double exponential distributions with density

$$f(x \mid \alpha, \beta) = \frac{1}{2\beta} \exp \left(\frac{-\mid x - \alpha \mid}{\beta} \right), \tag{5.22}$$

when α is unknown and β is known, and the negative exponential distributions with unknown location parameter

$$f(x \mid \theta) = e^{-(x-\theta)}\, I_{(\theta,\infty)}(x). \tag{5.23}$$

EXAMPLE 5. The Cauchy distribution $\mathcal{C}(\theta, 1)$ provides an example of a distribution *without* a monotone likelihood ratio, for the ratio

$$\frac{f(x \mid \theta_2)}{f(x \mid \theta_1)} = \frac{1 + (x - \theta_1)^2}{1 + (x - \theta_2)^2}$$

converges to one as $x \to +\infty$ or $x \to -\infty$.

The following theorem is due to Karlin and Rubin (1956).

Theorem 1. If the distribution of X has monotone likelihood ratio, any test of the form

$$\phi(x) = \begin{cases} 1 & \text{if } x > x_0, \\ \gamma & \text{if } x = x_0, \\ 0 & \text{if } x < x_0, \end{cases} \tag{5.24}$$

has nondecreasing power function. Any test of the form (5.24) is UMP of its size for testing $H_0{:}\theta \leq \theta_0$ against $H_1{:}\theta > \theta_0$ for any $\theta_0 \in \Theta$, provided its size is not zero. For every $0 \leq \alpha \leq 1$ and every $\theta_0 \in \Theta$, there exist numbers $-\infty \leq x_0 \leq +\infty$ and $0 \leq \gamma \leq 1$ such that the test (5.24) is UMP of size α, for testing $H_0{:}\theta \leq \theta_0$ against $H_1{:}\theta > \theta_0$.

Proof. Let θ_1 and θ_2 be any points of Θ with $\theta_1 < \theta_2$. By the Neyman-Pearson lemma any test of the form

$$\phi(x) = \begin{cases} 1 & \text{if } f(x \mid \theta_2) > kf(x \mid \theta_1) \\ \gamma(x) & \qquad\qquad = \\ 0 & \qquad\qquad < \end{cases} \tag{5.25}$$

for $0 \leq k < \infty$, or, corresponding to $k = \infty$,

$$\phi(x) = \begin{cases} 1 & \text{if } f(x \mid \theta_1) = 0 \\ 0 & \text{if } f(x \mid \theta_1) > 0 \end{cases} \tag{5.26}$$

is best of its size for testing $\theta = \theta_1$ against $\theta = \theta_2$. Because the distribution has monotone likelihood ratio, any test of the form (5.24) is also of the form (5.25), provided $E_{\theta_1}\phi(X) > 0$ (because $E_{\theta_1}\phi(X) > 0$

implies $k < \infty$). Thus, provided $\alpha = E_{\theta_1}\phi(X) > 0$, (5.24) is best of size α for testing $\theta = \theta_1$ against $\theta = \theta_2$. Because $\phi^*(x) \equiv \alpha$ is a size α test, it has no larger power at θ_2; that is,

$$E_{\theta_2}\phi(X) \geq E_{\theta_2}\phi^*(X) = \alpha = E_{\theta_1}\phi(X).$$

In other words, $\theta_1 < \theta_2$ and $E_{\theta_1}\phi(X) > 0$ imply $E_{\theta_1}\phi(X) \leq E_{\theta_2}\phi(X)$, which fact, in turn, implies that the power function is nondecreasing.

The rest of the proof follows the steps for the normal distribution, except that care must be taken when $E_{\theta_0}\phi(X) = 0$. It is left to the student.

Corollary 1. If the distribution of X has monotone likelihood ratio, then for every test ϕ and every $\theta_0 \in \Theta$, there exists a test ϕ' of the form (5.24) such that

$$E_\theta\,\phi'(X) \leq E_\theta\,\phi(X) \quad \text{for} \quad \theta \leq \theta_0$$

and

$$E_\theta\,\phi'(X) \geq E_\theta\,\phi(X) \quad \text{for} \quad \theta \geq \theta_0\,. \tag{5.27}$$

Proof. Let $\alpha = E_{\theta_0}\phi(X)$ and let ϕ' be UMP of size α of the form (5.24) for testing $H_0:\theta \leq \theta_0$ against $H_1:\theta > \theta_0$, so that $E_\theta\phi'(X) \geq E_\theta\phi(X)$ for $\theta \geq \theta_0$. By symmetry Theorem 1 implies that $1 - \phi'$ is UMP of its size for testing $H_0':\theta \geq \theta_0$ against $H_1':\theta < \theta_0$, provided $\alpha \neq 1$, so that, for $\theta \leq \theta_0$, $E_\theta(1 - \phi'(X)) \geq E_\theta(1 - \phi(X))$. The case $\alpha = 1$ follows from the case $\alpha = 0$ by symmetry, thus completing the proof.

We now strengthen this result by allowing the loss function to satisfy certain inequalities that describe what is intuitively meant by the one-sided hypothesis testing situation. Let $\mathcal{A} = \{a_0, a_1\}$ and let $L(\theta, a_i)$ be the loss if action a_i is taken and θ is the true value of the parameter, $i = 0, 1$. We assume that the loss function satisfies the inequalities

$$\begin{aligned} L(\theta, a_1) - L(\theta, a_0) &\geq 0 \quad \text{if} \quad \theta < \theta_0\,, \\ L(\theta, a_1) - L(\theta, a_0) &\leq 0 \quad \text{if} \quad \theta > \theta_0\,. \end{aligned} \tag{5.28}$$

Thus, if $\theta < \theta_0$, action a_0 is preferred, whereas, if $\theta > \theta_0$, action a_1 is preferred, so that a_0 represents intuitively the action "accept H_0" where $H_0:\theta < \theta_0$ and a_1, the action "accept H_1" where $H_1:\theta > \theta_0$. Note that no restrictions are put on the loss function where $\theta = \theta_0$; hypothesis H_0 could easily be $\theta \leq \theta_0$ or, alternatively, H_1 could be $\theta \geq \theta_0$.

Theorem 2. If the loss function satisfies the inequalities (5.28) and the distribution of X has monotone likelihood ratio, the class of one-sided tests (5.24) is essentially complete. If the set of points $\{x:f(x \mid \theta) > 0\}$ is independent of θ and if there exist numbers $\theta_1 \in \Theta$, $\theta_2 \in \Theta$ with $\theta_1 \leq \theta_0 \leq \theta_2$ such that $L(\theta_1, a_1) - L(\theta_1, a_0) > 0$ and $L(\theta_2, a_1) - L(\theta_2, a_0) < 0$, then any test of the form (5.24) is admissible.

Proof. Let ϕ be any test and find from Corollary 1 a test ϕ' of the form (5.24) that satisfies (5.27) for the θ_0 found in (5.28). The difference of the risks, according to (5.3), is

$$R(\theta, \phi) - R(\theta, \phi') = [L(\theta, a_1) - L(\theta, a_0)][E_\theta\, \phi(X) - E_\theta\, \phi'(X)]. \tag{5.29}$$

If $\theta > \theta_0$, (5.27) and (5.28) imply that both terms on the right of (5.29) are nonpositive; if $\theta < \theta_0$, both terms are nonnegative; and for $\theta = \theta_0$, $E_\theta\, \phi(X) = E_\theta\, \phi'(X)$, so that in all cases $R(\theta, \phi) - R(\theta, \phi') \geq 0$, which shows that the class of one-sided tests (5.24) is essentially complete.

Let ϕ' be a test in the form of (5.24) and suppose that ϕ is a test for which $R(\theta, \phi) \leq R(\theta, \phi')$ for all θ. To show admissibility of ϕ', we must show that $R(\theta, \phi) \geq R(\theta, \phi')$ for all θ. From (5.28) and (5.29) it follows that $E_{\theta_1}\, \phi(X) - E_{\theta_1}\, \phi'(X) \leq 0$ and $E_{\theta_2}\, \phi(X) - E_{\theta_2}\, \phi'(X) \geq 0$. If $E_{\theta_1}\, \phi'(X) > 0$, ϕ' is UMP of its size for testing $H_0:\theta \leq \theta_1$ against $H_1:\theta > \theta_1$ from Theorem 1. If $E_{\theta_1}\, \phi'(X) = 0$, it is UMP of its size for testing $H_0:\theta \leq \theta_1$ against $H_1:\theta > \theta_1$, because all such tests have zero power when $\{x:f(x \mid \theta) > 0\}$ is independent of θ. Because it is UMP in all cases, it follows that $E_\theta\, \phi(X) - E_\theta\, \phi'(X) \leq 0$ for all $\theta > \theta_1$. By symmetry $E_\theta\, \phi(X) - E_\theta\, \phi'(X) \geq 0$ for all $\theta < \theta_2$. This implies $R(\theta, \phi) - R(\theta, \phi') \geq 0$, completing the proof.

It is important to note that the test ϕ', which is as good as ϕ, does not depend on the loss function L, provided L is of the form (5.28) and θ_0 is given. Even θ_0 may not be uniquely determined by L.

An essentially complete class of decision rules, which contains only admissible rules, might be called *essentially minimal complete.* Such a class lies somewhere between a minimal complete class and a minimal essentially complete class. Theorem 2 in these terms states that for one-sided hypotheses testing problems (5.28) that satisfy the hypothesis of the theorem the class of one-sided tests (5.24) is essentially minimal complete.

Exercises

1. Complete the proof of Theorem 1.

2. Show that the uniform distribution $\mathcal{U}(\theta, \theta + 1)$ has monotone likelihood ratio. Find a counterexample to the second statement of Theorem 1 for a test of size zero.

3. Show that the family of distributions with density

$$f(x \mid \theta) = c(\theta)\, h(x) I_{(-\infty,\theta)}(x) \tag{5.30}$$

has monotone likelihood ratio. Show that the family of uniform distributions $\mathcal{U}(0, \theta)$ is a special case of (5.30).

4. Show that the double exponential distribution with density $f(x \mid \alpha, \beta)$ of formula (5.22) has monotone likelihood ratio when α is unknown and β is known.

5. Show that the negative exponential distribution with density $f(x \mid \theta)$ of (5.23) has monotone likelihood ratio.

6. Show that the Cauchy distribution $\mathcal{C}(0, \theta)$ has no monotone likelihood ratio but that the distribution of the sufficient statistic $|X|$ has monotone likelihood ratio.

7. Let $X_1, \cdots, X_n$ be a sample of size n from the uniform distribution $\mathcal{U}(0, \theta)$. Sufficiency reduces the problem to $T = \max_i X_i$.
(a) Find the class of all Neyman-Pearson best tests of $H_0{:}\theta = \theta_0$ against $H_1{:}\theta = \theta_1$, where $\theta_1 > \theta_0$.
(b) Find the subclass of the tests that are independent of θ_1. These are UMP tests of H_0 against $H_1'{:}\theta > \theta_0$.
(c) Show that the test

$$\phi(t) = \begin{cases} 1 & \text{if} \quad t > \theta_0, \\ \alpha & \text{if} \quad t \le \theta_0, \end{cases}$$

is UMP of size α for testing $H_0'{:}\theta \le \theta_0$ against $H_1'{:}\theta > \theta_0$ but that ϕ is not admissible.
(d) Show that

$$\phi(t) = \begin{cases} 1 & \text{if} \quad t > \theta_0 \quad \text{or} \quad t \le b, \\ 0 & \text{if} \quad b < t \le \theta_0, \end{cases}$$

where $b = \theta_0 \sqrt[n]{\alpha}$ is a UMP test of size α for testing $H_0{:}\theta = \theta_0$ against $H{:}\theta \ne \theta_0$.

(e) Generalize the result of (d) to the "nonregular" family of distributions (5.30).

8. Extend the game of bluffing defined in Section 1.1 to a decision-theory problem as follows. Suppose that if player I bets, player II is allowed to observe the number X of times that player I blinks his eyes in a certain length of time. Player II knows that X is a random variable with the Poisson distribution $X \in \mathcal{P}(\lambda)$. In fact, if player I is betting when he has a losing card, then $\lambda = \lambda_0$, whereas if player I is betting with a winning card, then $\lambda = \lambda_1 < \lambda_0$, where λ_0 and λ_1 are known. Let $\phi(x)$ denote a behavioral decision function for which $\phi(x)$ is the probability of folding when $X = x$ is observed, $0 \leq \phi(x) \leq 1$, $x = 0, 1, 2, \cdots$.

(a) Show that the risk function is

$$R(\text{bluff}, \phi) = (a + b)(2P - 1) - bPE_{\lambda_1}\phi(X) + (2a + b)(1 - P)E_{\lambda_0}\phi(X)$$

$$R(\text{honest}, \phi) = a(2P - 1) + bP - bPE_{\lambda_1}\phi(X).$$

(b) Suppose that $P < 1/2$. Show that the minimax strategy for player II is the ϕ that is a best test of size $b/(2a + b)$ for testing $H_0{:}\lambda = \lambda_0$ against $H_1{:}\lambda = \lambda_1$. Note that this rule is minimax, regardless of the value of $P < \frac{1}{2}$ and of $\lambda_1 < \lambda_0$.

Hint. Let ϕ_α be a best test of size α for testing H_0 against H_1. Show that attention may be restricted to the tests ϕ_α, and that $g_1(\alpha) = R(\text{bluff}, \phi_\alpha)$ is convex in α and $g_1(0) \leq g_1(b/(2a + b)) \leq g_1(1)$, while $g_2(\alpha) = R(\text{honest}, \phi_\alpha)$ is decreasing in α and

$$g_1(b/(2a + b)) = g_2(b/(2a + b)).$$

Note. This is the solution under the assumption that Player I realizes that he blinks his eyes at Poisson rates λ_0 and λ_1 depending on his strategy choice, but can do nothing about it. Under the more realistic assumption that he is unaware that he is giving away information by blinking his eyes, this is no longer a game in the von Neumann-Morgenstern sense, but is called instead, a game of incomplete information or a pseudo-game. The problem of how Player I may play a sequence of identical pseudo-games gaining knowledge about the structure of the information leak in order to minimize its effects has been studied by Baños (1967).

9. Show that the logistic distribution with location parameter θ, having density

$$f(x \mid \theta) = \frac{e^{(x-\theta)}}{(1 + e^{(x-\theta)})^2} = \frac{1}{2(1 + \cosh(x - \theta))},$$

has monotone likelihood ratio.

10. Show that the families of distributions in (3.61) have monotone likelihood ratio (θ real), provided $\pi_1(\theta)$ and $\pi_2(\theta)$ are nondecreasing functions of θ.

11. Let $X_1, \cdots, X_n$ be a sample of size n from $\mathcal{U}(\theta, \theta + 1)$, the uniform distribution on the interval $(\theta, \theta + 1)$. Find the joint distribution of the sufficient statistics $T_1 = \min X_i$ and $T_2 = \max X_i$. Show that a UMP size α test of $H_0{:}\theta \leq 0$ against $H_1{:}\theta > 0$ exists in the form

$$\phi_0(t_1, t_2) = \begin{cases} 0 & \text{if } t_1 < k \text{ and } t_2 < 1, \\ 1 & \text{otherwise.} \end{cases}$$

Find the value of k that makes this a size α test ($k = 1 - \alpha^{1/n}$).

5.3 Two-Sided Tests

In Section 5.2 we were concerned with tests of one-sided hypotheses, such as $H_0{:}\theta \leq \theta_0$ against $H_1{:}\theta > \theta_0$. Here we consider two-sided hypotheses, such as $H_0'{:}\theta = \theta_0$ against $H_1'{:}\theta \neq \theta_0$ or $H_0{:}\theta_1 \leq \theta \leq \theta_2$ (with $\theta_1 < \theta_2$) against $H_1{:}\theta < \theta_1$ or $\theta > \theta_2$. Just as distributions with monotone likelihood ratio are natural for one-sided problems, the Polya type-3 distributions are a natural set of distributions for two-sided problems. However, to get across the ideas involved in two-sided tests we treat only the special case of one-parameter exponential families of distributions, leaving the interested reader to consult Karlin (1957a) for the generalizations. It is shown that the two-sided tests form an essentially complete class for these problems. The proofs given here are constructive in the sense that for a given test ϕ a two-sided test ϕ', which is as good as ϕ, is directly exhibited.

In the proof of the existence of such tests we rely on the following lemma, which is of importance in itself. The corollary to Lemma 1 is quite useful in some nonparametric statistical problems. Let X be a random variable and let

$$F(x, \gamma) = P\{X < x\} + \gamma P\{X = x\}, \tag{5.31}$$

where $0 \le \gamma \le 1$ (so that $F(x, 1)$ is the distribution function of X). If we say that $(x, \gamma) < (x', \gamma')$ if, and only if, either $x < x'$, or $x = x'$ and $\gamma < \gamma'$, then it is clear that $F(x, \gamma)$ is a nondecreasing function of (x, γ). Furthermore, we can show, as in the proof of the existence part of the Neyman-Pearson lemma, that each value of α, $0 \le \alpha \le 1$, is assumed by the function F at some point (x, γ) $(x = \pm\infty$ is allowed).

Lemma 1. Let $V \in \mathcal{U}(0, 1)$ and let V be independent of X. The distribution of the random variable $W = F(X, V)$ is $\mathcal{U}(0, 1)$.

Proof. Let $0 \le w \le 1$ and find $x(w)$ and $\gamma(w)$ so that

$$F(x(w), \gamma(w)) = w.$$

Then

$$\begin{aligned} P\{W \le w\} &= P\{F(X, V) \le w\} \\ &= P\{(X, V) \le (x(w), \gamma(w))\} \\ &= P\{X < x(w)\} + P\{V \le \gamma(w)\} P\{X = x(w)\} \\ &= F(x(w), \gamma(w)) = w, \end{aligned} \tag{5.32}$$

as was to be shown.

Corollary 1. If the distribution function F of X is continuous then $F(X) \in \mathcal{U}(0, 1)$.

We now consider testing the hypothesis $H_0{:}\theta_1 \le \theta \le \theta_2$ against $H_1{:}\theta < \theta_1$ or $\theta > \theta_2$. We want to find a test ϕ' whose power function $E_\theta\phi'(X)$ is small when $\theta_1 \le \theta \le \theta_2$ and large when $\theta < \theta_1$ and $\theta > \theta_2$. We say that a test ϕ is a *two-sided test* if ϕ has the form

$$\phi(x) = \begin{cases} 1 & \text{if} \quad x < x_1 \quad \text{or} \quad x > x_2, \\ \gamma_i & \text{if} \quad x = x_i, \qquad i = 1, 2, \\ 0 & \text{if} \quad x_1 < x < x_2 \end{cases} \tag{5.33}$$

for some numbers $x_1 \le x_2$ and $0 \le \gamma_i \le 1$. [We allow the possibility $x_1 = x_2$; also, $x_1 = -\infty$ and $x_2 = +\infty$ are allowed, so that the class of two-sided tests contains the class of one-sided tests as a subset. The tests $\psi(x) = 1 - \phi(x)$ may also be considered as two-sided tests, good for testing H_1 against H_0, but in this section we consider only the problems for which the class of tests (5.33) is good.] In this section we restrict our attention to one-parameter exponential families of distributions of the form

$$f(x \mid \theta) = c(\theta)\, h(x) e^{\theta x} \tag{5.34}$$

for $\theta \in \Theta$, some interval on the real line.

Theorem 1. Suppose that the distribution of the random observable X has the form (5.34). Then, given any test ϕ and $\theta_1 \in \Theta$, $\theta_2 \in \Theta$ with $\theta_1 < \theta_2$, there exists a two-sided test ϕ' in the form of (5.33) for which

$$E_{\theta_1} \phi'(X) = E_{\theta_1} \phi(X) \quad \text{and} \quad E_{\theta_2} \phi'(X) = E_{\theta_2} \phi(X). \tag{5.35}$$

Moreover, for any such two-sided test ϕ'

$$E_\theta\, \phi'(X) - E_\theta\, \phi(X) \begin{cases} \le 0 & \text{for} \quad \theta_1 < \theta < \theta_2, \\ \ge 0 & \text{for} \quad \theta < \theta_1 \quad \text{and} \quad \theta > \theta_2. \end{cases} \tag{5.36}$$

Proof (Existence). Let $\alpha_1 = E_{\theta_1}\phi(X)$ and $\alpha_2 = E_{\theta_2}\phi(X)$. Let $\phi_w(x)$ be a one-sided test of size w at θ_1 of the form (5.33) with $x_2 = +\infty$. Then, for all $0 \le u \le \alpha_1$,

$$\phi'_u(x) = \phi_u(x) + 1 - \phi_{1-\alpha_1+u}(x) \tag{5.37}$$

is a test of size α_1 at θ_1 of the form (5.33). We are to show that there is a u between 0 and α_1 for which $E_{\theta_2}\phi'_u(X) = \alpha_2$.

The tests ϕ'_0 and ϕ'_{α_1} have the largest and smallest power at θ_2 in the class of all tests of size α_1 at θ_1; therefore,

$$E_{\theta_2}\phi'_{\alpha_1}(X) \le \alpha_2 \le E_{\theta_2}\phi'_0(X). \tag{5.38}$$

To complete the proof, it is sufficient to show that $E_{\theta_2}\phi'_u(X)$ is a continuous function of u. Let $g(w) = E_{\theta_2}\phi_w(X)$, so that $E_{\theta_2}\phi'_u(X) = g(u) + 1 - g(1 - \alpha_1 + u)$. It remains to show that g is continuous.

Let $V \in \mathcal{U}(0, 1)$ be independent of X, and let $W = F_{\theta_1}(X, V)$ as in Lemma 1, so that $W \in \mathcal{U}(0, 1)$ when X has distribution P_{θ_1}. Then, since

$\phi_w(x) = E\, I_{[o,w)}(F_{\theta_1}(x, V))$, g may be written

$$\begin{aligned} g(w) &= E_{\theta_2}\phi_w(X) \\ &= E_{\theta_1}\phi_w(X)\,\frac{c(\theta_2)}{c(\theta_1)}\exp\left[(\theta_2 - \theta_1)X\right] \\ &= E_{\theta_1} I_{[0,w)}(W)\,\frac{c(\theta_2)}{c(\theta_1)}\exp\left[(\theta_2 - \theta_1)X\right]. \end{aligned}$$

If g were not continuous at some point w_0, then $E_{\theta_1} I_{\{w_0\}}(W)(c(\theta_2)/c(\theta_1)) \exp((\theta_2 - \theta_1)X)$ would be positive. But this is impossible because $E_{\theta_1} I_{\{w_0\}}(W)$ is zero.

(*Moreover*). To prove the inequalities (5.36) we show that every two-sided test ϕ' in the form of (5.33) has maximum power at $\theta < \theta_1$ and $\theta > \theta_2$ and minimum power at θ between θ_1 and θ_2 out of all tests ϕ that satisfy (5.35).

First note that the one-sided tests with $x_1 = -\infty$ [that is, the tests ϕ_0' of (5.37)] are best of their size for testing $\theta = \theta_1$ against $\theta = \theta_2$. If ϕ is any other test of the same size at $\theta = \theta_1$, having the same maximum power at θ_2, then, by the uniqueness part of the Neyman-Pearson lemma, $\phi = \phi_0'$, except perhaps for a set of probability zero under P_{θ_1}; but, because any set that is null under P_{θ_1} is also null under P_θ for all $\theta \in \Theta$, ϕ and ϕ_0' must have the same power function, and inequalities (5.36) are automatically satisfied (with $\phi' = \phi_0'$). The one-sided tests with $x_2 = +\infty$ may be treated similarly. We may now suppose that both x_1 and x_2 are finite.

CASE 1. $\theta < \theta_1$. By the generalized Neyman-Pearson lemma (Exercise 5.1.2) any test ϕ_0 of the form

$$\phi_0(x) = \begin{cases} 1 & \text{if } f(x \mid \theta) > k_1 f(x \mid \theta_1) + k_2 f(x \mid \theta_2), \\ \gamma(x) & \text{if } f(x \mid \theta) = k_1 f(x \mid \theta_1) + k_2 f(x \mid \theta_2), \\ 0 & \text{if } f(x \mid \theta) < k_1 f(x \mid \theta_1) + k_2 f(x \mid \theta_2), \end{cases} \tag{5.39}$$

for finite k_1 and k_2 and $0 \le \gamma(x) \le 1$ maximizes the quantity $E_\theta\, \phi(X)$ out of all tests ϕ for which $E_{\theta_i}\, \phi(X) = E_{\theta_i}\, \phi_0(X)$ for $i = 1, 2$. We show that every two-sided test in the form of (5.33) with finite x_1 and x_2 is in the form of (5.39), when $\theta < \theta_1$, so that the lower inequality of (5.36) must be valid for $\theta < \theta_1$. The lower inequality of (5.39), except for some x for which $h(x) = 0$ is equivalent to the inequality

$$c(\theta)e^{\theta x} < k_1 c(\theta_1) \exp(\theta_1 x) + k_2 c(\theta_2) \exp(\theta_2 x),$$

which in turn is equivalent to

$$k_1' \exp(b_1 x) + k_2' \exp(b_2 x) > 1, \tag{5.40}$$

where $b_2 = (\theta_2 - \theta) > b_1 = (\theta_1 - \theta) > 0$, $k_1' = k_1 c(\theta_1)/c(\theta)$ and $k_2' = k_2 c(\theta_2)/c(\theta)$. First suppose that $x_1 < x_2$. Then, by choosing k_1' and k_2' to satisfy the simultaneous equations

$$\begin{aligned} k_1' \exp(b_1 x_1) + k_2' \exp(b_2 x_1) &= 1, \\ k_1' \exp(b_1 x_2) + k_2' \exp(b_2 x_2) &= 1, \end{aligned} \tag{5.41}$$

whose determinant is not zero, we find $k_1' > 0$, $k_2' < 0$, and inequality (5.40) is equivalent to $x_1 < x < x_2$. Thus every test in the form of (5.33) with $x_1 < x_2$ is in the form of (5.39). Now suppose $x_1 = x_2$. Then k_1' and k_2' may be found by solving the simultaneous equations

$$\begin{aligned} k_1' \exp(b_1 x_1) + k_2' \exp(b_2 x_1) &= 1, \\ k_1' b_1 \exp(b_1 x_1) + k_2' b_2 \exp(b_2 x_1) &= 0, \end{aligned} \tag{5.42}$$

the second of which reflects the requirement that the slope of

$$k_1' \exp(b_1 x) + k_2' \exp(b_2 x)$$

must be zero at x_1. Thus, again, every test in the form of (5.33) with $x_1 = x_2$ is in the form of (5.39).

CASE 2. $\theta > \theta_2$. This case follows from Case 1 by symmetry.

CASE 3. $\theta_1 < \theta < \theta_2$. According to the generalized Neyman-Pearson lemma with ϕ replaced by $1 - \phi$, any test ϕ_0 of the form

$$\phi_0(x) = \begin{cases} 0 & \text{if } f(x \mid \theta) > k_1 f(x \mid \theta_1) + k_2 f(x \mid \theta_2), \\ \gamma(x) & \text{if } f(x \mid \theta) = k_1 f(x \mid \theta_1) + k_2 f(x \mid \theta_2), \\ 1 & \text{if } f(x \mid \theta) < k_1 f(x \mid \theta_1) + k_2 f(x \mid \theta_2), \end{cases} \tag{5.43}$$

for finite k_1 and k_2 and $0 \le \gamma(x) \le 1$, minimizes the quantity $E_\theta\, \phi(X)$ out of all tests ϕ for which $E_{\theta_i}\, \phi(X) = E_{\theta_i}\, \phi_0(X)$, for $i = 1, 2$. We will be finished when we show that every two-sided test in the form of (5.33) with finite x_1 and x_2 is in the form of (5.43) when $\theta_1 < \theta < \theta_2$, so that the upper inequality of (5.36) must be valid. The upper inequality of (5.43) is equivalent, except for some x for which $h(x) = 0$, to

$$k_1' \exp(b_1 x) + k_2' \exp(b_2 x) < 1, \tag{5.44}$$

where $b_1 = (\theta_1 - \theta) < 0 < b_2 = (\theta_2 - \theta)$. This time, if $x_1 < x_2$, we may choose k_1' and k_2' to satisfy (5.41) and find that $k_1' > 0$ and $k_2' > 0$. Then, we see that inequality (5.44) is equivalent to $x_1 < x < x_2$. Thus every test in the form of (5.33) with $x_1 < x_2$ is in the form of (5.43). Finally, if $x_1 = x_2$, again k_1' and k_2' may be chosen to satisfy (5.42) as before, thus completing the proof.

We turn now to our other two-sided problem; namely testing $H_0{:}\theta = \theta_0$ against $H_1{:}\theta \neq \theta_0$. Recall from Section 3.5 that when the random observable X has a one-parameter exponential family of distributions the power function of any test ϕ is a differentiable, in fact, analytic, function of θ. Furthermore, the derivative may be passed beneath the integral sign so that (Exercise 3)

$$\begin{aligned}\frac{d}{d\theta} E_\theta\, \phi(X) &= \int \phi(x)\,(c'(\theta) + c(\theta)x)\, h(x)\, e^{\theta x}\, dx \\ &= E_\theta X\, \phi(X) - E_\theta\, \phi(X) E_\theta\, X,\end{aligned} \tag{5.45}$$

with the integral replaced by a summation sign in the discrete case. It is quite clear, moreover, that in order for a two-sided test ϕ' of the same size as ϕ at θ_0 to have nowhere a smaller power, $E_\theta\, \phi'(X) \geq E_\theta\, \phi(X)$, it is necessary (if θ_0 is an interior point of Θ) that the power function have the same slope at $\theta = \theta_0$. That this condition is also sufficient is the content of the following theorem.

Theorem 2. Suppose the distribution of the random observable X has the form (5.34). Then, given any test ϕ and $\theta_0 \in \Theta$, there exists a two-sided test ϕ' in the form of (5.33) for which

$$E_{\theta_0}\, \phi'(X) = E_{\theta_0}\, \phi(X) \tag{5.46}$$

and

$$\left.\frac{d}{d\theta} E_\theta\, \phi'(X)\right|_{\theta=\theta_0} = \left.\frac{d}{d\theta} E_\theta\, \phi(X)\right|_{\theta=\theta_0} \tag{5.47}$$

Moreover, for any such two-sided test ϕ' and for all $\theta \in \Theta$

$$E_\theta \, \phi'(X) \geq E_\theta \, \phi(X). \tag{5.48}$$

Proof (*Existence*). Let $\alpha = E_{\theta_0}\phi(X)$ and $\alpha' = (d/d\theta)E_\theta\phi(X)\,|_{\theta=\theta_0}$. Let $\phi_w(x)$ be a one-sided test of size w at θ_0 of the form (5.33) with $x_2 = +\infty$. Then, for all $0 \leq u \leq \alpha$,

$$\phi'_u(x) = \phi_u(x) + 1 - \phi_{1-\alpha+u}(x)$$

is a test of size α at θ_0 of the form (5.33). We are to show that there is a u between 0 and α for which $(d/d\theta)E_\theta\phi'_u(X)\,|_{\theta=\theta_0} = \alpha'$.

By the Neyman-Pearson lemma the tests ϕ'_α and ϕ'_0 are easily seen to have the smallest and largest values of the slope (5.45) at θ_0 in the class of tests of size α at θ_0; therefore,

$$\frac{d}{d\theta} E_\theta\phi'_\alpha(X)\,|_{\theta=\theta_0} \leq \alpha' \leq \frac{d}{d\theta} E_\theta\phi'_0(X)\,|_{\theta=\theta_0}. \tag{5.49}$$

To complete the proof, we need only to show that $(d/d\theta)E_\theta\phi'_u(X)\,|_{\theta=\theta_0}$ is a continuous function of u. Let $g(w) = (d/d\theta)E_\theta\phi_w(X)\,|_{\theta=\theta_0}$, so that $(d/d\theta)E_\theta\phi'_u(X)\,|_{\theta=\theta_0} = g(u) - g(1 - \alpha + u)$. It remains for us to show that g is continuous.

Let $V \in \mathcal{U}(0, 1)$ be independent of X, and let $W = F_{\theta_0}(X, V)$ as in Lemma 1, so that $W \in \mathcal{U}(0, 1)$ when X has distribution P_{θ_0}. Then, since $\phi_w(x) = E\, I_{[0,w)}(F_{\theta_0}(x, V))$, g may be written in the form (5.45)

$$\begin{aligned} g(w) &= E_{\theta_0}\phi_w(X)(X - E_{\theta_0}X) \\ &= E_{\theta_0}I_{[0,w)}(W)(X - E_{\theta_0}X). \end{aligned} \tag{5.50}$$

If g were not continuous at some point w_0, then $E_{\theta_0}I_{\{w_0\}}(W)(X - E_{\theta_0}X)$ would be nonzero. This is impossible because $E_{\theta_0}I_{\{w_0\}}(W)$ is zero.

(*Moreover*). We prove the inequality (5.48) by showing that every two-sided test ϕ' in the form of (5.33) has maximum power at each θ out of all tests ϕ that satisfy (5.46) and (5.47).

As in the proof of Theorem 1 the one-sided tests (with $x_1 = -\infty$ or $x_2 = +\infty$) are easily seen by the uniqueness part of the Neyman-Pearson lemma to be essentially unique out of all tests that satisfy (5.46) and (5.47) so that the power function of such tests are completely specified. We restrict attention to finite x_1 and x_2.

For definiteness, suppose that $\theta < \theta_0$; the case $\theta > \theta_0$ can be treated symmetrically. By the generalized Neyman-Pearson lemma any test ϕ' of the form

$$\phi'(x) = \begin{cases} 1 & \text{if} \quad f(x \mid \theta) > k_1 f(x \mid \theta_0) + k_2 \dfrac{\partial}{\partial \theta} f(x \mid \theta) \Big|_{\theta=\theta_0}, \\ \gamma(x) & \text{if} \quad = \\ 0 & \text{if} \quad < \end{cases} \tag{5.51}$$

has a maximum value of $E_\theta\, \phi(X)$ out of all tests ϕ for which (5.46) and (5.47) hold. We are to show that every two-sided test in the form of (5.33) with finite x_1 and x_2 is in the form of (5.51). The lower inequality of (5.51), except for some x for which $h(x) = 0$, is equivalent to the inequality

$$k_1' + k_2' x > \exp\left[(\theta - \theta_0)x\right]. \tag{5.52}$$

If $x_1 < x_2$, we may choose k_1' and k_2' to satisfy the simultaneous equations

$$\begin{aligned} k_1' + k_2' x_1 &= \exp\left[(\theta - \theta_0)x_1\right], \\ k_1' + k_2' x_2 &= \exp\left[(\theta - \theta_0)x_2\right], \end{aligned} \tag{5.53}$$

whereas, if $x_1 = x_2$, we may choose k_1' and k_2' so that

$$\begin{aligned} k_1' + k_2' x_1 &= \exp\left[(\theta - \theta_0)x_1\right], \\ k_2' &= (\theta - \theta_0) \exp\left[(\theta - \theta_0)x_1\right]. \end{aligned} \tag{5.54}$$

In either case the two-sided tests (5.33) with finite x_1 and x_2 are thus seen to be in the form of (5.51), thus completing the proof.

We now combine and extend Theorems 1 and 2 by allowing the loss function to satisfy certain general inequalities, similar to (5.27) for the one-sided problems. This gives a theorem of remarkable generality, which, however, is still a very special case of a theorem of Karlin (1957a). We assume that the loss functions satisfy the inequalities

$$\begin{aligned} L(\theta, a_1) - L(\theta, a_0) &\geq 0 \quad \text{if} \quad \theta_1 < \theta < \theta_2, \\ L(\theta, a_1) - L(\theta, a_0) &\leq 0 \quad \text{if} \quad \theta < \theta_1 \quad \text{or} \quad \theta > \theta_2, \end{aligned} \tag{5.55}$$

where $\theta_1 \leq \theta_2$ and where Θ is still presumed to be an interval on the real line. Such problems may be referred to as *two-sided problems*. Note that no restriction is placed on the loss functions for $\theta = \theta_1$ or $\theta = \theta_2$. We allow $\theta_1 = -\infty$ and $\theta_2 = +\infty$, so that these problems contain the one-sided problems as a particular case. If $\theta_1 < \theta < \theta_2$, we prefer to take action a_0, and, if $\theta < \theta_1$ or $\theta > \theta_2$, we prefer action a_1. The case $\theta_1 < \theta_2$ describes the situation in Theorem 1, whereas Theorem 2 can be interpreted as the case $\theta_1 = \theta_2$ in which $L(\theta_1, a_1) - L(\theta_1, a_0) > 0$.

Theorem 3. Suppose that the distribution of the random observable X has the form (5.34) and that the loss function satisfies the inequalities (5.55) for some $\theta_1 \leq \theta_2$. Then the class of two-sided tests (5.33) is essentially complete.

Proof. The difference in the risk functions of two tests ϕ' and ϕ is

$$R(\theta, \phi') - R(\theta, \phi) = [L(\theta, a_1) - L(\theta, a_0)][E_\theta \phi'(X) - E_\theta \phi(X)]. \tag{5.56}$$

Let ϕ be any test. If $\theta_1 < \theta_2$, the two-sided test ϕ' described in Theorem 1 is easily seen, using (5.36) and (5.56), to provide a test for which the difference (5.56) is nonpositive. If $\theta_1 = \theta_2$, the two-sided test ϕ' described in Theorem 2 provides a test for which (5.56) is nonpositive. In either case ϕ' is as good as ϕ, thus completing the proof.

It is worth noting that the test ϕ', which improves ϕ, depends on the loss function L only through the values of θ_1 and θ_2.

Theorem 4. Suppose that, in addition to the hypotheses of Theorem 3, there exist three numbers $\theta_1' < \theta_2' < \theta_3'$ such that $L(\theta_1', a_1) - L(\theta_1', a_0) < 0$, $L(\theta_2', a_1) - L(\theta_2', a_0) > 0$, and $L(\theta_3', a_1) - L(\theta_3', a_0) < 0$. Then, every two-sided test (5.33) is admissible.

The proof is an exercise.

Exercises

1. Let X have the Beta distribution $\mathcal{B}e(\theta, 1)$ and suppose that the loss function is in the form of (5.55), with $\theta_1 = 1$ and $\theta_2 = 2$. I am thinking of a test ϕ for which $E_{\theta_1} \phi(X) = .50$ and $E_{\theta_2} \phi(X) = .30$. Find a test ϕ' that is as good as the one I am thinking of.
2. Let X have the Beta distribution $\mathcal{B}e(\theta, 1)$ and let $\theta_1 = 1$. Show how to find a two-sided test ϕ for which $E_{\theta_1} \phi(X) = 0.10$ and

$$(d/d\theta) E_\theta \phi(X) \,|_{\theta=\theta_1} = 0.$$

3. Let X have a one-parameter exponential family of distributions (5.34). Let ϕ be any test. Verify (5.45). (See Exercise 3.5.2.)
4. Check that the following slight extension of the *moreover* part of Theorem 1 is valid. Assume that the distribution of the random observable X has form (5.34); let ϕ' be a two-sided test, let ϕ be any test, and let $\theta_1 < \theta_2$. Denote the difference in the power functions by $D(\theta) = E_\theta \phi'(X) - E_\theta \phi(X)$.
   (a) If $D(\theta_1) \leq 0$ and $D(\theta_2) \leq 0$, then $D(\theta) \leq 0$ for all θ between θ_1 and θ_2.
   (b) If $D(\theta_1) \leq 0$ and $D(\theta_2) \geq 0$, then $D(\theta) \geq 0$ for $\theta \geq \theta_2$.
   (c) If $D(\theta_1) \geq 0$ and $D(\theta_2) \leq 0$, then $D(\theta) \geq 0$ for $\theta \leq \theta_1$.
5. Using Exercise 4, prove Theorem 4.

5.4 Uniformly Most Powerful Unbiased Tests

The results of Section 5.3 may be applied to the classical problem of Neyman and Pearson, (1936, 1938) concerning uniformly most powerful unbiased tests. Here, we follow the approach of Lehmann and Scheffé (1950, 1955).

Definition 1. A size α test ϕ of $H_0:\theta \in \Theta_0$ against $H_1:\theta \in \Theta_1$ is said to be *unbiased* if

$$E_\theta \phi(X) \geq \alpha \quad \text{for all} \quad \theta \in \Theta_1. \tag{5.57}$$

Thus an unbiased test of size α has a power function less than or equal to α for all $\theta \in \Theta_0$ and greater than or equal to α for all $\theta \in \Theta_1$. Unbiasedness seems a reasonable requirement to place on a test: that the probability of rejecting H_0 when false is never smaller than the probability of rejecting H_0 when true. *A uniformly most powerful unbiased test of size* α is then a test which out of all unbiased tests of size α has a maximum value of the power at each $\theta \in \Theta_1$. In contrast to best unbiased estimates discussed in Section 3.6 uniformly most powerful unbiased tests are in general admissible (see Exercise 1).

We first treat the problem of finding a UMP unbiased test of a two-sided hypothesis concerning the parameter θ of a one-parameter exponential family of distributions (5.34). Consider testing $H_0:\theta_1 \leq \theta \leq \theta_2$ against $H_1:\theta < \theta_1$ or $\theta > \theta_2$. Because the power function of any test involving the exponential family of distributions is continuous (in fact analytic; see Lemma 3.5.3), any size α unbiased test of H_0 against H_1 must have power function equal to α at θ_1 and θ_2. Theorem 5.3.1 may then be interpreted to state that *any two-sided test* ϕ_0 *of the form*

$$\phi_0(x) = \begin{cases} 1 & \text{if} \quad x < x_1 \quad \text{or} \quad x > x_2, \\ \gamma_i & \text{if} \quad x = x_i, \qquad i = 1, 2, \\ 0 & \text{if} \quad x_1 < x < x_2, \end{cases} \tag{5.58}$$

where x_1, x_2, γ_1, *and* γ_2 *are chosen so that*

$$E_{\theta_1}\,\phi_0(X) = E_{\theta_2}\,\phi(X) = \alpha \tag{5.59}$$

is UMP unbiased. Such a test exists from Theorem 5.3.1, for $\phi(x) \equiv \alpha$ is an unbiased test.

Similarly, if we are testing $H_0':\theta = \theta_0$ against $H_1':\theta \neq \theta_0$, any unbiased test of size α must have a power function whose value at θ_0 is α and whose slope at θ_0 is zero. Theorem 5.3.2 implies that *any two-sided test* ϕ_0 *in the form of* (5.58), *where* x_1, x_2, γ_1, *and* γ_2 *are chosen so that*

$$E_{\theta_0}\,\phi(X) = \alpha \tag{5.60}$$

and

$$E_{\theta_0}X\,\phi(X) = \alpha E_{\theta_0}X, \tag{5.61}$$

is UMP unbiased. Such a test exists, because again $\phi(x) \equiv \alpha$ is an unbiased test. Equation 5.61 reflects the requirement that the power function have slope zero at θ_0. (See 5.45)

As an example, let $X \in \mathfrak{N}(\theta, 1)$. Because the distribution of X is continuous, the values of the γ_i in a test in the form of (5.58) have no effect on the power function; thus the γ_i may be neglected and tests in the form of (5.58) may be determined by specifying the x_i. Because the distribution of X is symmetric and θ is a location parameter, the power function of any test in the form of (5.58) must be symmetric about the point $(x_1 + x_2)/2$. Suppose that we are testing $H_0:\theta_1 < \theta < \theta_2$ against $H_1:\theta < \theta_1$ or $\theta > \theta_2$ and are seeking a UMP unbiased test of size α in the form (5.58). The test in the form of (5.58) for which $(x_1 + x_2)/2 = (\theta_1 + \theta_2)/2 = \bar{\theta}$ will have equal power at θ_1 and θ_2. Hence any test with $x_1 = \bar{\theta} - c$ and $x_2 = \bar{\theta} + c$ is UMP unbiased of its size. For a given value of α it is easy to find a number c so that this test has size α. Suppose, on the other hand, that we are testing $H_0':\theta = \theta_0$ against $H_1':\theta \neq \theta_0$. The power function of any test in the form of (5.58) will have slope zero at θ_0, provided $x_1 = \theta_0 - c$ and $x_2 = \theta_0 + c$. Such a test therefore is UMP unbiased of its size for testing H_0' against H_1'. Again, it is easy to choose c so that the test has size α.

UMP Unbiased Tests for k-Parameter Exponential Families. The analysis just given may be extended to situations in which the parameter space Θ is a subset of Euclidean k-dimensional space. Consider testing the hypothesis $H_0:\theta \in \Theta_0$ against $H_1:\theta \in \Theta_1$, where Θ_0 and Θ_1 are disjoint with $\Theta_0 \cup \Theta_1 = \Theta$. If the power function of every test is continuous (as for exponential families—see Lemma 3.5.3), every unbiased test of size α must have power function identically equal to α on the boundary $\Theta_B = \bar{\Theta}_0 \cap \bar{\Theta}_1$. (If $\theta \in \Theta_B$, then θ is a limit of points in Θ_0 and of points in Θ_1, so that for an unbiased test, ϕ, $E_\theta\,\phi(X)$ is a limit of numbers $\geq \alpha$ and of numbers $\leq \alpha$.) Such tests are called α-similar on the set Θ_B.

Definition 2. A test ϕ is said to be *α-similar on a set* Θ_B if $E_\theta\,\phi(X) = \alpha$ for all $\theta \in \Theta_B$. A test is said to be *similar on a set* Θ_B if it is α-similar on Θ_B for some α.

Thus, if the power function of every test is continuous, then every unbiased test of size α is α-similar on the boundary $\Theta_B = \bar{\Theta}_0 \cap \bar{\Theta}_1$.

Definition 3. A test ϕ_0 is said to be a *uniformly most powerful α-similar test* of $H_0:\theta \in \Theta_0$ against $H_1:\theta \in \Theta_1$, if out of all tests ϕ which are α-similar on the boundary $\Theta_B = \bar{\Theta}_0 \cap \bar{\Theta}_1$, ϕ_0 has maximum power at points $\theta \in \Theta_1$; that is, $E_\theta\,\phi_0(X) \geq E_\theta\,\phi(X)$ for all $\theta \in \Theta_1$.

Usually the task of finding UMP similar tests is easier analytically than the task of finding UMP unbiased tests. Yet it often turns out that tests which are UMP similar on the boundary are UMP unbiased also, as the Theorem 1 indicates.

Theorem 1. If for every test ϕ of $H_0:\theta \in \Theta_0$ against $H_1:\theta \in \Theta_1$ the power function $E_\theta\,\phi(X)$ is continuous in θ, if ϕ_0 is a UMP α-similar test of H_0 against H_1, and if ϕ_0 is of size α for testing H_0 against H_1, then ϕ_0 is UMP unbiased.

Proof. The test ϕ_0 is unbiased because it has size $\alpha (E_\theta\, \phi(X) \leq \alpha$ for all $\theta \in \Theta_0)$ and because it has power at least as great as the similar test $\phi(x) \equiv \alpha$; but, if ϕ is any unbiased test of size α, ϕ is α-similar on the boundary $\Theta_B = \overline{\Theta}_0 \cap \overline{\Theta}_1$, for its power function is continuous. Hence ϕ_0 is as good as ϕ, thus completing the proof.

We now turn our attention to the problem of finding UMP α-similar tests.

Suppose that T is a sufficient statistic for $\theta \in \Theta_B$ and that ϕ is a test for which $E(\phi(X) \mid T = t) = \alpha$ for almost all t. Then ϕ is an α-similar test on Θ_B, since for $\theta \in \Theta_B$, $E_\theta\, \phi(X) = E_\theta E(\phi(X) \mid T) = \alpha$.

Definition 4. Let T be a sufficient statistic for $\theta \in \Theta_B$. Then a test ϕ, which is α-similar on Θ_B, is said to have *Neyman structure* if

$$E(\phi(X) \mid T = t) = \alpha \tag{5.62}$$

for all t except perhaps for t in a set having probability zero under all distributions P_θ, $\theta \in \Theta_B$.

It is easy to construct similar tests with Neyman structure. Just define $\phi(x)$ for those x in the set $\{T(x) = t\}$ so that $E(\phi(x) \mid T = t) = \alpha$. If this is done for all values of t and the resulting function $\phi(x)$ is a test (that is, if it is measurable), then it is similar with Neyman structure. Sometimes, it is easy to show that a certain test is UMP out of the class of similar tests with Neyman structure, and therefore it is important to know when such a test is UMP out of the class of all similar tests. Theorem 2 provides an answer.

Theorem 2. If T is a boundedly complete sufficient statistic for $\theta \in \Theta_B$, then every test similar on Θ_B has Neyman structure.

Proof. If $E_\theta\, \phi(X) = \alpha$ for all $\theta \in \Theta_B$, then $E_\theta(E(\phi(X) \mid T)) = \alpha$ for all $\theta \in \Theta_B$. Hence, from the definition of bounded completeness (Section 3.6) $E(\phi(X) \mid T = t) = \alpha$ for almost all t, thus completing the proof.

The natural application of these theorems is to k-parameter exponential families of distributions having natural parametrization (3.59). Suppose that $X_1, \cdots, X_n$ is a sample from a k-parameter exponential family of distributions, $n \geq k$. We immediately reduce the problem to the consideration of the sufficient statistics $T_1, \cdots, T_k$ having density

$$f_{\mathbf{T}}(\mathbf{t} \mid \boldsymbol{\theta}) = c(\boldsymbol{\theta})\, h(\mathbf{t}) \exp\left(\sum_{i=1}^{k} \theta_i t_i\right), \tag{5.63}$$

as in Lemma 3.5.1. Consider the problem of finding a UMP unbiased test of the hypothesis $H_0 : \theta_1 \leq \theta_1^0$ against $H_1 : \theta_1 > \theta_1^0$. For each fixed θ_1, $(T_2, \cdots, T_k)$ is a sufficient statistic for $(\theta_2, \cdots, \theta_k)$ having density

$$f_{T_2,\cdots,T_k}(t_2, \cdots, t_k \mid \boldsymbol{\theta}) = c(\boldsymbol{\theta})\left[\int h(\mathbf{t}) \exp(\theta_1 t_1)\, dt_1\right] \exp\left(\sum_{i=2}^{k} \theta_i t_i\right). \tag{5.64}$$

The conditional density of T_1, given $T_2, \cdots, T_k$, is clearly

$$f_{T_1 \mid T_2 = t_2, \cdots, T_k = t_k}(t_1 \mid \theta_1) = \left[\int h(\mathbf{t}) \exp(\theta_1 t_1)\, dt_1\right]^{-1} h(\mathbf{t}) \exp(\theta_1 t_1), \tag{5.65}$$

a one-parameter exponential family of distributions independent of $\theta_2, \cdots, \theta_k$. This suggests finding the UMP (unbiased) test of $H_0 : \theta_1 \leq \theta_1^0$ against $H_1 : \theta_1 > \theta_1^0$ based on the conditional distribution (5.65) of T_1, given $T_2, \cdots, T_k$. From the results of Section 5.2 there exists a UMP size α test of H_0 against H_1, conditional on $T_2, \cdots, T_k$, of the form

$$\phi_0(t_1, \cdots, t_k) = \begin{cases} 1 & \text{if } t_1 > z(t_2, \cdots, t_k), \\ \gamma(t_2, \cdots, t_k) & \text{if } t_1 = z(t_2, \cdots, t_k), \\ 0 & \text{if } t_1 < z(t_2, \cdots, t_k), \end{cases} \tag{5.66}$$

where γ and z are functions of $t_2, \cdots, t_n$ such that

$$E_{\theta_1^0}(\phi_0(\mathbf{T}) \mid T_2, \cdots, T_k) = \alpha. \tag{5.67}$$

Given that $\phi_0(t_1, \cdots, t_k)$ is a test in the unconditional problem (that is, given that ϕ_0 is measurable), it is a UMP unbiased size α test of H_0 against H_1, as the following argument shows. Lemma 3.5.3 implies that the power function of any test is continuous, and, from Theorem 1 of this section, it is sufficient to show that ϕ_0 is a size α test that is UMP α-similar on the boundary $\Theta_B = \{\boldsymbol{\theta} : \theta_1 = \theta_1^0\}$. Let ϕ be any test α-similar on Θ_B. Then, because $(T_2, \cdots, T_k)$ is a complete (hence boundedly complete) sufficient statistic for $\boldsymbol{\theta} \in \Theta_B$, ϕ has Neyman structure; that is, $E_{\theta_1^0}(\phi(\mathbf{T}) \mid T_2, \cdots, T_k) = \alpha$, except perhaps for a set of probability zero. Out of all tests ϕ, for which $E_{\theta_1^0}(\phi(\mathbf{T}) \mid T_2, \cdots, T_k) = \alpha$, ϕ_0 has maximum conditional power for $\theta_1 > \theta_1^0$ and minimum conditional power for $\theta_1 \leq \theta_1^0$. Then, for $\theta_1 > \theta_1^0$

$$E_{\boldsymbol{\theta}}\, \phi_0(\mathbf{T}) = E_{\boldsymbol{\theta}}(E_{\theta_1}(\phi_0(\mathbf{T}) \mid T_2, \cdots, T_k))$$
$$\geq E_{\boldsymbol{\theta}}(E_{\theta_1}(\phi(\mathbf{T}) \mid T_2, \cdots, T_k)) = E_{\boldsymbol{\theta}}\, \phi(\mathbf{T}), \qquad (5.68)$$

showing that ϕ_0 is UMP α-similar. Since for $\theta_1 \leq \theta_1^0$,

$$E_{\boldsymbol{\theta}}\, \phi_0(\mathbf{T}) = E_{\boldsymbol{\theta}}(E_{\theta_1}(\phi_0(\mathbf{T}) \mid T_2, \cdots, T_k)) \leq \alpha,$$

ϕ_0 is a test of size α, implying that ϕ_0 is UMP unbiased.

As an example, consider testing the hypothesis $H_0 : \mu \leq 0$ against $H_1 : \mu > 0$, based on a sample $X_1, \cdots, X_n$ from a normal distribution with mean μ and variance σ^2:

$$f(x \mid \mu, \sigma^2) = \frac{1}{\sqrt{2\pi}\,\sigma} \exp\left[-(x - \mu)^2/2\sigma^2\right].$$

The joint distribution of the sufficient statistics $T_1 = \sum_1^n X_i$ and $T_2 = \sum_1^n X_i^2$ has form (5.63), with $k = 2$, $\theta_1 = \mu/\sigma^2$, and $\theta_2 = -(2\sigma^2)^{-1}$. Because the hypothesis H_0 is equivalent to the hypothesis that $\theta_1 \leq 0$, the UMP unbiased test has form (5.66), with $\gamma(t_2)$ omitted, for the distribution of T_1, given T_2, is continuous:

$$\phi_0(t_1, t_2) = \begin{cases} 1 & \text{if} \quad t_1 > z(t_2), \\ 0 & \text{if} \quad t_1 < z(t_2), \end{cases}$$

where $z(t_2)$ is chosen to satisfy (5.67) or, equivalently,

$$P_{\theta_1=0}\{T_1 > z(t_2) \mid T_2 = t_2\} = \alpha.$$

To find $z(t_2)$ note that when $\theta_1 = 0$ the conditional distribution of $X_1, \cdots, X_n$, given $T_2 = t_2$, is uniform on the sphere $\sum X_i^2 = t_2$, from the spherical symmetry of the joint distribution of $X_1, \cdots, X_n$. Thus $\sqrt{t_2}$ is a scale parameter for the conditional distribution of $\sum_1^n X_i$, given $T_2 = t_2$. It follows that $z(t_2)/\sqrt{t_2}$ must be constant; that is $z(t_2) = C\sqrt{t_2}$ for some constant C. Hence

$$\phi_0(t_1, t_2) = \begin{cases} 1 & \text{if} \quad \dfrac{t_1}{\sqrt{t_2}} > C, \\[2ex] 0 & \text{if} \quad \dfrac{t_1}{\sqrt{t_2}} < C, \end{cases}$$

where C may be chosen so that the unconditional test has size α. This test is, in fact, the usual t-test, more often written in the form

$$\phi_0(\mathbf{x}) = \begin{cases} 1 & \text{if} \quad \sqrt{n-1}\,\dfrac{\bar{x}}{s} > C', \\ 0 & \text{if} \quad \sqrt{n-1}\,\dfrac{\bar{x}}{s} < C', \end{cases} \tag{5.69}$$

where $\bar{x} = t_1/n$ and $s^2 = \sum(x_i - \bar{x})^2/n = (t_2/n) - \bar{x}^2$. To see this, we have merely to check that the inequality

$$\sqrt{n-1}\,\bar{x}/s = \frac{\sqrt{n-1}\,t_1}{\sqrt{nt_2 - t_1^2}} > C'$$

is equivalent to the inequality

$$\frac{t_1}{\sqrt{t_2}} > C = \frac{C'\sqrt{n}}{\sqrt{n-1+(C')^2}}.$$

The statistic $\sqrt{n-1}\,\bar{x}/s$ has a t_{n-1}-distribution when $\theta_1 = 0$, as may be seen by writing it in the form

$$\sqrt{n-1}\,\frac{\bar{x}}{s} = \frac{\sqrt{n}\,\bar{x}/\sigma}{\sqrt{n}\,s/\sqrt{n-1}\,\sigma},$$

where $\sqrt{n}\,\bar{x}/\sigma \in \mathfrak{N}(0, 1)$ and $(\sqrt{n}\,s/\sigma)^2 \in \chi^2_{n-1}$ are independent, and applying the definition of the t-distribution (Section 3.1). The constant C' in the test (5.69) may be chosen from the t-tables so that the size of the test is α.

UMP Unbiased Two-Sided Tests. We have seen that when testing $\theta_1 \le \theta_1^0$ against $\theta_1 > \theta_1^0$, where θ_1 is one parameter of a k-parameter exponential family of distributions (5.63), there exists a UMP unbiased test which may be described as follows. Pretend that $T_2, \cdots, T_k$ are given and treat the conditional problem of testing $\theta_1 \le \theta_1^0$ against $\theta_1 > \theta_1^0$ on the basis of the conditional distribution of T_1, given $T_2, \cdots, T_k$. This conditional distribution is a one-parameter exponential family of distributions so that a one-sided test in the form of (5.66) is UMP for each conditional problem. This test considered as a test in the unconditional problem is UMP unbiased (barring measurability difficulties).

This procedure of finding a UMP unbiased test works equally well on the two-sided problems of testing $H_0:\theta_1^{(1)} < \theta_1 < \theta_1^{(2)}$ against $H_1:\theta_1 < \theta_1^{(1)}$ or $\theta_1 > \theta_1^{(2)}$ or of testing $H_0':\theta_1 = \theta_1^{(0)}$ against $H_1':\theta_1 \neq \theta_1^{(0)}$. We pretend that $T_2, \cdots, T_k$ are given and treat the conditional problem of testing H_0 against H_1 or H_0' against H_1', based on the conditional distribution of T_1, given $T_2, \cdots, T_k$. Because this conditional distribution is a one-parameter exponential family, independent of $\theta_2, \cdots, \theta_k$, the two-sided test of the form

$$\phi_0(t_1, \cdots, t_n) = \begin{cases} 1 \quad \text{if} \quad t_1 < z_1(t_2, \cdots, t_k) \quad \text{or} \quad t_1 > z_2(t_2, \cdots, t_k), \\ \gamma_i(t_2, \cdots, t_k) \quad \text{if} \quad t_1 = z_i(t_2, \cdots, t_k), \qquad i = 1, 2 \\ 0 \quad \text{if} \quad z_1(t_2, \cdots, t_k) < t_1 < z_2(t_2, \cdots, t_k), \end{cases} \tag{5.70}$$

where z_1, z_2, γ_1, and γ_2 are chosen so that this test be unbiased for this conditional distribution, is UMP unbiased in the unconditional problem as well.

The supporting argument is an extension of the one given for the one-sided problem. First consider testing $H_0:\theta_1^{(1)} < \theta_1 < \theta_1^{(2)}$ against $H_1:\theta_1 < \theta_1^{(1)}$ or $\theta_1 > \theta_1^{(2)}$. If ϕ is any unbiased size α test of H_0 against H_1, ϕ is α-similar on the boundary $\Theta_B = \{\boldsymbol{\theta}:\theta_1 = \theta_1^{(1)} \text{ or } \theta_1 = \theta_1^{(2)}\}$. Any α-similar test on $\{\boldsymbol{\theta}:\theta_1 = \theta_1^{(i)}\}$ must have Neyman structure for $i = 1$ and 2, so that ϕ must satisfy

$$E_{\theta_1^{(1)}}(\phi(\mathbf{T}) \mid T_2, \cdots, T_k) = E_{\theta_1^{(2)}}(\phi(\mathbf{T}) \mid T_2, \cdots, T_k) = \alpha, \tag{5.71}$$

except perhaps for that set of probability zero. Out of all tests satisfying (5.71), however, the test ϕ_0 in the form of (5.70) has greatest conditional power for any $\boldsymbol{\theta} \in \Theta_1$ and smallest conditional power for any $\boldsymbol{\theta} \in \Theta_0$. This must then be true of the unconditional power, which proves that ϕ_0 is UMP unbiased.

For testing $H_0':\theta_1 = \theta_1^{(0)}$ against $H_1':\theta_1 \neq \theta_1^{(0)}$ any unbiased test ϕ of size α not only must be α-similar on the boundary $\Theta_B = \{\boldsymbol{\theta}:\theta_1 = \theta_1^{(0)}\}$ but must also have power function whose partial derivative with respect to θ_1 vanishes at $\theta_1 = \theta_1^{(0)}$. The condition that the partial derivative of $E_{\boldsymbol{\theta}}\,\phi(\mathbf{T})$ with respect to θ_1 must vanish at $\theta_1^{(0)}$ may be written as

$$E_{\boldsymbol{\theta}}((\phi(\mathbf{T}) - \alpha)T_1)\,|_{\theta_1=\theta_1^{(0)}} = 0 \tag{5.72}$$

(Exercise 2). Because $T_2, \cdots, T_k$ is a *complete* sufficient statistic for $\boldsymbol{\theta} \in \Theta_B$, any α-similar test ϕ on Θ_B, which also satisfies (5.72), must satisfy the equations

$$\begin{aligned} E_{\theta_1^{(0)}}(\phi(\mathbf{T}) \mid T_2, \cdots, T_k) &= \alpha, \\ E_{\theta_1^{(0)}}((\alpha - \phi(\mathbf{T}))T_1 \mid T_2, \cdots, T_k) &= 0, \end{aligned} \tag{5.73}$$

except perhaps for that set of probability zero, by the same argument used in Theorem 2. Out of all tests ϕ satisfying (5.73) the test ϕ_0 in the form of (5.70) has greatest conditional power for any $\boldsymbol{\theta} \in \Theta$. This must also be true of the unconditional power, proving that ϕ_0 is UMP unbiased for testing H_0' against H_1'.

Continuing the preceding example, consider testing $H_0'{:}\,\mu = 0$ against $H_1'{:}\,\mu \neq 0$ on the basis of a sample $X_1, \cdots, X_n$ from a normal distribution with unknown mean μ and unknown variance σ^2. The theory implies the existence of a UMP unbiased test of size α in the form of (5.70) with $k = 2$, $T_1 = \sum_1^n X_i$, and $T_2 = \sum_1^n X_i^2$, where $z_1(t_2)$ and $z_2(t_2)$ are chosen so that the conditions in (5.73) are satisfied (γ_1 and γ_2 may be put equal to zero). The lower condition in (5.73) may be satisfied when $z_1(t_2) = -z_2(t_2)$. The upper condition in (5.73) may be satisfied by choosing in addition $z_2(t_2) = C\sqrt{t_2}$ for some constant C. Analogous to the one-sided case, the resultant test is equivalent to the t-test:

$$\phi_0(\mathbf{x}) = \begin{cases} 1 & \text{if} \quad \left|\sqrt{n-1}\,\dfrac{\bar{x}}{s}\right| > C', \\[2ex] 0 & \text{if} \quad \left|\sqrt{n-1}\,\dfrac{\bar{x}}{s}\right| < C', \end{cases}$$

where C' is chosen to make the test have size α.

This general theory applies to the problem of finding UMP unbiased one-sided or two-sided tests for any linear combination of the parameters of a k-parameter exponential family. Thus, if $\vartheta = \sum_1^k C_i\theta_i$, where C_1, say, is not zero, then

$$\exp\left(\textstyle\sum_1^k \theta_i T_i\right) = \exp\left[\vartheta \frac{T_1}{C_1} + \textstyle\sum_2^k \theta_j \left(T_j - \dfrac{C_j}{C_1} T_1\right)\right],$$

so that by a change of variable from $(T_1, T_2, \cdots, T_k)$ to

$$[T_1/C_1,\ T_2 - (C_2/C_1)T_1,\ \cdots,\ T_k - (C_k/C_1)T_1]$$

we would deal with a k-parameter exponential family in which ϑ is one of the parameters. The following example, due to Hoel (1945), illustrates this idea.

Consider the problem of comparing the means of two Poisson distributions. Let $X \in \mathcal{P}(\lambda)$, $Y \in \mathcal{P}(\mu)$, with X and Y independent:

$$f_{X,Y}(x, y \mid \lambda, \mu) = \frac{1}{x!y!} \exp\left[-(\lambda + \mu) + (\log \lambda)x + (\log \mu)y\right]$$

for $x = 0, 1, \cdots$, and $y = 0, 1, \cdots$. We may find a UMP unbiased one-sided or two-sided test of a hypothesis concerning any linear combination, $C_1 \log \lambda + C_2 \log \mu$. In particular, we may find a UMP unbiased test of $\lambda = \mu$ (by testing $\log \lambda - \log \mu = 0$), of $\lambda = 2\mu$ ($\log \lambda - \log \mu = \log 2$), or of $\lambda \le \mu^2$ ($\log \lambda - 2 \log \mu \le 0$). This method, however, does not apply to finding, say, a UMP unbiased test of $\mu \le \lambda + 1$. To test $\lambda = k\mu$ for some constant k, we deal with the parameter $\vartheta = \log \lambda - \log \mu$ and test $\vartheta = \log k$. The preceding paragraph suggests transforming from (X, Y) to (X, U), where $U = X + Y$; therefore we search for the UMP unbiased test conditional on U. The marginal distribution of $X + Y$ is the Poisson distribution $\mathcal{P}(\lambda + \mu)$, and the conditional distribution of X, given $U = u$, is binomial $\mathcal{B}(u, p)$, where $p = \lambda/(\lambda + \mu)$,

$$f_{X\mid U=u}(x \mid p) = \frac{u!}{x!(u - x)!} p^x(1 - p)^{u-x}, \qquad x = 0, 1, \cdots, u.$$

The hypothesis $\lambda = \mu$ is equivalent to the hypothesis $p = \frac{1}{2}$ and the hypothesis $\lambda \le 2\mu$ is equivalent to the hypothesis $p \le \frac{2}{3}$. For either of these problems a UMP unbiased test which deals only with the above binomial distribution, given $X + Y = u$, may be found.

Exercises

1. A test is said to be unbiased if it is unbiased of some size α, $0 \le \alpha \le 1$. Suppose the loss function satisfies the inequalities $L(\theta, a_0) < L(\theta, a_1)$ for $\theta \in \Theta_0$ and $L(\theta, a_0) > L(\theta, a_1)$ for $\theta \in \Theta_1$. Show that if an unbiased test of $H_0{:}\theta \in \Theta_0$ against $H_1{:}\theta \in \Theta_1$ is admissible within the class of unbiased tests, it is admissible.

2. Let $\mathbf{T}$ have the k-parameter exponential distribution (5.63) and let ϕ be any test. Then

$$\frac{\partial}{\partial \theta_1} E_{\boldsymbol\theta}(\phi(\mathbf{T})) = E_{\boldsymbol\theta}(\phi(\mathbf{T})T_1) - E_{\boldsymbol\theta}\,\phi(\mathbf{T})E_{\boldsymbol\theta}\,T_1.$$

3. Let X and Y be random variables with joint density

$$f_{X,Y}(x, y \mid \lambda, \mu) = \lambda\mu e^{-\lambda x - \mu y} I_{(0,\infty)}(x)\, I_{(0,\infty)}(y).$$

Find a UMP unbiased test of size $\alpha = .20$ for testing
(a) $H_0:\lambda \le \mu + 1$ against $H_1:\lambda > \mu + 1$,
(b) $H_0:\lambda = \mu$ against $H_1:\lambda \ne \mu$,
(c) $H_0:\lambda \ge 2\mu$ against $H_1:\lambda < 2\mu$.

4. Let $X_1, \cdots, X_n$ be a sample of size n from $\mathcal{U}(\theta_1, \theta_2)$, the uniform distribution on the interval (θ_1, θ_2). Sufficient for θ_1 and θ_2 are $T_1 = \min X_i$ and $T_2 = \max X_i$. Consider testing $H_0:\theta_1 \le 0$ against $H_1:\theta_1 > 0$.
(a) Show that the conditional distribution of T_1, given $T_2 = t_2$, is the distribution of the minimum of a sample of size $n - 1$ from $\mathcal{U}(\theta_1, t_2)$.
(b) Find a UMP size α test of H_0 against H_1 based on the distribution in (a).
(c) Show that the unconditional test generated in (b) is a UMP unbiased size α test of H_0 against H_1.

5. Let X and Y be independent random variables with geometric distributions

$$f_{X,Y}(x, y \mid \theta_1, \theta_2) = (1 - \theta_1)(1 - \theta_2)\theta_1^x\theta_2^y,$$

$$x = 0, 1, \cdots, \quad y = 0, 1, \cdots,$$

where $0 < \theta_1 < 1$ and $0 < \theta_2 < 1$. Find a UMP unbiased test of size $\alpha = .20$ for testing
(a) $H_0:\theta_1 \le \theta_2$ against $H_1:\theta_1 > \theta_2$,
(b) $H_0:\theta_1 = \theta_2$ against $H_1:\theta_1 \ne \theta_2$.
For what functions $\varphi(\theta_1, \theta_2)$ do the methods of this section guarantee the existence of a UMP unbiased test of $H_0:\varphi(\theta_1, \theta_2) = 0$ against $H_1:\varphi(\theta_1, \theta_2) \ne 0$?

6. Let $X_1, \cdots, X_m$ and $Y_1, \cdots, Y_n$ be independent samples from $\mathfrak{N}(\mu, 1)$ and $\mathfrak{N}(\eta, 1)$, respectively. Find a UMP unbiased test of size α for testing $H_0:\mu \le \eta$ against $H_1:\mu > \eta$.

7. Let $X_1, \cdots, X_m$ and $Y_1, \cdots, Y_n$ be independent samples from $\mathfrak{N}(\mu, \sigma^2)$ and $\mathfrak{N}(\eta, \sigma^2)$, respectively. Show that the usual t-test: reject if

$$|\bar{X} - \bar{Y}| \ge s\sqrt{\frac{1}{m} + \frac{1}{n}}\, t_{m+n-2;\alpha/2},$$

where

$$s^2 = (m + n - 2)^{-1}\left(\sum_1^m (X_i - \bar{X})^2 + \sum_1^n (Y_i - \bar{Y})^2\right)$$

is UMP unbiased of size α for testing the hypothesis $\mu = \eta$.

5.5 Locally Best Tests

So far we have treated the testing of one-sided and two-sided problems for a real parameter when the distribution of the random observable X is sufficiently well behaved, that is, a one-parameter exponential family or a family with monotone likelihood ratio. In this section we devote a few words to the difficult problem of finding optimal (that is, admissible) tests for distributions without a monotone likelihood ratio. In such situations it is no longer true that the one-sided tests form an essentially complete class for the one-sided problems. The problem of finding optimal tests is not easily solved. Usually, Bayes procedures are admissible (Section 2.3) but difficult to compute. Also, the best test of some simple hypothesis, $h_0 \in H_0$, against some simple alternative, $h_1 \in H_1$, is usually admissible. Among the various other methods of arriving at admissible tests the locally best tests of Neyman and Pearson (1936, 1938) deserve special mention.

Locally Best Tests for One-Sided Problems. Consider the problem of testing the hypothesis $H_0:\theta \leq \theta_0$ against the alternative $H_1:\theta > \theta_0$. Assume that the distribution of the random observable X is such that the *power function*

$$\beta_\phi(\theta) = E_\theta\, \phi(X) \tag{5.74}$$

of any test ϕ admits one continuous derivative which may be passed beneath the integral (summation) sign:

$$\beta'_\phi(\theta) = \int \phi(x) \frac{\partial}{\partial\theta} f(x \mid \theta)\, dx. \tag{5.75}$$

In such a case a test ϕ_0 is said to be *a locally best test* for testing $H_0:\theta \leq \theta_0$ against $H_1:\theta > \theta_0$, if for any other test ϕ for which $\beta_\phi(\theta_0) = \beta_{\phi_0}(\theta_0)$ we have $\beta'_\phi(\theta_0) \leq \beta'_{\phi_0}(\theta_0)$. Thus a locally best test is one which has maximum slope at θ_0 out of all tests with the same size at θ_0.

A locally best test may be found with the help of the Neyman-Pearson lemma. According to this lemma, any test of the form

$$\phi_0(x) = \begin{cases} 1 & \text{if } \left.\dfrac{\partial}{\partial\theta} f(x \mid \theta)\right|_{\theta_0} > k\, f(x \mid \theta_0), \\ \gamma(x) & \text{if } \qquad = \\ 0 & \text{if } \qquad < \end{cases} \tag{5.76}$$

for some k and $0 \leq \gamma(x) \leq 1$, will have a maximum value of $\beta_\phi'(\theta_0)$ out of all tests ϕ for which $\beta_\phi(\theta_0) = \beta_{\phi_0}(\theta_0)$. Noting that

$$\frac{(\partial/\partial\theta)\, f(x \mid \theta) \mid_{\theta_0}}{f(x \mid \theta_0)} = \frac{\partial}{\partial\theta} \log f(x \mid \theta) \mid_{\theta_0} \tag{5.77}$$

for those values of x for which $f(x \mid \theta_0) \neq 0$, we see that any test of the form

$$\phi_0(x) = \begin{cases} 1 & \text{if } \left.\dfrac{\partial}{\partial\theta} \log f(x \mid \theta)\right|_{\theta_0} > k, \\ \gamma(x) & \text{if } \qquad\qquad = \\ 0 & \text{if } \qquad\qquad < \end{cases} \tag{5.78}$$

is also of the form (5.76), hence locally best at θ_0.

As an example, consider testing one-sided hypotheses for the median θ of a Cauchy distribution with known semi-interquartile range; that is, let $X_1, \cdots, X_n$ be a sample of size n from $\mathcal{C}(\theta, 1)$ and consider testing $H_0{:}\theta = 0$ (or $H_0{:}\theta \leq 0$) against $H_1{:}\theta > 0$. The joint density of $X_1, \cdots, X_n$ is

$$f(x_1, \cdots, x_n \mid \theta) = \frac{1}{\pi^n} \prod_{i=1}^{n} \frac{1}{1 + (x_i - \theta)^2}. \tag{5.79}$$

Using the Neyman-Pearson lemma, we can check that no best test of size α, $0 < \alpha < 1$, for testing $\theta = 0$ against $\theta = \theta_1 (\theta_1 > 0)$ can also be a best test of size α for testing $\theta = 0$ against $\theta = \theta_2$ $(\theta_2 > 0)$ when $\theta_2 \neq \theta_1$; in other words, a UMP test of size α, $0 < \alpha < 1$, for testing H_0 against H_1 does not exist. A locally best test may be found in (5.78), which in the present case says to reject H_0 if

$$\frac{\partial}{\partial\theta} \log f(x_1, \cdots, x_n \mid \theta) = \sum_{i=1}^{n} \frac{2(x_i - \theta)}{1 + (x_i - \theta)^2}, \tag{5.80}$$

evaluated at $\theta = 0$, is bigger than some constant k. The locally best test of size α is

$$\phi_0(x) = \begin{cases} 1 & \text{if } \displaystyle\sum_{i=1}^{n} \frac{2x_i}{1 + x_i^2} > k, \\ 0 & \qquad\qquad\qquad \leq \end{cases} \tag{5.81}$$

where k is chosen so that the test is of size α. To find the cutoff point k we can try to find the distribution of $\sum 2x_i/(1 + x_i^2)$. However, if n is large, we may use the normal approximation (see Exercise 1).

The optimum property of the test (5.81) may be stated as follows: given any other significantly different test ϕ_1 of the same size at θ_0, there exists an $\epsilon > 0$ such that $E_\theta\, \phi_0(X) > E_\theta\, \phi_1(X)$ for $0 < \theta < \epsilon$. This statement of the local optimality of the test (5.81) requires that the test satisfying (5.76) be unique, as it is in this case.

Unfortunately, the test (5.81) has what many would consider a serious drawback. Note that as $x_j \to \infty$, $x_j/(1 + x_j^2) \to 0$. Thus, if $\alpha < \frac{1}{2}$, so that the k in (5.81) is positive, the rejection region is a bounded set in n-dimensions. The probability that $(X_1, \cdots, X_n)$ will fall in any given bounded set goes to zero as $\theta \to \infty$, which implies that

$$\beta_{\phi_0}(\theta) \to 0 \quad \text{as} \quad \theta \to \infty.$$

Although ϕ_0 is good at detecting small departures from the null hypothesis, it is unsuccessful in detecting sufficiently large ones! In defense of ϕ_0, we can reiterate that a UMP test does not exist in this problem and that if a test is good for some θ it will probably be poor for some others.

Locally Best Unbiased Tests. The notion of a locally best test may be extended to testing the hypothesis $H_0'\!:\theta = \theta_0$ against $H_1'\!:\theta \neq \theta_0$ by specifying the slope of the power function at θ_0 and requiring the second derivative of the power function at θ_0 to be a maximum. Here we consider only unbiased tests with slope zero at θ_0.

Assume that the distribution of the random observable X is such that the power function $\beta_\phi(\theta) = E_\theta\, \phi(X)$ of any test ϕ admits two continuous derivatives which may be passed beneath the integral (summation) sign:

$$\begin{aligned} \beta_\phi'(\theta) &= \int \phi(x) \frac{\partial}{\partial\theta} f(x \mid \theta)\, dx, \\ \beta_\phi''(\theta) &= \int \phi(x) \frac{\partial^2}{\partial\theta^2} f(x \mid \theta)\, dx. \end{aligned} \tag{5.82}$$

In such situations a test ϕ_0 is said to be *locally best unbiased of size* α for testing $H_0':\theta = \theta_0$ against $H_1':\theta \neq \theta_0$, if out of all tests ϕ satisfying

$$\beta_\phi(\theta_0) = \alpha \tag{5.83}$$

and

$$\beta_\phi'(\theta_0) = 0, \tag{5.84}$$

the test ϕ_0 maximizes the value of the second derivative at θ_0, that is,

$$\beta_{\phi_0}''(\theta_0) \geq \beta_\phi''(\theta_0). \tag{5.85}$$

When such a test is unique, the optimum property of a locally best unbiased test may be stated as follows: If ϕ is any other unbiased test of size α, there exists a $\epsilon > 0$ such that $E_\theta\, \phi_0(X) > E_\theta\, \phi(X)$ for all θ for which $0 < |\theta - \theta_0| < \epsilon$.

A locally best unbiased test may be found by using the generalized Neyman-Pearson lemma. According to this lemma, any test of the form

$$\phi_0(x) = \begin{cases} 1 & \text{if } \left.\dfrac{\partial^2}{\partial\theta^2} f(x \mid \theta)\right|_{\theta_0} > k_1 f(x \mid \theta_0) + k_2 \left.\dfrac{\partial}{\partial\theta} f(x \mid \theta)\right|_{\theta_0}, \\ \gamma(x) & \text{if } \qquad = \\ 0 & \text{if } \qquad < \end{cases} \tag{5.86}$$

where k_1, k_2, and $0 \leq \gamma(x) \leq 1$ are chosen to satisfy (5.83) and (5.84), will be locally best unbiased of size α. Using (5.77) and noting that

$$\left.\frac{\partial^2}{\partial\theta^2}\frac{f(x \mid \theta)}{f(x \mid \theta_0)}\right|_{\theta_0} = \left.\frac{\partial^2}{\partial\theta^2}\log f(x \mid \theta)\right|_{\theta_0} + \left(\left.\frac{\partial}{\partial\theta}\log f(x \mid \theta)\right|_{\theta_0}\right)^2 \tag{5.87}$$

for those values of x for which $f(x \mid \theta_0) \neq 0$, we see that any test of the form

$$\phi_0(x) = \begin{cases} 1 & \text{if } \left.\dfrac{\partial^2}{\partial\theta^2}\log f(x \mid \theta)\right|_{\theta_0} + \left(\left.\dfrac{\partial}{\partial\theta}\log f(x \mid \theta)\right|_{\theta_0}\right)^2 \\ & \qquad > k_1 + k_2 \left.\dfrac{\partial}{\partial\theta}\log f(x \mid \theta)\right|_{\theta_0}, \\ \gamma(x) & \text{if } \qquad = \\ 0 & \text{if } \qquad < \end{cases} \tag{5.88}$$

is also in the form of (5.86), hence is locally best unbiased of size α provided (5.83) and (5.84) are satisfied.

Continuing the example of the Cauchy distribution, suppose we are testing $H_0'{:}\theta = 0$ against $H_1'{:}\theta \neq 0$ on the basis of a sample of size n, $X_1, \cdots, X_n$ from $\mathcal{C}(\theta, 1)$. To compute the locally best unbiased test of H_0' against H_1' we need, in addition to (5.80),

$$\frac{\partial^2}{\partial\theta^2} \log f(x_1, \cdots, x_n \mid \theta) = \sum_{i=1}^{n} 2\, \frac{(x_i - \theta)^2 - 1}{(1 + (x_i - \theta)^2)^2}. \tag{5.89}$$

Then (5.88) becomes

$$\phi_0(x) = \begin{cases} 1 & \text{if} \quad V > k_1 + k_2 U, \\ \gamma(x) & \text{if} \quad\quad = \\ 0 & \text{if} \quad\quad < \end{cases} \tag{5.90}$$

where

$$U = \sum_{i=1}^{n} \frac{2x_i}{1 + x_i^2},$$

$$V = \sum_{i=1}^{n} 2\, \frac{x_i^2 - 1}{(1 + x_i^2)^2} + \left(\sum_{i=1}^{n} \frac{2x_i}{1 + x_i^2}\right)^2. \tag{5.91}$$

It would appear difficult to find k_1 and k_2 so that (5.83) and (5.84) would be satisfied; however, a simple device may be used to show that equation (5.84) can be satisfied if, and essentially only if, $k_2 = 0$. This device uses the symmetry of the Cauchy distribution but is otherwise quite general. The key fact is that because of this symmetry the distribution of (U, V) in the plane is symmetric about the line $U = 0$: if R is a region in the half plane $U > 0$ and Я is its reflection across the line $U = 0$, $Я = \{(u, v){:}(-u, v) \in R\}$, then

$$P_0\{(U, V) \in R\} = P_0\{(U, V) \in Я\}$$

for if $(U(X_1, \cdots, X_n), V(X_1, \cdots, X_n)) \in R$, then

$$\begin{aligned}(U(-X_1, \cdots, -X_n),\ & V(-X_1, \cdots, -X_n)) \\ &= (-U(X_1, \cdots, X_n), V(X_1, \cdots, X_n)) \in Я,\end{aligned}$$

and the Cauchy distribution $\mathcal{C}(0, 1)$ is symmetric about 0. Now note that condition (5.84) for ϕ_0 may be written as

$$E_0\, \phi_0(X) U = 0. \tag{5.92}$$

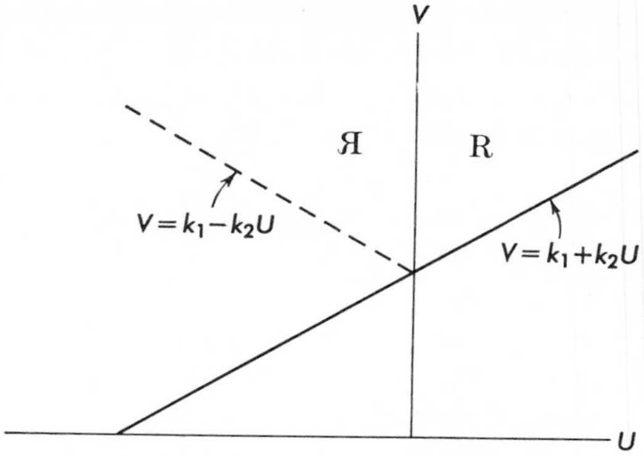

Fig. 5.3

If $k_2 = 0$, then obviously this equation is satisfied; for $E_0 \mid U \mid < \infty$ and $E_0\,\phi_0(X)U$ is the integral of U over the set $V > k_1$. Conversely, if $k_2 > 0$, the integral of U over the sets $\mathrm{R} = \{V > k_1 + k_2U,\ U > 0\}$ and Я $= \{V > k_1 - k_2U,\ U < 0\}$ cancel (Fig. 5.3), leaving $E_0\,\phi_0(X)U$ as the integral of U over the set

$$\{k_1 + k_2U < V \le k_1 - k_2U,\ U < 0\}.$$

On this set U is negative, so that $E_0\,\phi_0(X)U < 0$ unless

$$P\{k_1 + k_2U < V \le k_1 - k_2U,\ U < 0\} = 0;$$

but if this probability is zero, so that $E_0\,\phi_0(X)U = 0$, we may put $k_2 = 0$ in the test ϕ_0 without changing the size. Similarly, for $k_2 < 0$. In other words, if $E_0\,\phi_0(X)U = 0$, we may as well put $k_2 = 0$ in the test ϕ_0. Hence the test ϕ_0 (putting $\gamma(x) = 0$) becomes

$$\phi_0(x) = \begin{cases} 1 & \text{if} \quad V > k_1, \\ 0 & \text{if} \quad V \le k_1, \end{cases} \tag{5.93}$$

where k_1 is chosen so that (5.83) is satisfied.

Exercises

1. (a) Let $X \in \mathcal{C}(0, 1)$ and let $U = 2X/(1 + X^2)$. Show that $EU = 0$ and $\operatorname{Var} U = \frac{1}{2}$. *Hint.* Show that

$$EU^2 = \frac{8}{\pi}\int_0^{\pi/2} \cos^2\theta \sin^2\theta\, d\theta.$$

(b) For $n = 100$ and $\alpha = .05$ find the approximate cutoff point k for the size α test in the form of (5.81)

2. Let $X_1, \cdots, X_n$ be a sample from the double exponential distribution with location parameter θ:

$$f(x_1, \cdots, x_n \mid \theta) = \frac{1}{2^n} \exp\left(-\sum_{i=1}^{n} |x_i - \theta|\right).$$

One continuous derivative of the power function of any test exists and may be placed beneath the integral sign.

(a) Show that no UMP size α test of $H_0{:}\theta \leq 0$ against $H_1{:}\theta > 0$ exists for $n \geq 2$ and $0 < \alpha < 1$.

(b) Find the locally best test of H_0 against H_1. Show how to find the approximate cutoff point to achieve a test of size α. Show how to find the power function of the test.

(c) Find a test ϕ for which $\beta_\phi(\theta)$ does not have a second derivative at some point.

3. Let $X_1, \cdots, X_n$ be a sample from the logistic distribution with location parameter θ,

$$f(x_1, \cdots, x_n \mid \theta) = \prod_{i=1}^{n} [2(1 + \cosh(x_i - \theta))]^{-1}.$$

(a) Find the locally best test of $H_0{:}\theta = 0$ against $H_1{:}\theta > 0$.

(b) Find the locally best unbiased test of $H_0{:}\theta = 0$ against $\bar{H}{:}\theta \neq 0$.

4. Give a rough plot of the power function of the test (5.81) when $k > 0$; when $k < 0$; when $k = 0$.

5. Let X, Y have a bivariate Poisson distribution with probability mass function

$$f_{X,Y}(x, y \mid \theta) = \exp(\theta - \lambda_1 - \lambda_2) \sum_{k=0}^{\min(x,y)} \frac{\theta^k (\lambda_1 - \theta)^{x-k} (\lambda_2 - \theta)^{y-k}}{k!(x-k)!(y-k)!},$$

$$x = 0, 1, 2, \cdots, \quad y = 0, 1, 2, \cdots, \qquad (5.94)$$

where $\lambda_1 > 0$, $\lambda_2 > 0$ are assumed known, and $0 \leq \theta \leq \min(\lambda_1, \lambda_2)$. The marginal distributions of X and Y are Poisson, $\mathcal{P}(\lambda_1)$ and $\mathcal{P}(\lambda_2)$, respectively, and the hypothesis to be tested is $H_0{:}\theta = 0$, that X and Y are independent.

(a) Show that the locally best test of H_0 against $H_1{:}\theta > 0$ is of this form: reject H_0 if $(X - \lambda_1)(Y - \lambda_2)$ is too large.

(b) What is the form of the locally best test of H_0 against H_1 based on a sample of size n, (X_1, Y_1), (X_2, Y_2), $\cdots$, (X_n, Y_n), from the distribution (5.94)?

5.6 Invariance in Hypothesis Testing

The purpose of this section is to investigate the special problems that arise in the application of the invariance principle to hypothesis testing problems.

We assume that $\mathcal{A} = \{a_0, a_1\}$, and that there are subsets Θ_0 and Θ_1 of the parameter space Θ, with $\Theta_0 \cap \Theta_1 = \emptyset$ and $\Theta_0 \cup \Theta_1 = \Theta$, such that the loss function is in the form of

$$L(\theta, a_0) = \begin{cases} c_{00} & \text{if} \quad \theta \in \Theta_0, \\ c_{01} & \text{if} \quad \theta \in \Theta_1, \end{cases}$$

$$L(\theta, a_1) = \begin{cases} c_{10} & \text{if} \quad \theta \in \Theta_0, \\ c_{11} & \text{if} \quad \theta \in \Theta_1, \end{cases} \tag{5.95}$$

where $c_{00} < c_{10}$ and $c_{11} < c_{01}$. Suppose that the decision problem is invariant under a group $\mathcal{G}$ of transformations on the sample space $\mathfrak{X}$ and let $\bar{\mathcal{G}}$ and $\tilde{\mathcal{G}}$ be the groups of transformations induced on Θ and $\mathcal{A}$, respectively. This means first of all that *the distributions are invariant*; that is

$$E_\theta\, \phi(g(X)) = E_{\bar{g}(\theta)}\, \phi(X) \tag{5.96}$$

for every (measurable) real-valued function ϕ (4.2). The condition that *the loss be invariant*, $L(\bar{g}(\theta), \tilde{g}(a)) = L(\theta, a)$ for all $\theta \in \Theta$, $a \in \mathcal{A}$, and $g \in \mathcal{G}$ may be expressed in simpler terms. In the general case in which the c_{ij} are all different it is clear that the loss is invariant if and only if $\tilde{g}(a) = a$ for all $\tilde{g} \in \tilde{\mathcal{G}}$ and

$$\bar{g}(\Theta_0) = \Theta_0 \quad \text{and} \quad \bar{g}(\Theta_1) = \Theta_1 \tag{5.97}$$

for all $\bar{g} \in \bar{\mathcal{G}}$. In hypothesis testing problems we take $\tilde{\mathcal{G}}$ to consist of the identity transformation and replace the condition on the invariance of the loss function with (5.97). In other words, *a group $\mathcal{G}$ of transformations on $\mathfrak{X}$ leaves a hypothesis-testing problem invariant if $\mathcal{G}$ leaves both families of distributions $\{P_\theta, \theta \in \Theta_0\}$ and $\{P_\theta, \theta \in \Theta_1\}$ invariant.*

A behavioral decision rule (that is, *a test*) *ϕ is invariant if for all $x \in \mathfrak{X}$ and $g \in \mathcal{G}$*

$$\phi(g(x)) = \phi(x). \tag{5.98}$$

This is easily seen by applying the definition in Section 4.2 and recalling that $\tilde{g}(a) = a$ for hypothesis testing problems.

The main simplification in the invariance principle applied to hypothesis-testing problems (in fact, to any invariant decision problem for which $\tilde{\mathcal{G}}$ consists of the identity transformation) is that it is possible to describe the invariant decision rules as rules that are functions of a special statistic, the maximal invariant.

Definition. Let $\mathfrak{X}$ be a space and $\mathcal{G}$, a group of transformations on $\mathfrak{X}$. A function $T(x)$ on $\mathfrak{X}$ is said to be a *maximal invariant* with respect to $\mathcal{G}$ if

(a) (*Invariance*) $T(g(x)) = T(x)$ for all $x \in \mathfrak{X}$ and $g \in \mathcal{G}$;
(b) (*Maximality*) $T(x_1) = T(x_2)$ implies $x_1 = g(x_2)$ for some $g \in \mathcal{G}$.

Condition (a) says that T is constant on orbits, whereas condition (b) says that each orbit gets a different value, that is, T distinguishes orbits.

Examples of Maximal Invariants

1. *Location invariance.* Suppose that $\mathfrak{X} = E_n$ and that $\mathcal{G}$ consists of the translations

$$g_c(x_1, \cdots, x_n) = (x_1 + c, \cdots, x_n + c).$$

Then the $(n - 1)$-tuple $T(\mathbf{x}) = (x_1 - x_n, \cdots, x_{n-1} - x_n)$ is a maximal invariant. It is clearly invariant; that is, $T(g_c(\mathbf{x})) = T(\mathbf{x})$. If $T(\mathbf{x}) = T(\mathbf{x}')$, so that $x_i - x_n = x_i' - x_n'$ for all i, then $g_c(\mathbf{x}') = \mathbf{x}$, where $c = x_n - x_n'$, thus proving maximality.

2. *Scale invariance.* Suppose that $\mathfrak{X} = E_n$ and that $\mathcal{G}$ consists of scale changes

$$g_c(x_1, \cdots, x_n) = (cx_1, \cdots, cx_n),$$

where $c > 0$. Let

$$z^2 = \textstyle\sum_1^n x_i^2;$$

then

$$T(\mathbf{x}) = \begin{cases} 0 & \text{if } z = 0, \\ \left(\dfrac{x_1}{z}, \cdots, \dfrac{x_n}{z}\right) & \text{if } z \neq 0, \end{cases}$$

is a maximal invariant. It is clearly invariant. Suppose $T(\mathbf{x}) = T(\mathbf{x}')$; if $T(\mathbf{x}) = T(\mathbf{x}') = 0$, then $x_i = x_i' = 0$; otherwise, $T(\mathbf{x}) = T(\mathbf{x}') \neq 0$ implies $x_i/z = x_i'/z'$ for all i, so that $g_c(\mathbf{x}') = \mathbf{x}$ where $c = z/z'$.

3. *Location and scale invariance.* Suppose that $\mathfrak{X} = E_n$ and $\mathcal{G}$ is the group of location and scale changes,

$$g_{a,b}(x_1, \cdots, x_n) = (ax_1 + b, \cdots, ax_n + b)$$

with $a > 0$. Let

$$\bar{x} = n^{-1} \textstyle\sum_1^n x_i \quad \text{and} \quad s^2 = n^{-1} \sum_1^n (x_i - \bar{x})^2.$$

Then

$$T(\mathbf{x}) = \begin{cases} 0 & \text{if } s = 0, \\ \left(\dfrac{x_1 - \bar{x}}{s}, \ldots, \dfrac{x_n - \bar{x}}{s}\right) & \text{if } s \neq 0, \end{cases}$$

is a maximal invariant.

4. *The orthogonal group.* Suppose that $\mathfrak{X} = E_n$ and that $\mathcal{G}$ is the group of orthogonal transformations on $\mathfrak{X}$. A maximal invariant is $T(\mathbf{x}) = \sum_1^n x_i^2$.

5. *The permutation group.* Suppose that $\mathfrak{X} = E_n$ and $\mathcal{G}$ is the group of permutations of n symbols. Then, if $g \in \mathcal{G}$, $g(x_1, \cdots, x_n)$ just permutes the subscripts of the x_i. A maximal invariant is the vector of order statistics $T(\mathbf{x}) = (x_{(1)}, x_{(2)}, \cdots, x_{(n)})$, where $x_{(1)} =$ the smallest of the x_i, $x_{(2)} =$ the next smallest, $\cdots$, $x_{(n)} =$ the largest. Clearly, $T(\mathbf{x})$ is invariant, and if $T(\mathbf{x}) = T(\mathbf{x}')$ then $\mathbf{x}'$ is a permutation of $\mathbf{x}$, proving maximality.

6. *The group of functional transformations.* Suppose that $\mathfrak{X} = E_n$ and let $\mathcal{G}$ be the group of transformations $g_\varphi(x_1, \cdots, x_n) = (\varphi(x_1), \cdots, \varphi(x_n))$, where φ is a continuous increasing function from the real line onto the real line. A maximal invariant is the set of ranks $T(\mathbf{x}) = (r_1, \cdots, r_n)$, where r_i is the rank of x_i. The smallest of the x_i has rank 1, the next smallest has rank 2, $\cdots$, and the largest has rank n. If there are ties, the tied observations may be assigned the average of the available ranks. To be specific

$$r_i = \sum_{j=1}^n I_{(-\infty, x_i]}(x_j) - \frac{1}{2} \sum_{j=1, j\neq i}^n I_{(x_i]}(x_j).$$

Theorem 1. Let $\mathfrak{X}$ be a space, let $\mathcal{G}$ be a group of transformations on

$\mathfrak{X}$, and let $T(x)$ be a maximal invariant with respect to $\mathcal{G}$. A function $\phi(x)$ is invariant with respect to $\mathcal{G}$ if, and only if, ϕ is a function of $T(x)$.

Proof. *If.* If $\phi(x) = h(T(x))$, then

$$\phi(g(x)) = h(T(g(x))) = h(T(x)) = \phi(x),$$

so that ϕ is invariant.

Only if. Suppose that ϕ is invariant and that $T(x_1) = T(x_2)$. Then for some $g \in \mathcal{G}$, $x_1 = g(x_2)$, so that $\phi(x_1) = \phi(x_2)$, thus completing the proof.

The use of this theorem in hypothesis-testing problems is as follows. Suppose a problem is invariant under a group $\mathcal{G}$ and let T be a maximal invariant with respect to $\mathcal{G}$. Instead of restricting attention to the class of invariant tests, we may restrict attention to the conceptually simpler class of tests which are functions of T, for these two classes of tests are equivalent.

Theorem 4.2.1, which states that the risk function of an invariant rule is constant on orbits of Θ, may be given the following more precise formulation for hypothesis testing problems.

Theorem 2. Suppose a statistical decision problem is invariant under a group $\mathcal{G}$ and let $v(\theta)$ be a maximal invariant under $\bar{\mathcal{G}}$. Then if $T(x)$ is invariant under $\mathcal{G}$, the distribution of $T(X)$ depends only on $v(\theta)$.

Proof. If $\theta' = \bar{g}(\theta)$, then $P_{\theta'}\{T(X) \in A\} = P_\theta\{T(g(X)) \in A\} = P_\theta\{T(X) \in A\}$, so that $P_\theta\{T(X) \in A\}$ is constant on orbits. Any such function is a function of the maximal invariant from Theorem 1.

Examples of UMP Invariant Tests

1. *Location parameter.* Let $X_1, X_2, \cdots, X_n$ be random variables whose distribution depends on a parameter θ. We are interested in testing the hypothesis H_0 that the joint distribution function of $X_1, \cdots, X_n$ satisfies

$$H_0: F_{X_1,\cdots,X_n}(x_1, \cdots, x_n \mid \theta) = G_0(x_1 - \theta, \cdots, x_n - \theta)$$

against the hypothesis H_1 that

$$H_1: F_{X_1,\cdots,X_n}(x_1, \cdots, x_n \mid \theta) = G_1(x_1 - \theta, \cdots, x_n - \theta),$$

where G_0 and G_1 are known distribution functions. In each hypothesis θ may vary over the entire real line. This problem is invariant under change of location, $g_c(x_1, \cdots, x_n) = (x_1 + c, \cdots, x_n + c)$, and a maximal invariant is $\mathbf{Y} = (Y_1, \cdots, Y_{n-1})$, where $Y_i = X_i - X_n$, $i = 1, \cdots, n - 1$. The class of all invariant tests is then the class of all tests that are functions of $\mathbf{Y}$. The distribution of $\mathbf{Y}$ under hypothesis H_0 or H_1 is independent of θ. Therefore, in using the principle of invariance, we are testing a simple hypothesis against a simple alternative, a problem that can be solved by the Neyman-Pearson lemma. The resulting test will be UMP invariant. (See Exercises 1 and 2.)

2. *One-sided t-test.* Let $X_1, \cdots, X_n$ be a sample from the normal distribution with unknown mean μ and unknown variance σ^2 and consider testing the hypothesis $H_0: \mu \leq 0$ against the alternative $H_1: \mu > 0$. We have seen that the UMP unbiased test of H_0 against H_1 is the usual Student's t-test. This test is also UMP invariant, but the argument now depends on the noncentral t-distribution.

A sufficient statistic for the problem just stated is (U, V), where $U = \sqrt{n}\,\bar{X}$ and $V = \sum(X_i - \bar{X})^2$. The statistics U and V are independent, $U/\sigma \in \mathfrak{N}(\sqrt{n}\,\mu/\sigma, 1)$, and $V/\sigma^2 \in \chi^2_{n-1}$. This problem is invariant under changes of scale, hence under the group of transformations $g_c(U, V) = (cU, c^2V)$, where $c > 0$. A maximal invariant for this problem is $T = U/\sqrt{V/(n-1)}$, which, by definition, has a noncentral t-distribution with $n - 1$ degrees of freedom and noncentrality parameter $\delta = \sqrt{n}\,\mu/\sigma$. The problem of finding a UMP invariant test of H_0 against H_1 becomes the problem of finding a UMP test based on T of the hypothesis $H_0: \delta \leq 0$ against $H_1: \delta > 0$. It may be shown that the distribution of T has monotone likelihood ratio, hence that the usual Student's t-test, to reject H_0 when T is too large, is a UMP invariant test of H_0 against H_1. An alternative way of showing that this is the UMP invariant test is by the Neyman-Pearson lemma. It is sufficient to show that $f_T(t \mid \delta)/f_T(t \mid 0)$ is an increasing function of t when $\delta > 0$. This is left as an exercise (Exercise 4).

3. *Two-sided t-test.* Suppose, in the preceding example, that we are to test $H_0': \mu = 0$ against $H_1': \mu \neq 0$. The problem is now invariant under multiplication of all the observations by minus one, hence under the group of transformations $g_c(U, V) = (cU, c^2V)$, where $c \neq 0$. This time, a maximal invariant is $|T| = |U|/\sqrt{V/(n-1)}$, and the problem is to test $H_0': \delta = 0$ against $H_1': |\delta| > 0$. The density of $|T|$ is $f_{|T|}(t \mid \delta) = f_T(t \mid \delta) + f_T(-t \mid \delta)$ for $t > 0$. Again, it can be shown that $f_{|T|}(t \mid \delta)/f_{|T|}(t \mid 0)$ is increasing in t for $t > 0$ (Exercise 5), so that the usual two-sided t-test (reject H_0' when $|T|$ is too large) is a UMP invariant test of H_0' against H_1'.

Maximal Invariants in Invariant Constant Risk Decision Problems. With the notion of a maximal invariant, we may put into proper perspective the arguments of Section 4.7 which lead to the Pitman estimate. Consider a statistical decision problem invariant under a group $\mathcal{G}$ for which $\bar{\mathcal{G}}$ is transitive so that the risk function of an invariant rule is independent of the true state of nature, θ. Let $\mathbf{X}$ be the observable random quantity and let $\mathbf{Y}$ be a maximal invariant. Because the distribution of $\mathbf{Y}$ does not depend on θ, we may pretend that $\mathbf{Y}$ was observed first and then that $\mathbf{X}$ was chosen from the conditional distribution of $\mathbf{X}$, given $\mathbf{Y}$. This conditional decision problem, given $\mathbf{Y}$, is still invariant under $\mathcal{G}$. Suppose, for simplicity, that invariant randomized rules and invariant behavioral rules are equivalent so that attention may be restricted to the nonrandomized invariant rules (Lemma 4.5.1). Then, if we are lucky, we may be able to find, for (almost) all $\mathbf{y}$, a best invariant nonrandomized decision rule for the conditional problem, given $\mathbf{Y} = \mathbf{y}$, that is, an invariant rule d_0 that minimizes, among all invariant nonrandomized rules, d,

$$E_{\theta_0}(L(\theta_0, d(\mathbf{X})) \mid \mathbf{Y} = \mathbf{y}), \tag{5.99}$$

for some fixed θ_0. This should be relatively easy because, $\mathbf{Y}$ being a maximal invariant, the distribution of $\mathbf{X}$ given $\mathbf{Y}$ lies on a given orbit, and an invariant rule is determined on such an orbit once its value is chosen at one point. In other words, each of these conditional problems is a one-variable minimum problem. The resulting rule $d_0(\mathbf{X})$ is, in general, best invariant for the original unconditional problem as well. A heuristic proof may be given along the lines of (4.52). (See Exercise 12.)

Exercises

1. Let X_1 and X_2 be independent identically distributed random variables, let H_0 be the hypothesis that the underlying distribution is $\mathcal{N}(\theta, 1)$, a normal distribution with variance one, and let H_1 be the hypothesis that the underlying distribution is $\mathcal{C}(\theta, 1)$, a Cauchy distribution with semi-interquartile range 1. Show that

$$\phi(x_1, x_2) = \begin{cases} 1 & \text{if} \quad |x_1 - x_2| > K, \\ 0 & \text{if} \quad |x_1 - x_2| < K, \end{cases}$$

is a UMP invariant test of H_0 against H_1.

2. Let $X_1, \cdots, X_n$ be independent identically distributed, let H_0 be the hypothesis that the common distribution is $\mathcal{N}(\theta, 1)$, and let H_1

be the hypothesis that the common distribution has density

$$f(x \mid \theta) = \exp\{-\exp[(x-\theta)] + (x-\theta)\}.$$

Find the form of the UMP invariant tests of H_0 against H_1.

3. Let $X_1, \cdots, X_n$ be independent identically distributed $\mathfrak{N}(\theta, \sigma^2)$, and consider testing $H_0: \sigma^2 \geq 1$ against $H_1: \sigma^2 < 1$. Show that

$$\phi(\mathbf{x}) = \begin{cases} 1 & \text{if} \quad \sum (x_i - \bar{x})^2 < K, \\ 0 & \text{if} \quad \sum (x_i - \bar{x})^2 > K, \end{cases}$$

is a UMP invariant test of H_0 against H_1.

4. Show that the ratio $f_T(t \mid \delta)/f_T(t \mid 0)$ is a nondecreasing function of t when $\delta > 0$, where $f_T(t \mid \delta)$ is the density of the noncentral t-distribution with ν degrees of freedom and noncentrality parameter δ.

5. Let $f_{|T|}(t \mid \delta)$ denote the density of the distribution of $|T|$ where T has a noncentral t-distribution with ν degrees of freedom and noncentrality parameter δ. Show that $f_{|T|}(t \mid \delta)/f_{|T|}(t \mid 0)$ is an increasing function of t for $t > 0$.

6. Let $X_1, \cdots, X_m$ and $Y_1, \cdots, Y_n$ be independent samples from populations with distributions $\mathfrak{N}(\mu, \sigma^2)$ and $\mathfrak{N}(\eta, \sigma^2)$, respectively. Find UMP invariant tests for testing $H_0: \mu \leq \eta$ against $H_1: \mu > \eta$ and for testing $H_0': \mu = \eta$ against $H_1': \mu \neq \eta$.

7. Let X and Y have joint distribution given in Exercise 5.4.3 and let $\theta = \lambda/\mu$. (a) Show that the problem of testing $H_0: \theta \leq 1$ against $H_1: \theta > 1$ is invariant under the group $\mathcal{G}$ of transformations $g_c(x, y) = (cx, cy)$, $c > 0$, and find a UMP invariant test of size α. (b) Show that the problem of testing $H_0': \theta = 1$ against $H_1': \theta \neq 1$ is invariant *in addition* under the transformation $g(x, y) = (y, x)$, and find a UMP invariant test of size α.

Ans. (a) Reject if $T = Y/X > (1 - \alpha)/\alpha$; (b) reject if $\max(T, 1/T) > (2 - \alpha)/\alpha$.

8. Let X and Y be independent with densities $f_X(x \mid \lambda) = f(x - \lambda)$ and $f_Y(y \mid \mu) = f(y - \mu)$. Let $Z = X - Y$ and $\theta = \lambda - \mu$. (a) Suppose that, for $\theta > 0$, $f_Z(z \mid \theta)/f_Z(z \mid 0)$ is nondecreasing in z and find a UMP invariant size α test of $H_0: \theta \leq \theta_0$ against $H_1: \theta > \theta_0$. (b) Suppose that, for $\theta \neq 0$, $f_{|Z|}(z \mid \theta)/f_{|Z|}(z \mid 0)$ is nondecreasing in z and find a UMP invariant size α test of $H_0': \theta = \theta_0$ against $H_1': \theta \neq \theta_0$.

9. Let $\Theta = \{(\Delta, \nu): \Delta \text{ real}, 1 \le \nu \le n, \nu \text{ an integer}\}$ and let the distribution of $X_1, \cdots, X_n$, given $\theta = (\Delta, \nu)$, be as independent random variables with $X_i \in \mathfrak{N}(0, 1)$ for $i \neq \nu$ and $X_\nu \in \mathfrak{N}(\Delta, 1)$. Test the hypothesis $H_0: \Delta = 0$ against alternatives $H_1: \Delta > 0$ or $\bar{H}: \Delta \neq 0$.
(a) Show that this problem is invariant under the group of permutations of $(X_1, \cdots, X_n)$ and that the distribution of the maximal invariant $Y_1 = X_{(1)}, \cdots, Y_n = X_{(n)}$ (the order statistics) has density

$$f_{Y_1,\cdots,Y_n}(y_1, \cdots, y_n \mid \Delta)$$

$$= (2\pi)^{-n/2} \exp\left(-\tfrac{1}{2}\sum y_i^2 - \tfrac{1}{2}\Delta^2\right)(n-1)! \sum_{\nu=1}^{n} \exp(\Delta y_\nu)$$

for $y_1 < y_2 < \cdots < y_n$ and zero elsewhere.
(b) Show that the locally best invariant test of H_0 against H_1 is to reject H_0 if $\sum_1^n X_i$ is too large.
(c) Show that the locally best unbiased invariant test of H_0 against $\bar{H}$ is to reject H_0 if $\sum_1^n X_i^2$ is too large.

10. Let $\Theta = \{(\Delta, (\nu_1, \cdots, \nu_n)): \Delta \ge 0, (\nu_1, \cdots, \nu_n) \text{ is a permutation of } (1, \cdots, n)\}$ and let the distribution of $X_1, \cdots, X_n$, given $\theta = (\Delta, (\nu_1, \cdots, \nu_n))$, be as independent random variables with gamma distributions, $X_i \in \mathcal{G}[\alpha, \beta \exp(\Delta b_{\nu_i})]$, where $\alpha > 0, \beta > 0$, and $b_1, \cdots, b_n$ are known real numbers, such that $\sum_1^n b_i > 0$. Test the hypothesis $H_0: \Delta = 0$ against the alternative $H_1: \Delta > 0$.
(a) Show that this problem is invariant under the group of permutations of $(X_1, \cdots, X_n)$ and that the distribution of the maximal invariant $Y_1 = X_{(1)}, \cdots, Y_n = X_{(n)}$ (the order statistics) has density

$$f_{Y_1,\cdots,Y_n}(y_1, \cdots, y_n \mid \Delta)$$

$$= \frac{(\prod y_i)^{\alpha-1} \exp(-\alpha\Delta \sum b_i)}{\Gamma(\alpha)^n \beta^{n\alpha}} \sum{}^{*} \exp\left\{-\frac{1}{\beta}\sum_{i=1}^{n} y_i \exp(-\Delta b_{\nu_i})\right\}$$

for $y_1 < y_2 < \cdots < y_n$ and zero elsewhere, where $\sum^*$ denotes the sum over all permutations $(\nu_1, \cdots, \nu_n)$ of $(1, \cdots, n)$.
(b) Show that the locally best invariant test of H_0 against H_1 is to reject H_0 when $\sum_1^n X_i$ is too large. Note that the local optimality of this test is uniform in the b_i, provided $\sum_1^n b_i > 0$.

11. Let $X_1, \cdots, X_n$ be as in Exercise 10, with β unknown (so that $\Theta = \{(\Delta, \beta, (\nu_1, \cdots, \nu_n)): \Delta \text{ real}, \beta > 0, (\nu_1, \cdots, \nu_n) \text{ a permutation of } (1, \cdots, n)\}$. Suppose we are interested in testing $H_0: \Delta = 0$ against $\bar{H}: \Delta \neq 0$. In this problem it is assumed without loss of generality that $\sum b_i = 0$.

(a) Show that this problem is invariant under permutations of the subscripts and under change of scale.

(b) Let $Y_1, \cdots, Y_n$ be the order statistics and make the change of variable $Z_1 = Y_1/Y_n, \cdots, Z_{n-1} = Y_{n-1}/Y_n, W = Y_n$. Show that the distribution of $Z_1, \cdots, Z_{n-1}, W$ has density

$$f(z_1, \cdots, z_{n-1}, w) = \frac{(\prod z_i)^{\alpha-1} w^{n\alpha-1}}{\Gamma(\alpha)^n \beta^{n\alpha}} \sum{}^* \exp\left\{-\frac{w}{\beta}\sum_1^n z_i \exp(-\Delta b_{\nu_i})\right\}$$

for $w > 0$ and $0 < z_1 < \cdots < z_{n-1} < 1$, and zero elsewhere, where for notational convenience we let $z_n = 1$. Show that the distribution of $Z_1, \cdots, Z_{n-1}$ has density

$$h(z_1, \cdots, z_{n-1}) = \frac{\Gamma(n\alpha)}{\Gamma(\alpha)^n} (\prod z_i)^{\alpha-1} \sum{}^* [\sum_1^n z_i \exp(-\Delta b_{\nu_i})]^{-n\alpha}$$

for $0 < z_1 < \cdots < z_{n-1} < 1$ and zero elsewhere.

(c) Show that any invariant test has slope zero at $\Delta = 0$ by showing that $\partial h/\partial\Delta\,|_{\Delta=0} \equiv 0$.

(d) Show that in such circumstances the locally best invariant test of H_0 against $\bar{H}$ rejects H_0 when $\partial^2 h/\partial\Delta^2\,|_{\Delta=0} > kh|_{\Delta=0}$ for some constant k. Derive this test and show that it is independent of $b_1, \cdots, b_n$, provided $\sum_1^n b_i^2 \neq 0$. *Ans.* Reject H_0 if $\sum_1^n X_i^2/(\sum_1^n X_i)^2$ is too large. For extensions, see Ferguson (1961).

12. Consider a statistical decision problem in which $\Theta = \{-1, +1\}$, $\mathcal{A}$ is the real line, and $L(\theta, a) = (\theta - a)^2$. The observable random variables X_1 and X_2 are independent with Poisson distributions, $X_1 \in \mathcal{P}(1)$ and $X_2 \in \mathcal{P}(2)$ if $\theta = -1$, whereas $X_1 \in \mathcal{P}(2)$ and $X_2 \in \mathcal{P}(1)$ if $\theta = +1$. Show that this problem is invariant under the group $\mathcal{G}$ generated by $g(x_1, x_2) = (x_2, x_1)$, and find $\bar{\mathcal{G}}$ and $\tilde{\mathcal{G}}$. Find a maximal invariant under $\mathcal{G}$. Using the method of the last paragraph of this section, find a best invariant decision rule. *Ans.* $d(x_1, x_2) = (2^{x_1} - 2^{x_2})/(2^{x_1} + 2^{x_2})$.

5.7 The Two-Sample Problem

The two-sample problem may be stated roughly as follows. We are given samples $X_1, \cdots, X_m$ and $Y_1, \cdots, Y_n$ from two populations with respective distribution functions F and G. We wish to test the hypothesis $H_0:F(x) \geq G(x)$ for all x or $H_0':F(x) = G(x)$ for all x against the alternatives $H_1:F(x) \leq G(x)$ for all x but $F \neq G$. In other words, we wish to test the hypothesis that the distribution of the X_i is situated to the *left* of the distribution of the Y_i (or that these distributions are equal) against the alternative that the distribution of the X_i is situated to the *right* of the distribution of the Y_i. If it is known that the two distributions are normal with a common variance, we may use the UMP invariant test (Exercise 5.6.6), which is also UMP unbiased, the usual Student's t-test. However, if the distributions are not normal, then this test may be very poor.

Here we consider the nonparametric two-sample problem in which we do not make specific assumptions on the distributions F and G other than continuity.

We assume that H_0 contains all pairs (F, G) of continuous distribution functions for which $F(x) \geq G(x)$ for all x and H_1 contains all pairs (F, G) of continuous distribution functions for which $F(x) \leq G(x)$ for all x, yet $F(x) \neq G(x)$ for some x. We restrict attention to tests that are functions of the sufficient statistics for the problem $(X_{(1)}, \cdots, X_{(m)})$ and $(Y_{(1)}, \cdots, Y_{(n)})$, the two vectors of order statistics for the two samples. Under this formulation the problem is invariant under the group $\mathcal{G}$ of all transformations $g_\varphi(x_{(1)}, \cdots, x_{(m)}, y_{(1)}, \cdots, y_{(n)}) = (\varphi(x_{(1)}), \cdots, \varphi(x_{(m)}), \varphi(y_{(1)}), \cdots, \varphi(y_{(n)}))$, where φ is a continuous increasing function from the real line onto the real line. We have seen in Section 5.6 that a maximal invariant is the set of ranks $(R_1, \cdots, R_m, S_1, \cdots, S_n)$, where $R_1 < R_2 < \cdots < R_m$ are the ranks of the order statistics of the X_i (in the total sample of $m + n$ observations) and where $S_1, \cdots, S_n$ are the ranks of the order statistics of the Y_i. Because $S_1, \cdots, S_n$ is determined from knowledge of $R_1, \cdots, R_m$, a simpler maximal invariant is $(R_1, \cdots, R_m)$. The principle of invariance has led us to the consideration of *rank tests*, that is, tests that depend only on the ranks $(R_1, \cdots, R_m)$. For this problem the terms "invariant test" and "rank test" are synonymous.

As will be seen, there does not exist a UMP rank test of H_0 (or H_0') against H_1. The two tests most commonly applied to this problem are the following:

1. The Wilcoxon Test. Reject H_0 if $\sum_{i=1}^m R_i$ is too large. This is commonly called the rank-sum or the Mann-Whitney test.

2. The Fisher-Yates Test. Reject H_0 if $\sum_{i=1}^{m} \mathfrak{N}_{(Ri)}^{m+n}$ is too large, where $\mathfrak{N}_{(r)}^{N}$ is the expected value of the rth order statistic of a sample of size N from the distribution $\mathfrak{N}(0, 1)$.

We now show that these tests have certain optimum properties in the class of all rank (invariant) tests. (Admissibility of these tests in the class of all tests is a difficult and unsolved problem.) Under the hypothesis $H_0'{:}F(x) = G(x)$ for all x, all $\binom{m+n}{m}$ permutations of the X's among the Y's are equally likely so that

$$P\{R_1 = r_1, \cdots, R_m = r_m \mid H_0'\} = \binom{m+n}{m}^{-1}. \tag{5.100}$$

For the problems we are considering it is necessary to compute the joint distribution of $R_1, \cdots, R_m$ under the general hypothesis of arbitrary, absolutely continuous F and G. This is done by considering the following more general problem. Let $Z_1, Z_2, \cdots, Z_N$ be independent, absolutely continuous random variables with densities $f_1, f_2, \cdots, f_N$, respectively, and let $Q_1, Q_2, \cdots, Q_N$ denote the ranks of $Z_1, Z_2, \cdots, Z_N$; then

$$P\{Q_1 = q_1, \cdots, Q_N = q_N\} = \int_S \cdots \int f_1(z_1) \cdots f_N(z_N)\, dz_1 \cdots dz_N, \tag{5.101}$$

where S is the set of $(z_1, \cdots, z_N)$ for which the rank of z_i is q_i, $i = 1, \cdots, N$. Letting $(\nu_1, \cdots, \nu_N)$ denote the inverse permutation of $(q_1, \cdots, q_N)$, so that the rank of z_{ν_i} is i, $i = 1, \cdots, N$, we may write $S = \{(z_1, \cdots, z_N){:}z_{\nu_1} < z_{\nu_2} < \cdots < z_{\nu_N}\}$. The transformation $v_i = z_{\nu_i}$ or $z_i = v_{q_i}$ changes the integral (5.101) into the form

$$P\{Q_1 = q_1, \cdots, Q_N = q_N\} = \int_S \cdots \int f_1(v_{q_1}) \cdots f_N(v_{q_N})\, dv_1 \cdots dv_N, \tag{5.102}$$

where $S = \{(v_1, \cdots, v_N){:}v_1 < v_2 < \cdots < v_N\}$. Let h be any density on the real line that is positive whenever any of the f_i is positive. Because the joint distribution of the order statistics $V_{(1)}, \cdots, V_{(N)}$ of a sample of size N from a population with density h has the form

$$f_{V_{(1)},\cdots,V_{(N)}}(v_1, \cdots, v_N) = N!\, h(v_1) \cdots h(v_N)\, I_S(v_1, \cdots, v_N),$$

where $S = \{(v_1, \cdots, v_N){:}v_1 < v_2 < \cdots < v_N\}$, probability (5.102) may be written

$$P\{Q_1 = q_1, \cdots, Q_N = q_N\}$$

$$= \int_S \cdots \int \frac{f_1(v_{q1}) \cdots f_N(v_{qN})}{N!h(v_{q1}) \cdots h(v_{qN})} N!h(v_1) \cdots h(v_N)\, dv_1 \cdots dv_N$$

$$= \frac{1}{N!} E \frac{f_1(V_{(q1)}) \cdots f_N(V_{(qN)})}{h(V_{(q1)}) \cdots h(V_{(qN)})}, \tag{5.103}$$

where the expectation is taken with respect to the distribution for which $V_{(1)}, \cdots, V_{(N)}$ are order statistics from a sample of size N from density h. If we specialize this result to the problem, $N = m + n$, $f_1 = \cdots = f_m = f$, and $f_{m+1} = \cdots = f_{m+n} = g$, f and g being the densities of the distributions F and G, respectively, and take $h = g$, as we may do, provided $g(x) = 0$ implies $f(x) = 0$, we arrive at the following theorem, a special case of a result of Hoeffding (1951).

Theorem 1. Let $X_1, \cdots, X_m$ and $Y_1, \cdots, Y_n$ be samples from populations with densities f and g, respectively, and let $R_1, \cdots, R_m$ denote the ordered ranks of the X's in the total sample of $m + n$ observations. Then, if $g(x) = 0$ implies $f(x) = 0$,

$$P\{R_1 = r_1, \cdots, R_m = r_m\} = \binom{m+n}{m}^{-1} E \prod_{i=1}^{m} \frac{f(V_{(r_i)})}{g(V_{(r_i)})}, \tag{5.104}$$

where $V_{(1)}, \cdots, V_{(m+n)}$ are the order statistics of a sample of size $m + n$ from a population with density g.

Under the null hypothesis: $f = g$, (5.104) reduces to (5.100). If we use (5.104), the power function of any test may be computed at any alternative for which $g(x) = 0$ implies $f(x) = 0$. Under this condition, we may write $F(x) = \psi(G(x))$, where ψ is an absolutely continuous nondecreasing function from the interval $[0, 1]$ onto the interval $[0, 1]$. The densities satisfy the relation $f(x) = \psi'(G(x))\, g(x)$, so that (5.104) becomes

$$P\{R_1 = r_1, \cdots, R_m = r_m\} = \binom{m+n}{m}^{-1} E \prod_{i=1}^{m} \psi'(G(V_{(r_i)})). \tag{5.105}$$

The distribution of $G(V_{(1)}), \cdots, G(V_{(m+n)})$ is the same as the distribution of $U_{(1)}, \cdots, U_{(m+n)}$, the order statistics of a sample of size $m + n$ from a uniform distribution on the interval $[0, 1]$ (see Corollary 5.3.1).

Hence the distribution of $R_1, \cdots, R_m$ depends on F and G only through the function ψ. The power of any rank test is constant on the set

$$\{(F, G): F(x) = \psi(G(x))\}.$$

We shall derive optimum rank tests against specific alternatives, but it should be remembered that the resulting tests are optimum against a much larger class of alternatives.

We proceed to find tests that are admissible out of the class of rank tests by finding tests that are optimal against specific alternatives. Consider, first, a fixed one-parameter exponential family of distributions with density

$$f(x \mid \theta) = c(\theta)\, h(x)\, e^{\theta x}, \tag{5.106}$$

and let us find the best test based on the ranks of the simple hypothesis $f(x) = g(x) = f(x \mid \theta_0)$ against the simple alternative $f(x) = f(x \mid \theta_1)$, $g(x) = f(x \mid \theta_0)$, where $\theta_0 < \theta_1$. The best rank tests, given by the Neyman-Pearson lemma, are found to be

$$\phi(\mathbf{r}) = \begin{cases} 1 & \text{if} \quad E \exp\left[(\theta_1 - \theta_0) \sum_{i=1}^{m} V_{(r_i)}\right] > K, \\ \gamma & \qquad\qquad\qquad\qquad\qquad\qquad = \\ 0 & \qquad\qquad\qquad\qquad\qquad\qquad < \end{cases} \tag{5.107}$$

Unfortunately, the rejection region (5.107) is difficult to find analytically. In the case of the exponential distribution it can be found rather easily (Exercises 1 and 2).

Another method of finding admissible rank tests is to search for locally optimal tests. Suppose we fix $g(x) = f(x \mid \theta_0)$, using the one-parameter exponential family of (5.106) and consider testing $f(x) = f(x \mid \theta_0)$ against $f(x) = f(x \mid \theta)$ for values of θ tending to θ_0 from above. The locally best test may be found by using (5.78) in the form

$$\phi(\mathbf{r}) = \begin{cases} 1 & \text{if} \quad \left.\dfrac{\partial}{\partial \theta} \log P_\theta\{R_1 = r_1, \cdots, R_m = r_m\}\right|_{\theta=\theta_0} > K, \\ \gamma & \qquad\qquad\qquad\qquad\qquad\qquad\qquad\qquad = \\ 0 & \qquad\qquad\qquad\qquad\qquad\qquad\qquad\qquad < \end{cases} \tag{5.108}$$

or, equivalently,

$$\phi(\mathbf{r}) = \begin{cases} 1 & \text{if} \quad \sum_{i=1}^{m} EV_{(r_i)} > K, \\ \gamma & \qquad\qquad\qquad = \\ 0 & \qquad\qquad\qquad < \end{cases} \tag{5.109}$$

If $f(x \mid \theta_0)$ is taken to be the density of the normal distribution with mean $\theta_0 = 0$ and variance one, this test immediately specializes to the Fisher-Yates test. To obtain the Wilcoxon test, we need a distribution $f(x \mid \theta_0)$ for which the expected value of the rth order statistic $EV_{(r)}$ is linear in r. This is true for the uniform distribution; in fact, if $V_{(r)}$ is the rth order statistic of a sample of size N from the $\mathcal{U}(0, 1)$ distribution $EV_{(r)} = r/(N + 1)$ (Exercise 3). Hence the Wilcoxon test is locally best for testing $\theta = 0$ against $\theta > 0$ for the truncated exponential distribution, $f(x \mid \theta) = c(\theta)\, e^{\theta x} I_{(a,b)}(x)$, since $f(x \mid 0)$ is the uniform distribution, $\mathcal{U}(a, b)$.

An alternative procedure for obtaining locally best rank tests is to consider location parameter families of distributions. Suppose we fix $g(x)$ and consider testing $f(x) = g(x)$ against $f(x) = g(x - \theta)$ for values of θ tending to zero from above. The locally best test may be found by (5.108), which in this case leads to the test in the form of

$$\phi(\mathbf{r}) = \begin{cases} 1 & \text{if} \quad -\sum_{i=1}^{m} E\left[\dfrac{g'(V_{(r_i)})}{g(V_{(r_i)})}\right] > K \\ \gamma & \qquad\qquad\qquad\qquad = \\ 0 & \qquad\qquad\qquad\qquad < \end{cases} \tag{5.110}$$

For the normal distribution $g(x) = (2\pi)^{1/2} \exp(-x^2/2)$ this again leads to the Fisher-Yates test. For the logistic distribution $g(x) = e^x/(1 + e^x)^2$ this test becomes the Wilcoxon test (Exercise 4).

Exercises

1. Let $X_1, X_2, \cdots, X_N$ be a sample from a distribution with density $f(x \mid \theta) = \theta e^{\theta x} I_{(-\infty,0)}(x)$ and let $V_{(1)} < V_{(2)} < \cdots < V_{(N)}$ denote the order statistics. Show that $Y_1 = V_{(1)} - V_{(2)}$, $Y_2 = V_{(2)} - V_{(3)}, \cdots,$ $Y_{N-1} = V_{(N-1)} - V_{(N)}$, $Y_N = V_{(N)}$ are completely independent random variables and that Y_j has density $j\theta e^{j\theta x} I_{(-\infty,0)}(x)$ for $j = 1, 2, \cdots, N$.

2. Using Exercise 1, the expectation in (5.107) may be computed by noticing that $\sum_{i=1}^{m} V_{(r_i)}$ may be written as a linear combination of the Y_j and using the fact that the expectation of the product of independent random variables is the product of the expectations. Show, in the particular case $\theta_0 = 1$ and $\theta_1 = 2$, that (5.107) becomes

$$\phi(\mathbf{r}) = \begin{cases} 1 & \text{if } r_1(r_2+1)\cdots(r_m+m-1) > K, \\ \gamma & \qquad\qquad\qquad\qquad\qquad\quad = \\ 0 & \qquad\qquad\qquad\qquad\qquad\quad < \end{cases}$$

Show that this test is a most powerful rank test of $H_0: F = G$ against the alternatives $H_1: F = G^2$. (Use the remark after Theorem 1.)

3. Let $V_{(1)} < V_{(2)} < \cdots < V_{(N)}$ be the order statistics of a sample of size N from the uniform distribution on the interval (0, 1). Show that $EV_{(r)} = r/(N+1)$.

4. Show for the logistic distribution $g(x) = e^x/(1 + e^x)^2$ that $-g'(x)/g(x) = 2G(x) - 1$. Hence the test (5.110) becomes

$$\phi(\mathbf{r}) = \begin{cases} 1 & \text{if } \sum_{i=1}^{m} E\,G(V_{(r_i)}) > K, \\ \gamma & \qquad\qquad\qquad\quad = \\ 0 & \qquad\qquad\qquad\quad < \end{cases}$$

But, because $G(V_{(1)}), G(V_{(2)}), \cdots, G(V_{(m+n)})$ are distributed as the order statistics from a $\mathcal{U}(0, 1)$-distribution, the preceding exercise reduces this test to the Wilcoxon test.

5. Find the locally best rank test of the hypothesis

$$f(x) = g(x) = e^{-|x|}/2$$

(the double exponential distribution) against alternatives

$$g(x) = \frac{e^{-|x|}}{2} \qquad \text{and} \qquad f(x) = \frac{e^{-|x-\theta|}}{2}, \ \theta > 0.$$

For $m + n = N$ large show that this test is approximately

$$\phi(\mathbf{r}) = \begin{cases} 1 & \text{if} \quad \sum_{i=1}^{m} \Phi\left(\frac{r_i - N/2}{\sqrt{N/4}}\right) > K, \\ \gamma & \qquad\qquad = \\ 0 & \qquad\qquad < \end{cases}$$

where Φ is the cumulative distribution function of the standard normal distribution.

6. Consider the family of distributions with density

$$f(x \mid \theta) = c(\theta)\, h(x) I_{(\theta,\infty)}(x).$$

Show that the uniformly (in $h(x)$ and θ) most powerful rank test of the hypothesis

$$f(x) = g(x) = f(x \mid \theta_0)$$

against the alternatives

$$g(x) = f(x \mid \theta_0) \qquad \text{and} \qquad f(x) = f(x \mid \theta),\ \theta > \theta_0,$$

is

$$\phi(\mathbf{r}) = \begin{cases} 1 & \text{if} \quad r_1 > K \\ \gamma & \qquad = \\ 0 & \qquad < \end{cases}$$

7. What is the locally best rank test of $F = G$ against $F = (e^{\theta G} - 1)/(e^{\theta} - 1)$, $\theta > 0$? Of $F = G$ against $F = G/(e^{\theta}(1 - G) + G)$, $\theta > 0$?

5.8 Confidence Sets

In this section the direct relation between families of tests and confidence sets is examined.

Definition 1. Let $\{S(x)\}$ be a family of subsets of the parameter space Θ, where $x \in \mathfrak{X}$, the sample space. $\{S(x)\}$ is said to be a *family of confidence sets at confidence level* $1 - \alpha$, if

$$P_\theta\{S(X) \text{ contains } \theta\} = 1 - \alpha \qquad \text{for all } \theta \in \Theta. \tag{5.111}$$

We may easily construct families of confidence sets at a given level $1 - \alpha$ by the following method. Let $A(\theta_0)$ be the acceptance region of a size α nonrandomized test φ of the hypothesis $H_0{:}\theta = \theta_0$ against any alternative; that is,

$$\varphi(x) = \begin{cases} 1 & \text{if} \quad x \notin A(\theta_0), \\ 0 & \text{if} \quad x \in A(\theta_0), \end{cases}$$

where $P_{\theta_0}\{X \in A(\theta_0)\} = 1 - \alpha$. If we consider the sets $A(\theta)$, so formed for every $\theta \in \Theta$, we have a family of acceptance regions, each a subset of $\mathfrak{X}$, such that $P_\theta\{X \in A(\theta)\} = 1 - \alpha$. We then define the sets

$$S(x) = \{\theta{:}x \in A(\theta)\}, \tag{5.112}$$

so that, whenever θ is in $S(x)$, x is in $A(\theta)$ and vice versa. Then

$$P_\theta\{S(X) \text{ contains } \theta\} = P_\theta\{X \in A(\theta)\} = 1 - \alpha,$$

so that $\{S(x)\}$ is a family of confidence sets at level $1 - \alpha$.

Conversely, given a family $\{S(x)\}$ of confidence sets at level $1 - \alpha$, if one defines a family of subsets $A(\theta)$ of $\mathfrak{X}$ by

$$A(\theta) = \{x{:}\theta \in S(x)\}, \tag{5.113}$$

then $A(\theta_0)$ is the acceptance region of a size α test of the hypothesis $H_0{:}\theta = \theta_0$.

We have defined what should perhaps be called nonrandomized confidence sets and have seen the equivalence to a family of nonrandomized tests. More generally, we could define randomized confidence sets as certain objects equivalent to a family of randomized tests. For our purposes, this is an unnecessary refinement. The main ideas are contained in the simpler but less general notion of nonrandomized confidence sets, to which we restrict attention.

When choosing a family of confidence sets among the superabundance of such, a suitable criterion to use is to prefer small probability of covering false values; that is, $P_{\theta'}\{S(X) \text{ covers } \theta\}$ is to be small if $\theta' \neq \theta$. Optimal properties of the family of tests generally carry over to optimal properties of the family of confidence sets, as in the following theorem.

Theorem 1. Let $A(\theta_0)$ be the acceptance region of a UMP test of size α of the hypothesis $H_0:\theta = \theta_0$ against $H_1:\theta \in \bar{H}(\theta_0)$. Then $S(x)$ defined by (5.112) minimizes the probability $P_{\theta'}\{S(X)$ contains $\theta\}$ for all $\theta' \in \bar{H}(\theta)$, among all level $1 - \alpha$ families of confidence sets.

Proof. Let $\{S'(x)\}$ be any other level $1 - \alpha$ family of confidence sets and let $A'(\theta) = \{x:\theta \in S'(x)\}$. Then $P_\theta\{X \in A'(\theta)\} = P_\theta\{S'(X)$ contains $\theta\} = 1 - \alpha$, so that, for every $\theta_0 \in \Theta$, $A'(\theta_0)$ is the acceptance region of a size α test of the hypothesis $H_0:\theta = \theta_0$. Because $A(\theta_0)$ is the acceptance region of a UMP size α test of $H_0:\theta = \theta_0$ against $H_1:\theta \in \bar{H}(\theta_0)$, then, for any $\theta' \in \bar{H}(\theta_0)$, $P_{\theta'}\{S(X)$ contains $\theta_0\} = P_{\theta'}\{X \in A(\theta_0)\} \leq P_{\theta'}\{X \in A'(\theta_0)\} = P_{\theta'}\{S'(X)$ contains $\theta_0\}$, thus completing the proof.

Example 1. *Monotone likelihood ratio.* Let the distribution of X be continuous and have monotone likelihood ratio. The acceptance region of a UMP size α test of $H_0:\theta = \theta_0$ against $H_1:\theta > \theta_0$ is in the form of $A(\theta_0) = (-\infty, u(\theta_0)]$. The function $u(\theta)$ may be chosen nondecreasing; for

$$P_{\theta'}\{X \geq u(\theta')\} = \alpha = P_\theta\{X \geq u(\theta)\} \leq P_{\theta'}\{X \geq u(\theta)\}$$

if $\theta \leq \theta'$ (power of UMP test $\geq$ size). The family of confidence sets $S(x) = \{\theta:x \in A(\theta)\}$ is in the form of $S(x) = [u^{-1}(x), \infty)$ or $(u^{-1}(x), \infty)$, where u^{-1} is defined as $u^{-1}(x) = \text{glb}\ \{\theta:u(\theta) \geq x\}$ (Fig. 5.4). By Theorem 1 the family of confidence sets $\{S(x)\}$ minimizes the probability $P_{\theta'}\{S(X)$ contains $\theta\}$ for all $\theta < \theta'$ out of all level $1 - \alpha$ families of confidence sets.

Example 2. *The nonregular family.* The only examples we have of

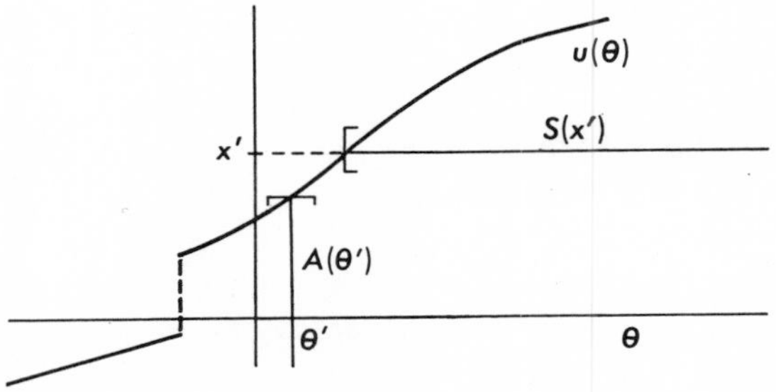

Fig. 5.4

UMP two-sided tests are for the "nonregular" families of distribution, admitting sufficient statistics,

$$f(x \mid \theta) = c(\theta)\, h(x)\, I_{(-\infty,\theta)}(x) \tag{5.114}$$

(Exercise 5.2.7). A UMP size α test of $H_0{:}\theta = \theta_0$ against $H_1{:}\theta \neq \theta_0$ has acceptance region of the form $A(\theta_0) = (u(\theta_0), \theta_0]$. As before, it can be shown that $u(\theta)$ is a nondecreasing function of θ. The associated family of confidence sets is therefore in the form of $S(x) = [x, u^{-1}(x))$ or $[x, u^{-1}(x)]$, with u^{-1} defined as before. This family of confidence sets has the optimum property that out of all level $1 - \alpha$ families of confidence sets $\{S(x)\}$ minimizes the probability of covering the wrong value $P_{\theta'}\{S(X) \text{ contains } \theta\}$ for $\theta' \neq \theta$.

This optimum property is so marvelous (though admittedly rare) that it deserves a name.

Definition 2. A family $\{S(x)\}$ of confidence sets at confidence level $1 - \alpha$ is said to be a *uniformly most accurate* (UMA) family of confidence sets at level $1 - \alpha$ if for any other family $\{S'(x)\}$ of confidence sets at level $1 - \alpha$, and all θ and θ',

$$P_{\theta'}\{S(X) \text{ contains } \theta\} \leq P_{\theta'}\{S'(X) \text{ contains } \theta\}.$$

For the parameter θ of the nonregular family (5.114) there exists a UMA family of confidence sets at any level $1 - \alpha$; furthermore, these confidence sets are intervals. Because the number of occasions on which UMP two-sided tests exist is rather small, we appeal to the principles of unbiasedness and invariance. It turns out that, in general, UMP unbiased or invariant two-sided tests lead to UMA unbiased or invariant confidence sets.

Unbiased Confidence Sets. Let $\Theta = \{(\theta, \mu)\}$ and consider the problem of obtaining confidence sets for θ alone.

Definition 3. $\{S(x)\}$ is an *unbiased family of confidence sets* for θ at confidence level $1 - \alpha$ if $P_{\theta,\mu}\{S(X) \text{ contains } \theta\} = 1 - \alpha$ for all θ and μ, and $P_{\theta'\mu}\{S(X) \text{ contains } \theta\} \leq 1 - \alpha$ for all θ, θ', and μ.

Theorem 2. If $A(\theta_0)$ is the acceptance region of a UMP unbiased test of size α of $H_0:\theta = \theta_0$ against $H_1:\theta \neq \theta_0$ for each θ_0, then $S(x) = \{\theta : x \in A(\theta)\}$ is a UMA unbiased family of confidence sets in that $S(x)$ is unbiased and $P_{\theta'\mu}\{S(X) \text{ contains } \theta\}$ is a minimum for all θ, θ', and μ among unbiased families of confidence sets at level $1 - \alpha$.

Proof. $\{S(x)\}$ is *unbiased:*

$$P_{\theta,\mu}\{S(X) \text{ contains } \theta\} = P_{\theta',\mu}\{X \in A(\theta)\} \leq 1 - \alpha.$$

$\{S(x)\}$ is *UMA unbiased:* Let $S'(x)$ be any other unbiased family of confidence sets at level $1 - \alpha$ and let $A'(\theta) = \{x:\theta \in S'(x)\}$. Then $P_{\theta'\mu}\{X \in A'(\theta)\} = P_{\theta'\mu}\{S'(X) \text{ contains } \theta\} \leq 1 - \alpha$, so that $A'(\theta)$ is the acceptance region of an unbiased test of size α. Hence $P_{\theta'\mu}\{S'(X)$ contains $\theta\} = P_{\theta'\mu}\{X \in A'(\theta)\} \geq P_{\theta'\mu}\{X \in A(\theta)\} = P_{\theta'\mu}\{S(X)$ contains $\theta\}$, for $A(\theta)$ is the acceptance region of a UMP unbiased test, thus completing the proof.

UMP unbiased two-sided tests exist for any parameter or linear combination of parameters of an exponential family of distributions, with density, say,

$$f_{\mathbf{T}}(\mathbf{t} \mid \boldsymbol{\theta}) = c(\boldsymbol{\theta})\, h(\mathbf{t}) \exp\{\sum_1^k \theta_i t_i\}.$$

The UMP unbiased test of $H_0:\theta_1 = \theta_1^0$ against $H_1:\theta_1 \neq \theta_1^0$ has acceptance region in the form of an interval about t_1

$$A(\theta_1^0) = \{\mathbf{t}: z_1(\theta_1^0; t_2, \cdots, t_k) < t_1 < z_2(\theta_1^0; t_2, \cdots, t_k)\}.$$

The associated UMA unbiased confidence sets are $S(\mathbf{t}) = \{\theta_1 : \mathbf{t} \in A(\theta_1)\}$. The functions $z_1(\theta_1; t_2, \cdots, t_k)$ and $z_2(\theta_1; t_2, \cdots, t_k)$ are increasing functions of θ_1 for fixed $t_2, \cdots, t_k$ (Exercise 1). Hence $S(\mathbf{t})$ is an interval of the form $(u_1(\mathbf{t}), u_2(\mathbf{t}))$.

Invariant Confidence Sets. Let $\Theta = \{(\theta, \mu)\}$ and suppose that for each θ_0 there exists a group $\mathcal{G}_{\theta_0}$ of transformations that leaves the problem of testing $H_0:\theta = \theta_0$ against $H_1:\theta \neq \theta_0$ invariant.

Theorem 3. Suppose for each θ_0 that $A(\theta_0)$ is the acceptance region of a UMP invariant test of size α of $H_0:\theta = \theta_0$ against $H_1:\theta \neq \theta_0$ under the group $\mathcal{G}_{\theta_0}$ and let $S(x) = \{\theta : x \in A(\theta)\}$.

(a) Assume for all θ_0, (θ, μ) and $g \in \mathcal{G}_{\theta_0}$, that if $\bar{g}(\theta, \mu) = (\theta', \mu')$ then θ' depends only on g and θ and not on μ, so that $\bar{g}$ induces a transformation on the space of θ.

(b) Assume, for all $g \in \bigcup_\theta \mathcal{G}_\theta$, that $S(g(x)) = \bar{g}(S(x))$ for all $x \in \mathfrak{X}$. Then, out of all $1 - \alpha$ level families of confidence sets satisfying (b), $S(x)$ minimizes $P_{\theta'\mu}\{S(X) \text{ contains } \theta\}$ for all θ, θ', μ.

Proof. Let $S'(x)$ be any other family of confidence sets at level $1 - \alpha$ satisfying (b) and let $A'(\theta) = \{x : \theta \in S'(x)\}$. Then $A'(\theta)$ is invariant under $\mathcal{G}_\theta$:

$$\begin{aligned} g(A'(\theta)) = \{g(x) : \theta \in S'(x)\} &= \{x : \theta \in S'(g^{-1}(x))\} \\ &= \{x : \theta \in \bar{g}^{-1}S'(x)\} = \{x : \bar{g}(\theta) \in S'(x)\} \end{aligned}$$

and if $g \in \mathcal{G}_\theta$, then $\bar{g}(\theta) = \theta$, so that

$$g(A'(\theta)) = \{x : \theta \in S'(x)\} = A'(\theta).$$

Hence $A'(\theta)$ is invariant and

$$\begin{aligned} P_{\theta'\mu}\{S'(X) \text{ contains } \theta\} &= P_{\theta'\mu}\{X \in A'(\theta)\} \\ &\geq P_{\theta'\mu}\{X \in A(\theta)\} = P_{\theta'\mu}\{S(X) \text{ contains } \theta\}, \end{aligned}$$

thus completing the proof.

In problems in which the groups $\mathcal{G}_\theta$ satisfy condition (a) we may define a family of confidence sets $\{S(x)\}$ as *invariant* under the group $\mathcal{G}$ generated by $\bigcup_\theta \mathcal{G}_\theta$ if condition (b) holds for $\{S(x)\}$. Then the content of Theorem 3 is that the family of confidence sets associated with a family of UMP invariant tests is UMA invariant under $\mathcal{G}$, provided it is invariant under $\mathcal{G}$ [that is, satisfies condition (b)].

Condition (b) is more easily checked in applications if it is expressed in terms of the defining relation for $S(x)$ as follows:

(b') For all θ_0 and all $g \in \mathcal{G}_{\theta_0}$,

$$\{\theta : g(x) \in A(\bar{g}(\theta))\} = \{\theta : x \in A(\theta)\}.$$

This expresses the condition that $\bar{g}^{-1}\, S(g(x)) = S(x)$; for

$$\begin{aligned} \bar{g}^{-1}\, S(g(x)) &= \bar{g}^{-1}\{\theta : g(x) \in A(\theta)\} \\ &= \{\bar{g}^{-1}(\theta) : g(x) \in A(\theta)\} \\ &= \{\theta : g(x) \in A(\bar{g}(\theta))\}. \end{aligned}$$

As an example of these considerations, we take the two-sided t-test, observed in Section 5.6 to be UMP invariant for testing that the mean of a normal distribution is zero. When $X_1, \cdots, X_n$ is a sample from $\mathfrak{N}(\theta, \sigma^2)$, the test

$$\phi(\mathbf{x}) = \begin{cases} 1 & \text{if} \quad \sqrt{n-1}\,\dfrac{|\bar{x}|}{s} > K, \\[2ex] 0 & \text{if} \quad \sqrt{n-1}\,\dfrac{|\bar{x}|}{s} < K, \end{cases}$$

of $H_0{:}\theta = 0$ against $H_1{:}\theta \neq 0$ is UMP invariant under the group of scale changes $g_c(\mathbf{x}) = c\mathbf{x}$, $c > 0$. If we test the hypothesis $H_0{:}\theta = \theta_0$ against $H_1{:}\theta \neq \theta_0$, then by considering the variables $X_1 - \theta_0, \cdots, X_n - \theta_0$ we are reduced to the preceding problem. The corresponding t-test is

$$\phi(\mathbf{x}) = \begin{cases} 1 & \text{if} \quad \sqrt{n-1}\,\dfrac{|\bar{x} - \theta_0|}{s} > K, \\[2ex] 0 & \text{if} \quad \sqrt{n-1}\,\dfrac{|\bar{x} - \theta_0|}{s} < K. \end{cases}$$

This test is UMP invariant under the group $\mathcal{G}_{\theta_0}$ of transformations $g_c(\mathbf{x}) = (c(\mathbf{x} - \theta_0\mathbf{1}) + \theta_0\mathbf{1})$. Since for $g_c \in \mathcal{G}_{\theta_0}$,

$$\bar{g}_c(\theta, \sigma^2) = (c(\theta - \theta_0) + \theta_0, c^2\sigma^2),$$

condition (a) of Theorem 3 is satisfied, and we may write

$$\bar{g}_c(\theta) = c(\theta - \theta_0) + \theta_0 \qquad \text{for } g_c \in \mathcal{G}_{\theta_0}.$$

Because

$$A(\theta) = \{\mathbf{x}{:}\sqrt{n-1}\,|\bar{x} - \theta| < sK\},$$

the family of confidence sets

$$S(\mathbf{x}) = \{\theta{:}\mathbf{x} \in A(\theta)\} = \{\theta{:}\sqrt{n-1}\,|\bar{x} - \theta| < sK\}$$

is UMA invariant under the group $\mathcal{G}$ generated by $\bigcup_\theta \mathcal{G}_\theta$, provided it is invariant under $\mathcal{G}$. To see that condition (b') holds, let $g_c \in \mathcal{G}_{\theta_0}$ for some θ_0; then

$$\begin{aligned} \{\theta{:}g_c(\mathbf{x}) \in A(\bar{g}_c(\theta))\} &= \{\theta{:}\sqrt{n-1}\,|c(\bar{x} - \theta_0) + \theta_0 - \bar{g}_c(\theta)| < csK\} \\ &= \{\theta{:}\sqrt{n-1}\,|\bar{x} - \theta| < sK\} = \{\theta{:}\mathbf{x} \in A(\theta)\}. \end{aligned}$$

Therefore $S(\mathbf{x})$ satisfies condition (b') and is UMA invariant.

Exercises

1. Let X have density $f(x \mid \theta) = c(\theta)\, h(x) e^{\theta x}$ and let $A(\theta_0) = (z_1(\theta_0), z_2(\theta_0))$ be the acceptance region of a UMP unbiased test of $H_0{:}\theta = \theta_0$ against $H_1{:}\theta \neq \theta_0$. Show that $z_1(\theta)$ and $z_2(\theta)$ are increasing functions of θ.

2. Let Θ be the real line and let X be $\mathfrak{N}(\theta - 1, 1)$ if $\theta < 0$, $\mathfrak{N}(0, 1)$ if $\theta = 0$ and $\mathfrak{N}(\theta + 1, 1)$ if $\theta > 0$. Show that X has monotone likelihood ratio and find a level $1 - \alpha$ family of confidence sets, $S(x)$, as in Example 1, for which $P_{\theta'}\{S(X) \text{ contains } \theta\}$ is uniformly minimum for $\theta < \theta'$.

3. Let $X_1, \cdots, X_n$ be a sample from the uniform distribution $\mathfrak{U}(0, \theta)$. Find a UMA family of confidence sets for θ, at level $1 - \alpha$ (see Exercise 5.2.7).

4. Let $X_1, \cdots, X_m$ be a sample from $\mathfrak{N}(\lambda, \sigma^2)$ and let $Y_1, \cdots, Y_n$ be a sample from $\mathfrak{N}(\mu, \sigma^2)$, independent of the X_i's. Find a UMA unbiased family of confidence intervals at level $1 - \alpha$ for $\theta = \lambda - \mu$. (Use the result of Exercise 5.4.7.)

5. Let X and Y be independent with respective densities $f_X(x \mid \lambda) = f(x - \lambda)$ and $f_Y(y \mid \mu) = f(y - \mu)$. Let $T = |X - Y|$ and $\theta = \lambda - \mu$ and suppose that $f_T(t \mid \theta)/f_T(t \mid 0)$ is nondecreasing in t. Find a UMP invariant family of confidence sets for θ at level $1 - \alpha$.

6. Let X and Y have joint distribution given in Exercise 5.4.3.

(a) Find a UMA unbiased family of confidence sets for $\theta = \lambda/\mu$, at level $1 - \alpha$.

(b) Show that this family of confidence sets is also UMA invariant.

Ans. $S(x, y) = (t/K, tK)$, where $t = y/x$ and $K = (2 - \alpha)/\alpha$.

7. Let $\mathfrak{NE}(\alpha, \beta)$ denote the negative exponential distribution with density of (3.65). Let X be a random variable whose distribution given θ is $\mathfrak{NE}(\theta - 1, 1)$ if $\theta < 0$, $\mathfrak{NE}(0, 1)$ if $\theta = 0$, and $\mathfrak{NE}(\theta + 1, 1)$ if $\theta > 0$. Show that $A(\theta_0) = (\theta_0 - 1, \theta_0 - 1 - \log \alpha)$ if $\theta_0 < 0$, $A(\theta_0) = (0, -\log \alpha)$ if $\theta_0 = 0$, and $A(\theta_0) = (\theta_0 + 1, \theta_0 + 1 - \log \alpha)$ if $\theta_0 > 0$ is a family of UMP size α tests of the hypothesis $H_0{:}\theta = \theta_0$ against $H_1{:}\theta \neq \theta_0$. Find the associated UMA family of confidence sets for θ at level $1 - \alpha$. Consider, in particular, the case $\alpha = e^{-1/2}$.

5.9 The General Linear Hypothesis

Another application of the principle of invariance in hypothesis testing is to the general linear hypothesis. One form of the general (univariate) linear hypothesis concerns independent random variables $X_1, \cdots, X_n$ having normal distributions with common variance σ^2. The vector of means $\boldsymbol{\xi} = E\mathbf{X}$ is assumed to lie in some k-dimensional subspace $\mathcal{L}$ of E_n, where $k < n$. Usually this assumption is put in the form of $E\mathbf{X} = \mathbf{A}\boldsymbol{\theta}$, where $\mathbf{A}$ is a known $n \times k$ matrix of full rank $k < n$ and $\boldsymbol{\theta}$ is an arbitrary k-dimensional vector; in this case, $\mathcal{L}$ is the space spanned by the column vectors of $\mathbf{A}$. In short, our assumptions are

$$\mathbf{X} \in \mathfrak{N}(\mathbf{A}\boldsymbol{\theta}, \sigma^2\mathbf{I}), \tag{5.115}$$

where $\mathbf{I}$ represents the $n \times n$ identity matrix. The hypothesis to be tested is that the vector of means $\boldsymbol{\xi} = E\mathbf{X}$ lie in $\mathcal{L}_1$, some $(k - r)$-dimensional subspace of $\mathcal{L}$. Usually this hypothesis is expressed as restricting the parameters $\boldsymbol{\theta}$ to satisfy r constraints,

$$H_0: \mathbf{B}\boldsymbol{\theta} = \mathbf{0}, \tag{5.116}$$

where $\mathbf{B}$ is a given $r \times k$ matrix of full rank r, $1 \leq r \leq k$.

Example 1. *One-way classification.* Let X_{ij} for $j = 1, \cdots, n_i$ and $i = 1, \cdots, I$ be independent, normally distributed random variables with common variance σ^2 and suppose that $EX_{ij} = \theta_i$. Thus $X_{i1}, \cdots, X_{in_i}$ is a sample of size n_i from $\mathfrak{N}(\theta_i, 1)$. In the notation of the preceding paragraph $n = n_1 + \cdots + n_I$ and $k = I$. The traditional null hypothesis is that $\theta_1 = \theta_2 = \cdots = \theta_I$, so that $r = I - 1$. If $\mathbf{X}$ denotes the vector $(X_{11}, \cdots, X_{1n_1}, X_{21}, \cdots, X_{2n_2}, \cdots, X_{I1}, \cdots, X_{In_I})^T$, the matrix $\mathbf{A}$ is given by (a_{ij}), where, for $i = 1, \cdots, n$ and $j = 1, \cdots, k$, $a_{ij} = 1$ if $n_1 + \cdots + n_{j-1} < i \leq n_1 + \cdots + n_j$ and $a_{ij} = 0$ otherwise. The space $\mathcal{L}$ is generated by the columns of $\mathbf{A}$, the space $\mathcal{L}_1$, by the n-vector with all components one.

Example 2. *Regression.* Independent normally distributed observations $X_1, \cdots, X_n$ are taken at "levels" $z_1, \cdots, z_n$, respectively, with $EX_i = \xi_i = \beta_0 + \beta_1 z_i + \beta_2 z_i^2 + \cdots + \beta_m z_i^m$ where $m + 1 < n$, and with common variance σ^2. The matrix $\mathbf{A} = (a_{ij})$ has components $a_{ij} = z_i^{j-1}$, $i = 1, \cdots, n$, $j = 1, \cdots, m + 1$. This matrix $\mathbf{A}$ has full rank $m + 1$, provided there are at least $m + 1$ distinct numbers among $z_1, \cdots, z_n$. With this assumption, $k = m + 1$. Usually, we are interested in testing the hypothesis that

$$\beta_{s+1} = \beta_{s+2} = \cdots = \beta_m = 0$$

(that is, that the regression is a polynomial of degree at most s). If this is so, then $r = m - s$.

EXAMPLE 3. *Two-way classification, one observation per cell.* Let X_{ij}, $i = 1, \cdots, I, j = 1, \cdots, J$, be independent, normally distributed random variables with common variance, and let $\xi_{ij} = EX_{ij}$. It is assumed that the mean of X_{ij} is a linear sum of an effect due to the subscript i, called the row effect, and an effect due to the subscript j, called the column effect, $\xi_{ij} = \mu_i' + \eta_j'$. Written in this form, the parameters μ_i' and η_j' are not identifiable, for if $\xi_{ij} = \mu_i' + \eta_j'$ then $\xi_{ij} = (\mu_i' + c) + (\eta_j' - c)$; that is, increasing all μ_i' by a constant c and decreasing the η_j' by the same constant leaves the distribution of the X_{ij} unaltered. Therefore it is customary to write

$$\xi_{ij} = \xi + \mu_i + \eta_j, \tag{5.117}$$

where

$$\sum \mu_i = 0 \qquad \text{and} \qquad \sum \eta_j = 0$$

$$(\xi = \bar{\mu}' + \bar{\eta}', \; \mu_i = \mu_i' - \bar{\mu}', \; \eta_j = \eta_j' - \bar{\eta}').$$

In this formulation there are $I + J + 1$ parameters and two restrictions, so that the ξ_{ij} are restricted to lie in a $k = (I + J - 1)$-dimensional space. The null hypothesis may be that of no column effect

$$H_0: \eta_1 = \eta_2 = \cdots = \eta_J = 0. \tag{5.118}$$

This puts $r = J - 1$ new restrictions on $\eta_1, \cdots, \eta_J$, so that $\mathfrak{L}_1$ is a $k - r = I$-dimensional space.

Canonical form. By a change of variable from $\mathbf{X}$ to $\mathbf{Y}$ the general linear hypothesis can be put into the following much simpler form, called the canonical form of the general linear hypothesis. In this form $Y_1, \cdots, Y_n$ are independent random variables having normal distributions with a common variance σ^2 and with means $EY_i = \mu_i$ for $i = 1, \cdots, k$, and $EY_i = 0$ for $i = k + 1, \cdots, n$. In matrix notation

$$\mathbf{Y} \in \mathfrak{N}(\boldsymbol{\mu}, \sigma^2\mathbf{I}), \tag{5.119}$$

where $\mu_{k+1} = \cdots = \mu_n = 0$. In canonical form the hypothesis to be tested is simply

$$H_0: \mu_1 = \cdots = \mu_r = 0. \tag{5.120}$$

This is a great simplification in the statement of the problem over the original form (5.115) and (5.116). To transform the original problem into canonical form we let $\mathbf{Y} = \mathbf{RX}$, where $\mathbf{R}$ is an orthogonal matrix with the following properties. The last $(n - k)$ rows of $\mathbf{R}$ are perpen-

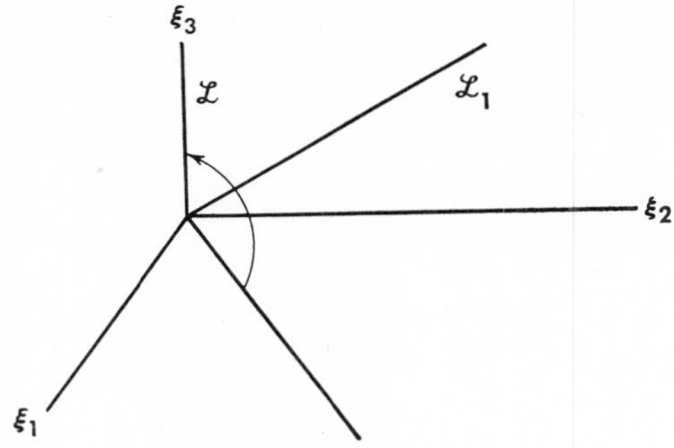

Fig. 5.5

dicular to the space $\mathcal{L}$ (that is, perpendicular to the columns of **A**). The first r rows of **R** are vectors in $\mathcal{L}$ perpendicular to $\mathcal{L}_1$; and rows $r + 1$ through k of **R** span the space $\mathcal{L}_1$. Then, clearly, $\mathbf{Y} \in \mathfrak{N}(\mathbf{RA\theta}, \sigma^2\mathbf{I})$, for the matrix **R** is orthogonal. Furthermore, if $\mu = \mathbf{RA\theta}$, then $\mu_{k+1} = \cdots = \mu_n = 0$; for the last $n - k$ rows of **R** are perpendicular to $\mathcal{L}$. Under the null hypothesis $\mu_1 = \cdots = \mu_r = 0$, for the first r rows of **R** are perpendicular to $\mathcal{L}_1$.

As an example, consider the two-sample problem based on observations X_1, X_2, X_3 with means $EX_1 = \theta_1$, $EX_2 = \theta_1$, $EX_3 = \theta_2$, where the null hypothesis is that $\theta_1 = \theta_2$. The space $\mathcal{L}$ is that subspace of E_3 in which $\xi_1 = \xi_2$ (Fig. 5.5). The matrix **A** is

$$\mathbf{A} = \begin{pmatrix} 1 & 0 \\ 1 & 0 \\ 0 & 1 \end{pmatrix},$$

which is of full rank 2, so that $\mathcal{L}$ is generated by the vectors $(1, 1, 0)^T$ and $(0, 0, 1)^T$. The space $\mathcal{L}_1$ is that subspace of $\mathcal{L}$ in which $\xi_1 = \xi_2 = \xi_3$. Hence the second row of **R** is $(1, 1, 1)$ normalized to have length one, namely, $(1/\sqrt{3}, 1/\sqrt{3}, 1/\sqrt{3})$. The first row of **R** is a linear combination of $(1, 1, 0)$ and $(0, 0, 1)$ perpendicular to $(1, 1, 1)$. Because

$$(a, a, b) \cdot (1, 1, 1) = 2a + b,$$

$(1, 1, -2)$ is such a vector. Normalized, the first row of **R** is $(1/\sqrt{6}, 1/\sqrt{6}, -2/\sqrt{6})$. For the third row of **R** we seek a vector perpendicular to $(1, 1, 0)$ and $(0, 0, 1)$. If (a, b, c) is such a vector, then $a + b = 0$ and

$c = 0$. The third row of $\mathbf{R}$ may be taken, therefore, as $(1/\sqrt{2}, -1/\sqrt{2}, 0)$. In full

$$\mathbf{R} = \begin{pmatrix} \dfrac{1}{\sqrt{6}} & \dfrac{1}{\sqrt{6}} & \dfrac{-2}{\sqrt{6}} \\ \\ \dfrac{1}{\sqrt{3}} & \dfrac{1}{\sqrt{3}} & \dfrac{1}{\sqrt{3}} \\ \\ \dfrac{1}{\sqrt{2}} & \dfrac{-1}{\sqrt{2}} & 0 \end{pmatrix}.$$

If $\mathbf{Y} = \mathbf{RX}$, then $\mathbf{Y} \in \mathfrak{N}(\boldsymbol{\mu}, \sigma^2\mathbf{I})$, where $\boldsymbol{\mu} = \mathbf{RA}\boldsymbol{\theta}$:

$$\boldsymbol{\mu} = \begin{pmatrix} \dfrac{2(\theta_1 - \theta_2)}{\sqrt{6}} \\ \\ \dfrac{(2\theta_1 + \theta_2)}{\sqrt{3}} \\ \\ 0 \end{pmatrix}.$$

The null hypothesis $\theta_1 = \theta_2$ becomes $\mu_1 = 0$.

Invariance. The canonical form of the general linear hypothesis is invariant under the group generated by the following three groups of transformations:

$\mathcal{G}_1$ $g(\mathbf{Y}) = \mathbf{Y}'$, where $Y_i' = Y_i + c_i$ for $i = r + 1, \cdots, k$,

$Y_i' = Y_i$ otherwise

$\mathcal{G}_2$ the group of orthogonal transformations on $Y_1, \cdots, Y_r$.

$\mathcal{G}_3$ scale change: $g(\mathbf{Y}) = b\mathbf{Y}$, where $b > 0$.

Sufficiency reduces the problem to consideration of $Y_1, \cdots, Y_k$, $\sum_{k+1}^{n} Y_i^2$. We further restrict attention to invariant rules. Invariance under $\mathcal{G}_1$ restricts attention to $Y_1, \cdots, Y_r, \sum_{k+1}^{n} Y_i^2$. Invariance under $\mathcal{G}_2$ further restricts attention to $\sum_1^r Y_i^2$ and $\sum_{k+1}^{n} Y_i^2$. Invariance under $\mathcal{G}_3$ further restricts attention to the ratio $\sum_1^r Y_i^2 / \sum_{k+1}^{n} Y_i^2$. We consider instead the equivalent invariant statistic

$$F = \frac{(1/r) \sum_{1}^{r} Y_i^2}{(1/(n-k)) \sum_{k+1}^{n} Y_i^2}, \tag{5.121}$$

which under the general linear hypothesis has a noncentral $\mathfrak{F}$-distribution $\mathfrak{F}_{r,n-k}(\gamma^2)$, with r and $n - k$ degrees of freedom and noncentrality parameter $\gamma^2 = \sum_1^r \mu_i^2/\sigma^2$. It is easily seen that this statistic is maximal invariant (Exercise 1).

The null hypothesis, when dealing with the maximal invariant F becomes the simple hypothesis $H_0 : \gamma^2 = 0$ and the alternative $H_1 : \gamma^2 > 0$. It may be shown that the distribution $\mathfrak{F}_{r,n}(\gamma^2)$ has monotone likelihood ratio in γ^2, so that from Theorem 5.2.1 any test that rejects H_0 when $F > c$ is *UMP* of its size based on F. However, it is easier to show that the ratio $f(x \mid \gamma^2)/f(x \mid 0)$ is nondecreasing in x, where $f(x \mid \gamma^2)$ is the density of $\mathfrak{F}_{r,n}(\gamma^2)$ of (3.19) (Exercise 2). Then the Neyman-Pearson lemma implies that the test, which rejects when $F > c$, is UMP of its size based on F. Hence any such test is UMP invariant for the original problem. The constant c may be chosen from the central $\mathfrak{F}$-distribution so that the size of the test is a given constant α.

Least squares. In application it is usually more convenient to carry out the computations in terms of the observables X_i rather than to find the transformation to canonical form. We proceed to express the test statistic F in terms of the original variables. Note that $\sum_{i=k+1}^{n} Y_i^2$ is the unrestricted minimum of the sum of squares

$$S^2 = \sum_{1}^{k} (Y_i - \mu_i)^2 + \sum_{i=k+1}^{n} Y_i^2.$$

But orthogonal transformations leave distances unchanged so that

$$S^2 = \sum_{1}^{n} (Y_i - EY_i)^2 = \sum_{1}^{n} (X_i - \xi_i)^2.$$

The minimum of $\sum_1^n (Y_i - EY_i)^2$ under variation of $\mu_1, \cdots, \mu_k$ is therefore the minimum of $\sum_1^n (X_i - \xi_i)^2$ under variation of $\boldsymbol{\xi}$ in $\mathcal{L}$. Hence

$$\sum_{k+1}^{n} Y_i^2 = \sum_{1}^{n} (X_i - \hat{\xi}_i)^2 = \min_{\xi \epsilon \mathcal{L}} \sum_{1}^{n} (X_i - \xi_i)^2, \tag{5.122}$$

where $\hat{\boldsymbol{\xi}}$ represents the least squares estimates of $\boldsymbol{\xi}$ under the general hypothesis.

Similarly, the minimum of

$$\sum_{1}^{n} (Y_i - EY_i)^2 = \sum_{1}^{n} (X_i - \xi_i)^2$$

under the null hypothesis H_0 is on the left side $\sum_1^r Y_i^2 + \sum_{k+1}^n Y_i^2$ and on the right side

$$\sum_{1}^{n} (X_i - \hat{\hat{\xi}}_i)^2 = \min_{\xi \epsilon \mathcal{L}_1} \sum_{1}^{n} (X_i - \xi_i)^2,$$

where $\hat{\hat{\boldsymbol{\xi}}}$ is the least squares estimate of $\boldsymbol{\xi}$ under H_0. Hence

$$\sum_{1}^{r} Y_i^2 = \sum_{1}^{n} (X_i - \hat{\hat{\xi}}_i)^2 - \sum_{1}^{n} (X_i - \hat{\xi}_i)^2. \tag{5.123}$$

Equations 5.122 and 5.123 give the denominator and numerator sum of squares of the F statistic.

There is a simplification in (5.123) that may be viewed geometrically (Fig. 5.6). The vectors $\hat{\boldsymbol{\xi}}$ and $\hat{\hat{\boldsymbol{\xi}}}$ are the projections of the vector $\mathbf{X}$ onto the spaces $\mathcal{L}$ and $\mathcal{L}_1$, respectively. Hence the vectors $\mathbf{X} - \hat{\boldsymbol{\xi}}$ and $\hat{\boldsymbol{\xi}} - \hat{\hat{\boldsymbol{\xi}}}$

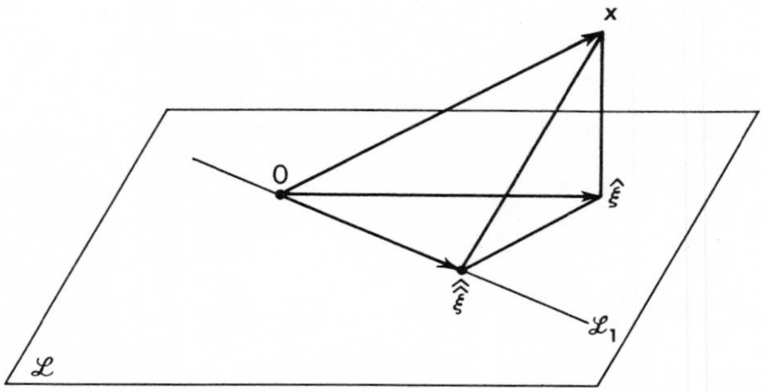

Fig. 5.6

are orthogonal. By the theorem of Pythagoras

$$\sum(X_i - \hat{\hat{\xi}}_i)^2 = \sum(X_i - \hat{\xi}_i)^2 + \sum(\hat{\xi}_i - \hat{\hat{\xi}}_i)^2.$$

Hence

$$\sum_1^r Y_i^2 = \sum_1^n (\hat{\xi}_i - \hat{\hat{\xi}}_i)^2, \tag{5.124}$$

so that the statistic F may be written

$$F = \frac{(1/r)\sum(\hat{\xi}_i - \hat{\hat{\xi}}_i)^2}{1/(n-k)\sum(X_i - \hat{\xi}_i)^2}. \tag{5.125}$$

To find the noncentrality parameter γ^2 in terms of $\boldsymbol{\xi}$ the trick is to note that, in the transformation $\mathbf{Y} = \mathbf{RX}$ and $\boldsymbol{\mu} = \mathbf{R}\boldsymbol{\xi}$, $\boldsymbol{\xi}$ undergoes the same transformation as $\mathbf{X}$. If we express $\hat{\boldsymbol{\xi}}$ and $\hat{\hat{\boldsymbol{\xi}}}$ as functions of $\mathbf{X}$, say $\hat{\boldsymbol{\xi}}(\mathbf{X})$ and $\hat{\hat{\boldsymbol{\xi}}}(\mathbf{X})$, then in analogy to (5.124)

$$\gamma^2 = \frac{1}{\sigma^2}\sum_1^r \mu_i^2 = \frac{1}{\sigma^2}\sum_1^n (\hat{\xi}_i(\boldsymbol{\xi}) - \hat{\hat{\xi}}_i(\boldsymbol{\xi}))^2. \tag{5.126}$$

As an example of these considerations, we take the two-way classification with one observation per cell, as in Example 3. The sum of squares S^2 may be written

$$S^2 = \sum_i \sum_j (X_{ij} - \xi_{ij})^2 = \sum\sum (X_{ij} - \xi - \mu_i - \eta_j)^2. \tag{5.127}$$

This problem, and the general problem of finding least squares estimates, may be solved by the method of Lagrange multipliers. However, in the present problem additional insight is gained by writing the sum of squares in the form

$$\begin{aligned} S^2 = {} & \sum\sum (X_{ij} - \bar{X}_{i.} - \bar{X}_{.j} + \bar{X}_{..})^2 + \sum\sum (\bar{X}_{i.} - \bar{X}_{..} - \mu_i)^2 \\ & + \sum\sum (\bar{X}_{.j} - \bar{X}_{..} - \eta_j)^2 + \sum\sum (\bar{X}_{..} - \xi)^2, \end{aligned} \tag{5.128}$$

where

$$\bar{X}_{i.} = \sum_{j=1}^J X_{ij}/J, \qquad \bar{X}_{.j} = \sum_{i=1}^J X_{ij}/I,$$

and

$$\bar{X}_{..} = \sum\sum X_{ij}/IJ \qquad \text{(Exercise 3)}.$$

It follows that

$$\hat{\xi} = \bar{X}_{..},\ \hat{\mu}_i = \bar{X}_{i.} - \bar{X}_{..},\ \hat{\eta}_j = \bar{X}_{.j} - \bar{X}_{..}$$

and that

$$\sum \sum (X_{ij} - \hat{\xi}_{ij})^2 = \sum \sum (X_{ij} - \bar{X}_{i.} - \bar{X}_{.j} + \bar{X}_{..})^2.$$

If we are testing no column effect, $H_0: \eta_1 = \eta_2 = \cdots = \eta_J$, it follows from (5.128) that $\hat{\hat{\xi}} = \bar{X}_{..}$, $\hat{\hat{\mu}}_i = \bar{X}_{i.} - \bar{X}_{..}$, so that

$$\sum \sum (\hat{\xi}_{ij} - \hat{\hat{\xi}}_{ij})^2 = \sum \sum (\bar{X}_{.j} - \bar{X}_{..})^2 = I \sum_{j=1}^{J} (\bar{X}_{.j} - \bar{X}_{..})^2. \tag{5.129}$$

From this the statistic F of (5.125) may easily be computed. It has an $\mathfrak{F}_{J-1,IJ-I-J+1}(\gamma^2)$-distribution, where the noncentrality parameter γ^2 may be computed from (5.126) and (5.129) by replacing X_{ij} by $\xi + \mu_i + \eta_j$:

$$\gamma^2 = I \sum_{j=1}^{J} \frac{\eta_j{}^2}{\sigma^2}.$$

A UMP invariant test of H_0 is therefore to reject H_0 when $F > c$, where c is chosen from the central $\mathfrak{F}$-distribution to obtain a given size. The power of this test may be found from the tables of the noncentral $\mathfrak{F}$-distribution.

Exercises

1. Show that (5.121) is a maximal invariant under the group generated by $\mathcal{G}_1$, $\mathcal{G}_2$, and $\mathcal{G}_3$.
2. Let $f(x \mid \gamma^2)$ be the density given by (3.19). Show that $f(x \mid \gamma^2)/f(x \mid 0)$ is for fixed $\gamma^2 > 0$ an increasing function of x.
3. Show that (5.128) is valid. *Hint.* Write

$$\sum\sum(X_{ij} - \xi - \mu_i - \eta_j)^2 = \sum\sum[(X_{ij} - \bar{X}_{i.} - \bar{X}_{.j} + \bar{X}_{..}) + (\bar{X}_{i.} - \bar{X}_{..} - \mu_i) + (\bar{X}_{.j} - \bar{X}_{..} - \eta_j) + (\bar{X}_{..} - \xi)]^2$$

and in squaring keep the terms in parenthesis as units.

4. *Two-way classification K observations per cell.* Let X_{ijk}, $i = 1, \cdots, I$, $j = 1, \cdots, J$, $k = 1, \cdots, K$, satisfy the general linear hypothesis with

$$EX_{ijk} = \xi + \mu_i + \eta_j + \delta_{ij},$$

where $\sum \mu_i = 0$, $\sum \eta_j = 0$, $\sum_i \delta_{ij} = 0$ for all j, and $\sum_j \delta_{ij} = 0$ for all i. (δ_{ij} is called the interaction effect of the ith row and jth column.)
(a) Show that

$$\begin{aligned} S^2 &= \sum\sum\sum (X_{ijk} - \xi - \mu_i - \eta_j - \delta_{ij})^2 \\ &= \sum\sum\sum (X_{ijk} - \bar{X}_{ij.})^2 + \sum\sum\sum (\bar{X}_{ij.} - \bar{X}_{i..} - \bar{X}_{.j.} \\ &\quad + \bar{X}_{...} - \delta_{ij})^2 + \sum\sum\sum (\bar{X}_{i..} - \bar{X}_{...} - \mu_i)^2 \\ &\quad + \sum\sum\sum (\bar{X}_{.j.} - \bar{X}_{...} - \eta_j)^2 + \sum\sum\sum (\bar{X}_{...} - \xi)^2, \end{aligned}$$

where $\bar{X}_{ij.} = \sum_k X_{ijk}/K$, and so on.
(b) Find the UMP invariant test of the hypothesis of no row effect $H_0: \mu_1 = \cdots = \mu_I = 0$. What is the distribution of the test statistic under the general linear hypothesis (including the noncentrality parameter)?
(c) Find the UMP invariant test of the hypothesis of no interaction effect $H_0: \delta_{ij} = 0$ for all i, j. What is the distribution of the test statistic under the general linear hypothesis?

5. Consider the regression problem where $X_1, \cdots, X_n$ satisfy the general linear hypothesis with $EX_i = \beta_0 + \beta_1 z_i$. We assume for simplicity that $\sum z_i = 0$ and $\sum z_i^2 \neq 0$.
(a) Find the least squares estimates $\hat{\beta}_0$ and $\hat{\beta}_1$.
(b) Under the hypothesis $H_0: \beta_1 = 0$ find the least squares estimate $\hat{\hat{\beta}}_0$.
(c) Find the UMP invariant test of H_0. What is the distribution of the test statistic under the general linear hypothesis?

6. Consider a two-way classification X_{ij}, $i = 1, \cdots, I$, $j = 1, \cdots, J$, with the assumptions of the general linear hypothesis for which $EX_{ij} = \alpha + \beta z_i + \eta_j$, where α, β, and η_j are unknown parameters subject to the restriction $\sum \eta_j = 0$, and where z_i are known numbers for which $\sum z_i = 0$ and $\sum z_i^2 = 1$.
(a) Find the UMP invariant test of the hypothesis

$$H_0: \eta_1 = \eta_2 = \cdots = \eta_J = 0.$$

(b) What is the distribution of the test statistic under the general linear hypothesis?

7. Show that the least squares estimates $\hat{\xi}$ of ξ are also the maximum likelihood estimates under the general linear hypothesis.

8. *One-way classification.* Let X_{ij} be as in Example 1. Find the UMP invariant test of the hypothesis $H_0: \theta_1 = \cdots = \theta_I$. What is the distribution of the test statistic under the general hypothesis?

9. *Three-way classification, one observation per cell.* Let X_{ijk}, $i = 1, \cdots, I$, $j = 1, \cdots, J$, and $k = 1, \cdots, K$, satisfy the general linear hypothesis with $EX_{ijk} = \xi_{ijk} = \xi + \lambda_i + \mu_j + \eta_k$, subject to the restrictions $\sum \lambda_i = 0$, $\sum \mu_j = 0$, and $\sum \eta_k = 0$.
(a) Find a decomposition of the sum of squares

$$S^2 = \sum\sum\sum (X_{ijk} - \xi_{ijk})^2$$

similar to that of (5.128).
(b) Find the UMP invariant test of $H_0: \lambda_1 = \cdots = \lambda_I = 0$. What is the distribution of the test statistic under the general hypothesis?

10. *Latin square.* Let X_{ij}, $i = 1, \cdots, m$ and $j = 1, \cdots, m$ satisfy the general linear hypothesis with $EX_{ij} = \xi_{ij} = \xi + \lambda_i + \mu_j + \eta_{\nu(i,j)}$, where $\lambda_1, \cdots, \lambda_m$, $\mu_1, \cdots, \mu_m$, and $\eta_1, \cdots, \eta_m$ satisfy the restrictions $\sum \lambda_i = 0$, $\sum \mu_j = 0$, and $\sum \eta_\nu = 0$, and where $\nu(i, j)$ has the properties.
(a) For each fixed i, $(\nu(i, 1), \cdots, \nu(i, m))$ is a permutation of $(1, \cdots, m)$, and
(b) For each fixed j, $(\nu(1, j), \cdots, \nu(m, j))$ is a permutation of $(1, \cdots, m)$. Find the UMP invariant test of the hypothesis $H_0: \lambda_1 = \cdots = \lambda_m = 0$. What is the distribution of the test statistic under the general hypothesis?

11. Let X_i, $i = 1, \cdots, I$ and Y_j, $j = 1, \cdots, J$ satisfy the general linear hypothesis with $EX_i = \alpha_1 + \beta_1 u_i$ and $EY_j = \alpha_2 + \beta_2 v_j$, where the u_i and v_j are known and $\sum u_i = \sum v_j = 0$, $\sum u_i^2 = I$ and $\sum v_j^2 = J$.
(a) Find the UMP invariant test of $H_0: \beta_1 = \beta_2$.
(b) Find the UMP invariant test of $H_0': \alpha_1 = \alpha_2$ and $\beta_1 = \beta_2$.

5.10 Confidence Ellipsoids and Multiple Comparisons

The UMP invariant tests of the general linear hypothesis developed in Section 5.9 may be used to obtain UMA invariant confidence sets for the parameters. First we consider the canonical form of the general linear hypothesis $\mathbf{Y} \in \mathfrak{N}(\boldsymbol{\mu}, \sigma^2\mathbf{I})$, where $\mu_{k+1} = \cdots = \mu_n = 0$. The UMP invariant size α test of the hypothesis $\mu_i = 0$ for $i = 1, \cdots, r$, has acceptance region $A = \{\mathbf{y}: \sum_1^r y_i^2 \le r\hat{\sigma}^2 F_{r,n-k;\alpha}\}$, where

$$\hat{\sigma}^2 = \sum_{k+1}^{n} y_i^2/(n - k)$$

(called the "residual" estimate of the variance) and where $F_{r,n-k;\alpha}$ is the αth cutoff point for the $\mathfrak{F}_{r,n-k}$-distribution, that is, the probability that a random variable with an $\mathfrak{F}_{r,n-k}$-distribution is greater than $F_{r,n-k;\alpha}$ is α. The problem of testing $\mu_i = \mu_i^0$ for $i = 1, \cdots, r$ may be transformed into canonical form by subtracting μ_i^0 from Y_i, $i = 1, \cdots, r$. The UMP invariant size α test of the hypothesis $\mu_i = \mu_i^0$, $i = 1, \cdots, r$ therefore has acceptance region

$$A\{\mu_1^0, \cdots, \mu_r^0\} = \{\mathbf{y}: \sum_1^r (y_i - \mu_i^0)^2 \le c\hat{\sigma}^2\}, \tag{5.130}$$

where $c = rF_{r,n-k;\alpha}$.

The associated family of confidence sets for $\mu_1, \cdots, \mu_r$ is

$$S(\mathbf{y}) = \{(\mu_1, \cdots, \mu_r): \sum_1^r (y_i - \mu_i)^2 \le c\hat{\sigma}^2\}. \tag{5.131}$$

These are spheres centered at $(y_1, \cdots, y_r)$. In the following paragraph we show that this family of confidence sets is UMA invariant by checking the conditions of Theorem 5.8.3.

To describe the group $\mathcal{G}_{\mu_1^0, \cdots, \mu_r^0}$, with respect to which the test with acceptance region (5.130) is UMP invariant we use the notation $\mathbf{y}_r$ to denote the vector $(y_1, \cdots, y_r)^T$ and $\boldsymbol{\mu}_r$[resp. $\boldsymbol{\mu}_r^0$] to denote the vector $(\mu_1, \cdots, \mu_r)^T$[resp. $(\mu_1^0, \cdots, \mu_r^0)^T$]. The group $\mathcal{G}_{\boldsymbol{\mu}_r^0}$ is generated by the following three groups of transformations:

1. If $g(\mathbf{y}) = \mathbf{y}'$, then $y_i' = y_i + c_i$ for $r < i \le k$ and $y_i' = y_i$ for $i \le r$ and $i > k$.
2. If $g(\mathbf{y}) = \mathbf{y}'$, then $\mathbf{y}_r' = \mathbf{P}(\mathbf{y}_r - \boldsymbol{\mu}_r^0) + \boldsymbol{\mu}_r^0$ and $y_i' = y_i$ for $i > r$, where $\mathbf{P}$ is an $r \times r$ orthogonal matrix.
3. If $g(\mathbf{y}) = \mathbf{y}'$, then $y_i' = b(y_i - \overset{0}{\mu_i}) + \overset{0}{\mu_i}$ for $i \le r$ and $y_i' = by_i$ for $i > r$, where $b > 0$.

To check condition (a) of Theorem 5.8.3 we must find the group $\bar{\mathcal{G}}_{\boldsymbol{\mu}_r^0}$. This group is generated by the corresponding three groups:

1. If $\bar{g}(\boldsymbol{\mu}, \sigma) = (\boldsymbol{\mu}', \sigma')$, then $\mu_i' = \mu_i + c_i$ for $r < i \le k$, $\mu_i' = \mu_i$ for $i \le r$ and $i > k$, and $\sigma' = \sigma$.
2. If $\bar{g}(\boldsymbol{\mu}, \sigma) = (\boldsymbol{\mu}', \sigma')$, then $\boldsymbol{\mu}_r' = \mathbf{P}(\boldsymbol{\mu}_r - \boldsymbol{\mu}_r^0) + \boldsymbol{\mu}_r^0$, $\mu_i' = \mu_i$ for $i > r$ and $\sigma' = \sigma$.
3. If $\bar{g}(\boldsymbol{\mu}, \sigma) = (\boldsymbol{\mu}', \sigma')$, then $\mu_i' = b(\mu_i - \mu_i^0) + \mu_i^0$ for $i \le r$, $\mu_i' = b\mu_i$ for $i > r$, and $\sigma' = b\sigma$.

Because the μ_i' for $i \leq r$ are functions only of the μ_i for $i \leq r$ and not of σ and μ_i for $i > r$, condition (a) of Theorem 5.8.3 is satisfied. To check condition (b) it is sufficient to show that $\bar{g}^{-1} S(g(\mathbf{y})) = S(\mathbf{y})$ for any set of generators in any of the groups $\mathcal{G}_{\boldsymbol{\mu}_{r}0}$. The generators in (1) do not change $S(\mathbf{y})$. For g satisfying (2)

$$S(g(y)) = \{\boldsymbol{\mu}_r: (g\mathbf{Y}_r - \boldsymbol{\mu}_r)^T(g\mathbf{Y}_r - \boldsymbol{\mu}_r) \leq c\hat{\sigma}^2\},$$

so that

$$\begin{aligned}
\bar{g}^{-1} S(g(\mathbf{y})) &= \{\bar{g}^{-1}\boldsymbol{\mu}_r: (g\mathbf{Y}_r - \boldsymbol{\mu}_r)^T(g\mathbf{Y}_r - \boldsymbol{\mu}_r) \leq c\hat{\sigma}^2\} \\
&= \{\boldsymbol{\mu}_r: (g\mathbf{Y}_r - \bar{g}\boldsymbol{\mu}_r)^T(g\mathbf{Y}_r - \bar{g}\boldsymbol{\mu}_r) \leq c\hat{\sigma}^2\} \\
&= \{\boldsymbol{\mu}_r: (\mathbf{Y}_r - \boldsymbol{\mu}_r)^T(\mathbf{Y}_r - \boldsymbol{\mu}_r) \leq c\hat{\sigma}^2\} \\
&= S(\mathbf{y}).
\end{aligned}$$

For g satisfying (3),

$$\begin{aligned}
\bar{g}^{-1} S(g(\mathbf{y})) &= \{\bar{g}^{-1}\boldsymbol{\mu}_r: (g\mathbf{Y}_r - \boldsymbol{\mu}_r)^T(g\mathbf{Y}_r - \boldsymbol{\mu}_r) \leq cb^2\hat{\sigma}^2\} \\
&= \{\boldsymbol{\mu}_r: (g\mathbf{Y}_r - \bar{g}\boldsymbol{\mu}_r)^T(g\mathbf{Y}_r - \bar{g}\boldsymbol{\mu}_r) \leq cb^2\sigma^2\} \\
&= \{\boldsymbol{\mu}_r: b^2(\mathbf{Y}_r - \boldsymbol{\mu}_r)^T(\mathbf{Y}_r - \boldsymbol{\mu}_r) \leq cb^2\hat{\sigma}^2\} \\
&= S(\mathbf{y}).
\end{aligned}$$

The conditions in Theorem 5.8.3 are seen to be fulfilled; therefore the confidence regions are UMA invariant.

Using the methods of Section 5.9, we may translate the confidence sets (5.131) out of canonical form into terms of $\mathbf{X}$ and $\boldsymbol{\xi}$ of the original problem. The confidence sets (5.131) become

$$S(\mathbf{x}) = \{\boldsymbol{\xi}: \sum_{i=1}^{n} (\hat{\xi}_i(\mathbf{X} - \boldsymbol{\xi}) - \hat{\hat{\xi}}_i(\mathbf{X} - \boldsymbol{\xi}))^2 \leq c\hat{\sigma}^2\}, \qquad (5.132)$$

where $c = rF_{r,n-k;\alpha}$ and where $\hat{\sigma}^2$ in terms of $\mathbf{X}$ is

$$\hat{\sigma}^2 = \sum_{1}^{n} (X_i - \hat{\xi}_i(\mathbf{X}))^2/(n - k).$$

Because $\hat{\xi}_i(\mathbf{X})$ and $\hat{\hat{\xi}}_i(\mathbf{X})$ are linear functions of their arguments, these sets are ellipsoids in E_n. The more usual form of the confidence ellipsoids is in terms of $\boldsymbol{\theta}$; this may be obtained by replacing $\boldsymbol{\xi}$ with $\mathbf{A}\boldsymbol{\theta}$.

As an example, consider the two-way classification with one observation per cell, X_{ij}, $i = 1, \cdots, I$, $j = 1, \cdots, J$, where

$$EX_{ij} = \xi_{ij} = \xi + \mu_i + \eta_j,$$

with $\sum \mu_i = 0$ and $\sum \eta_j = 0$. As observed in Section 5.9, $n = IJ$, $k = I + J - 1$, and

$$\hat{\sigma}^2 = \sum\sum (X_{ij} - \bar{X}_{i.} - \bar{X}_{.j} + \bar{X}_{..})^2/(IJ - I - J + 1).$$

If we are testing $H_0: \mu_1 = \mu_2 = \cdots = \mu_I = 0$, so that $r = I - 1$, then

$$\sum\sum (\hat{\xi}_{ij} - \hat{\hat{\xi}}_{ij})^2 = \sum\sum (\bar{X}_{i.} - \bar{X}_{..})^2.$$

Replacing each X_{ij} with $X_{ij} - \xi_{ij} = X_{ij} - \xi - \mu_i - \eta_j$, we find that $\bar{X}_{i.} - \bar{X}_{..}$ is replaced by $\bar{X}_{i.} - \bar{X}_{..} - \mu_i$, and the UMA invariant family of confidence sets (5.132) is

$$S(\mathbf{x}) = \{\boldsymbol{\mu}: \sum\sum (\bar{X}_{i.} - \bar{X}_{..} - \mu_i)^2 \le (I - 1)\hat{\sigma}^2 F_{(I-1),IJ-I-J+1;\alpha}\}. \tag{5.133}$$

In interpreting the sets (5.133), we should remember that $\boldsymbol{\mu}$ is further restricted so that $\sum \mu_i = 0$. By symmetry the UMA invariant family of confidence sets for the η_j must be

$$S(\mathbf{x}) = \{\boldsymbol{\eta}: \sum\sum (\bar{X}_{.j} - \bar{X}_{..} - \eta_j)^2 \le (J - 1)\hat{\sigma}^2 F_{(J-1),IJ-I-J+1;\alpha}\}. \tag{5.134}$$

If we want a confidence ellipsoid for $\mu_1, \cdots, \mu_I, \eta_1, \cdots, \eta_J$, we must first find the numerator of the F-statistic for testing the hypothesis

$$H_0: \mu_1 = \cdots = \mu_I = \eta_1 = \cdots = \eta_J = 0.$$

This numerator turns out to be

$$[\sum\sum (\bar{X}_{i.} - \bar{X}_{..})^2 + \sum\sum (\bar{X}_{.j} - \bar{X}_{..})^2]/(I + J - 2).$$

Hence the UMA invariant family of confidence sets for $\boldsymbol{\mu}$ and $\boldsymbol{\eta}$ is

$$S(\mathbf{x}) = \{(\boldsymbol{\mu}, \boldsymbol{\eta}): \sum\sum (\bar{X}_{i.} - \bar{X}_{..} - \mu_i)^2 + \sum\sum (\bar{X}_{.j} - \bar{X}_{..} - \eta_j)^2 \le (I + J - 2)\hat{\sigma}^2 F_{(I+J-2),IJ-I-J+1;\alpha}\}. \tag{5.135}$$

Multiple Comparisons. To introduce the subject of multiple comparisons we consider the one-way classification with J observations per cell. Let X_{ij} satisfy the general linear hypothesis with $\xi_{ij} = EX_{ij} = \theta_i$ for $i = 1, \cdots, I$ and $j = 1, \cdots, J$. If we test the hypothesis

$$H_0: \theta_1 = \cdots = \theta_I$$

and reject it, we should, in general, like to know what it is that is making us do so. Perhaps one of the θ_i is much larger than the others and we should like to know which. More typically, we should like to make a tentative ranking of the θ_i and to indicate which of the differences $\theta_i - \theta_j$ seem to be significantly large as evidenced by the data. This can be considered as equivalent to making many confidence interval statements. One simple way of making multiple confidence statements is to make each statement considered by itself at the 95 percent level. If each of n statements is made separately at the 95 percent level, the expected number of true statements will be $0.95n$. We now describe another method of making these multiple comparisons, due to Scheffé (1953) or see Scheffé (1959), which has the property that the probability is at least 0.95 that *all* confidence interval statements made will be correct.

We describe the multiple comparison method for the canonical form of the general linear hypothesis and later make the usual transformation to general form. The following lemma is fundamental.

Multiple Comparisons Lemma. Let $\mathbf{y} \in E_r$, $\boldsymbol{\mu} \in E_r$, and let $c_0 \geq 0$. Then $\| \mathbf{y} - \boldsymbol{\mu} \|^2 \leq c_0$ if, and only if, for all $\mathbf{a} \in E_r$, $(\mathbf{a}^T\mathbf{y} - \mathbf{a}^T\boldsymbol{\mu})^2 \leq c_0 \| \mathbf{a} \|^2$.

Proof. *If.* Let $\mathbf{a} = (\mathbf{y} - \boldsymbol{\mu})$. The inequality $(\mathbf{a}^T(\mathbf{y} - \boldsymbol{\mu}))^2 \leq c_0 \| \mathbf{a} \|^2$ becomes $\| \mathbf{y} - \boldsymbol{\mu} \|^4 \leq c_0 \| \mathbf{y} - \boldsymbol{\mu} \|^2$. If $\mathbf{y} \neq \boldsymbol{\mu}$, $\| \mathbf{y} - \boldsymbol{\mu} \|^2$ may be canceled, thus yielding $\| \mathbf{y} - \boldsymbol{\mu} \|^2 \leq c_0$. If $\mathbf{y} = \boldsymbol{\mu}$, the inequality $\| \mathbf{y} - \boldsymbol{\mu} \|^2 \leq c_0$ is obvious.

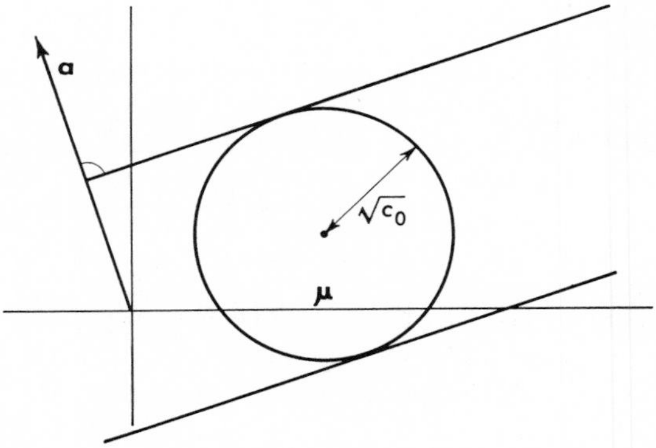

Fig. 5.7

Only if. By the Schwarz inequality $(\mathbf{a}^T(\mathbf{y} - \boldsymbol{\mu}))^2 \leq \|\mathbf{a}\|^2 \|\mathbf{y} - \boldsymbol{\mu}\|^2$. Hence $(\mathbf{a}^T\mathbf{y} - \mathbf{a}^T\boldsymbol{\mu})^2 \leq c_0 \|\mathbf{a}\|^2$, thus completing the proof.

This lemma may be interpreted geometrically (Fig. 5.7). The inequality $\|\mathbf{y} - \boldsymbol{\mu}\|^2 \leq c_0$ states that $\mathbf{y}$ lies in the sphere of radius $\sqrt{c_0}$ centered at $\boldsymbol{\mu}$. If $\mathbf{a}$ is a unit vector ($\|\mathbf{a}\|^2 = 1$), then $(\mathbf{a}^T(\mathbf{y} - \boldsymbol{\mu}))^2 \leq c_0$ means that $\mathbf{y}$ lies between the two parallel planes tangent to the sphere $\|\mathbf{y} - \boldsymbol{\mu}\|^2 \leq c_0$ and perpendicular to the vector $\mathbf{a}$. Thus the lemma could be restated geometrically: a closed sphere is equal to the intersection of all regions formed between (and on) parallel tangent planes.

This lemma may be used in the canonical problem, $\mathbf{Y} \in \mathfrak{N}(\boldsymbol{\mu}, \sigma^2\mathbf{I})$, where $\mu_{k+1} = \cdots = \mu_n = 0$, as follows. The usual confidence statement for $(\mu_1, \cdots, \mu_r)$ is

$$P\{\sum_1^r (Y_i - \mu_i)^2 \leq c\hat{\sigma}^2\} = 1 - \alpha, \tag{5.136}$$

where $c = rF_{r,n-k;\alpha}$ and where $\hat{\sigma}^2 = \sum_{k+1}^n Y_i^2/(n-k)$. This, by the lemma, is equivalent to

$$P\{(\sum_1^r a_iY_i - \sum_1^r a_i\mu_i)^2 \leq \|\mathbf{a}\|^2 \hat{\sigma}^2 c \quad \text{for all} \quad \mathbf{a} \in E_r\} = 1 - \alpha. \tag{5.137}$$

The important point to note is that the words "for all $\mathbf{a} \in E_r$" appear on the inside of the probability statement. Thus, if we make all confidence interval statements

$$\sum_1^r a_iY_i - \|\mathbf{a}\| \hat{\sigma}\sqrt{c} \leq \sum_1^r a_i\mu_i \leq \sum_1^r a_iY_i + \|\mathbf{a}\| \hat{\sigma}\sqrt{c}, \tag{5.138}$$

the probability that all such statements are true is $1 - \alpha$. Furthermore, these multiple statements are equivalent to the F-test of the hypothesis $H_0: \mu_1 = \cdots = \mu_r = 0$. If the F-test accepts H_0, all the confidence intervals (5.138) contain the origin. If the F-test rejects the hypothesis H_0, the confidence sphere (5.131) does not contain the origin and there is some linear combination $\sum_1^r a_i\mu_i$, whose corresponding confidence interval (5.138) does not contain zero.

The statement (5.137) may be written more concisely. Let L be a linear function of $\mu_1, \cdots, \mu_r$ and let $\hat{L}$ be the least squares estimate of L obtained by replacing $(\mu_1, \cdots, \mu_r)$ in L with $(Y_1, \cdots, Y_r)$. If $L(\boldsymbol{\mu}) = \sum_1^r a_i\mu_i$, the variance of $\hat{L}$ may be written var $(\hat{L}) = \sigma^2\sum_1^r a_i^2$. An estimate of the variance of $\hat{L}$ is $\widehat{\text{var}}\ (\hat{L}) = \hat{\sigma}^2 \sum_1^r a_i^2$. Hence (5.137) may be written

$$P\{(\hat{L} - L)^2 \leq c\ \widehat{\text{var}}\ (\hat{L}) \quad \text{for all} \quad L \in \mathfrak{L}_1\} = 1 - \alpha, \qquad (5.139)$$

where $c = rF_{r,n-k;\alpha}$ and $\mathfrak{L}_1$ is the space of all linear functions of $(\mu_1, \cdots, \mu_r)$. This equation is particularly appropriate for transformation to the general form.

In the general form $\mathbf{X} \in \mathfrak{N}(\mathbf{A}\boldsymbol{\theta}, \sigma^2\mathbf{I})$ with $\mathbf{A}$ of full rank k. Suppose we are interested in making confidence statements about a certain class of linear combinations of the θ_i which span a space $\mathfrak{L}_1$ of dimension r. We may make the usual orthogonal transformation to canonical form in which the space $\mathfrak{L}_1$ becomes the space of linear combinations of the first r transformed variables. Therefore statement (5.139) applies equally well to the general form.

Consider again the one-way classification with J observations per cell, $X_{ij} \in \mathfrak{N}(\xi_{ij}, \sigma^2)$, where $\xi_{ij} = \theta_i$ for $i = 1, \cdots, I$, $j = 1, \cdots, J$. Suppose we are interested in making confidence statements about all the differences of the means. This spans a space $\mathfrak{L}_1$ of dimension $I - 1$, $\mathfrak{L}_1 = \{L = \sum a_i\theta_i, \text{ with } \sum a_i = 0\}$, so that (5.139) may be applied with $r = I - 1$ and $n - k = IJ - I$. Because $\hat{\theta}_i = \bar{X}_{i.}$, we have $\hat{L} = \sum a_i\bar{X}_{i.}$ and $\text{var}\ (\hat{L}) = \sum a_i^2\sigma^2/J$. Hence the probability is $1 - \alpha$ that for all $\mathbf{a}$ for which $\sum a_i = 0$

$$\left(\sum a_i\bar{X}_{i.} - \sum a_i\theta_i\right)^2 \leq \frac{c\sum a_i^2\hat{\sigma}^2}{J} \qquad (5.140)$$

where $\hat{\sigma}^2 = \sum\sum(X_{ij} - \bar{X}_{i.})^2/(IJ - I)$ and $c = (I - 1)F_{I-1,IJ-I;\alpha}$. In comparing θ_1 and θ_2, for instance, we take $a_1 = 1$, $a_2 = -1$, $a_3 = \cdots = a_I = 0$. The confidence interval (5.140) becomes

$$\bar{X}_{1.} - \bar{X}_{2.} - \left(\frac{2\hat{\sigma}^2c}{J}\right)^{1/2} \leq \theta_1 - \theta_2 \leq \bar{X}_{1.} - \bar{X}_{2.} + \left(\frac{2\hat{\sigma}^2c}{J}\right)^{1/2}. \qquad (5.141)$$

In the one-way classification it is customary to summarize the multiple comparison procedure by listing the means in increasing order of their estimates, with accompanying statements regarding which of the various confidence intervals contain the origin. If $I = 6$, $J = 4$, $\bar{X}_{1.} = 2$, $\bar{X}_{2.} = 4$, $\bar{X}_{3.} = -1$, $\bar{X}_{4.} = 6$, $\bar{X}_{5.} = 3$, $\bar{X}_{6.} = 10$, $\hat{\sigma}^2 = 2.4$, and $\alpha = 0.10$, then $F_{5,18;\alpha} = 2.20$, $(2\hat{\sigma}^2c/J)^{1/2} = 3.63, \cdots$, and we rank

$$\underline{\theta_3 < \theta_1} < \underline{\theta_5 < \theta_2 < \theta_4} < \theta_6, \qquad (5.142)$$

where a bar jointly beneath θ_i and θ_j means that the confidence interval for $\theta_i - \theta_j$ contains the origin. It may happen that the usual F-test rejects the hypothesis $H_0:\theta_1 = \cdots = \theta_I$ and that in the ranking all confidence intervals for the differences contain zero. This is possible; what must be satisfied when the F-test rejects H_0 is that there must be some linear combination $\sum a_i\theta_i$ with $\sum a_i = 0$ whose confidence interval does not contain zero.

If we were interested in finding in addition a confidence interval for, say, θ_6 alone, then $\mathfrak{L}_1$ becomes the I-dimensional space of all linear combinations of the θ_i and statement (5.140) would be changed to state: the probability is $1 - \alpha$ that for all $(a_1, \cdots, a_I)$ $(\sum a_i\theta_i - \sum a_i\hat{\theta}_i)^2 \leq \sum a_i^2\hat{\sigma}^2 c$, where here $c = IF_{I,IJ-I;\alpha}$.

As a final example, consider the two-way classification with one observation per cell, X_{ij}, $i = 1, \cdots, I$, $j = 1, \cdots, J$, where $EX_{ij} = \xi + \mu_i + \eta_j$ with $\sum \mu_i = 0$ and $\sum \eta_j = 0$. If we are interested in obtaining simultaneous confidence intervals for $\mu_1, \cdots, \mu_I$, we note that because $\mu_I = -\sum_1^{I-1}\mu_i$ this is equivalent to obtaining simultaneous confidence intervals for $\mu_1, \cdots, \mu_{I-1}$, which spans a space $\mathfrak{L}_1$ of dimension $r = I - 1$. Arbitrary linear combinations $\sum_1^{I-1} a_i\mu_i$, however, are equivalent to linear combinations $\sum_1^I a_i'\mu_i$, whose coefficients sum to zero: $\sum_1^I a_i' = 0$. To see this, let $\Delta = \sum_1^{I-1} a_i/I$ and define $a_i' = a_i - \Delta$ for $i = 1, \cdots, I - 1$, and $a_I' = -\Delta$. Then $\sum_1^I a_1' = 0$ and

$$\sum_1^I a_i'\mu_i = \sum_1^{I-1} (a_i - \Delta)\mu_i - \Delta\mu_I = \sum_1^{I-1} a_i\mu_i.$$

Hence

$$\mathfrak{L}_1 = \{\sum_1^I a_i\mu_i : \sum_1^I a_i = 0\}.$$

Now, $\hat{\mu}_i = \bar{X}_{i.} - \bar{X}_{..}$, so that for $L \in \mathfrak{L}_1$

$$\text{Var}\,\hat{L} = \text{Var}\,(\sum a_i(\bar{X}_{i.} - \bar{X}_{..})) = \text{Var}\,(\sum a_i\bar{X}_{i.}) = \sum a_i^2\sigma^2/J.$$

Hence statement (5.139) becomes the probability is $1 - \alpha$ that

$$(\sum a_i\bar{X}_{i.} - \sum a_i\mu_i)^2 \leq \frac{\sum a_i^2\hat{\sigma}^2 c}{J} \tag{5.143}$$

for all $\mathbf{a} \in E_I$ for which $\sum a_i = 0$, where

$$\hat{\sigma}^2 = \sum\sum(X_{ij} - \bar{X}_{i.} - \bar{X}_{.j} + \bar{X}_{..})/(IJ - I - J + 1)$$

and

$$c = (I - 1)F_{I-1,IJ-I-J+1;\alpha}.$$

Exercises

1. Prove the following extensions of the multiple comparisons lemma. Let $\mathbf{x} \in E_r$, $\boldsymbol{\theta} \in E_r$, $c_0 \geq 0$ and let $\boldsymbol{\Sigma}$ be an $r \times r$ symmetric positive definite matrix.

(a) $(\mathbf{x} - \boldsymbol{\theta})^T\boldsymbol{\Sigma}^{-1}(\mathbf{x} - \boldsymbol{\theta}) \leq c_0$ if, and only if, $(\mathbf{a}^T(\mathbf{x} - \boldsymbol{\theta}))^2 \leq c_0\mathbf{a}^T\boldsymbol{\Sigma}\mathbf{a}$, for all $\mathbf{a} \in E_r$.

(b) Suppose $\mathbf{1}^T(\mathbf{X} - \boldsymbol{\theta}) = 0$ where $\mathbf{1}^T = (1, 1, \cdots, 1)$. Then

$$||\mathbf{X} - \boldsymbol{\theta}||^2 \leq c_0$$

if, and only if, $(\mathbf{a}^T(\mathbf{X} - \boldsymbol{\theta}))^2 \leq c_0||\mathbf{a}||^2$ for all $\mathbf{a} \in E_r$ such that $\mathbf{1}^T\mathbf{a} = 0$.

(c) Suppose $\mathbf{1}^T\boldsymbol{\Sigma}^{-1}(\mathbf{X} - \boldsymbol{\theta}) = 0$. Then $(\mathbf{X} - \boldsymbol{\theta})^T\boldsymbol{\Sigma}^{-1}(\mathbf{X} - \boldsymbol{\theta}) \leq c_0$ if, and only if, $(\mathbf{a}^T(\mathbf{X} - \boldsymbol{\theta}))^2 \leq c_0\mathbf{a}^T\boldsymbol{\Sigma}\mathbf{a}$ for all $\mathbf{a} \in E_r$ such that $\mathbf{1}^T\mathbf{a} = 0$.

2. Using Exercise 1(b), show how the multiple confidence statement (5.143) follows directly from the confidence set (5.133).

3. Let X_{ij} be observations from a one-way classification with differing numbers of observations per cell,

$$EX_{ij} = \theta_i, \quad i = 1, \cdots, I, \quad j = 1, \cdots, n_i.$$

(a) Find the UMA invariant confidence ellipsoid for the parameters θ_i, $i = 1, \cdots, I$. Derive the multiple confidence interval statements for all linear combinations of $\theta_1, \cdots, \theta_I$ directly from this confidence ellipsoid, using Exercise 1(a).

(b) Write $\theta_i = \xi + \mu_i$ where $\xi = \bar{\theta}$ and $\mu_i = \theta_i - \bar{\theta}$ so that $\sum \mu_i = 0$. Find the UMA invariant confidence ellipsoid for the parameters μ_i, $i = 1, \cdots, I$. Derive the multiple confidence interval statements for all linear combinations $\sum a_i\theta_i$ for which $\sum a_i = 0$ directly from this confidence ellipsoid, using Exercise 1(c).

4. Suppose that in a one-way classification with $I = 4$ and $n_i = 6$ observations per cell it is observed that $\bar{X}_{1.} = 0$, $\bar{X}_{2.} = 4$, $\bar{X}_{3.} = 5$, $\bar{X}_{4.} = 7$, and $\hat{\sigma}^2 = 10$.

(a) Find the ordering similar to (5.142), using $\alpha = 0.10$. ($F_{3,20;.10} = 2.38$.)

(b) Find the ordering similar to (5.142), using $\alpha = 0.01$ ($F_{3,20;.10} = 4.94$.)

(c) Show that the F-test of the hypothesis $H_0{:}\theta_1 = \cdots = \theta_4$ at level $\alpha = 0.01$ rejects that hypothesis.

(d) Find a linear combination $\sum a_i\theta_i$ with $\sum a_i = 0$, whose associated confidence interval in the multiple comparison with $\alpha = 0.01$ does not contain the origin.

5. Let X_i be the observations in a regression model in which $\xi_i = EX_i = \beta_0 + \beta_1 z_i + \beta_2 z_i^2$, for $i = 1, \cdots, n$, and suppose that $\sum z_i = 0$, $\sum z_i^2 = 1$, and $\sum z_i^3 = 0$.
(a) Find the least squares estimates $\hat{\beta}_0$, $\hat{\beta}_1$, and $\hat{\beta}_2$.
(b) Find the UMA invariant confidence ellipsoid for β_1 and β_2.
(c) Find the multiple confidence interval statements for all linear combinations of β_1 and β_2.

6. In the two-way classification with one observation per cell ($EX_{ij} = \xi + \mu_i + \eta_j$, $i = 1, \cdots, I$, $j = 1, \cdots, J$, $\sum \mu_i = 0$, $\sum \eta_j = 0$) find the multiple confidence interval statements for all linear combinations of μ_i and η_j.

7. The UMA invariant confidence ellipsoids are also best invariant (under a slightly different group) set estimates (see Exercises 4.6.9 and 4.6.10). Let $\Theta = \{(\mu_1, \cdots, \mu_k, \sigma), \sigma > 0\}$, let $\mathcal{A}$ be the set of all open (or measurable) subsets of E_r, let $L(\theta, a) = cl_r(a)\sigma^{-1} - I_a(\mu_1, \cdots, \mu_r)$, where l_r is Lebesgue measure on E_r, and let

$$\mathbf{Y} \in \mathfrak{N}(\boldsymbol{\mu}, \sigma^2\mathbf{I}),$$

where

$$\mathbf{Y} = (Y_1, \cdots, Y_n)^T, \qquad \boldsymbol{\mu} = (\mu_1, \cdots, \mu_n)^T,$$

and

$$\mu_{k+1} = \cdots = \mu_n = 0.$$

A sufficient statistic for this problem is $(Y_1, \cdots, Y_k, \sum_{k+1}^n Y_i^2)$.
(a) Show that the problem is invariant under the group generated by the groups $\mathcal{G}_1: g(Y_1, \cdots, Y_k, \sum_{k+1}^n Y_i^2) = (Y_1 + b_1, \cdots, Y_k + b_k, \sum_{k+1}^n Y_i^2)$, and $\mathcal{G}_2: g(Y_1, \cdots, Y_k, \sum_{k+1}^n Y_i^2) = (KY_1, \cdots, KY_k, K^2 \sum_{k+1}^n Y_i^2)$, where $K > 0$.
(b) Find the form of the nonrandomized invariant decision rules.
(c) Show that a best invariant decision rule is given by (5.131) (with strict inequality if $\mathcal{A}$ contains only open sets).

CHAPTER 6

Multiple Decision Problems

6.1 Monotone Multiple Decision Problems

Multiple decision problems were defined in Section 1.3 as those decision problems in which the space $\mathcal{A}$ of actions of the statistician consists of a finite number ($k \geq 3$) of points, $\mathcal{A} = \{a_1, \cdots, a_k\}$. There are two distinct types of multiple-decision problem that seem to arise in practice. In one the parameter space Θ is partitioned into k subsets $\Theta_1, \cdots, \Theta_k$, according to the increasing value of a single real parameter $\gamma(\theta)$, and action a_i is preferred if $\theta \in \Theta_i$. For example, if an experimenter is comparing two treatments with means θ_1 and θ_2, he might have available to him only a finite number of actions $a_1, \cdots, a_k$, among which he has preferences based on the magnitude of the differences of the means $\theta_2 - \theta_1$. An important particular case occurs when he may choose from the three alternatives: (a) decide treatment 1 is better, (b) decide treatment 2 is better, and (c) withhold judgment until more evidence is available. This type of multiple decision problem is called monotone and is treated in this section. The other type arises when the actions have symmetric effects; for example, when the problem is to choose one of k drugs to use on a certain patient or to choose one of k beauty contestants to receive a prize. This type of problem is treated in the next two sections.

For multiple decision problems a nonrandomized decision rule may be represented by a measurable partition $(\omega_1, \cdots, \omega_k)$ of the sample space $\mathfrak{X}$ (that is, $\omega_i \subset \mathfrak{X}$, ω_i are pairwise disjoint and $\bigcup \omega_i = \mathfrak{X}$), with the understanding that if X falls in ω_i action a_i is taken. Randomized decision rules therefore choose at random a partition from the class of all measurable partitions of $\mathfrak{X}$. This involves a rather complicated random mechanism. Because behavioral decision rules, on the contrary, are rather simple to describe, we deal with them in the following. A behavioral decision rule ϕ gives for each $x \in \mathfrak{X}$ a probability distribution over $\mathfrak{A}$. We use the notation $\phi(i \mid x)$ to represent the probability that a decision rule ϕ chooses action a_i when $X = x$ is observed. Thus a behavioral decision rule may be represented as

$$\phi: (\phi(1 \mid x), \cdots, \phi(k \mid x)), \tag{6.1}$$

where, for each i, $\phi(i \mid x)$ is a nonnegative measurable function of x, and $\sum_1^k \phi(i \mid x) = 1$ for all x. It may be noticed that, for each i, $\phi(i \mid x)$ is a test, because $0 \leq \phi(i \mid x) \leq 1$. Using this notation, we may write the risk function for a given decision rule ϕ as

$$\hat{R}(\theta, \phi) = \sum_{i=1}^{k} L(\theta, a_i) E_\theta\, \phi(i \mid X). \tag{6.2}$$

Thus the risk function depends on ϕ only through the values of $E_\theta\, \phi(i \mid X)$. In the following we drop the notation $\hat{R}$ and use R to denote the risk function, remembering that we are assuming that behavioral rules and randomized rules are equivalent, as in Section 1.5.

Monotone multiple decision problems are a generalization of the one-sided hypothesis-testing problems discussed in Section 5.2 to the case $\mathfrak{A} = \{a_1, \cdots, a_k\}$, $k \geq 3$. The following definition makes this precise. Note that Θ is assumed here to be a subset of the real line.

Definition 1. A multiple decision problem with Θ a subset of the real line is said to be *monotone* if for some ordering of $\mathfrak{A}$, say $\mathfrak{A} = \{a_1, \cdots, a_k\}$, there exist numbers $\theta_1 \leq \theta_2 \leq \cdots \leq \theta_{k-1}$ ($\theta_i \in \Theta$ for $i = 1, \cdots, k-1$) such that the loss function satisfies

$$\begin{aligned} L(\theta, a_i) - L(\theta, a_{i+1}) &\leq 0 \quad \text{for} \quad \theta < \theta_i, \\ L(\theta, a_i) - L(\theta, a_{i+1}) &\geq 0 \quad \text{for} \quad \theta > \theta_i, \end{aligned} \tag{6.3}$$

for $i = 1, 2, \cdots, k-1$.

This is clearly a generalization of the inequalities (5.27) which define a one-sided hypothesis-testing problem. The inequalities (6.3) imply for $i = 1, \cdots, k$, that if $\theta \in \Theta$ and $\theta_{i-1} < \theta < \theta_i$ (where $\theta_0 = -\infty$ and $\theta_k = +\infty$), then action a_i is preferred, since, for $\theta_{i-1} < \theta < \theta_i$, $L(\theta, a_i) = \inf_j L(\theta, a_j)$. Note that no restriction is put on the difference $L(\theta, a_i) - L(\theta, a_{i+1})$ when $\theta = \theta_i$. We shall extend Theorem 5.2.2 to the multiple decision case. Therefore we define a class of decision rules, analogous to the one-sided tests of (5.24), which may be called monotone decision rules and which we hope will form an essentially complete class under some assumption regarding the distribution of the random observable X. As in Section 5.2, we assume that X is a random variable with a monotone (nondecreasing) likelihood ratio in θ, as in Definition 5.2.3. For extensions, see Karlin and Rubin (1956). The sample space $\mathfrak{X}$ is thus taken to be the real line, and monotone decision rules may be defined as follows:

Definition 2. For a given monotone multiple decision problem, as in Definition 1, with sample space the real line, a decision rule ϕ is said to be *monotone* if there exist numbers x_i, γ_i, and γ_i' for $i = 1, \cdots, k-1$, with

$$-\infty = x_0 \le x_1 \le x_2 \le \cdots \le x_k = +\infty, \; 0 \le \gamma_i \le 1,$$

and $0 \le \gamma_i' \le 1$, such that for $i = 1, 2, \cdots, k$,

$$\phi(i \mid x) = \begin{cases} 0 & \text{if } x < x_{i-1}, \\ \gamma_{i-1}' & \text{if } x = x_{i-1}, \\ 1 & \text{if } x_{i-1} < x < x_i, \\ \gamma_i & \text{if } x = x_i, \\ 0 & \text{if } x > x_i. \end{cases} \tag{6.4}$$

This generalizes the notion of a one-sided test.

Theorem 1. If the distribution of the random observable has monotone likelihood ratio, the class of monotone decision rules is essentially complete for a monotone multiple decision problem.

Proof. Let $\psi: (\psi(1 \mid x), \cdots, \psi(k \mid x))$ be any decision rule and define

$$\psi_j(x) = \sum_{i=j+1}^{k} \psi(i \mid x) \qquad \text{for } j = 0, 1, \cdots, k.$$

Since $0 \leq \psi_j(x) \leq 1$, ψ_j may be considered as a test for testing, say, $H_0: \theta = \theta_j$. Therefore by Corollary 5.2.1 there exists a one-sided test $\phi_j(x)$ in the form of (5.24),

$$\phi_j(x) = \begin{cases} 1 & \text{if } x > x_j, \\ \gamma_j'' & \text{if } x = x_j, \\ 0 & \text{if } x < x_j, \end{cases} \tag{6.5}$$

for $j = 1, 2, \cdots, k - 1$, such that

$$E_\theta \phi_j(X) - E_\theta \psi_j(X) \begin{cases} \leq 0 & \text{for } \theta < \theta_j, \\ = 0 & \text{for } \theta = \theta_j, \\ \geq 0 & \text{for } \theta > \theta_j. \end{cases} \tag{6.6}$$

In addition, let $\phi_0(x) \equiv 1$ and $\phi_k(x) \equiv 0$. Now, because

$$E_{\theta_{j-1}} \phi_{j-1}(X) = E_{\theta_{j-1}} \psi_{j-1}(X) \geq E_{\theta_{j-1}} \psi_j(X) \geq E_{\theta_{j-1}} \phi_j(X),$$

we may take (changing x_j and γ_j'' if necessary) $\phi_{j-1}(x) \geq \phi_j(x)$ for all x, that is, $x_1 \leq x_2 \leq \cdots \leq x_k$, and so on. Hence we may define $\phi(j \mid x) = \phi_{j-1}(x) - \phi_j(x)$ for $j = 1, \cdots, k$, so that $\phi: (\phi(1 \mid x), \cdots, \phi(k \mid x))$ is a monotone decision rule. We show that ϕ is as good as ψ. From (6.2)

$$\begin{aligned} R(\theta, \psi) - R(\theta, \phi) &= \sum_{i=1}^{k} L(\theta, a_i) \{E_\theta \psi(i \mid X) - E_\theta \phi(i \mid X)\} \\ &= \sum_{i=1}^{k} L(\theta, a_i) \{[E_\theta \phi_i(X) - E_\theta \psi_i(X)] \\ &\qquad - [E_\theta \phi_{i-1}(X) - E_\theta \psi_{i-1}(X)]\} \\ &= \sum_{i=1}^{k-1} [E_\theta \phi_i(X) - E_\theta \psi_i(X)][L(\theta, a_i) - L(\theta, a_{i+1})]. \end{aligned} \tag{6.7}$$

Each term of this sum is nonnegative, for, if $\theta < \theta_i$, then $L(\theta, a_i) - L(\theta, a_{i+1}) \leq 0$ and $E_\theta\, \phi_i(X) - E_\theta\, \psi_i(X) \leq 0$, whereas, if $\theta > \theta_i$, then $L(\theta, a_i) - L(\theta, a_{i+1}) \geq 0$ and $E_\theta\, \phi_i(X) - E_\theta\, \psi_i(X) \geq 0$ and, if $\theta = \theta_i$, then $E_\theta \phi_i(X) - E_\theta\, \psi_i(X) = 0$. Hence for all $\theta \in \Theta$, $R(\theta, \phi) \leq R(\theta, \psi)$, showing that ϕ is as good as ψ and completing the proof.

It is interesting to note that the rule ϕ, which is as good as ψ, does not depend on the loss function L, provided that L has the form (6.3) and $\theta_1, \cdots, \theta_{k-1}$ are given (see Exercise 1).

Without further rather strong restrictions, not every monotone procedure is admissible (see Exercises 2 and 3). Thus the last statement of Theorem 5.2.2 does not extend easily to multiple decision problems. A study of conditions under which all monotone rules are admissible may be found in Karlin (1957b).

As an example of the use of Theorem 1, consider the multiple decision problem in which $\mathcal{A} = \{a_1, a_2, a_3\}$, Θ is the real line and X has a normal distribution with mean θ and variance 1. The loss function is given by Table 6.1.

Table 6.1

	a_1	a_2	a_3
$\theta < -\theta_0$	0	1	L
$-\theta_0 \leq \theta \leq \theta_0$	1	0	1
$\theta > \theta_0$	L	1	0

$L(\theta, a)$ (6.8)

We assume that $L \geq 1$ and therefore that the problem is monotone. Thus, if $\Theta_1 = (-\infty, -\theta_0)$, $\Theta_2 = [-\theta_0, \theta_0]$, and $\Theta_3 = (\theta_0, \infty)$, then a_i might be taken to represent the action "accept the hypothesis that $\theta \in \Theta_i$". If the true hypothesis is accepted, there is no loss. The loss is L if a_1 is taken when $\theta \in \Theta_3$ or if a_3 is taken when $\theta \in \Theta_1$, and the loss is one in the remaining cases.

Because the distribution of X is continuous, a monotone decision rule, ϕ, is determined by two numbers, $x_1 \le x_2$, where, say,

$$\begin{aligned} \phi(1 \mid x) &= I_{(-\infty, x_1)}(x), \\ \phi(2 \mid x) &= I_{[x_1, x_2]}(x), \\ \phi(3 \mid x) &= I_{(x_2, \infty)}(x). \end{aligned} \tag{6.9}$$

Because the distribution of X has monotone likelihood ratio, we may restrict attention to monotone rules.

Now suppose we are interested in finding a minimax rule. Because $P_\theta\{X \le x\} = P_{-\theta}\{-X \le x\}$ for all θ and x, the problem is invariant under the transformation $g(x) = -x$, with $\bar{g}(\theta) = -\theta$, $\tilde{g}(a_1) = a_3$, $\tilde{g}(a_2) = a_2$, and $\tilde{g}(a_3) = a_1$. By Theorem 4.3.1 we may restrict attention to the invariant rules, that is, rules for which $\phi(1 \mid -x) = \phi(3 \mid x)$ and $\phi(2 \mid x) = \phi(2 \mid -x)$. This suggests that perhaps we can restrict attention to the invariant monotone rules, those rules of the form (6.9) for which $x_1 = -x_2$ and $x_2 \ge 0$. Indeed, this is correct (Exercise 4). The risk function of invariant rules is symmetric about zero, and in the search for minimax rules among such rules we need consider the risk function for $\theta \ge 0$ only. The risk function for rule (6.9), with $x_1 = -x_2$, is

$$R(\theta, \phi) = \begin{cases} P_\theta\{X < -x_2\} + P_\theta\{X > x_2\} & \text{for } 0 \le \theta \le \theta_0, \\ LP_\theta\{X < -x_2\} + P_\theta\{-x_2 < X < x_2\} & \text{for } \theta > \theta_0. \end{cases}$$

This function is easily seen to be increasing in θ for $0 \le \theta \le \theta_0$ and decreasing in θ for $\theta > \theta_0$. Hence

$$\sup_\theta R(\theta, \phi) = \max [P_{\theta_0}\{X < -x_2\} + P_{\theta_0}\{X > x_2\}, (L-1)P_{\theta_0}\{X < -x_2\} + P_{\theta_0}\{X < x_2\}] \tag{6.10}$$

The minimax rule is obtained by finding that value of x_2 which minimizes this expression, a relatively straightforward numerical problem. If $L = 2$, this expression achieves its minimum when $P_{\theta_0}\{X > x_2\} = P_{\theta_0}\{X < x_2\}$, namely, when $x_2 = \theta_0$. If $L = 1$, this expression achieves its minimum when $P_{\theta_0}\{X < -x_2\} + P_{\theta_0}\{X > x_2\} = P_{\theta_0}\{X < x_2\}$; this gives a point x_2 somewhat larger than θ_0. For L sufficiently large the second term in the right side of (6.10) dominates, and the minimum of this expression is achieved when $(\partial/\partial x_2)[(L-1)P_{\theta_0}\{X < -x_2\} + P_{\theta_0}\{X < x_2\}] = 0$; using the normal distribution function in this expression, we find that $x_2 = (2\theta_0)^{-1} \log (L-1)$.

This analysis is valid for other distributions besides the normal. See Exercise 3 for the logistic distribution.

Exercises

1. Suppose that X has the binomial distribution $\mathcal{B}(5, \theta)$, with $\theta \in [0, 1]$; let $\mathcal{a} = \{a_1, a_2, a_3\}$ and let the loss function satisfy (6.3) with $\theta_1 = 1/3$ and $\theta_2 = 2/3$. Find a monotone decision rule which is as good as the decision rule ψ, which takes a_i if $X \equiv i - 1 \pmod 3$:

$$\psi(1 \mid x) = I_{\{0,3\}}(x), \quad \psi(2 \mid x) = I_{\{1,4\}}(x), \quad \psi(3 \mid x) = I_{\{2,5\}}(x).$$

2. Suppose X has a normal distribution with mean θ and variance one; let $\mathcal{a} = \{a_1, a_2, a_3\}$ and let $L(\theta, a_1) = (\theta + 1)^2$, $L(\theta, a_2) = \theta^2$, and $L(\theta, a_3) = (\theta - 1)^2$.
(a) Suppose that $\Theta = [-1, 1]$ and consider the monotone rule ϕ:

$$\phi(1 \mid x) = I_{(-\infty, -.1)}(x),$$
$$\phi(2 \mid x) = I_{[-.1, .1]}(x),$$
$$\phi(3 \mid x) = I_{(.1, \infty)}(x).$$

Show that ϕ is not admissible.
(b) Show that if $\Theta = (-\infty, \infty)$ then every monotone rule is admissible. *Hint.* It is sufficient to show that for a given monotone ϕ there does not exist a better monotone ψ. From (6.7) this can happen only if $E_\theta\phi_1(X) < E_\theta\psi_1(X)$ and $E_\theta\phi_2(X) > E_\theta\psi_2(X)$ for all θ. If this is so, a contradiction can be obtained by letting $\theta \to \pm\infty$.

3. In the example in the text suppose that X has the logistic distribution $\mathcal{L}(\theta, 1)$ (3.15).
(a) Find the minimax rule if $L = 1$.
(b) Find the minimax rule if $L = 2$.
(c) Show that the rule ϕ of (6.9) with $x_1 = -x_2$ is not admissible if $L > 1 + e^{x_2}$.

4. Consider the monotone multiple decision problem of Definition 1, let X have monotone likelihood ratio, and assume that the problem is invariant under the group generated by the transformation $gx = -x$, with $\bar{g}\theta = -\theta$, and $\tilde{g}_i = k + 1 - i$. Let ψ be any invariant rule. Show that there exists an invariant monotone rule ϕ that is as good as ψ. *Hint.* Show that the rule ϕ in the proof of Theorem 1 can be chosen to be invariant.

5. Let X have a normal distribution with mean θ and variance 1, let Θ be the real line, let $\mathcal{A} = \{a_1, a_2, a_3, a_4\}$, and let the loss be as in Table 6.2.

Table 6.2

	a_1	a_2	a_3	a_4
$\theta \le -1$	0	1	1	1
$-1 < \theta < 0$	1	0	1	1
$0 \le \theta < 1$	1	1	0	1
$\theta \ge 1$	1	1	1	0

$L(\theta, a)$

Find the minimax rule.
Ans. $\phi(1 \mid x) = I_{(-\infty,-b)}(x)$, $\phi(2 \mid x) = I_{(-b,0)}(x)$, $\phi(3 \mid x) = I_{(0,b)}(x)$, and $\phi(4 \mid x) = I_{(b,\infty)}(x)$, where b is determined by the equation,

$$P_0(X < b - 1) + P_0(X < b) = \tfrac{3}{2}, \qquad b = 1.262 \cdots.$$

6.2 Bayes Rules in Multiple Decision Problems

The features involved in treating nonmonotone multiple decision problems may be illustrated by considering a class of problems called *classification problems.* In problems of this type it is desired to classify an observation as coming from one of k known distributions. Specifically, we have $\Theta = \{\theta_1, \cdots, \theta_k\}$, $\mathcal{A} = \{a_1, \cdots, a_k\}$, and a random observable X, whose distribution, when θ_i is the true state of nature, has density (or probability mass function) $f_i(x)$, $i = 1, \cdots, k$. The loss is zero if a correct classification is made and one if an incorrect classification is made. If a_j represents the action "classify the observation as coming from distribution $f_j(x)$", the loss function may be written

$$L(\theta_i, a_j) = \begin{cases} 1 & \text{if} \quad i \neq j, \\ 0 & \text{if} \quad i = j. \end{cases} \tag{6.11}$$

For a given multiple decision rule $\phi : \phi(1 \mid x), \cdots, \phi(k \mid x)$ the risk function is

$$\begin{aligned} R(\theta_i, \phi) &= \sum_{j=1}^{k} L(\theta_i, a_j)\, E_{\theta_i}\, \phi(j \mid X) \\ &= 1 - E_{\theta_i}\, \phi(i \mid X). \end{aligned} \tag{6.12}$$

We search for a Bayes rule with respect to a prior distribution τ, giving probability p_i to θ_i, where $p_i \geq 0$ and $\sum_1^k p_i = 1$. The Bayes risk with respect to τ is

$$r(\tau, \phi) = 1 - \sum_{i=1}^{k} p_i E_{\theta_i}\, \phi(i \mid X). \tag{6.13}$$

The risk (6.13) may be interpreted as the probability of making a misclassification, using the rule ϕ, when it is known that the distribution from which X was taken was chosen according to τ. This situation occurs, for example, when one individual is drawn at random from a population stratified into k strata, with known proportion p_i of individuals in the ith stratum. One observation, X (X may as well be a vector), is made on this individual, and the distribution of X depends on the stratum this individual inhabits. In such a case the problem of choosing ϕ to minimize the risk (6.13) is equivalent to the problem of choosing ϕ to maximize the probability of a correct classification. The following theorem, due to Hoel and Peterson (1949), gives an answer to this problem.

Theorem 1. Any rule ϕ for which

$$\phi(i \mid x) = 0 \qquad \text{whenever } p_i f_i(x) < \max_j p_j f_j(x) \tag{6.14}$$

for $i = 1, \cdots, k$, is Bayes with respect to τ.

Remark. Because $\sum_1^k \phi(i \mid x) = 1$ for all x, (6.14) implies that when $p_i f_i(x) > p_j f_j(x)$ for all $j \neq i$, then $\phi(i \mid x) = 1$. More generally, the Bayes rule (6.14) allows any randomization among those i for which $p_i f_i(x) = \max_j p_j f_j(x)$. This amounts to choosing any a_i for which θ_i is most probable under the posterior distribution given the observations.

Proof. Let ϕ' be any other rule. We are to show that $r(\tau, \phi') - r(\tau, \phi) \geq 0$, where ϕ is any rule of the form (6.14). We treat only the absolutely continuous case, the discrete case being completely analogous.

$$r(\tau, \phi') - r(\tau, \phi) = \sum_{i=1}^{k} p_i \int \phi(i \mid x)\, f_i(x)\, dx - \sum_{i=1}^{k} p_j \int \phi'(j \mid x)\, f_j(x)\, dx$$

$$= \sum_{i=1}^{k} \sum_{j=1}^{k} \int \phi(i \mid x)\, \phi'(j \mid x)[p_i f_i(x) - p_j f_j(x)]\, dx.$$

This quantity cannot be negative, for whenever $p_i f_i(x) - p_j f_j(x)$ is negative, $\phi(i \mid x)$ is zero, thus completing the proof.

EXAMPLE 1. Suppose that $f_i(x) = \beta_i^{-1} \exp(-x/\beta_i) I_{(0,\infty)}(x)$ for $i = 1, \cdots, k$, where $\beta_1 < \beta_2 < \cdots < \beta_k$. The rule (6.14) requires us, if $X = x$ is observed, to take action a_i for some integer i for which $p_i\beta_i^{-1} \exp(-x/\beta_i)$ is a maximum. Suppose that $p_i = 1/k$ for $i = 1, 2, \cdots, k$; then, for $i < j$,

$$p_i\beta_i^{-1} \exp(-x/\beta_i) > p_j\beta_j^{-1} \exp(-x/\beta_j)$$

if, and only if,

$$x < \frac{\beta_i\beta_j}{\beta_j - \beta_i} (\log \beta_j - \log \beta_i).$$

It is easy to show that this is an increasing function of β_j for fixed $\beta_i < \beta_j$ and an increasing function of β_i for fixed $\beta_j > \beta_i$ (Exercise 1). This, combined with the fact that $f_i(x)$ is decreasing in x for $x > 0$, implies that

$$\phi(i \mid x) = \begin{cases} 1 & \text{if} \quad x_{i-1} \leq x < x_i, \\ 0 & \text{otherwise,} \end{cases}$$

is a Bayes rule, where $x_0 = 0$, $x_k = +\infty$, and for $i = 1, \cdots, k-1$,

$$x_i = \frac{\beta_i \beta_{i+1}}{\beta_{i+1} - \beta_i} (\log \beta_{i+1} - \log \beta_i).$$

If the p_i are proportional to β_i (that is, $p_i = \beta_i / \sum_1^k \beta_i$), then, because $\exp(-x/\beta_i) < \exp(-x/\beta_j)$ for $i < j$, the rule $\phi(i \mid x) \equiv 0$ for $i \neq k$ and $\phi(k \mid x) \equiv 1$ is Bayes. In this case we do not need to observe X.

EXAMPLE 2. Suppose that $\mathbf{X}$ is an n-dimensional random vector and that $f_i(\mathbf{x})$ is the density of a multivariate normal distribution with known mean $\boldsymbol{\mu}_i$ and nonsingular covariance matrix $\boldsymbol{\Sigma}$. We assume that $\boldsymbol{\Sigma}$ is the identity matrix $\mathbf{I}$, for otherwise we could treat $\boldsymbol{\Sigma}^{-1/2}\mathbf{X}$ which has a multivariate normal distribution with covariance matrix $\mathbf{I}$. The inequality $p_i f_i(\mathbf{x}) > p_j f_j(\mathbf{x})$ becomes

$$p_i \exp\left[-\tfrac{1}{2}(\mathbf{x} - \boldsymbol{\mu}_i)^T(\mathbf{x} - \boldsymbol{\mu}_i)\right] > p_j \exp\left[-\tfrac{1}{2}(\mathbf{x} - \boldsymbol{\mu}_j)^T(\mathbf{x} - \boldsymbol{\mu}_j)\right]$$

or equivalently, for $p_i \neq 0$ and $p_j \neq 0$,

$$(\boldsymbol{\mu}_i - \boldsymbol{\mu}_j)^T\mathbf{x} > \tfrac{1}{2}(\boldsymbol{\mu}_i^T\boldsymbol{\mu}_i - \boldsymbol{\mu}_j^T\boldsymbol{\mu}_j) + \log\frac{p_j}{p_i}.$$

When $\boldsymbol{\mu}_i \neq \boldsymbol{\mu}_j$, this inequality is satisfied for $\mathbf{x}$ in a half-space bounded by a hyperplane perpendicular to $\boldsymbol{\mu}_i - \boldsymbol{\mu}_j$. If, in addition, $p_i = p_j$, this hyperplane is the perpendicular bisector of the line segment from $\boldsymbol{\mu}_i$ to $\boldsymbol{\mu}_j$. Taking into consideration the inequalities $p_i f_i(\mathbf{x}) > p_j f_j(\mathbf{x})$ for all i and j, we see that the set on which $\phi(i \mid \mathbf{x})$ is one, being the intersection of $(k-1)$ half-spaces, is a convex polyhedron with at most $(k-1)$ faces.

If $p_i = 1/k$ for $i = 1, \cdots, k$, then $\mathbf{X}$ is classified as belonging to that distribution whose mean $\boldsymbol{\mu}_i$ is closest to $\mathbf{X}$. For arbitrary covariance matrix $\boldsymbol{\Sigma}$ this also holds true, provided that distance is measured as $(\mathbf{x} - \boldsymbol{\mu}_i)\boldsymbol{\Sigma}^{-1}(\mathbf{x} - \boldsymbol{\mu}_i)$.

EXAMPLE 3. RANKING. Consider n normal populations with known means $\mu_1 < \mu_2 < \cdots < \mu_n$. One observation is taken from each of these populations. These observations are put in random order and called $X_1, \cdots, X_n$. The statistician's problem is to say which observation came from which population. More precisely, let Θ and $\mathcal{A}$ be the set of all permutations $(\nu_1, \nu_2, \cdots, \nu_n)$ of $(1, 2, \cdots, n)$, and let $L(\theta, a)$ be zero if $a = \theta$ and one if $a \neq \theta$. The random variables $X_1, \cdots, X_n$ are independent and $X_i \in \mathfrak{N}(\mu_{\nu_i}, 1)$, where $\theta = (\nu_1, \cdots, \nu_n)$, so that

$$f_{X_1,\cdots,X_n}(x_1, \cdots, x_n \mid \theta) = (2\pi)^{-n/2} \exp\left[-\tfrac{1}{2} \sum_{i=1}^{n} (x_i - \mu_{\nu_i})^2\right]. \tag{6.15}$$

The assumption that the observations were put in random order is equivalent to saying that the prior distribution of θ is uniform over the $n!$ elements of Θ. According to Theorem 1, the Bayes rule with respect to the uniform prior distribution ($p_i = (n!)^{-1}$ for all i) chooses that permutation $(\nu_1, \cdots, \nu_n)$ for which (6.15) achieves its maximum or, equivalently, for which $\sum(x_i - \mu_{\nu_i})^2$ is minimized. The answer is as might be expected; namely, we choose ν_i as the rank of X_i so that the smallest of the X_i is associated with μ_1, the next smallest with μ_2, and so on. To see this, let $(\nu_1, \cdots, \nu_n)$ be the ranks of $X_1, \cdots, X_n$ and let $(b_1, \cdots, b_n)$ be the inverse permutation, that is, $X_{b_1} < X_{b_2} < \cdots < X_{b_n}$. Then for any other permutation $(\nu_1', \cdots, \nu_n')$ with inverse $(b_1', \cdots, b_n')$

$$\begin{aligned}\sum_{i=1}^{n} (X_i - \mu_{\nu_i'})^2 - \sum_{i=1}^{n} (X_i - \mu_{\nu_i})^2 \\ &= \sum_{i=1}^{n} (X_{b_i'} - \mu_i)^2 - \sum_{i=1}^{n} (X_{b_i} - \mu_i)^2 \\ &= 2 \sum_{i=1}^{n} (X_{b_i} - X_{b_i'})\mu_i \\ &= 2 \sum_{i=2}^{n} \left(\sum_{j=i}^{n} (X_{b_j} - X_{b_j'})\right)(\mu_i - \mu_{i-1}) \geq 0,\end{aligned}$$

using the fact that $\sum X_{b_i}^2 = \sum X_{b_i'}^2$. The final inequality follows, since $\mu_i > \mu_{i-1}$ for all i and $\sum_{j=i}^{n} X_{b_j} \geq \sum_{j=i}^{n} X_{b_j'}$ for all i for any permutation $(b_1', \cdots, b_n')$. This shows that $\sum(X_i - \mu_{\nu_i})^2$ is minimized by choosing $(\nu_1, \cdots, \nu_n)$ as the ranks of $(X_1, \cdots, X_n)$.

This result, that $(\nu_1, \cdots, \nu_n)$ are taken to be the ranks of $(X_1, \cdots, X_n)$, depends on the assumption of normality. If the populations are Cauchy with location parameters μ_i, different results occur (Exercise 4).

EXAMPLE 4. DISCRIMINATION. Let $X_1, \cdots, X_k, Y$ be independent random variables, let $X_i \in \mathfrak{N}(\theta_i, 1)$ for $i = 1, \cdots, k$, and let $Y \in \mathfrak{N}(\theta_\nu, 1)$ for some integer ν, $1 \leq \nu \leq k$. The statistician is to guess at ν. He loses nothing if his guess is correct, but he loses one if his guess is incorrect. Thus $\Theta = \{(\theta_1, \cdots, \theta_k, \nu); \theta_i \text{ real}, \nu \text{ integral } 1 \leq \nu \leq k\}$, $\mathcal{A} = \{1, 2, \cdots, k\}$, and

$$L((\theta_1, \cdots, \theta_k, \nu), a) = \begin{cases} 1 & \text{if} \quad a \neq \nu, \\ 0 & \text{if} \quad a = \nu. \end{cases}$$

We shall obtain a Bayes rule with respect to the prior distribution over Θ, for which $\theta_1, \cdots, \theta_k, \nu$ are distributed independently, $\theta_i \in \mathfrak{N}(0, \sigma^2)$ for $i = 1, \cdots, k$, and $p_i = P\{\nu = i\}$ for $i = 1, \cdots, k$.

We search for a nonrandomized decision rule $d(x_1, \cdots, x_k, y)$ which chooses the element of $\mathcal{A}$ that minimizes the conditional expected loss, given the observations $X_1 = x_1, \cdots, X_k = x_k, Y = y$. In other words, $d(x_1, \cdots, x_k, y)$ is the integer $a \in \mathcal{A}$ that minimizes $E\{L((\theta_1, \cdots, \theta_k, \nu), a) \mid X_1 = x_1, \cdots, X_k = x_k, Y = y\}$, where the expectation is taken, using the posterior distribution of $(\theta_1, \cdots, \theta_k, \nu)$, given the observations. The joint density of $X_1, \cdots, X_k, Y, \theta_1, \cdots, \theta_k, \nu$ is

$$f(x_1, \cdots, x_k, y, \theta_1, \cdots, \theta_k, \nu) = (2\pi)^{-(2k+1)/2} \sigma^{-k}$$
$$\times \exp\left\{-\frac{1}{2}\left[\sum (x_i - \theta_i)^2 + (y - \theta_\nu)^2 + \frac{1}{\sigma^2} \sum \theta_i^2\right]\right\} p_\nu,$$
$$\nu = 1, 2, \cdots, k.$$

After completing the squares of the terms involving the θ_i, we find

$$f(x_1, \cdots, x_k, y, \theta_1, \cdots, \theta_k, \nu)$$
$$= (2\pi)^{-(2k+1)/2} \sigma^{-k} \exp\left\{-\frac{1}{2}\left[\sum \frac{x_i^2}{\sigma^2 + 1} - \frac{\sigma^2 + 1}{2\sigma^2 + 1} y^2\right]\right\}$$
$$\times \exp\left\{-\frac{1}{2}\left[\sum_{i \neq \nu} \frac{\sigma^2 + 1}{\sigma^2}\left(\theta_i - \frac{x_i \sigma^2}{\sigma^2 + 1}\right)^2 + \frac{2\sigma^2 + 1}{\sigma^2}\left(\theta_\nu - \frac{(x_\nu + y)\sigma^2}{2\sigma^2 + 1}\right)^2\right]\right\}$$
$$\times p_\nu \exp\left\{\frac{\sigma^2 x_\nu}{(4\sigma^2 + 2)}\left[2y - \frac{\sigma^2}{\sigma^2 + 1} x_\nu\right]\right\}, \qquad \nu = 1, 2, \cdots, k.$$

Because the posterior distribution of $\theta_1, \cdots, \theta_k, \nu$ is given by this density normalized by a function of $x_1, \cdots, x_k, y$, we see that this posterior distribution may be described as follows. If p_i' denotes the posterior probability that $\nu = i$, then p'_i is proportional to

$$p'_i \sim p_i \exp\left[\frac{\sigma^2 x_i(2y - \sigma^2 x_i(\sigma^2 + 1)^{-1})}{4\sigma^2 + 2}\right],$$

and, given ν, the variables $\theta_1, \cdots, \theta_k$ are independent, with

$$\theta_i \in \mathfrak{N}(x_i\sigma^2/(\sigma^2 + 1),\ \sigma^2/(\sigma^2 + 1)) \quad \text{for} \quad i \neq \nu$$

and $\theta_\nu \in \mathfrak{N}((x_\nu + y)\sigma^2/(2\sigma^2 + 1),\ \sigma^2/(2\sigma^2 + 1))$. The conditional expected loss, given $X_1 = x_1, \cdots, X_k = x_k$, $Y = y$, is therefore

$$1 - P\{\nu = a \mid X_1 = x_1, \cdots, X_k = x_k, Y = y\} = 1 - p'_a.$$

Hence the Bayes rule $d(x_1, \cdots, x_k, y)$ is that integer a for which p'_a is largest. Let

$$x'_i = \frac{\sigma^2}{\sigma^2 + 1} x_i.$$

Then p'_i is proportional to $p_i \exp[(\sigma^2 + 1)(2x'_i y - x'^2_i)/(4\sigma^2 + 2)]$ or to $p_i \exp[-(\sigma^2 + 1)(x'_i - y)^2/(4\sigma^2 + 2)]$. Thus $d(x_1, \cdots, x_k, y)$ is that integer a for which $(x'_a - y)^2 - [(4\sigma^2 + 2) \log p_a]/(\sigma^2 + 1)$ is smallest. If all p_i are equal, then $d(x_1, \cdots, x_k, y)$ is that a for which $(x'_a - y)^2$ is smallest. As $\sigma \to \infty$, $d(x_1, \cdots, x_k, y)$ becomes that a for which $(x_a - y)^2$ is smallest.

Exercises

1. Let $f(x, y) = xy(\log y - \log x)/(y - x)$. Show that $f(x, y)$ is increasing in y for fixed $x < y$ and increasing in x for fixed $y > x$. *Hint.* Use the fact that $1 - z + \log z \geq 0$ for all $z > 0$.
2. Let $\mu_1 < \mu_2 < \cdots < \mu_k$ and let $f_i(x) = \exp[-(x - \mu_i)] I_{(\mu_i, \infty)}(x)$ for $i = 1, \cdots, k$.
   (a) Find the Bayes rule (6.14) with respect to the distribution for which $p_i = 1/k$, $i = 1, \cdots, k$.
   (b) Describe the class of rules which are Bayes with respect to the distribution for which p_i is proportional to $\exp(-\mu_i)$.
3. Consider Example 2 in which $\mathbf{X}$ is two-dimensional and $k = 5$, with $\boldsymbol{\mu}_1^T = (0, 0)$, $\boldsymbol{\mu}_2^T = (1, 0)$, $\boldsymbol{\mu}_3^T = (0, 1)$, $\boldsymbol{\mu}_4^T = (-1, 0)$, $\boldsymbol{\mu}_5^T = (0, -1)$, and $\boldsymbol{\Sigma} = \mathbf{I}$. Plot on the plane the regions in which $\phi(i \mid \mathbf{x}) = 1$ for $i = 1, \cdots, 5$, where ϕ is a Bayes rule with respect to the distributions

(a) $p_1 = 1/3$, $p_2 = p_3 = p_4 = p_5 = 1/6$ ($\log 2 = 0.69315\cdots$),
(b) $p_1 = p_2 = p_3 = p_4 = p_5 = 1/5$,
(c) $p_1 = 1/(1 + 4e^{1/2})$, $p_2 = p_3 = p_4 = p_5 = e^{1/2}/(1 + 4e^{1/2})$.

4. Let Θ, $\mathcal{A}$, and L be as in Example 3, with $n = 2$. However, let X_1 and X_2 be independent random variables with Cauchy distributions, $X_i \in \mathcal{C}(\mu_{\nu_i}, 1)$, where (ν_1, ν_2) is $(1, 2)$ or $(2, 1)$ and where μ_1 and μ_2 are known numbers $\mu_1 < \mu_2$. Find the Bayes rule with respect to the distribution giving weight $1/2$ to both points of Θ.

Ans. If $x_1 < x_2$, then $d(x_1, x_2) = (1, 2)$ if, and only if, $(x_1 - \mu_1)(x_2 - \mu_2) + (x_1 - \mu_2)(x_2 - \mu_1) < 2$.
Plot the set on which $d(x_1, x_2) = (1, 2)$ in the plane when $\mu_1 = -1$ and $\mu_2 = +1$.

5. Let Θ, $\mathcal{A}$, and L be as in Example 3 and let $X_1, \cdots, X_n$ be independent random variables. Let $\mu_1 < \mu_2 < \cdots < \mu_n$ and suppose that X_i has density (probability mass function) $f(x_i \mid \mu_{\nu_i})$, where $f(x \mid \mu)$ is the density (probability mass function) of a family of distributions with monotone likelihood ratio which satisfies inequality (5.21) (with θ replaced by μ). Show that a Bayes rule with respect to the prior distribution, giving equal weight to all points of Θ, is $d(x_1, \cdots, x_n) = (\nu_1, \cdots, \nu_n)$, where ν_i is the rank of X_i and where ties are decided by assigning any of the tied ranks to the tied variables. *Hint.* Solve the problem first for $n = 2$.

6. Let Θ, $\mathcal{A}$, and L be as in Example 3 with $n = 2$ and let $\mathbf{X}_1$ and $\mathbf{X}_2$ be independent m-dimensional random vectors with $\mathbf{X}_i \in \mathcal{N}(\boldsymbol{\mu}_{\nu_i}, \mathbf{I})$. Find a Bayes rule with respect to the uniform prior distribution. *Hint.* Solve the problem first for $\boldsymbol{\mu}_1 = \mathbf{0}$ and $\boldsymbol{\mu}_2 = (\lambda, 0, \cdots, 0)^T$. Then, by a change of location and a rotation, obtain the original problem.

7. Let $\mathbf{X}_1, \cdots, \mathbf{X}_k$, $\mathbf{Y}$ be independent n-dimensional vectors $\mathbf{X}_i \in \mathcal{N}(\boldsymbol{\theta}_i, \mathbf{I})$, $\mathbf{Y} \in \mathcal{N}(\boldsymbol{\theta}_\nu, \mathbf{I})$ for some integer ν, $1 \le \nu \le k$. Let

$$\Theta = \{(\boldsymbol{\theta}_1, \cdots, \boldsymbol{\theta}_k, \nu)\}, \qquad \mathcal{A} = \{1, \cdots, k\},$$

and $L((\boldsymbol{\theta}_1, \cdots, \boldsymbol{\theta}_k, \nu), a)$ be zero if $\nu = a$ and one otherwise. Let the prior distribution of $\boldsymbol{\theta}_1, \cdots, \boldsymbol{\theta}_k, \nu$ be such that $\boldsymbol{\theta}_1, \cdots, \boldsymbol{\theta}_k$, and ν are independent, $\boldsymbol{\theta}_i \in \mathcal{N}(\mathbf{0}, \boldsymbol{\Sigma})$ for $i = 1, \cdots, k$, and p_i is the probability that ν is equal to i.
(a) Show that the posterior distribution of ν (given $\mathbf{X}_1 = \mathbf{x}_1, \cdots, \mathbf{X}_k = \mathbf{x}_k$, $\mathbf{Y} = \mathbf{y}$) is such that if p_i' is the posterior probability that

ν is equal to i then p_i' is proportional to

$$p_i \exp\left[-\tfrac{1}{2}(\hat{\mathbf{x}}_\nu - \mathbf{y})^T \mathbf{A}^{-1}(\hat{\mathbf{x}}_\nu - \mathbf{y})\right\},$$

where $\hat{\mathbf{x}}_\nu = \{(\mathbf{\Sigma}^{-1} + 2\mathbf{I})(\mathbf{\Sigma}^{-1} + \mathbf{I})^{-1} - \mathbf{I}]\mathbf{x}_\nu$, and

$$\mathbf{A} = (\mathbf{\Sigma}^{-1} + 2\mathbf{I})(\mathbf{\Sigma}^{-1} + \mathbf{I})^{-1}(\mathbf{\Sigma}^{-1} + 2\mathbf{I}) - (\mathbf{\Sigma}^{-1} + 2\mathbf{I});$$

given ν, the posterior distribution of $\boldsymbol{\theta}_1, \cdots, \boldsymbol{\theta}_k$ is such that $\boldsymbol{\theta}_1, \cdots, \boldsymbol{\theta}_k$ are independent, $\boldsymbol{\theta}_i \in \mathfrak{N}((\mathbf{\Sigma}^{-1} + \mathbf{I})^{-1}\mathbf{x}_i, (\mathbf{\Sigma}^{-1} + \mathbf{I})^{-1})$ for $i \neq \nu$, $\boldsymbol{\theta}_\nu \in \mathfrak{N}((\mathbf{\Sigma}^{-1} + 2\mathbf{I})^{-1}(\mathbf{x}_\nu + \mathbf{y}), (\mathbf{\Sigma}^{-1} + 2\mathbf{I})^{-1})$.
(b) Show that when $p_i = 1/k$ for all i the Bayes rule chooses that integer ν for which $(\hat{\mathbf{x}}_\nu - \mathbf{y})^T \mathbf{A}^{-1}(\hat{\mathbf{x}}_\nu - \mathbf{y})$ is a minimum. Specialize to the case $\mathbf{\Sigma} = \sigma^2\mathbf{I}$.

6.3 Slippage Problems

As an application of the theorem of Section 6.2, we consider the *k-sample slippage problem,* which, in essence, is the problem of deciding whether one of k populations has "slipped" to the right of the rest and, if so, which one. To be specific, we consider random variables $X_1, \cdots, X_k$ and $k + 1$ hypotheses $H_0, H_1, \cdots, H_k$. Under hypothesis H_0, $X_1, \cdots, X_k$ are independent and identically distributed with common density (probability mass function) $f_0(x)$. Under hypothesis H_i, $X_1, \cdots, X_k$ are independent, $X_1, \cdots, X_{i-1}, X_{i+1}, \cdots, X_k$ are identically distributed with common density $f_0(x)$, and x_i has density $f_1(x)$, $i = 1, \cdots, k$.

This problem is invariant under the group $\mathcal{G}$ of permutations of the observations in the sense that under any $g \in \mathcal{G}$ hypothesis H_0 is unaltered and hypotheses $H_1, \cdots, H_k$ are permuted among themselves. Thus we could use the usual invariance methods to treat this problem. For variety, we try a different approach that involves looking at decision rules which are Bayes with respect to an invariant prior distribution. Given a decision problem invariant under a group $\mathcal{G}$, *a prior distribution τ is said to be invariant under $\bar{\mathcal{G}}$ if for every measurable subset Θ_1, of Θ, and for all* $\bar{g} \in \bar{\mathcal{G}}$, $\tau(\bar{g}\Theta_1) = \tau(\Theta_1)$. Instead of looking at invariant rules directly and trying to find good rules among them, we look at the class of rules which are Bayes with respect to some invariant prior distribution. This class of rules should contain all good invariant rules as indicated by Theorem 4.3.3.(a).

Returning to the k-sample slippage problem, let us find Bayes decision rules with respect to the invariant prior distributions. A prior distribution τ may be described by giving the probabilities $\tau\{H_i\}$ for $i = 0, 1, \cdots, k$. The prior distributions invariant under permutations of $H_1, \cdots, H_k$ among themselves give equal weight to $H_1, \cdots, H_k$. Therefore the invariant distributions are in the form of τ_p, where $\tau_p(H_0) = 1 - kp$, $\tau_p(H_i) = p$ for $i = 1, \cdots, k$ and $0 \leq p \leq 1/k$. Using the loss function of Section 6.2, namely loss zero if the correct hypothesis is accepted and loss one otherwise, we may find the Bayes rules with respect to τ_p by Theorem 6.2.1. Under H_0 the density of $X_1, \cdots, X_k$ is $\prod_1^k f_0(x_j)$, whereas under H_i it is $(\prod_1^k f_0(x_j))(f_1(x_i)/f_0(x_i))$. Therefore any rule ϕ for which

$$\phi(0 \mid \mathbf{X}) = 0 \quad \text{whenever} \quad \frac{1 - kp}{p} < \max_j \frac{f_1(X_j)}{f_0(X_j)} \tag{6.16}$$

and for $i = 1, \cdots, k$

$$\phi(i \mid \mathbf{X}) = 0$$

whenever either

$$\frac{f_1(X_i)}{f_0(X_i)} < \max_j \frac{f_1(X_j)}{f_0(X_j)}$$

or

$$\frac{1 - kp}{p} > \max_j \frac{f_1(X_j)}{f_0(X_j)} \tag{6.17}$$

is Bayes with respect to τ_p.

Instead of classifying these rules according to the variable p, we could, as in hypothesis-testing problems, classify them according to the variable α, $0 \leq \alpha \leq 1$, where

$$\alpha = 1 - P\{\text{accept } H_0 \mid H_0\} = 1 - E_{H_0}\,\phi(0 \mid \mathbf{X}).$$

With such a classification we might refer to these problems as testing multiple hypotheses. If we define the random variable

$$V = \max_j \left[f_1(X_j)/f_0(X_j) \right],$$

there exist numbers c and γ, $0 \leq c \leq \infty$ and $0 \leq \gamma \leq 1$, such that $E_{H_0}\,\phi(0 \mid \mathbf{X}) = 1 - \alpha$, where

$$\phi(0 \mid \mathbf{X}) = \begin{cases} 1 & \text{if} \quad V < c, \\ \gamma & \text{if} \quad V = c, \\ 0 & \text{if} \quad V > c. \end{cases} \tag{6.18}$$

In defining $\phi(i \mid \mathbf{X})$ for $i = 1, \cdots, k$, we take only those rules among the rules (6.17) that are invariant. Define ν as the number of subscripts i for which $f_1(X_i)/f_0(X_i) = V$. Then we take

$$\phi(i \mid \mathbf{X}) = \begin{cases} \dfrac{1}{\nu} & \text{if } \dfrac{f_1(X_i)}{f_0(X_i)} = V > c, \\[2ex] \dfrac{1-\gamma}{\nu} & \text{if } \dfrac{f_1(X_i)}{f_0(X_i)} = V = c, \\[2ex] 0 & \text{otherwise.} \end{cases} \tag{6.19}$$

The rule defined by (6.18) and (6.19) is Bayes with respect to τ_p, where $c = (1 - kp)/p$. For these invariant rules the probabilities $P\{\text{accept } H_i \mid H_i\}$ do not depend on i for $i \neq 0$. When $c = \infty$ (because $p = 0$), we should take $\gamma = 0$, since $\alpha = 0$ for any γ and P $\{$accept $H_i \mid H_i\}$ is nonincreasing in γ.

In this form the optimal property of these invariant Bayes rules may be stated as in Theorem 1. For simplicity, we restrict attention to the case $\alpha > 0$. Consider all decision rules ϕ with the properties (a) $P\{\text{accept } H_0 \mid H_0\} \geq 1 - \alpha$, where $0 < \alpha \leq 1$; (b) $P\{\text{accept } H_i \mid H_i\} = P\{\text{accept } H_j \mid H_j\}$ for all $i \neq 0, j \neq 0$.

Out of this class of rules we search for those rules for which the common value of $P\{\text{accept } H_i \mid H_i\}$, $i \neq 0$, is a maximum.

Theorem 1. Out of the class of all rules satisfying (a) and (b) the rule ϕ of (6.18) and (6.19), where c and γ are chosen so that $P\{\text{accept } H_0 \mid H_0\} = 1 - \alpha$, maximizes the common value of $P\{\text{accept } H_i \mid H_i\}$ for $i \neq 0$.

Proof. Let ϕ' be any other decision rule satisfying (a) and (b) and suppose that $E_{H_i} \phi'(i \mid \mathbf{X}) > E_{H_i} \phi(i \mid \mathbf{X})$ for $i \neq 0$. Then, because $\alpha > 0$ implies that $\tau_p(H_1) = p > 0$ where $c = (1 - kp)/p$, we have

$$r(\tau_p, \phi') - r(\tau_p, \phi) = \sum_{i=0}^{k} \tau_p(H_i)[E_{H_i} \phi(i \mid \mathbf{X}) - E_{H_i} \phi'(i \mid \mathbf{X})] < 0,$$

contradicting the fact that ϕ is Bayes with respect to τ_p and completing the proof.

The idea of this theorem and its proof is taken from Paulson (1952).

EXAMPLE 1. Suppose we are given k coins, all presumably fair but admitting the possibility that one of them is biased toward probability of heads greater than 1/2. Each coin is tossed n times, and X_i is used to denote the number of heads appearing in the n tosses of the ith coin, $i = 1, \cdots, k$. We assume that this problem fits the description given in the first paragraph of this section, with $f_0(x) = \binom{n}{x} 2^{-n}$, $f_1(x) = \binom{n}{x} \theta^x (1-\theta)^{n-x}$, $x = 0, 1, \cdots, n$, where θ is known and $\theta > 1/2$.

Because

$$\frac{f_1(x)}{f_0(x)} = \left[\frac{1-\theta}{2}\right]^n \left[\frac{\theta}{1-\theta}\right]^x,$$

we have

$$V = \left[\frac{1-\theta}{2}\right]^n \left[\frac{\theta}{1-\theta}\right]^{\max_j X_j},$$

and the inequality $V < c$ is equivalent to $\max_j X_j < c'$, using the assumption $\theta > 1/2$. Hence the rule ϕ of (6.18) and (6.19) becomes

$$\phi(0 \mid \mathbf{X}) = \begin{cases} 1 & \text{if } \max X_j < c', \\ \gamma & \text{if } \max X_j = c', \\ 0 & \text{if } \max X_j > c', \end{cases}$$

$$\phi(i \mid \mathbf{X}) = \begin{cases} \dfrac{1}{\nu} & \text{if } X_i = \max X_j > c', \\ \dfrac{(1-\gamma)}{\nu} & \text{if } X_i = \max X_j = c', \\ 0 & \text{otherwise}, \end{cases} \tag{6.20}$$

for $i = 1, \cdots, k$, where c' and γ are chosen so that $E_{H_0}\,\phi(0 \mid \mathbf{X}) = 1 - \alpha$, where α is given in advance. This implies that c' and γ, hence the test

ϕ, do not depend on θ, provided $\theta > \frac{1}{2}$. In other words, this test ϕ satisfies the optimal property of Theorem 1 uniformly in $\theta > \frac{1}{2}$. This example extends to families of distributions with monotone likelihood ratio (Exercise 2).

EXAMPLE 2. *The k-sample slippage problem for normal distributions with unknown means.* Let $X_1, \cdots, X_k$ be independent random variables and consider the $k + 1$ hypotheses

$$H_0: X_j \in \mathfrak{N}(\mu, 1), \qquad j = 1, \cdots, k,$$

$$H_i: X_j \in \mathfrak{N}(\mu, 1) \quad \text{for} \quad j \neq i \quad \text{and} \quad X_i \in \mathfrak{N}(\mu + \Delta, 1), \qquad i = 1, \cdots, k,$$

where μ is an unknown parameter and Δ is a known positive number, $\Delta > 0$. This problem is invariant under a change in location in the sense that adding the same constant to each X_i leaves each of the hypotheses H_i, $i = 0, 1, \cdots, k$ unchanged. Thus it seems reasonable to require that a multiple decision rule ϕ, designed to distinguish among the $k + 1$ hypotheses should be invariant under a change of location. Therefore we require (c) $\phi(i \mid x_1, \cdots, x_k) = \phi(i \mid x_1 + c, \cdots, x_k + c)$ for all $i, x_1, \cdots, x_k$, and c. For such rules, the probabilities $P\{\text{accept } H_i \mid H_j\}$ do not depend on μ. Therefore we may further require (a) and (b) of Theorem 1. We search for that rule which, out of the class of rules satisfying (a), (b), and (c), maximizes the common value of $P\{\text{accept } H_i \mid H_i\}$, $i \neq 0$. We proceed by considering rules that are functions of the maximal invariant $Y_1, \cdots, Y_{k-1}$, where $Y_1 = X_1 - X_k, \cdots, Y_{k-1} = X_{k-1} - X_k$, and by applying the methods of Theorem 1. To find the joint distribution of $Y_1, \cdots, Y_{k-1}$ we make the transformation from $X_1, \cdots, X_k$ to $Y_1, \cdots, Y_{k-1}, Z$, where $Z = X_k$. The inverse transformation is $X_1 = Z + Y_1, \cdots, X_{k-1} = Z + Y_{k-1}, X_k = Z$, whose Jacobian is one. Because the distribution of $Y_1, \cdots, Y_{k-1}$ does not depend on μ, we put $\mu = 0$ from the start to simplify the computations. Under H_i, $i \neq 0$, the distribution of $X_1, \cdots, X_k$ has density

$$f_{X_1,\cdots,X_k}(x_1, \cdots, x_k) = (2\pi)^{-k/2} \exp\left[-\frac{1}{2}\left(\sum_{j=1}^{k} x_j^2\right) + \Delta x_i - \frac{\Delta^2}{2}\right]. \tag{6.21}$$

Therefore the density of $Y_1, \cdots, Y_{k-1}, Z$ is

$$f_{Y_1,\cdots,Y_{k-1},Z}(y_1, \cdots, y_{k-1}, z) = (2\pi)^{-k/2}$$
$$\times \exp\left[-\frac{1}{2} \sum_{j=1}^{k} (z + y_j)^2 + \Delta(z + y_i) - \frac{\Delta^2}{2}\right],$$

where the dummy variable y_k represents zero, $y_k = 0$. By integrating out z we find

$$f_{Y_1,\cdots,Y_{k-1}}(y_1, \cdots, y_{k-1}) = (2\pi)^{-k/2} \exp\left[-\frac{1}{2}\sum_1^k y_j^2 + \Delta y_i - \frac{\Delta^2}{2}\right]$$

$$\times \int_{-\infty}^{\infty} \exp\left[-\frac{kz^2}{2} - z\left(\sum y_j - \Delta\right)\right] dz \qquad (6.22)$$

$$= k^{-1/2}(2\pi)^{-(k-1)/2} \exp\left[-\frac{1}{2}\sum_1^k y_j^2 + \Delta y_i - \frac{\Delta^2}{2} + \frac{k}{2}\left(\bar{y} - \frac{\Delta}{k}\right)^2\right]$$

$$= k^{-1/2}(2\pi)^{-(k-1)/2} \exp\left\{-\frac{k}{2}\left[\frac{1}{k}\sum (y_i - \bar{y})^2\right] - \frac{\Delta^2(k-1)}{2k} + \Delta(y_i - \bar{y})\right\},$$

where $\bar{y} = \sum_1^k y_j/k$. The distribution of $Y_1, \cdots, Y_{k-1}$ under H_0 is obtained by putting $\Delta = 0$. From this, a Bayes rule with respect to the invariant prior distribution giving probability $1 - kp$ to H_0 and p to $H_1, \cdots, H_k$, may be found as

$$\phi(0 \mid \mathbf{x}) = \begin{cases} 1 & \text{if} \quad \dfrac{(1 - kp)}{p} \geq \max_i \exp\left[\Delta(y_i - \bar{y})\right], \\ 0 & \qquad\qquad < \end{cases}$$

and so forth. But, since $y_i - \bar{y} = x_i - \bar{x}$ and $\Delta > 0$,

$$\phi(0 \mid \mathbf{x}) = \begin{cases} 1 & \text{if} \quad \max_j (x_j - \bar{x}) \leq c, \\ 0 & \text{if} \qquad\qquad > c, \end{cases}$$

and

$$\phi(i \mid \mathbf{x}) = \begin{cases} 1 & \text{if} \quad (x_i - \bar{x}) = \max_j (x_j - \bar{x}) > c, \\ 0 & \text{otherwise,} \end{cases}$$

for $i = 1, \cdots, k$. The constant c is chosen so that $P\{\text{accept } H_0 \mid H_0\} = 1 - \alpha$. That this rule has maximum value of $P\{\text{accept } H_i \mid H_i\}$ for

$i \neq 0$, out of the class of all rules satisfying (a), (b), and (c), no matter what the value of $\Delta > 0$, may be demonstrated by using the proof of Theorem 1.

EXAMPLE 3. *The k-sample slippage problem for the normal distribution with unknown mean and variance.* Let $X_1, \cdots, X_k$ be independent random variables and consider the $k + 1$ hypotheses

$$H_0: X_j \in \mathfrak{N}(\mu, \sigma^2), \qquad j = 1, \cdots, k,$$

$$H_i: X_j \in \mathfrak{N}(\mu, \sigma^2), \quad \text{for} \quad j \neq i \quad \text{and} \quad X_i \in \mathfrak{N}(\mu + \Delta, \sigma^2), \qquad i = 1, \cdots, k,$$

where μ and $\sigma^2 > 0$ are unknown parameters and Δ is a known positive number, $\Delta > 0$. This problem is invariant under change in location and scale, and we restrict attention to rules invariant under change of location and scale,

$$\text{(d)} \quad \phi(i \mid x_1, \cdots, x_k) = \phi\left(i \mid \frac{x_1 - c}{b}, \cdots, \frac{x_k - c}{b}\right) \quad \text{for all} \quad i, \quad x_1, \cdots, x_k, c, \quad \text{and} \quad b > 0.$$

For such rules the probabilities $P\{\text{accept } H_i \mid H_i\}$ do not depend on μ or σ^2. Therefore we may further require (a) and (b) of Theorem 1 and proceed by the method of Example 2 to find a rule that, out of the class of all rules satisfying (a), (b), and (d), maximizes $P\{\text{accept } H_i \mid H_i\}$ for $i \neq 0$. The maximal invariant for this problem is the set of ratios of differences

$$R_1 = \frac{X_1 - X_k}{X_{k-1} - X_k}, \cdots, \qquad R_{k-2} = \frac{X_{k-2} - X_k}{X_{k-1} - X_k}.$$

It is necessary to find the joint distribution of $R_1, \cdots, R_{k-2}$. Because this distribution does not depend on μ or σ^2, we put $\mu = 0$ and $\sigma^2 = 1$ to simplify the computations in deriving it. Under H_i, $i \neq 0$ the joint distribution of $Y_1, \cdots, Y_{k-1}$ is given by (6.22). To find the distribution of $R_1, \cdots, R_{k-2}$ we make the transformation $R_1 = Y_1/Y_{k-1}, \cdots, R_{k-2} = Y_{k-2}/Y_{k-1}$, $U = Y_{k-1}$, whose inverse transformation is $Y_1 = UR_1, \cdots, Y_{k-2} = UR_{k-2}$, $Y_{k-1} = U$ with Jacobian $| U |^{k-2}$. Therefore under $H_i, i \neq 0$

$$f_{R_1, \cdots, R_{k-2}, U}(r_1, \cdots, r_{k-2}, u) = k^{-1/2}(2\pi)^{-(k-1)/2} \mid u \mid^{k-2}$$

$$\times \exp\left\{-\frac{ku^2}{2}\left[\frac{1}{k}\sum_{j=1}^{k}(r_j - \bar{r})^2\right] - \frac{\Delta^2}{2}\frac{(k-1)}{k} + \Delta u(r_i - \bar{r})\right\},$$

where $r_{k-1} = 1$, $r_k = 0$, and $\bar{r} = \sum_1^k r_i/k$. By integrating out u and letting $s_r^2 = (1/k) \sum_1^k (r_j - \delta)^2$

$$f_{R_1,\cdots,R_{k-2}}(r_1, \cdots, r_{k-2}) = k^{-1/2}(2\pi)^{-(k-1)/2} \exp\left[-\frac{\Delta^2(k-1)}{2k}\right]$$

$$\times \int_{-\infty}^{\infty} \exp\left[-\frac{ku^2}{2} s_r^2 + \Delta u(r_i - \bar{r})\right] | u |^{k-2} \, du$$

$$= k^{-1/2}(2\pi)^{-(k-1)/2} s_r^{-(k-1)} \exp\left[-\frac{\Delta^2(k-1)}{2k}\right]$$

$$\times \int_{-\infty}^{\infty} \exp\left[-\frac{kv^2}{2} + \Delta \frac{(r_i - \bar{r})}{s_r} v\right] | v |^{k-2} \, dv,$$

where $v = us_r$. Under H_0 the distribution of $R_1, \cdots, R_{k-2}$ is given as above, with $\Delta = 0$. The function

$$h(\delta) = \int_{-\infty}^{\infty} \exp\left(-\frac{kv^2}{2} + \delta v\right) | v |^{k-2} \, dv$$

$$= \int_0^{\infty} \exp\left(-\frac{kv^2}{2}\right) v^{k-2} \cosh \delta v \, dv$$

is an increasing function of δ, so that $\max_i h(\Delta(r_i - \bar{r})/s_r) < c$ is equivalent to $\max_i (r_i - \bar{r})/s_r < c'$. Because $(r_i - \bar{r})/s_r = (x_i - \bar{x})/s_x$, the Bayes rule is

$$\phi(0 \mid \mathbf{x}) = \begin{cases} 1 & \text{if} \quad \max_j \dfrac{x_j - \bar{x}}{s_x} \le c', \\ 0 & \qquad\qquad\qquad > c', \end{cases}$$

$$\phi(i \mid x) = \begin{cases} 1 & \text{if} \quad \dfrac{x_i - \bar{x}}{s_x} = \max_j \dfrac{x_j - \bar{x}}{s_x} > c', \\ 0 & \text{otherwise}, \end{cases}$$

for $i = 1, 2, \cdots, k$.

Exercises

1. Show that the problem in Exercise 6.2.5 is invariant under the group of all permutations of $(X_1, \cdots, X_n)$. Using the result of Exercise 6.2.5, describe an invariant rule that is Bayes with respect to the prior distribution, giving equal weight to all points of Θ [see Theorem 4.3.3(b)]. Prove that this rule is best invariant. Conclude that because this rule is independent of $\mu_1, \cdots, \mu_n$ it is uniformly best invariant for the problem in which $\mu_1 < \cdots < \mu_n$ are unknown parameters.

2. Let $f(x \mid \theta)$ be the density (probability mass function) of a family of distributions with monotone likelihood ratio satisfying Definition 5.2.3, let $\theta_0 < \theta_1$, and let $f_0(x) = f(x \mid \theta_0)$ and $f_1(x) = f(x \mid \theta_1)$. Show that the rules of equations (6.18) and (6.19) have form (6.20), hence that these rules satisfy the optimal property of Theorem 1 uniformly in $\theta_1 > \theta_0$ (see Karlin and Truax (1960)).

3. Let $X_1, \cdots, X_k$ be independent and consider the $k + 1$ hypotheses

$$H_0: X_j \in \mathfrak{N}(\mu, 1) \quad \text{for} \quad j = 1, \cdots, k,$$

$$H_i: X_j \in \mathfrak{N}(\mu, 1) \quad \text{for} \quad j \neq i$$

and

$$X_i \in \mathfrak{N}(\mu + \Delta, 1), \quad \text{for} \quad i = 1, \cdots, k,$$

where μ is unknown and where Δ may take one of the two values $-\Delta_0$ and $+\Delta_0$, with Δ_0 known. This problem is invariant under change of location and under multiplication by -1.

(a) Find the density of the maximal invariant $|Y_1|$, $Y_2 \operatorname{sgn} Y_1, \cdots, Y_{k-1} \operatorname{sgn} Y_1$, where $Y_i = X_i - X_k$, $i = 1, \cdots, k - 1$ and $\operatorname{sgn} Y_1 = |Y_1|/Y_1$.

(b) Find the decision rule which maximizes $P\{\text{accept } H_i \mid H_i\}$, $i \neq 0$, out of the class of all invariant rules satisfying (a) and (b) of Theorem 1 and show that this rule is independent of Δ_0.

4. Let $X_1, \cdots, X_k$ be independent random variables with gamma distributions and consider the $k + 1$ hypotheses

$$H_0: X_j \in \mathcal{G}(\alpha, \beta) \quad \text{for} \quad j = 1, \cdots, k,$$

$$H_i: X_j \in \mathcal{G}(\alpha, \beta) \quad \text{for} \quad j \neq i \quad \text{and} \quad X_i \in \mathcal{G}(\alpha, \lambda\beta),$$

where $\beta > 0$ is an unknown parameter and $\alpha > 0$ and $\lambda > 1$ are known numbers. This problem is invariant under scale changes.

(a) Find the distribution of the maximal invariant $Z_1, \cdots, Z_{k-1}$, where $Z_i = X_i/X_k$, $i = 1, \cdots, k-1$.
(b) Find the decision rule which maximizes $P\{\text{accept } H_i \mid H_i\}$, $i \neq 0$, out of the class of all invariant rules satisfying (a) and (b) of Theorem 1 and show that this rule is independent of $\lambda > 1$.

5. Let $X_1, \cdots, X_k$ be independent and consider the $k(k-1)/2 + 1$ hypotheses

$$H_0: X_j \in \mathfrak{N}(0, 1) \quad \text{for} \quad j = 1, \cdots, k,$$

$$H_{h,i}: X_j \in \mathfrak{N}(0, 1) \quad \text{for} \quad j \neq i \quad \text{and } j \neq h,$$

$$X_h \in \mathfrak{N}(\Delta, 1),\ X_i \in \mathfrak{N}(\Delta, 1), \qquad 1 \leq h < i \leq k.$$

Find the decision rule which out of the class of all rules satisfying (a) $P\{\text{accept } H_0 \mid H_0\} \geq 1 - \alpha$, and (b) $P\{\text{accept } H_{h,i} \mid H_{h,i}\}$ is independent of h and i, maximizes $P\{\text{accept } H_{h,i} \mid H_{h,i}\}$, and show that this rule is independent of Δ.

CHAPTER 7

Sequential Decision Problems

7.1 Sequential Decision Rules

Sequential decision problems differ from the decision problems we have discussed so far in that the statistician is given the freedom to look at a sequence of observations one at a time and to decide after each observation whether to stop sampling and take an action immediately or to continue sampling and take an action sometime later. We assume here that the order of the sequence of random variables which the statistician may observe is prescribed in advance. Decision problems in which the statistician has, in addition, a choice of experiments to perform at each stage are generally of a higher order of difficulty. Such problems are called sequential design problems.

The elements of a sequential decision problem are as follows.

1. Θ, *the parameter space,* or space of states of nature,

2. $\mathcal{A}$, the action space, called *the space of terminal actions of the statistician,*

3. L, a real-valued function defined on $\Theta \times \mathcal{A}$, called *the loss function,* where $L(\theta, a)$ represents the loss incurred by the statistician if θ is the true state of nature and a is the terminal action chosen by him. We assume throughout this chapter that L is bounded below; then, without loss of generality, we assume that L is nonnegative

$$L(\theta, a) \geq 0.$$

4. $\mathbf{X} = (X_1, X_2, \cdots)$, *the random variables available to the statistician for observation.* These random variables need not be identically distributed or independent. The sample space of the random variable X_j is denoted by $\mathfrak{X}_j$, and $\mathfrak{X}$ denotes $\mathfrak{X}_1 \times \mathfrak{X}_2 \times \cdots$, the set of all points $\mathbf{x} = (x_1, x_2, \cdots)$, with $x_j \in \mathfrak{X}_j$. The distribution of $\mathbf{X}$ depends on the true state of nature θ. It is assumed that for every $\theta \in \Theta$, and every integer j, the statistician knows the distribution of $X_1, \cdots, X_j$.

5. $\{c_j(\theta, x_1, \cdots, x_j), j = 0, 1, 2, \cdots\}$, *the cost functions,* a sequence of real-valued functions with c_j defined on $\Theta \times \mathfrak{X}_1 \times \cdots \times \mathfrak{X}_j$. The number $c_j(\theta, x_1, \cdots, x_j)$ represents the cost to the statistician of taking observations $X_1 = x_1, \cdots, X_j = x_j$ and stopping when θ is the true value of the parameter. For simplicity it is assumed that the cost of each observation is always positive, that is, $c_0(\theta) \geq 0$ and

$$c_j(\theta, x_1, \cdots, x_j) < c_{j+1}(\theta, x_1, \cdots, x_j, x_{j+1})$$

and that the cost of sampling increases to infinity as the number of samples taken tends to infinity, that is,

$$c_j(\theta, x_1, \cdots, x_j) \to \infty \quad \text{as} \quad j \to \infty$$

for all $\theta \in \Theta$ and all $\mathbf{x} \in \mathfrak{X}$. In particular, it is assumed that the cost of an infinite number of observations is always infinite. In most applications the cost of each observation will be some constant c so that $c_j(\theta, x_1, \cdots, x_j) = jc$.

Given these five elements, the statistician is to sample the X_j sequentially and decide when to stop sampling and what terminal action to take when he does stop. His objective is to minimize the over-all expected value of his cost plus his loss. A smaller sample costs less, but it entails a larger expected loss.

We now describe the decision rule of the statistician. We prefer to describe the behavioral decision rule at the start, letting the nonrandomized decision rule be a particular case. The decision rule of the statistician is conveniently divided into two parts, the stopping rule and the terminal decision rule.

The Stopping Rule, $\boldsymbol{\varphi}$. A stopping rule is a sequence of functions $\boldsymbol{\varphi}(\mathbf{x}) = (\varphi_0, \varphi_1(x_1), \varphi_2(x_1, x_2), \cdots)$ with φ_j defined on $\mathfrak{X}_1 \times \cdots \times \mathfrak{X}_j$ and $0 \leq \varphi_j \leq 1$ for all j. The function $\varphi_j(x_1, \cdots, x_j)$ represents the conditional probability that the statistician will cease sampling, given that he has taken j observations and $X_1 = x_1, \cdots, X_j = x_j$ (φ_0 is a constant representing the probability of taking no observations at all). Alternatively, we may define a stopping rule by the sequence of functions $\boldsymbol{\psi}(\mathbf{x}) = (\psi_0, \psi_1(x_1), \psi_2(x_1, x_2), \cdots)$, where $\psi_j(x_1, \cdots, x_j)$ represents the conditional probability of not stopping after the first $j - 1$ observations and then stopping after the jth observation, given that $X_1 = x_1, \cdots, X_j = x_j$. The functions $\boldsymbol{\varphi}$ and $\boldsymbol{\psi}$ are related by the formulas $\psi_0 = \varphi_0$ and

$$\psi_j(x_1, \cdots, x_j) = (1 - \varphi_0)(1 - \varphi_1(x_1)) \cdots (1 - \varphi_{j-1}(x_1, \cdots, x_{j-1})) \times \varphi_j(x_1, \cdots, x_j) \quad \text{for} \quad j = 1, 2, \cdots. \tag{7.1}$$

The probability of stopping at time n is given by $E_\theta \psi_n(X_1, \cdots, X_n)$; hence the probability of stopping eventually (or, equivalently, the probability of a finite sample size) is

$$\sum_{n=0}^{\infty} E_\theta \psi_n(X_1, \cdots, X_n). \tag{7.2}$$

To obtain a finite expected cost it must be assumed (at least) that this probability is one for all $\theta \in \Theta$.

For a given stopping rule, $\boldsymbol{\varphi}$ or $\boldsymbol{\psi}$, we may define a random variable N to represent *the random stopping time*. The conditional distribution of N, given $\mathbf{X} = \mathbf{x}$, is defined for any $\theta \in \Theta$ as independent of θ by the formula

$$P_\theta\{N = n \mid \mathbf{X} = \mathbf{x}\} = \psi_n(x_1, \cdots, x_n), \qquad n = 0, 1, \cdots, \tag{7.3}$$

and

$$P_\theta\{N = \infty \mid \mathbf{X} = \mathbf{x}\} = 1 - \sum_{j=0}^{\infty} \psi_j(x_1, \cdots, x_j).$$

The function φ_j may be written in terms of the random stopping time as

$$\varphi_j(x_1, \cdots, x_j) = P_\theta\{N = j \mid N \geq j, \mathbf{X} = \mathbf{x}\}. \tag{7.4}$$

The Terminal Decision Rule, $\boldsymbol{\delta}$. A terminal decision rule is a sequence of functions $\boldsymbol{\delta}(\mathbf{x}) = (\delta_0, \delta_1(x_1), \delta_2(x_1, x_2), \cdots)$, where, for all j, δ_j is a behavioral decision rule for the statistical decision problem $(\Theta, \mathcal{A}, L)$, based on the fixed sample size $X_1, \cdots, X_j$; that is, $\boldsymbol{\delta}_j(x_1, \cdots, x_j)$ is a probability distribution over a σ-field of subsets of $\mathcal{A}$ for which the expected loss $E_\theta L(\theta, \delta_j(X_1, \cdots, X_j))$ exists and is finite for all $\theta \in \Theta$. The interpretation of $\boldsymbol{\delta}$ as a terminal decision rule is as follows. If it is decided to stop after observing $X_1 = x_1, \cdots, X_j = x_j$, then a point in $\mathcal{A}$ is chosen at random according to the distribution $\delta_j(x_1, \cdots, x_j)$.

For a given stopping rule, $\boldsymbol{\varphi}$, δ_j needs to be defined only for those $(x_1, \cdots, x_j)$ for which $\psi_j(x_1, \cdots, x_j) > 0$. It is useful, however, to define a terminal decision rule so that it is independent of the stopping rule, as we shall see.

A (behavioral) sequential decision rule is a pair $(\boldsymbol{\varphi}, \boldsymbol{\delta})$, in which $\boldsymbol{\varphi}$ is a stopping rule and $\boldsymbol{\delta}$ is a terminal decision rule.

The risk function of a sequential decision rule $(\boldsymbol{\varphi}, \boldsymbol{\delta})$ is the expected value, using this rule, of the loss plus the cost when θ is the true value of the parameter. If the probability of a finite sample size (7.2) is not equal to one, this risk is defined as infinite; otherwise

$$\hat{R}(\theta, (\boldsymbol{\varphi}, \boldsymbol{\delta})) = \sum_{j=0}^{\infty} E_\theta\{\psi_j(X_1, \cdots, X_j)[L(\theta, \delta_j(X_1, \cdots, X_j)) + c_j(\theta, X_1, \cdots, X_j)]\}, \quad (7.5)$$

where $L(\theta, \delta_j)$ is defined as in Section 1.5 as $EL(\theta, Z)$, where Z is a random variable taking values in $\mathcal{A}$ whose distribution is given by δ_j. In terms of the random stopping time N, the risk function may be written

$$\hat{R}(\theta, (\boldsymbol{\varphi}, \boldsymbol{\delta})) = E_{\theta,\boldsymbol{\varphi}}\{L(\theta, \delta_N(X_1, \cdots, X_N)) + c_N(\theta, X_1, \cdots, X_N)\}, \quad (7.6)$$

provided $P_{\theta,\boldsymbol{\varphi}}\{N \neq \infty\}$ of formula (7.2) is equal to one. In this notation $E_{\theta,\boldsymbol{\varphi}}$ represents the expectation given θ as the true value of the parameter and, given the stopping rule, $\boldsymbol{\varphi}$. The stopping rule $\boldsymbol{\varphi}$ determines the distribution of N. Any expectation or probabilities in which N is involved must have a stopping rule appended as a subscript to show what distribution is given to N.

Nonrandomized terminal decision rules are denoted by $\mathbf{d}$. If $\mathbf{d} = (d_0, d_1, \cdots)$, then d_j is a mapping from $\mathfrak{X}_1 \times \cdots \times \mathfrak{X}_j$ into $\mathcal{A}$. This may be considered as a special case of a behavioral decision rule, $\boldsymbol{\delta}$, such that for all j and $(x_1, \cdots, x_j)$, $\delta_j(x_1, \cdots, x_j)$ is a degenerate probability distribution on $\mathcal{A}$. A nonrandomized stopping rule is a stopping rule, $\boldsymbol{\varphi}$, for which $\varphi_j(x_1, \cdots, x_j)$ is zero or one for all j and $(x_1, \cdots, x_j)$. If $\boldsymbol{\varphi}$ is nonrandomized, then $\psi_j(x_1, \cdots, x_j)$ is zero or one for all j and $(x_1, \cdots, x_j)$ also. No special notation is used for nonrandomized stopping rules.

A randomized sequential decision rule is a probability distribution over the space of nonrandomized sequential decision rules $\{(\boldsymbol{\varphi}, \mathbf{d}), \boldsymbol{\varphi}$ and $\mathbf{d}$ nonrandomized$\}$. As for nonsequential decision problems, it is important to know whether randomized and behavioral decision rules are equivalent in the sense that for every behavioral rule there is a randomized rule with identical risk and conversely. In Wald and Wolfowitz (1951) it is shown that there is an affirmative answer to this question when $\mathcal{A}$ is a separable complete metric space. An argument similar to that found in Section 1.5 and rigorous for the discrete case may be given here also. This we leave to interested students. In the following we assume that behavioral rules and randomized rules are equivalent in the above sense.

We deal almost exclusively with behavioral decision rules in the following. We drop the notation $\hat{R}$ and use R instead to denote the risk from now on.

7.2 Bayes and Minimax Sequential Decision Rules

In this section we investigate methods for finding Bayes rules with respect to a given prior distribution, and we apply these methods for finding minimax sequential decision rules to a few specific problems. A worthwhile supplementary reference is the original paper of Arrow, Blackwell, and Girshick (1949), or the subsequent book by Blackwell and Girshick (1954).

As in Section 1.6, Θ^* denotes a space of probability distributions over a fixed σ-field of subsets of Θ. We suppose that for each $\tau \in \Theta^*$ the Bayes risk

$$r(\tau, (\boldsymbol{\varphi}, \boldsymbol{\delta})) = ER(T, (\boldsymbol{\varphi}, \boldsymbol{\delta})) = \int R(\theta, (\boldsymbol{\varphi}, \boldsymbol{\delta}))\, d\tau(\theta) \qquad (7.7)$$

(where T is a random variable on Θ with distribution τ) is well defined for all $(\boldsymbol{\varphi}, \boldsymbol{\delta}) \in \mathfrak{D}$. To gain generality, we extend the notion of a Bayes rule in fixed sample problems for which the Bayes risk is infinite. We take it as a definition that a Bayes (ϵ-Bayes) rule for a statistical decision problem $(\Theta, \mathcal{A}, L)$ based on $X_1, \cdots, X_j$, minimizes (comes within ϵ of minimizing) the conditional Bayes expected loss

$$E\{L(T, \delta_j(x_1, \cdots, x_j)) \mid X_1 = x_1, \cdots, X_j = x_j\}$$

except perhaps for $(x_1, \cdots, x_j)$ in a set of probability zero. If in a sequential problem the minimum conditional Bayes expected loss is $+\infty$ for some $(x_1, \cdots, x_j)$, we expect a good stopping rule to assign probability zero to stopping after $X_1 = x_1, \cdots, X_j = x_j$ is observed (see Exercise 2). The problem is to find for a fixed $\tau \in \Theta^*$ a rule $(\boldsymbol{\varphi}, \boldsymbol{\delta})$ which minimizes $r(\tau, (\boldsymbol{\varphi}, \boldsymbol{\delta}))$.

The problem may be divided into two parts. First, we fix both τ and

$\boldsymbol{\varphi}$ and try to find $\boldsymbol{\delta}$ which minimizes $r(\tau, (\boldsymbol{\varphi}, \boldsymbol{\delta}))$. Then we try to find a stopping rule $\boldsymbol{\varphi}$ which minimizes $\inf_{\boldsymbol{\delta}} r(\tau, (\boldsymbol{\varphi}, \boldsymbol{\delta}))$. The first part of this program is easily carried out. The result is the following theorem.

Theorem 1. Let $\delta_j^0(X_1, \cdots, X_j)$ be a Bayes (ϵ-Bayes) rule for the statistical decision problem $(\Theta, \mathcal{A}, L)$ based on a fixed sample $X_1, \cdots, X_j$, with respect to a prior distribution $\tau \in \Theta^*$. Then, for any fixed stopping rule $\boldsymbol{\varphi}$, $r(\tau, (\boldsymbol{\varphi}, \boldsymbol{\delta}))$ is minimized (within ϵ of being minimized) by $\boldsymbol{\delta}^0 = (\delta_0^0, \delta_1^0, \delta_2^0, \cdots)$.

Remark. The noteworthy point of this theorem is that $\boldsymbol{\delta}^0$ is independent of $\boldsymbol{\varphi}$! This not only facilitates the computation of Bayes rules but also has interesting implications regarding the behavior of the Bayesian in decision problems. Suppose that an experimenter comes to the statistician (a Bayesian) with his data already collected and asks for an estimate of a parameter based on this data, using some given loss function. The experimenter has already paid the cost of obtaining the data and has no doubt used a rather complicated stopping rule, depending on a hasty glance at the data as they were being collected and how he "felt" about them. Lack of knowledge concerning the stopping rule used by the experimenter does not hinder the statistician in arriving at a terminal decision rule. He acts as if he were faced with a fixed sample decision problem and chooses a Bayes rule with respect to someone's subjective prior distribution. This leads to the possibility of use by the experimenter of a stopping rule that estimates with a strong bias. For example, in estimating the probability that a coin will come up heads with loss $L(\theta, a) = (\theta - a)^2/(\theta(1 - \theta))$, the Bayes rule with respect to the uniform distribution on $(0, 1)$ based on n tosses of the coin is $\hat{\theta}_n = S_n/n$; S_n is the number of heads. The experimenter may use the stopping rule $\boldsymbol{\varphi}$, which stops after the first toss if it is heads and after the second toss if the first toss is tails. This rule has positive bias: $E_{\theta,\boldsymbol{\varphi}}\hat{\theta}_N = \theta + \frac{1}{2}\theta(1 - \theta)$. There are stopping rules $\boldsymbol{\varphi}$ for which $E_{\theta,\boldsymbol{\varphi}}\hat{\theta}_N$ evaluated at $\theta = \frac{1}{2}$ is greater than .79 [see Chow and Robbins (1965)].

Proof. The Bayes risk may be written, using (7.7) and (7.5),

$$r(\tau, (\boldsymbol{\varphi}, \boldsymbol{\delta})) = \sum_{j=0}^{\infty} E\{\psi_j(X_1, \cdots, X_j)\, L(T, \delta_j(X_1, \cdots, X_j))\}$$

$$+ \sum_{j=0}^{\infty} E\{\psi_j(X_1, \cdots, X_j)\, c_j(T, X_1, \cdots, X_j)\}. \tag{7.8}$$

Because both τ and $\boldsymbol{\varphi}$ (hence $\boldsymbol{\psi}$) are given, we may minimize (7.8) by choosing δ_j for each j to minimize

$$E\{\psi_j(X_1, \cdots, X_j)\, L(T, \delta_j(X_1, \cdots, X_j))\}$$
$$= E\{\psi_j(X_1, \cdots, X_j)\, E(L(T, \delta_j(X_1, \cdots, X_j)) \mid X_1, \cdots, X_j)\},$$

which may be minimized by choosing δ_j to minimize $E(L(T, \delta_j) \mid X_1, \cdots, X_j)$, that is, the Bayes rule with respect to τ. If these Bayes rules δ_j^0 exist, then $\boldsymbol{\delta}^0 = (\delta_0^0, \delta_1^0, \cdots)$ clearly minimizes $r(\tau, (\boldsymbol{\varphi}, \boldsymbol{\delta}))$. If δ_j^0 is ϵ-Bayes for $j = 0, 1, \cdots$, then $\boldsymbol{\delta}^0 = (\delta_0^0, \delta_1^0, \cdots)$ comes within ϵ of minimizing $r(\tau, (\boldsymbol{\varphi}, \boldsymbol{\delta}))$, as is easily checked, thus completing the proof.

The first part of our program is finished. To find Bayes terminal decision rules, we use the data as if we were dealing with a fixed sample size problem and find the corresponding Bayes decision rule. The other part of our program is more difficult. Our attack is as follows. First we truncate the sequential decision problem at some integer J (i.e., we restrict the statistician to observing at most $X_1, \cdots, X_J$) and find the Bayes rule for this finitized problem. Then we let $J \to \infty$, hoping that the general sequential decision problem will be approximated well by the problem truncated at J when J is large.

A sequential decision problem is said to be *truncated at J* if the statistician is restricted to stopping rules $\boldsymbol{\varphi}$ (or $\boldsymbol{\psi}$) for which

$$\varphi_J(x_1, \cdots, x_J) \equiv 1 \qquad \left(\text{or } \sum_{j=0}^{J} \psi_j(x_1, \cdots, x_j) \equiv 1\right). \tag{7.9}$$

We use a method of backward induction for computing a Bayes stopping rule for a truncated sequential decision problem. This may be described roughly as follows. Whenever we stop, we use a Bayes terminal decision rule, which for the purposes of the present argument we assume exists. If we reach J, we know we must stop. Therefore, if we reach $J - 1$ and have observed $X_1 = x_1, \cdots, X_{J-1} = x_{J-1}$, we should stop if the conditional expected loss plus cost, given $X_1 = x_1, \cdots, X_{J-1} = x_{J-1}$ of stopping immediately, is less than the conditional expected loss plus cost, given $X_1 = x_1, \cdots, X_{J-1} = x_{J-1}$ of taking one more observation and then stopping. We would take one more observation if the reverse inequality were true. Thus we have found φ_{J-1} of the Bayes stopping rule $\boldsymbol{\varphi}$. Inductively, if we know $\varphi_j, \varphi_{j+1}, \cdots, \varphi_J$ of the Bayes stopping rule $\boldsymbol{\varphi}$, we may find φ_{j-1} as follows: if we reach $j - 1$ and have observed $X_1 = x_1, \cdots, X_{j-1} = x_{j-1}$, we stop if and only if the conditional expected loss plus cost, given $X_1 = x_1, \cdots, X_{j-1} = x_{j-1}$ of stopping immediately is less than the conditional expected loss plus cost, given

$X_1 = x_1, \cdots, X_{j-1} = x_{j-1}$ of taking one more observation and then using the Bayes stopping rule $\varphi_j, \cdots, \varphi_J$ from there on. We formalize this argument by introducing the following notation.

Let the minimum conditional Bayes expected loss given $X_1 = x_1, \cdots, X_j = x_j$ be denoted by $\rho_j(x_1, \cdots, x_j)$:

$$\begin{aligned}\rho_j(x_1, \cdots, x_j) &= \inf_{\delta_j} E\{L(T, \delta_j) \mid X_1 = x_1, \cdots, X_j = x_j\} \\ &= \inf_{a \in \mathfrak{A}} E\{L(T, a) \mid X_1 = x_1, \cdots, X_j = x_j\}.\end{aligned}$$

We assume that $\rho_j(x_1, \cdots, x_j)$ is integrable. If $\mathfrak{A}$ or $\mathfrak{X}$ is discrete, this follows automatically. The minimum conditional expected loss plus cost of stopping after observing $X_1 = x_1, \cdots, X_j = x_j$ is denoted by $U_j(x_1, \cdots, x_j)$:

$$\begin{aligned}U_j(x_1, \cdots, x_j) &= \rho_j(x_1, \cdots, x_j) \\ &\quad + E\{c_j(T, x_1, \cdots, x_j) \mid X_1 = x_1, \cdots, X_j = x_j\}.\end{aligned}$$

Now suppose that $X_1 = x_1, \cdots, X_{J-1} = x_{J-1}$ has been observed. If the statistician stops without observing X_J, his loss plus cost is $U_{J-1}(x_1, \cdots, x_{J-1})$. If he does look at X_J, his expected loss plus cost is

$$E\{U_J(x_1, \cdots, x_{J-1}, X_J) \mid X_1 = x_1, \cdots, X_{J-1} = x_{J-1}\}.$$

Thus he should stop if the former is smaller than the latter and continue if the latter is smaller than the former. In the case of equality it does not matter which he does. Thus $\overset{0}{\varphi}_{J-1}$ of the Bayes stopping rule is

$$\begin{aligned}&\overset{0}{\varphi}_{J-1}(x_1, \cdots, x_{J-1}) \\ &= \begin{cases} 1 & \text{if} \quad U_{J-1}(x_1, \cdots, x_{J-1}) \\ & \qquad < E\{U_J(x_1, \cdots, x_{J-1}, X_J) \mid X_1 = x_1, \cdots, X_{J-1} = x_{J-1}\}, \\ \text{any} & \qquad = \\ 0 & \qquad > \end{cases}\end{aligned}$$

The minimum conditional Bayes risk, given $X_1 = x_1, \cdots, X_{J-1} = x_{J-1}$, is

$$\begin{aligned}&V_{J-1}^{(J)}(x_1, \cdots, x_{J-1}) \\ &\qquad = \min [U_{J-1}(x_1, \cdots, x_{J-1}), \\ &\qquad\qquad E\{U_J(x_1, \cdots, x_{J-1}, X_J) \mid X_1 = x_1, \cdots, X_{J-1} = x_{J-1}\}].\end{aligned}$$

If $X_1 = x_1, \cdots, X_{J-2} = x_{J-2}$ has been observed, the statistician would stop if

$$U_{J-2}(x_1, \cdots, x_{J-2}) < E\{V_{J-1}^{(J)}(x_1, \cdots, x_{J-2}, X_{J-1}) \mid X_1 = x_1, \cdots, X_{J-2} = x_{J-2}\},$$

and so on. The minimum conditional Bayes risk, given $X_1 = x_1, \cdots, X_{j-1}, = x_{j-1}$, may be defined inductively as $V_J^{(J)} = U_J$ and

$$V_{j-1}^{(J)}(x_1, \cdots, x_{j-1}) = \min [U_{j-1}(x_1, \cdots, x_{j-1}), E\{V_j^{(J)}(x_1, \cdots, x_{j-1}, X_j) \mid X_1 = x_1, \cdots, X_{j-1} = x_{j-1}\}] \quad (7.10)$$

for $j = 1, \cdots, J$. The constant $V_0^{(J)}$ denotes the minimum Bayes risk of the problem truncated at J. The Bayes stopping rule is then $\boldsymbol{\varphi}^0 = (\varphi_0^0, \cdots, \varphi_J^0)$, where

$$\varphi_{j-1}^0(x_1, \cdots, x_{j-1}) = \begin{cases} 1 & \text{if} \quad U_{j-1}(x_1, \cdots, x_{j-1}) < E\{V_j^{(J)}(x_1, \cdots, x_{j-1}, X_j) \mid X_1 = x_1, \cdots, X_{j-1} = x_{j-1}\}, \\ \text{any} & \qquad = \\ 0 & \qquad > \end{cases} \quad (7.11)$$

for $j = 1, 2, \cdots, J$ (of course, $\varphi_J^0(x_1, \cdots, x_J) \equiv 1$).

Theorem 2. Let $\boldsymbol{\delta}^0$ be as in Theorem 1 and let $\boldsymbol{\varphi}^0$ be defined by (7.11). Then $(\boldsymbol{\varphi}^0, \boldsymbol{\delta}^0)$ is a Bayes (ϵ-Bayes) decision rule with respect to τ for the sequential decision problem truncated at J.

A formal proof involves showing $r(\tau, (\boldsymbol{\varphi}^0, \boldsymbol{\delta}^0)) \le r(\tau, (\boldsymbol{\varphi}, \boldsymbol{\delta}))$ (plus ϵ) for any rule $(\boldsymbol{\varphi}, \boldsymbol{\delta})$ in the problem truncated at J. It is left to the interested student.

We say that any such stopping rule $\boldsymbol{\varphi}^0$ of (7.11) is Bayes with respect to π in the truncated problem, whether or not there exists a Bayes terminal decision rule with respect to π. In general, Bayes stopping rules exist in truncated decision problems.

To complete the program we need a theorem which implies that the general problem (nontruncated) can be approximated by the problem truncated at a sufficiently large J. Sufficient conditions for this result are found in the following theorem, a version of a theorem found in Hoeffding

(1960). Let $V_0^{(\infty)}$ denote the minimum Bayes risk for the general problem

$$V_0^{(\infty)} = \inf_{(\boldsymbol{\varphi},\boldsymbol{\delta})} r(\tau, (\boldsymbol{\varphi}, \boldsymbol{\delta})).$$

Because $V_0^{(0)} \geq V_0^{(1)} \geq \cdots \geq V_0^{(J)} \geq \cdots$, the sequence $V_0^{(J)}$ converges. If the limit of $V_0^{(J)}$ as J tends to infinity is equal to $V_0^{(\infty)}$, the nontruncated decision problem can be well approximated by the problem truncated at a sufficiently large J. Note that $E\rho_J(X_1, \cdots, X_J)$ denotes the minimum Bayes expected loss for the fixed sample problem based on $X_1, \cdots, X_J$.

Theorem 3. If $E\rho_J(X_1, \cdots, X_J) \to 0$, as $J \to \infty$, or if L is bounded, $V_0^{(J)} \to V_0^{(\infty)}$ as $J \to \infty$.

Proof. Let $(\boldsymbol{\varphi}^\epsilon, \boldsymbol{\delta}^\epsilon)$ be an ϵ-Bayes rule for the general problem and modify $\boldsymbol{\delta}^\epsilon$, if necessary, so that $\delta_j{}^\epsilon$ is ϵ-Bayes for the fixed sample size problem based on $X_1, \cdots, X_j$, as in Theorem 1. Let $\boldsymbol{\varphi}^{\epsilon,J}$ be $\boldsymbol{\varphi}^\epsilon$ truncated at J; that is, $\varphi_j^{\epsilon,J} = \varphi_j{}^\epsilon$ for $j < J$ and $\varphi_J^{\epsilon,J} \equiv 1$. Then

$$r(\tau, (\boldsymbol{\varphi}^{\epsilon,J}, \boldsymbol{\delta}^\epsilon)) - r(\tau, (\boldsymbol{\varphi}^\epsilon, \boldsymbol{\delta}^\epsilon))$$

$$= \sum_{j-J}^{\infty} E\{\psi_j{}^\epsilon(X_1, \cdots, X_j)[L(T, \delta_J{}^\epsilon(X_1, \cdots, X_J)) - L(T, \delta_j{}^\epsilon(X, \cdots, X_j))$$
$$+ C_J(T, X_1, \cdots, X_J) - C_j(T, X_1, \cdots, X_j)]\}$$
$$\leq E\{\psi_J^{\epsilon,J}(X_1, \cdots, X_J)\, L(T, \delta_J{}^\epsilon(X_1, \cdots, X_J))\}.$$

If $E\rho_J(X_1, \cdots, X_J) \to 0$ or L is bounded, this last term can be made less than 2ϵ for J sufficiently large [since $E\{L(T, \delta_J{}^\epsilon(X_1, \cdots, X_J))\} \leq E\rho_J(X_1, \cdots, X_J) + \epsilon$]. Then, because $r(\tau, (\boldsymbol{\varphi}^\epsilon, \boldsymbol{\delta}^\epsilon)) \leq V_0^{(\infty)} + \epsilon$, we have for J sufficiently large

$$V_0^{(J)} \leq r(\tau, (\boldsymbol{\varphi}^{\epsilon,J}, \boldsymbol{\delta}^\epsilon)) \leq V_0^{(\infty)} + 3\epsilon.$$

Because ϵ is arbitrary, this completes the proof.

An example in which $V_0^{(J)}$ does not converge to $V_0^{(\infty)}$ is found in Exercise 2.

Example 1. Suppose that $X_1, X_2, \cdots$ are independent Bernoulli trials, $\mathcal{B}(1, \theta)$, where $\theta \in \Theta = [0, 1] = \mathcal{A}$. The loss is proportional to squared error, $L(\theta, a) = K(\theta - a)^2$, and there is a constant cost, c, for each observation, $c_j(\theta, x_1, \cdots, x_j) = jc$. The problem is to find a Bayes sequential estimate of θ when the prior distribution of θ is $\mathcal{U}(0, 1)$, the uniform distribution on the interval $(0, 1)$.

The Bayes terminal decision rule is easily found by using Theorem 1. In the fixed sample size problem, based on $X_1, \cdots, X_j$, $S_j = X_1 + \cdots + X_j$ is a sufficient statistic for θ having the binomial distribution, $\mathcal{B}(j, \theta)$ (see Exercise 1.8.8). The posterior distribution of θ, given $X_1 = x_1, \cdots, X_j = x_j$, is the Beta distribution with parameters $\alpha = s_j + 1$ and $\beta = j - s_j + 1$, where $s_j = x_1 + \cdots + x_j$. This has mean $(s_j + 1)/(j + 2)$ and variance $(s_j + 1)(j - s_j + 1)(j + 2)^{-2}(j + 3)^{-1}$. Thus the Bayes terminal decision rule is the nonrandomized rule $\mathbf{d}^0 = (d_0^0, d_1^0, \cdots)$, where for all j

$$d_j^0(x_1, \cdots, x_j) = \frac{s_j + 1}{j + 2},$$

and the minimum conditional Bayes expected loss is

$$\rho_j(x_1, \cdots, x_j) = K(s_j + 1)(j - s_j + 1)(j + 2)^{-2}(j + 3)^{-1}.$$

Hence

$$U_j(x_1, \cdots, x_j) = K(s_j + 1)(j - s_j + 1)(j + 2)^{-2}(j + 3)^{-1} + jc.$$

We approximate this problem by the problem truncated at J. To compute the minimum conditional Bayes risks $V_j^{(J)}$ we need the conditional distribution of X_j, given $X_1, \cdots, X_{j-1}$ (unconditional on θ). The unconditional distribution of $X_1, \cdots, X_j$ is

$$f_{X_1,\cdots,X_j}(x_1, \cdots, x_j) = \frac{\Gamma(s_j + 1)\Gamma(j - s_j + 1)}{\Gamma(j + 2)}$$

$$x_i = 0, 1, \qquad i = 1, \cdots, j.$$

Hence the conditional distribution of X_j, given $X_1, \cdots, X_{j-1}$, is

$$f_{X_j|X_1=x_1,\cdots,X_{j-1}=x_{j-1}}(x_j)$$

$$= \frac{\Gamma(s_j + 1)\Gamma(j - s_j + 1)\Gamma(j + 1)}{\Gamma(j + 2)\Gamma(s_{j-1} + 1)\Gamma(j - s_{j-1})}, \qquad x_j = 0, 1,$$

$$(7.12)$$

$$= \begin{cases} \dfrac{j - s_{j-1}}{j + 1}, & \text{if } x_j = 0, \\[2ex] \dfrac{s_{j-1} + 1}{j + 1} & \text{if } x_j = 1. \end{cases}$$

Using this distribution, we can compute the conditional expectation $E(V_j^{(J)}(x_1, \cdots, x_{j-1}, X_j) \mid X_1 = x_1, \cdots, X_{j-1} = x_{j-1})$ necessary for the inductive computation of the $V_j^{(J)}$.

To carry out the computations in a specific instance we choose $J = 5$, $K = 200$, and $c = 1$, that is, $L(\theta, a) = 200(\theta - a)^2$ and constant cost $c = 1$. Table 7.1 shows the results of this computation. The Bayes sequential rule turns out to be observe X_1, X_2, and X_3. If $S_3 = 0$ or 3, stop and estimate θ as $(S_3 + 1)/5$; otherwise observe X_4. If $S_4 = 1$ or 3, stop and estimate θ as $(S_4 + 1)/6$; otherwise observe X_5 and estimate θ as $(S_5 + 1)/7$.

EXAMPLE 2. Consider Example 1, except that the loss function is now changed to $L(\theta, a) = K(\theta - a)^2/(\theta(1 - \theta))$. Given $X_1, \cdots, X_j$ the posterior distribution of θ is a Beta distribution as before, but this time the Bayes estimate of θ is the usual unbiased estimate of θ,

$$d_j{}^0(x_1, \cdots, x_j) = \frac{s_j}{j}, \qquad j = 1, \cdots, J.$$

For $j = 0$ any estimate has infinite Bayes risk; therefore we are certain to take at least one observation. The minimal conditional Bayes loss given $X_1 = x_1, \cdots, X_j = x_j$ is

$$\rho_j(x_1, \cdots, x_j) = \frac{K\Gamma(j + 2)}{\Gamma(s_j + 1)\Gamma(j - s_j + 1)} \int_0^1 \left(\theta - \frac{s_j}{j}\right)^2 \theta^{s_j - 1}(1 - \theta)^{j - s_j - 1} d\theta$$

$$= \frac{K}{j}, \qquad j = 1, \cdots, J,$$

so that

$$U_j(x_1, \cdots, x_j) = \frac{K}{j} + jc, \qquad j = 1, \cdots, J.$$

This function is independent of $x_1, \cdots, x_j$, hence $V_j^{(J)}$, defined inductively by (7.10), is independent of $x_1, \cdots, x_j$. Equations 7.10 and 7.11 become

$$V_{j-1}^{(J)} = \min \{U_{j-1}, U_j, \cdots, U_J\}$$

$$\varphi_{j-1} = \begin{cases} 1 & \text{if} \quad U_{j-1} < V_j^{(J)}, \\ \text{any} & \qquad\quad = \\ 0 & \qquad\quad > \end{cases}$$

Table 7.1

S_5	d_5^0	Cost	ρ_5	$U_5 = V_5^{(5)}$
0	1/7	5	3.0612	8.0612
1	2/7	5	5.1020	10.1020
2	3/7	5	6.1224	11.1224
3	4/7	5	6.1224	11.1224
4	5/7	5	5.1020	10.1020
5	6/7	5	3.0612	8.0612

S_4	d_4^0	Cost	ρ_4	U_4	$E(V_5^{(5)} \mid X_1 \cdots X_4)$	$V_4^{(5)}$	φ_4^0
0	1/6	4	3.9683	7.9683	8.4013	7.9683	1
1	1/3	4	6.3492	10.3492	10.4421	10.3492	1
2	1/2	4	7.1429	11.1429	11.1224	11.1224	0
3	2/3	4	6.3492	10.3492	10.4421	10.3492	1
4	5/6	4	3.9683	7.9683	8.4013	7.9683	1

S_3	d_3^0	Cost	ρ_3	U_3	$E(V_4^{(5)} \mid X_1 \cdots X_3)$	$V_3^{(5)}$	φ_3^0
0	1/5	3	5.3333	8.3333	8.4445	8.3333	1
1	2/5	3	8.0000	11.0000	10.6585	10.6585	0
2	3/5	3	8.0000	11.0000	10.6585	10.6585	0
3	4/5	3	5.3333	8.3333	8.4445	8.3333	1

S_2	d_2^0	Cost	ρ_2	U_2	$E(V_3^{(5)} \mid X_1, X_2)$	$V_2^{(5)}$	φ_2^0
0	1/4	2	7.5000	9.5000	8.9146	8.9146	0
1	1/2	2	10.0000	12.0000	10.6585	10.6585	0
2	3/4	2	7.5000	9.5000	8.9146	8.9146	0

S_1	d_1^0	Cost	ρ_1	U_1	$E(V_2^{(5)} \mid X_1)$	$V_1^{(5)}$	φ_1^0
0	1/3	1	11.1111	12.1111	9.4959	9.4959	0
1	2/3	1	11.1111	12.1111	9.4959	9.4959	0

S_0	d_0^0	Cost	ρ_0	U_0	$EV_1^{(5)}$	$V_0^{(5)}$	φ_0^0
0	1/2	0	16.6666	16.6666	9.4959	9.4959	0

Thus the Bayes sequential decision rule for the truncated problem is a fixed sample size rule: observe $X_1, \cdots X_n$ when $n \leq J$ minimizes $Kn^{-1} + nc$, and estimate θ to be S_n/n. (If $K = 200$ and $c = 1$, then $n = 14$.) The risk function corresponding to this rule is the same as for the fixed sample size problem, plus the cost, namely, $R(\theta, (\varphi^0, \delta^0)) = Kn^{-1} + nc$ (see Exercise 2.11.8). Because $E\rho_J(X_1, \cdots, X_J) = KJ^{-1} \to 0$ as $J \to \infty$, $V_0^{(J)} \to V_0^{(\infty)}$ by Theorem 3; but if the n_0 which minimizes $Kn^{-1} + nc$ is less than J, then $V_0^{(J)} = Kn_0^{-1} + n_0c$, so that $V_0^{(\infty)} = Kn_0^{-1} + n_0c$. The Bayes sequential rule for the general problem is a fixed sample size rule also. Furthermore, because it has constant risk, it is minimax. The generalization of this example is in Theorem 4.

Theorem 4. If, for all j, $U_j(x_1, \cdots, x_j)$ is a constant independent of $x_1, \cdots, x_j$, then the Bayes sequential decision rule for the truncated problem is a fixed sample size rule. If, in addition $V_0^{(J)} \to V_0^{(\infty)}$ as $J \to \infty$, the Bayes sequential decision rule for the general problem is a fixed sample size rule.

The proof is an exercise.

The Bayes rule for the nontruncated problem of Example 1 may also be found, for it turns out that this Bayes rule is a truncated rule. It often turns out that a Bayes rule is a truncated rule. Theorem 5 gives a sufficient condition for this to happen and also an upper bound on the maximum number of observations required by the Bayes rule.

Theorem 5. If $V_0^{(J)} \to V_0^{(\infty)}$, as $J \to \infty$, and if for all $j > J_0$

$$\begin{aligned}\rho_{j-1}(x_1, \cdots, x_{j-1}) - E\{\rho_j(x_1, \cdots, x_{j-1}, X_j) \mid X_1 = x_1, \cdots, X_{j-1} = x_{j-1}\} \\ \leq E\{c_j(T, x_1, \cdots, x_{j-1}, X_j) \\ - c_{j-1}(T, x_1, \cdots, x_{j-1}) \mid X_1 = x_1, \cdots, X_{j-1} = x_{j-1}\} \quad (7.13)\end{aligned}$$

(except perhaps for $(x_1, \cdots, x_{j-1})$ in a set of probability zero), then $V_0^{(J_0)} = V_0^{(\infty)}$.

Proof. Let $J > J_0$ and consider the sequential decision problem truncated at J. Equation 7.13 is just another way of saying $U_{j-1} \leq E(U_j \mid X_1, \cdots, X_{j-1})$ for $j > J_0$ (with probability one). Hence $V_{J-1}^{(J)} =$

U_{J-1} (with probability one) and, by induction, $V_j^{(J)} = U_j$ for $j = J_0, \cdots, J$ (with probability one). This implies that $V_{J_0}^{(J)} = V_{J_0}^{(J_0)}$ for all $J > J_0$ (with probability one), hence that $V_0^{(J)} = V_0^{(J_0)}$ for all $J > J_0$. Then, because $V_0^{(J)} \to V_0^{(\infty)}$ as $J \to \infty$, $V_0^{(J_0)} = V_0^{(\infty)}$, as was to be proved.

Remark. The right side of (7.13) is the conditional expected cost of taking one more observation given $X_1 = x_1, \cdots, X_{j-1} = x_{j-1}$. The left side of (7.13) represents the conditional expected reduction in the loss, saved by taking one more observation, given $X_1 = x_1, \cdots, X_{j-1} = x_{j-1}$. The conclusion, $V_0^{(J_0)} = V_0^{(\infty)}$, means that we can do as well using rules truncated at J_0 as we can using nontruncated rules, in the sense that we obtain the same Bayes risk. Thus Theorem 5 says roughly: if $V_0^{(J)} \to V_0^{(\infty)}$, as $J \to \infty$, and for all $j \geq J_0$ the conditional expected saving in the loss is less than the conditional expected cost of taking one more observation, given that $X_1 = x_1, \cdots, X_{j-1} = x_{j-1}$ have been observed so far, a Bayes rule for the decision problem truncated at J_0 is also Bayes for the general problem.

EXAMPLE 1 (Continued). Because the loss is bounded, Theorem 3 implies that $V_0^{(J)} \to V_0^{(\infty)}$, as $J \to \infty$. We may compute from (7.12)

$$E(\rho_j(x_1, \cdots, x_{j-1}, X_j) \mid X_1 = x_1, \cdots, X_{j-1} = x_{j-1})$$

$$= \frac{K}{(j+2)^2(j+3)} E((S_j + 1)(j - S_j + 1) \mid X_1 = x_1, \cdots, X_{j-1} = x_{j-1})$$

$$= \frac{K(s_{j-1} + 1)(j - s_{j-1})}{(j+1)(j+2)^2}. \tag{7.14}$$

Hence

$$\rho_{j-1}(x_1, \cdots, x_{j-1}) - E(\rho_j(x_1, \cdots, x_{j-1}, X_j) \mid X_1 = x_1, \cdots, X_{j-1} = x_{j-1}) = \frac{K(s_{j-1} + 1)(j - s_{j-1})}{(j+1)^2(j+2)^2}.$$

This takes on its largest value when $s_{j-1} = [(j - 1)/2]$ (largest integer $\leq (j-1)/2$).

$$\frac{K(s_{j-1} + 1)(j - s_{j-1})}{(j+1)^2(j+2)^2} \leq \frac{K[(j+1)/2][(j+2)/2]}{(j+1)^2(j+2)^2}.$$

It is easily checked that this is decreasing in j for $j = 1, 2, \cdots$. Hence

J_0 is the largest integer j for which $K[(j+1)/2][(j+2)/2] \geq c(j+1)^2(j+2)^2$. In the numerical example in which $K = 200$ and $c = 1$ we find that $J_0 = 5$. Thus the Bayes rule we computed there for the problem truncated at 5 is a Bayes rule for the general nontruncated problem as well.

One final point must be cleared up before the approximation to a Bayes rule by a Bayes truncated rule becomes a computational reality. In order to know whether an integer J is a truncation point large enough to effect the approximation to within a desired degree of accuracy, it is necessary to find the number $V_0^{(\infty)}$ or at least to be able to approximate $V_0^{(\infty)}$ as closely as desired. We already have a sequence of upper bounds $V_0^{(1)} \geq V_0^{(2)} \geq \cdots \geq V_0^{(\infty)}$, which, under the assumptions of Theorem 3, converge to $V_0^{(\infty)}$. We may define a sequence of lower bounds $W_0^{(1)} \leq W_0^{(2)} \leq \cdots \leq V_0^{(\infty)}$ as follows. Consider the decision problem truncated at J with the single modification that if J is reached the loss is waived (i.e., put equal to zero). This modified truncated problem may be solved by the same method used to solve the truncated problem. We let $W_0^{(J)}$ represent the minimum Bayes risk in the modified truncated problem.

Theorem 6. $W_0^{(1)} \leq W_0^{(2)} \leq \cdots \leq V_0^{(\infty)}$. If $E\rho_J(X_1, \cdots, X_J) \to 0$ as $J \to \infty$, or if L is bounded, then $W_0^{(J)} \to V_0^{(\infty)}$ as $J \to \infty$.

Proof. Any rule used in the modified problem truncated at $J+1$, when used in the modified problem truncated at J with enforced stopping if J is reached, gives smaller risk; hence $W_0^{(J)} \leq W_0^{(J+1)}$. Similarly, $W_0^{(J)} \leq V_0^{(\infty)}$, proving the first statement. To prove the second, let $r_J(\tau, (\boldsymbol{\varphi}, \boldsymbol{\delta}))$ represent the Bayes risk in the modified truncated problem, truncated at J,

$$r_J(\tau, (\boldsymbol{\varphi}, \boldsymbol{\delta})) = \sum_{j=0}^{J-1} E\{\psi_j(X_1, \cdots, X_j)[L(T, \delta_j(X_1, \cdots, X_j)) + c_j(T, X_1, \cdots, X_j)]\} + E\{\psi_J(X_1, \cdots, X_J)\, c_J(T, X_1, \cdots, X_J)\}.$$

Let $(\boldsymbol{\varphi}^{(J)}, \boldsymbol{\delta}^{(J)})$ be an ϵ-Bayes rule for the modified truncated problem, where $\delta_J^{(J)}$ is chosen as ϵ-Bayes for the fixed sample size problem based on $X_1, \cdots, X_J$. Then

$$V_0^{(\infty)} - W_0^{(J)} \leq r(\tau, (\boldsymbol{\varphi}^{(J)}, \boldsymbol{\delta}^{(J)})) - r_J(\tau, (\boldsymbol{\varphi}^{(J)}, \boldsymbol{\delta}^{(J)})) + \epsilon$$
$$= E\{\psi_J^{(J)}(X_1, \cdots, X_J)\, L(T, \delta_J^{(J)}(X_1, \cdots, X_J))\} + \epsilon. \quad (7.15)$$

As in the proof of Theorem 3, the hypothesis $E\rho_J(X_1, \cdots, X_J) \to 0$, as $J \to \infty$, immediately implies that the expectation in (7.15) goes to zero, and the conclusion of the theorem. Suppose now that L is bounded by B, so that $V_0^{(\infty)} - W_0^{(J)} \le BP_{\varphi^{(J)}}(N = J) + \epsilon$. We will be finished when we show that $P_{\varphi^{(J)}}(N = J) \to 0$ as $J \to \infty$. Let B' be an arbitrary large number and let $S_J = \{(T, X_1, \cdots, X_J) : c_J(T, X_1, \cdots, X_J) < B'\}$. Then

$$\begin{aligned} V_0^{(\infty)} + \epsilon \ge W_0^{(J)} + \epsilon &\ge r_J(\tau, (\varphi^{(J)}, \boldsymbol{\delta}^{(J)})) \\ &\ge E\{\psi_J(X_1, \cdots, X_J)\, c_J(T, X_1, \cdots, X_J)\} \\ &\ge B'E\{\psi_J(X_1, \cdots, X_J) I_{S_J{}^c}(T, X_1, \cdots, X_J)\} \\ &\ge B'\{P_{\varphi^{(J)}}(N = J) - E\psi_J(X_1, \cdots, X_J) I_{S_J}(T, X_1, \cdots, X_J)\} \\ &\ge B'\{P_{\varphi^{(J)}}(N = J) - P(S_J)\}. \end{aligned}$$

Hence

$$P_{\varphi^{(J)}}(N = J) \le \frac{V_0^{(\infty)} + \epsilon}{B'} + P(S_J).$$

Since S_J tends monotonely to the empty set as $J \to \infty$, we have $P(S_J) \to 0$ as $J \to \infty$, hence $\lim_{J\to\infty} P_{\varphi^{(J)}}(N = J) \le (V_0^{(\infty)} + \epsilon)/B'$. Because B' is arbitrary, the proof is complete.

In computing $W_0^{(J)}$, we may define

$$W_J^{(J)} = E\{c_J(T, X_1, \cdots, X_J) \mid X_1, \cdots, X_J\}$$

and then define $W_j^{(J)}$ inductively for $j < J$ by (7.10) with V replaced by W. It is easy then to compute $W_j^{(J)}$ while $V_j^{(J)}$ is being computed by adding two columns to the tables for the computation of $V_j^{(J)}$, namely, $E\{W_{j+1}^{(J)}(| X_1, \cdots, X_j\}$ and $W_j^{(J)}$. An improvement of the lower bounds $W_0^{(J)}$ may be found in Hoeffding (1960), from which the proof of Theorem 6 is taken.

Exercises

1. Prove Theorem 2.

2. Let $X_1, X_2, \cdots$ be independent, with common distribution

$$P_\theta\{X_i = -1\} = 1/2,\ P_\theta\{X_i = 0\} = (1 - \theta)/2,\ P_\theta\{X_i = 1\} = \theta/2,$$

where $0 < \theta < 1$. Let $L(\theta, a) = (\theta - a)^2/(\theta(1 - \theta))$, let the cost of observation be constant, $c_j = jc$, and let the prior distribution of θ be uniform, $\mathcal{U}(0, 1)$. Show that $V_0^{(J)} = +\infty$ for all J but that $V_0^{(\infty)} < \infty$.

3. Prove Theorem 4.

4. Let $X_1, X_2, \cdots$ be independent with common Poisson distribution, $\mathcal{P}(\theta)$, $\theta > 0$. Let $L(\theta, a) = K(\theta - a)^2$, and let the prior distribution of θ be negative exponential $\mathcal{G}(1, 1)$.

(a) Show

$$\rho_j(x_1, \cdots, x_j) = \frac{K(s_j + 1)}{(j + 1)^2},$$

where $s_j = \sum_1^j x_i$.

(b) Show

$$E(\rho_j(x_1, \cdots, x_{j-1}, X_j) \mid X_1 = x_1, \cdots, X_{j-1} = x_{j-1}) = \frac{K(s_{j-1} + 1)}{j(j + 1)}.$$

(c) Find the Bayes sequential decision rule for the decision problem truncated at $J = 3$, when $K = 12$ and there is a constant cost $c = 1$ per observation.

5. Let $X_1, X_2, \cdots$ be independent Bernoulli trials, $\mathcal{B}(1, \theta)$, $0 < \theta < 1$. Consider the hypothesis testing problem, $\mathcal{A} = \{a_1, a_2\}$, for which $L(\theta, a_1) = K\theta$, $L(\theta, a_2) = K(1 - \theta)$. Find the Bayes sequential decision rule with respect to the uniform prior distribution $\mathcal{U}(0, 1)$ for the decision problem truncated at $J = 5$, when there is a constant cost $c = 1$ per observation and $K = 56$. *Note.* Theorem 5 gives $J_0 = 25$. However, S. N. Ray (1965) has derived an upper bound for the truncation point of a Bayes sequential decision rule, never larger than J_0 and which in the present problem is 5, so that $V_0^{(5)} = V_0^{(\infty)}$ and the rule is Bayes for the general problem as well ($V_0^{(5)} = 19.0667$).

6. Let $X_1, X_2, \cdots$ be independent $\mathcal{N}(\theta, 1)$, let $L(\theta, a) = (\theta - a)^2$, let the prior distribution τ be $\mathcal{N}(0, \sigma^2)$, where σ^2 is known, and let the cost of observation be constant.

(a) Show that the distribution of θ given $X_1, \cdots, X_j$ is $\mathcal{N}(\sigma^2 S_j/(1 + j\sigma^2), \sigma^2/(1 + j\sigma^2))$, where $S_j = \sum_1^j X_i$. Hence

$$\rho_j(x_1, \cdots, x_j) = \frac{\sigma^2}{1 + j\sigma^2}.$$

(b) Find the Bayes decision rule with respect to τ.

(c) Find the minimax sequential decision rule.

7. Let $X_1, X_2, \cdots$ be independent Poisson, $\mathcal{P}(\theta)$, let $L(\theta, a) = (\theta - a)^2/\theta$, let θ have a prior gamma distribution, $\mathcal{G}(\alpha, 1/\lambda)$, and let the cost of observation be constant $c_j = jc$.
(a) Show that the distribution of θ, given $X_1, \cdots, X_j$, is $\mathcal{G}(\alpha + S_j, 1/(\lambda + j))$, where $S_j = \sum_1^j X_i$.
(b) Show that $\rho_j(x_1, \cdots, x_j) = 1/(\lambda + j)$.
(c) Find the Bayes sequential decision rule.
(d) Show that the following rule is minimax: find the integer j that minimizes $jc + j^{-1}$, take a sample of fixed size j, and estimate θ to be $\bar{X} = S_j/j$.

8. Let $X_1, X_2, \cdots$ be independent negative exponential, $\mathcal{G}(1, 1/\theta)$, $\theta > 0$, let $L(\theta, a) = (\theta - a)^2/\theta^2$, let θ have a prior gamma distribution, $\mathcal{G}(\alpha, 1/\lambda)$, and let the cost of observation be constant, $c_j = jc$.
(a) Show that the distribution of θ, given $X_1, \cdots, X_j$, is $\mathcal{G}(\alpha + j, 1/(\lambda + S_j))$, where $S_j = \sum_1^j X_i$.
(b) Show that $\rho_j(x_1, \cdots, x_j) = 1/(\alpha + j + 1)$.
(c) Find the Bayes sequential decision rule.
(d) Show that the following rule is minimax: find the integer j that minimizes $jc + (j - 1)^{-1}$, take a sample of fixed size j, and estimate θ to be $(j - 2)/S_j$.

9. Let $X_1, X_2, \cdots$ be independent negative exponential, $\mathcal{G}(1, \theta), \theta > 0$, let $L(\theta, a) = (\theta - a)^2/\theta^2$, let θ have a prior reciprocal gamma distribution, $\mathcal{G}^{-1}(\alpha, 1/\lambda)$, with density

$$f(\theta) = \frac{\lambda^\alpha}{\Gamma(\alpha)} \exp(-\lambda/\theta)\theta^{-(\alpha+1)} I_{(0,\infty)}(\theta),$$

with $\alpha > 0$, $\lambda > 0$, and let the cost of observation be constant, $c_j = jc$.
(a) Show that the distribution of θ, given $X_1, \cdots, X_j$, is also reciprocal gamma, $\mathcal{G}^{-1}(\alpha + j, 1/(\lambda + S_j))$, where $S_j = \sum_1^j X_i$.
(b) Show that $\rho_j(x_1, \cdots, x_j) = 1/(\alpha + j + 1)$.
(c) Find the Bayes sequential decision rule.
(d) Show that the following rule is minimax: find the integer j which minimizes $jc + (j + 1)^{-1}$, take a sample of fixed size j, and estimate θ to be $S_j/(j + 1)$.

10. Let $X_1, X_2, \cdots$ be independent geometric, $\mathcal{NB}(1, \theta/(1 + \theta))$, $\theta > 0$, let $L(\theta, a) = (\theta - a)^2/(\theta(1 + \theta))$, where $\mathcal{A} = [0, \infty)$, let θ have prior distribution $\tau_{\alpha,\beta}$ described in Exercise 2.11.12, and let the cost of observation be constant, $c_j = jc$.

(a) Show that the distribution of θ, given $X_1, \cdots, X_j$, is $\tau_{\alpha+S_j,\beta+j}$ where $S_j = \sum_1^j X_i$.
(b) Show that $\rho_j(x_1, \cdots, x_j) = 1/(\beta + j + 1)$.
(c) Find the Bayes sequential decision rule.
(d) Show that the following rule is minimax: find the integer j that minimizes $jc + (j+1)^{-1}$, take a sample of fixed size j, and estimate θ to be $S_j/(j+1)$.

11. (Dependent observations). Let $X_1, \cdots, X_M$ represent dependent Bernoulli trials which arise from sampling a population of M items without replacement, θ of which are "defective" and $M - \theta$ "nondefective." Thus X_i is one or zero according as the ith item sampled is defective or nondefective. We take $\mathcal{A}$ to be the real line, $\Theta = \{0, 1, \cdots, M\}$, $L(\theta, a) = (\theta - a)^2$, and constant cost c per observation. Show that the Bayes sequential decision rule with respect to the distribution τ, where $\tau \in \mathcal{B}(M, p)$, is a fixed sample size rule.

12. Let $X_1, \cdots, X_M$, Θ, $\mathcal{A}$, L, and c be as in Exercise 11. This time let θ have the $\mathcal{BB}(\alpha, \beta, M)$-distribution.
(a) Find the Bayes terminal decision rule and show

$$\rho_j(x_1, \cdots, x_j) = \frac{(M-j)(s_j+\alpha)(j+\beta-s_j)(M+\alpha+\beta)}{(j+\alpha+\beta)^2(j+\alpha+\beta+1)},$$

where $s_j = \sum_1^j x_i$. (See Exercise 2.11.14.)
(b) Show

$$P\{X_j = x_j \mid X_1 = x_1, \cdots, X_{j-1} = x_{j-1}\} = \begin{cases} \dfrac{s_{j-1}+\alpha}{j+\alpha+\beta-1} & \text{if } x_j = 1, \\[2ex] \dfrac{j+\beta-s_{j-1}-1}{j+\alpha+\beta-1} & \text{if } x_j = 0. \end{cases}$$

(c) Show

$$\rho_{j-1}(x_1, \cdots, x_{j-1}) - E(\rho_j(x_1, \cdots, x_{j-1}, X_j) \mid X_1 = x_1, \cdots, X_{j-1} = x_{j-1})$$

$$= \frac{(M+\alpha+\beta)^2(s_{j-1}+\alpha)(j+\beta-s_{j-1}-1)}{(j+\alpha+\beta)^2(j+\alpha+\beta-1)^2}.$$

(d) Show that J_0 of Theorem 5 is the largest integer j for which

$$(M + \alpha + \beta)^2[(j + \alpha + \beta)/2][(j + \alpha + \beta - 1)/2] \geq c(j + \alpha + \beta)^2(j + \alpha + \beta - 1)^2.$$

Take $\alpha = \beta = 1$ (the uniform distribution on Θ), $M = 40$, and $c = 8$ and find the Bayes sequential decision rule (that is, show $J_0 = 5$ and construct a table similar to the table in Example 1). $V_0^{(\infty)} = 73.72$.

7.3 Convex Loss and Sufficiency

In this section we investigate the sequential analogues of the theorems on the essential completeness of the class of nonrandomized rules when the loss is convex, and of the class of rules based on a sufficient statistic (Theorems 2.8.1, 3.4.1, and 3.4.3). First, the analogue of Theorem 2.8.1.

Theorem 1. Suppose that $\mathcal{A}$ is a convex subset of E_k and that the loss function $L(\theta, a)$ is convex in $a \in \mathcal{A}$ for each $\theta \in \Theta$. Assume also that $L(\theta, a) \to \infty$ as $|a| \to \infty$. Then the class of decision rules $(\varphi, \mathbf{d})$, in which $\mathbf{d}$ is a nonrandomized terminal decision rule, forms an essentially complete class of sequential decision rules.

The proof is left to the student.

EXAMPLE 1. Let $X_1, X_2, \cdots, X_J$ be a sequence of J independent Bernoulli trials with probability θ of success and consider the problem of estimating θ sequentially with loss $L(\theta, a) = (\theta - a)^2/(\theta(1 - \theta))$ and constant cost c per observation. We show that the class of extended Bayes rules is essentially complete for this problem, using Theorem 1 and Theorem 2.10.3. We choose to use the ψ form for stopping rules. Theorem 1 states that the class C of all decision rules $(\psi, \mathbf{d})$, in which $\mathbf{d}$ is nonrandomized, is essentially complete. The class C of rules

$$(\psi, \mathbf{d}) = ((\psi_0, \psi_1(x_1), \cdots, \psi_J(x_1, \cdots, x_J)), (d_0, d_1(x_1), \cdots, d_J(x_1, \cdots, x_J)))$$

is a compact subset of the space $C' = [0, 1]^{(2^{J+1}-1)2}$ [considering ψ without the restriction $\sum_0^J \psi_j(x_1, \cdots, x_j) = 1$], itself a compact subset of $E_{2^{J+2}-2}$. The risk

$$R(\theta, (\varphi, \mathbf{d})) = \sum_{j=0}^{J} E_\theta\{\psi_j(X_1, \cdots, X_j)[L(\theta, d_j(X_1, \cdots, X_j)) + jc]\} \tag{7.16}$$

is obviously continuous in C', hence in C. From Theorem 2.10.3 the class of extended Bayes rules containing a nonrandomized terminal decision rule is essentially complete.

Sufficient Sequences. Consider a sequential decision problem with observations $X_1, X_2, \cdots$ and a sequence of statistics $T_1, T_2, \cdots$ such that T_j is a function of $X_1, X_2, \cdots, X_j$.

Definition 1. $\{T_j\}$ is said to be a *sufficient sequence* for θ if, for each j, T_j is sufficient for θ based on $X_1, \cdots, X_j$. A terminal decision rule $\boldsymbol{\delta}$ [a stopping rule φ or a decision rule $(\varphi, \boldsymbol{\delta})$] is said to be based on $\{T_j\}$ if, δ_j [φ_j or both] is a function of T_j for $j = 1, 2, \cdots$.

Given a sufficient sequence $\{T_j\}$ and a sequential decision rule $(\varphi, \boldsymbol{\delta})$, we ask whether it is possible to find a decision rule $(\varphi^0, \boldsymbol{\delta}^0)$ based on $\{T_j\}$ such that $(\varphi^0, \boldsymbol{\delta}^0)$ is as good as $(\varphi, \boldsymbol{\delta})$. We shall see that this is not always true. However, we can prove that, given any sequential decision rule $(\varphi, \boldsymbol{\delta})$, there exists a terminal decision rule $\boldsymbol{\delta}^0$ based on $\{T_j\}$ such that the sequential decision rule $(\varphi, \boldsymbol{\delta}^0)$ has risk identical to that of $(\varphi, \boldsymbol{\delta})$. The proof of this assertion is based on the following lemma. Recall that choosing a stopping rule is equivalent to choosing a distribution for the random stopping time N.

Lemma 1. Let $\{T_j\}$ be a sufficient sequence for θ. For a given stopping rule φ, (N, T_N) is a sufficient statistic for θ in the sense that the conditional distribution of $(X_1, \cdots, X_N)$, given (N, T_N), is independent of θ.

Proof. Let A be any (measurable) subset of E_n. Then

$$\begin{aligned} P_{\theta,\varphi}\{(X_1, \cdots, X_n) \in A \mid N = n, T_n\} \\ &= P_{\theta,\varphi}\{(X_1, \cdots, X_n) \in A, N = n \mid T_n\}/P_{\theta,\varphi}\{N = n \mid T_n\} \\ &= E\{I_A(X_1, \cdots, X_n)\psi_n(X_1, \cdots, X_n) \mid T_n\}/E\{\psi_n(X_1, \cdots, X_n) \mid T_n\} \end{aligned}$$

which is independent of θ, for T_n is sufficient for θ based on $X_1, \cdots, X_n$.

This leads immediately to the following theorem.

Theorem 2. If $\{T_j\}$ is a sufficient sequence for θ, the class of rules $\{(\varphi, \boldsymbol{\delta})\}$ in which $\boldsymbol{\delta}$ is based on $\{T_j\}$ is essentially complete.

Proof. Let $(\varphi, \boldsymbol{\delta})$ be any rule. Then

$$\begin{aligned} R(\theta, (\varphi, \boldsymbol{\delta})) &= E_{\theta,\varphi}\{L(\theta, \delta_N(X_1, \cdots, X_N)) + c_N(\theta, X_1, \cdots, X_N)\} \\ &= E_{\theta,\varphi}\{E_\varphi(L(\theta, \delta_N(X_1, \cdots, X_N)) \\ &\qquad\qquad + c_N(\theta, X_1, \cdots, X_N) \mid N, T_N)\} \\ &= E_{\theta,\varphi}\{E_\varphi(L(\theta, \delta_N{}^0(T_N)) + c_N(\theta, X_1, \cdots, X_N) \mid N, T_N)\} \\ &= E_{\theta,\varphi}\{L(\theta, \delta_N{}^0(T_N)) + c_N(\theta, X_1, \cdots, X_N)\} \\ &= R(\theta, (\varphi, \boldsymbol{\delta}^0)), \end{aligned}$$

where $\boldsymbol{\delta}^0$ is the terminal decision rule based on $\{T_j\}$ defined by $\delta_j{}^0(T_j) = E_\varphi(\delta_j(X_1, \cdots, X_j) \mid N = j, T_j)$, as in the proof of Theorem 3.4.1, thus completing the proof.

The sequential analogue of the Rao-Blackwell Theorem is as follows.

Theorem 3. Let $\mathcal{A}$ be a convex subset of E_k, let $L(\theta, a)$ be a convex function of $a \in \mathcal{A}$ for each $\theta \in \Theta$, and suppose that $\{T_j\}$ is a sufficient sequence for θ. If $(\varphi, \mathbf{d})$ is a sequential decision rule with $\mathbf{d}$ nonrandomized, then the rule $(\varphi, \mathbf{d}^0)$ with $\mathbf{d}^0$ nonrandomized and based on $\{T_j\}$ where

$$d_n{}^0(T_n) = E_\varphi\{d_n(X_1, \cdots, X_n) \mid N = n, T_n\}, \tag{7.17}$$

provided that this expectation exists, is as good as $(\varphi, \mathbf{d})$.

Proof. By Jensen's inequality

$$\begin{aligned} E_\varphi(L(\theta, d_N(X_1, \cdots, X_N)) \mid N, T_N) &\geq L(\theta, E_\varphi(d_N(X_1, \cdots, X_N) \mid N, T_N)) \\ &= L(\theta, d_N{}^0(T_N)) \end{aligned}$$

for all θ, so that

$$\begin{aligned} R(\theta, (\varphi, \mathbf{d})) &= E_{\theta,\varphi}\{L(\theta, d_N(X_1, \cdots, X_N)) + c_N(\theta, X_1, \cdots, X_N)\} \\ &= E_{\theta,\varphi}\{E_\varphi(L(\theta, d_N(X_1, \cdots, X_N) \mid N, T_N)\} \\ &\qquad\qquad + E_{\theta,\varphi}\{c_N(X_1, \cdots, X_N)\} \\ &\geq E_{\theta,\varphi}\{L(\theta, d_N{}^0(T_N)) + c_N(X_1, \cdots, X_N)\} \\ &= R(\theta, (\varphi, \mathbf{d}^0)), \end{aligned}$$

thus completing the proof.

When Blackwell (1947) originally suggested the use of (7.17), he intended it partly as a method which may often be used to construct for a given stopping rule, φ, an unbiased sequential estimate, $\mathbf{d}^0$, of θ. The only requirement is that $\mathbf{d}$ itself be an unbiased sequential estimate of θ, that is, $E_{\theta,\varphi}\{d_N(X_1, \cdots, X_N)\} \equiv \theta$, for then

$$E_{\theta,\varphi}\{d_N{}^0(T_N)\} = E_{\theta,\varphi}\{E_\varphi(d_N(X_1, \cdots, X_N) \mid N, T_N)\}$$
$$= E_{\theta,\varphi}\{d_N(X_1, \cdots, X_N)\} \equiv \theta. \tag{7.18}$$

If $\varphi_0 = 0$ and there exists a function of X_1, say $\hat{\theta}(X_1)$, which is an unbiased estimate of θ, $E_\theta\hat{\theta}(X_1) \equiv \theta$, then $\mathbf{d}$, defined by $d_n(X_1, \cdots, X_n) = \hat{\theta}(X_1)$, is clearly an unbiased sequential estimate of θ. So then is $\mathbf{d}_0$, defined by (7.17).

Example 2. Let X_1, X_2, X_3 be independent Bernoulli trials with probability θ of success and consider the stopping rule $\varphi_0 = 0$, $\varphi_1(x_1) = 0$, $\varphi_2(x_1, x_2) = x_1$, $\varphi_3(x_1, x_2, x_3) = 1$. Clearly, $\{T_j\}$ with $T_j = \sum_1^j X_i$ is a sufficient sequence and $d_j(X_1, \cdots, X_j) = X_1$ is an unbiased sequential estimate. If the loss function is convex, an improved unbiased sequential estimate may be found from (7.17). This gives $d_2{}^0(1) = E_\varphi(X_1 \mid N = 2, T_2 = 1) = 1$, for, if $N = 2$, then $X_1 = 1$. Similarly,

$$d_2{}^0(2) = E_\varphi(X_1 \mid N = 2, T_2 = 2) = 1$$

and

$$d_3{}^0(0) = d_3{}^0(1) = d_3{}^0(2) = 0.$$

It should be noted in this example that $d_j(X_1, \cdots, X_j) = X_2$, $j \geq 2$, is also an unbiased estimate of θ. The estimate given by (7.17) in this case is $d_2{}^0(1) = 0$, $d_2{}^0(2) = 1$, $d_3{}^0(0) = 0$, $d_3{}^0(1) = \frac{1}{2}$, $d_3{}^0(2) = 1$, which differs from the estimate in the preceding paragraph. This shows that even if T_j is a complete sufficient statistic for θ, based on $X_1, \cdots, X_j$ for each j, there may be, for a fixed stopping rule, more than one unbiased estimate based on $\{T_j\}$. Furthermore, it may be seen that the estimate of this paragraph is better than the estimate of the preceding paragraph when the loss is squared error (Exercise 2). This shows that the estimate of the preceding paragraph is not admissible within the class of unbiased estimates for this fixed φ.

In spite of this example, the spirit of the theory of minimum variance unbiased estimation found in Section 3.6 can be recaptured in the binomial case when supplementary conditions are placed on the stopping rule: that it be nonrandomized, based on $\{T_j\}$, closed and simple. This theory was introduced by Girschick, Mosteller, and Savage (1946), Wolfowitz (1946), and Savage (1947). A summary is contained in DeGroot (1959). For the complications that arise in the general (nonbinomial) case, see Lehmann and Stein (1950).

We now consider the more delicate question of the essential completeness of the class of rules in which both $\boldsymbol{\varphi}$ and $\boldsymbol{\delta}$ are based on a sufficient sequence. In this treatment we follow Bahadur (1954). First we give an example in which there is a sufficient sequence $\{T_j\}$ and a rule $(\boldsymbol{\varphi}, \boldsymbol{\delta})$ for which there is no rule $(\boldsymbol{\varphi}^0, \boldsymbol{\delta}^0)$ based on $\{T_j\}$ such that

$$R(\theta, (\boldsymbol{\varphi}^0, \boldsymbol{\delta}^0)) \leq R(\theta, (\boldsymbol{\varphi}, \boldsymbol{\delta}))$$

for all $\theta \in \Theta$.

Example 3. Consider the sequential decision problem truncated at $J = 2$, in which $\mathcal{A} = \Theta = (0, 1)$, $L(\theta, a) = (\theta - a)^2$, and there is a constant cost c per observation. Let $X_1 \in \mathcal{B}(1, 1/2)$, let the conditional distribution of X_2, given $X_1 = 1$, be $\mathcal{B}(1, 1/2)$, and let the conditional distribution of X_2, given $X_1 = 0$, be $\mathcal{B}(1, \theta)$. The sequence $\{T_j\}$, with $T_1 \equiv 0$ and $T_2 = (X_1, X_2)$, is a sufficient sequence. Consider the rule $(\boldsymbol{\varphi}, \boldsymbol{\delta})$: observe X_1. If $X_1 = 1$, stop and estimate θ to be 1/2. If $X_1 = 0$, observe X_2 and estimate θ to be 1 if $X_2 = 1$ and 0 if $X_2 = 0$. A simple computation shows that for all θ

$$R(\theta, (\boldsymbol{\varphi}, \boldsymbol{\delta})) = 1/8 + (3/2)c. \tag{7.19}$$

Now consider sequential decision rules based on the sufficient sequence $\{T_j\}$. The only restriction on these rules is that φ_1 and δ_1 cannot be functions of X_1. This is equivalent to a one-stage sequential decision problem in which either no observations are taken or both X_1 and X_2 are observed. The rule $(\boldsymbol{\varphi}^0, \boldsymbol{\delta}^0)$ "observe X_1 and X_2, estimate θ to be $\frac{1}{2}$ if $X_1 = 1$, estimate θ to be 1 if $X_1 = 0$, $X_2 = 1$, and estimate θ to be 0 if $X_1 = 0$, $X_2 = 0$" has constant risk

$$R(\theta, (\boldsymbol{\varphi}^0, \boldsymbol{\delta}^0)) = \tfrac{1}{8} + 2c. \tag{7.20}$$

Furthermore, it is easily checked by the methods of Section 7.2 that this rule is Bayes for this one-stage problem with respect to the prior distribution giving probability 1/2 to $\theta = 0$ and $\theta = 1$, provided $c \leq 1/16$ (Exercises 3 and 4). Hence $(\boldsymbol{\varphi}^0, \boldsymbol{\delta}^0)$ is minimax in this problem. Because the maximum risk of a rule based on $\{T_j\}$ is no less than $1/8 + 2c$, it is clear that no rule based on $\{T_j\}$ can have risk everywhere less than or equal to the risk of (7.19), so that rules based on $\{T_j\}$ are not essentially complete.

Bahadur shows that a sufficient condition for the rules based on $\{T_j\}$ to be essentially complete, when the cost $c_j(\theta, X_1, \cdots, X_j)$ depends on $X_1, \cdots, X_j$ only through the values of T_j, is that the sufficient sequence $\{T_j\}$ be transitive.

Definition 2. A sufficient sequence $\{T_j\}$ is said to be *transitive* if, for every $j = 1, 2, \cdots$, and all bounded integrable functions $f(x_1, \cdots, x_j)$,

$$E(f(X_1, \cdots, X_j) \mid T_j, T_{j+1}) = E(f(X_1, \cdots, X_j) \mid T_j), \tag{7.21}$$

except perhaps for a set of probability zero.

In other words, $\{T_j\}$ is transitive if, by using the conditional distribution given T_j, T_{j+1} is stochastically independent of $(X_1, \cdots, X_j)$ for $j = 1, 2, \cdots$. This condition is not satisfied for the sufficient sequence of Example 3 when $j = 1$.

Lemma 2. If a sufficient sequence $\{T_j\}$ is transitive, then for every $j = 1, 2, \cdots$ and all bounded integrable functions $f(x_1, \cdots, x_j)$

$$E\{E\{f(X_1, \cdots, X_j) \mid T_j\} \mid T_{j+1}\} = E\{f(X_1, \cdots, X_j) \mid T_{j+1}\}, \tag{7.22}$$

except perhaps for a set of probability zero.

Proof. For any bounded integrable function $f(x_1, \cdots, x_j)$

$$\begin{aligned} E\{f(X_1, \cdots, X_j) \mid T_{j+1}\} &= E\{E\{f(X_1, \cdots, X_j) \mid T_j, T_{j+1}\} \mid T_{j+1}\} \\ &= E\{E\{f(X_1, \cdots, X_j) \mid T_j\} \mid T_{j+1}\}, \end{aligned}$$

thus completing the proof.

The conclusion of Lemma 2 is sufficient as well as necessary for transitivity (Exercise 12). Bahadur shows that if the X_i are independent, a minimal sufficient sequence is transitive. This result is contained in Lemma 3, which suffices for most of the applications.

Lemma 3. If $X_1, X_2, \cdots$ are independent and $\{T_j\}$ is a sufficient sequence such that for all $j = 1, 2, \cdots$, T_{j+1} is a function of T_j and X_{j+1}, then $\{T_j\}$ is transitive.

Proof. For any bounded integrable function $f(x_1, \cdots, x_j)$

$$\begin{aligned} E\{ f(X_1, \cdots, X_j) \mid T_j, T_{j+1}\} \\ &= E\{E\{ f(X_1, \cdots, X_j) \mid T_j, X_{j+1}, T_{j+1}\} \mid T_j, T_{j+1}\} \\ &= E\{E\{ f(X_1, \cdots, X_j) \mid T_j, X_{j+1}\} \mid T_j, T_{j+1}\} \\ &= E\{E\{ f(X_1, \cdots, X_j) \mid T_j\} \mid T_j, T_{j+1}\} \\ &= E\{ f(X_1, \cdots, X_j) \mid T_j\}, \end{aligned}$$

thus completing the proof.

For example, if $X_1, X_2, \cdots$ is a sample from a one-parameter exponential family, $f(x \mid \theta) = c(\theta)\, h(x) \exp\{\theta x\}$, then

$$T_{j+1} = \sum_1^{j+1} X_i = T_j + X_{j+1},$$

so that $\{T_j\}$ is a transitive sufficient sequence.

The main property enjoyed by a transitive sufficient sequence is the following:

Lemma 4. Let $\boldsymbol{\varphi}$ be a given stopping rule and suppose that $\{T_j\}$ is a transitive sufficient sequence for θ. Then there exists a stopping rule $\boldsymbol{\varphi}^0$ based on $\{T_j\}$ such that the distribution of (N, T_N), using $\boldsymbol{\varphi}^0$, is identical to the distribution of (N, T_N), using $\boldsymbol{\varphi}$, for all $\theta \in \Theta$.

Proof. Let $\boldsymbol{\varphi} = (\varphi_0, \varphi_1, \cdots)$ and define for $n = 0, 1, 2, \cdots$,

$$\begin{aligned} \alpha_n(x_1, \cdots, x_n) &= \prod_{j=0}^{n} (1 - \varphi_j(x_1, \cdots, x_j)) \\ &= P_{\boldsymbol{\varphi}}(N \geq n + 1 \mid X_1 = x_1, \cdots, X_n = x_n). \qquad (7.23) \end{aligned}$$

Define $\varphi_0{}^0 = \varphi_0$, and for $n = 1, 2, \cdots$ define

$$\varphi_n{}^0(T_n) = \frac{E(\alpha_{n-1}(X_1, \cdots, X_{n-1})\ \varphi_n(X_1, \cdots, X_n) \mid T_n)}{E(\alpha_{n-1}(X_1, \cdots, X_{n-1}) \mid T_n)}$$

$$= \frac{P_{\boldsymbol{\varphi}}(N = n \mid T_n)}{P_{\boldsymbol{\varphi}}(N \geq n \mid T_n)} = P_{\boldsymbol{\varphi}}(N = n \mid N \geq n,\ T_n). \qquad (7.24)$$

We must show that for all $n = 1, 2, \cdots$ and all measurable sets A in the range space of T_n,

$$P_{\theta,\boldsymbol{\varphi}^0}(N = n,\ T_n \in A) = P_{\theta,\boldsymbol{\varphi}}(N = n,\ T_n \in A) \qquad (7.25)$$

for all $\theta \in \Theta$. To do this, we first establish that

$$E(\alpha_{n-1}^0(X_1, \cdots, X_{n-1}) \mid T_n) = E(\alpha_{n-1}(X_1, \cdots, X_{n-1}) \mid T_n) \qquad (7.26)$$

by induction on n, where

$$\alpha_{n-1}^0(X_1, \cdots, X_{n-1}) = \prod_{j=0}^{n-1} (1 - \varphi_j{}^0(T_j)),$$

analogous to (7.23). For $n = 1$ both sides of (7.26) are equal to $(1 - \varphi_0)$. If (7.26) is valid for a fixed n, then

$$E(\alpha_n{}^0(X_1, \cdots, X_n) \mid T_{n+1})$$

$$= E(\alpha_{n-1}^0(X_1, \cdots, X_{n-1})\,(1 - \varphi_n{}^0(T_n)) \mid T_{n+1})$$

$$= E((1 - \varphi_n{}^0(T_n))\ E(\alpha_{n-1}^0(X_1, \cdots, X_{n-1}) \mid T_n) \mid T_{n+1})$$

(transitivity, Lemma 2)

$$= E((1 - \varphi_n{}^0(T_n))\ E(\alpha_{n-1}(X_1, \cdots, X_{n-1}) \mid T_n) \mid T_{n+1})$$

(7.26)

$$= E(E(\alpha_{n-1}(X_1, \cdots, X_{n-1})\,(1 - \varphi_n(X_1, \cdots, X_n)) \mid T_n) \mid T_{n+1})$$

(7.24)

$$= E(\alpha_{n-1}(X_1, \cdots, X_{n-1})\,(1 - \varphi_n(X_1, \cdots, X_n)) \mid T_{n+1})$$

(transitivity, Lemma 2)

$$= E(\alpha_n(X_1, \cdots, X_n) \mid T_{n+1}),$$

proving that (7.26) is valid for n increased by one and completing the induction.

Returning to the proof of (7.25), we see that

$$P_{\theta,\boldsymbol{\varphi}^0}(N = n,\ T_n \in A)$$

$$= E_\theta \alpha_{n-1}^0(X_1, \cdots, X_{n-1})\ \varphi_n{}^0(T_n)\ I_A(T_n)$$

$$= E_\theta E(\alpha_{n-1}^0(X_1, \cdots, X_{n-1}) \mid T_n)\ \varphi_n{}^0(T_n)\ I_A(T_n)$$

$$= E_\theta E(\alpha_{n-1}(X_1, \cdots, X_{n-1}) \mid T_n)\ \varphi_n{}^0(T_n)\ I_A(T_n) \qquad (7.26)$$

$$= E_\theta E(\alpha_{n-1}(X_1, \cdots, X_{n-1})\ \varphi_n(X_1, \cdots, X_n) \mid T_n)\ I_A(T_n) \qquad (7.24)$$

$$= E_\theta \alpha_{n-1}(X_1, \cdots, X_{n-1})\ \varphi_n(X_1, \cdots, X_n)\ I_A(T_n)$$

$$= P_{\theta,\boldsymbol{\varphi}}(N = n,\ T_n \in A),$$

thus completing the proof.

We may now show that transitivity implies the essential completeness of the class of rules based on a sufficient sequence.

Theorem 4. If $\{T_j\}$ is a transitive sufficient sequence and the cost $c_j(\theta, x_1, \cdots, x_j)$ depends on $x_1, \cdots, x_j$ only through T_j, for $j = 1, 2, \cdots$, then the class of rules based on $\{T_j\}$ is essentially complete.

Proof. Let $(\boldsymbol{\varphi}, \boldsymbol{\delta})$ be any rule. Find $\boldsymbol{\delta}^0$ based on $\{T_j\}$, as in Theorem 2, for which $R(\theta, (\boldsymbol{\varphi}, \boldsymbol{\delta})) = R(\theta, (\boldsymbol{\varphi}, \boldsymbol{\delta}^0))$ for all $\theta \in \Theta$. Then find $\boldsymbol{\varphi}^0$ based on $\{T_j\}$ as in Lemma 4 such that the distribution of (N, T_N), using $\boldsymbol{\varphi}^0$, is the same as it is using $\boldsymbol{\varphi}$. Then

$$\begin{aligned} R(\theta, (\boldsymbol{\varphi}, \boldsymbol{\delta})) &= R(\theta, (\boldsymbol{\varphi}, \boldsymbol{\delta}^0)) \\ &= E_{\theta,\boldsymbol{\varphi}}(L(\theta, \delta_N{}^0(T_N)) + c_N(\theta, T_N)) \\ &= E_{\theta,\boldsymbol{\varphi}^0}(L(\theta, \delta_N{}^0(T_N)) + c_N(\theta, T_N)) \\ &= R(\theta, (\boldsymbol{\varphi}^0, \boldsymbol{\delta}^0)), \end{aligned}$$

thus completing the proof.

Example 4. For the stopping rule $\boldsymbol{\varphi}$ of Example 2 let us find a stopping rule $\boldsymbol{\varphi}^0$ based on $\{T_j\}$ for which the distributions of (N, T_N), using $\boldsymbol{\varphi}$ and $\boldsymbol{\varphi}^0$, are identical. Using (7.24), we find $\varphi_0{}^0 = \varphi_0$, $\varphi_1{}^0 = \varphi_1$, $\varphi_2{}^0(T_2) = E(\varphi_2(X_1, X_2) \mid T_2) = E(X_1 \mid T_2) = T_2/2$, $\varphi_3{}^0 = 1$. If $\mathbf{d}^0$ is an unbiased estimate of θ based on $\{T_j\}$, using φ, then it is also unbiased using φ^0. Furthermore, if the cost $c_j(\theta, X_1, \cdots, X_j) = c_j(\theta, T_j)$ for all j, then the risk function of any rule using φ^0 is the same as the risk function using φ. Thus the difficulties arising from the multiplicity of unbiased estimates of θ based on a sufficient sequence $\{T_j\}$ are not removed by restricting attention to stopping rules based on $\{T_j\}$.

Exercises

1. Prove Theorem 1.

2. Consider the distributions and stopping rule of Example 2 when the loss is squared error, $L(\theta, a) = (\theta - a)^2$. (The cost is immaterial, for the expected cost depends only on φ and φ is fixed.)
(a) Find the expectation of an arbitrary nonrandomized estimate **d** based on $\{T_j\}$.
(b) Find the class of all unbiased nonrandomized estimates **d** based on $\{T_j\}$. (Denote $d_3(2)$ by z.)
(c) Find the expected loss for estimates in this class as a function of θ and z.
(d) Show that no estimate of this class is admissible within this class unless $1/2 \leq z \leq 2/3$.

3. Verify (7.19) and (7.20).

4. Show that the rule $(\varphi^0, \boldsymbol{\delta}^0)$ of (7.20) is Bayes with respect to the prior distribution, giving mass 1/2 to $\theta = 0$ and $\theta = 1$ in the problem in which we observe Y_1, Y_2, where $Y_1 \equiv 0$ and $Y_2 = (X_1, X_2)$, with (X_1, X_2) distributed as in Example 3, provided the cost c of each stage is no larger than 1/16.

5. Let X_1, X_2, X_3, X_4 be independent Bernoulli trials with probability θ of success and consider the stopping rule φ, where $\varphi_0 = \varphi_1 = 0$, $\varphi_2(x_1, x_2) = x_2$, $\varphi_3(x_1, x_2, x_3) = x_3$, $\varphi_4 = 1$. Let **d** denote the nonrandomized terminal decision rule $d_j(x_1, \cdots, x_j) = x_j$, $j = 2, 3, 4$, and suppose that the cost $c_j(\theta, X_1, \cdots, X_j) = c_j(\theta, T_j)$, where $T_j = \sum_1^j X_i$. (Note that $\{T_j\}$ is a transitive sufficient sequence.)
(a) Find $(\varphi^0, \boldsymbol{\delta}^0)$ based on $\{T_j\}$ with the same risk function as $(\varphi, \mathbf{d})$, as in Theorem 4.
(b) If the loss $L(\theta, a)$ is strictly convex in (a), find $(\varphi^0, \mathbf{d}^0)$, depending on $\{T_j\}$, with improved risk, where $\mathbf{d}^0$ is nonrandomized.

6. Let $X_1, X_2, \cdots$ be a sequence of independent Bernoulli trials with probability θ of success and let $\{T_j\}$ be the sufficient sequence $T_j = \sum_1^j X_i$.

(a) Consider the stopping rule φ: stop whenever the number of successes exceeds the number of failures by 2 or the number of failures exceeds the number of successes by 3. Find the unbiased sequential estimate of θ, $d_n{}^0 = E_\varphi(X_1 \mid N = n, T_n)$.

Ans: $d_{n+1}^0 = F_{n-2}/F_n$, where F_n is the Fibonacci sequence, $F_0 = 0$, $F_1 = 1$, $F_n = F_{n-1} + F_{n-2}$.

(b) Consider the stopping rule φ': stop whenever the difference between the number of successes and the number of failures is in absolute value equal to 3. Find the unbiased sequential estimates of θ,

$$d_n' = E_{\varphi'}(X_1 \mid N = n, T_n = t)$$

and

$$d_n''(t) = E_{\varphi'}((X_1 + X_2)/2 \mid N = n, T_n = t).$$

Ans: $d_n'((n - 3)/2) = (1 - 3^{-(n-3)/2})/2$, $d_n'((n + 3)/2) = 1 - d_n'((n - 3)/2)$, $d_n''(t) = d_n'(t)$.

7. Let $X_1, X_2, \cdots$ be a sample from the Poisson distribution $\mathcal{P}(\theta)$.
(a) Show that the conditional distribution of $X_1, X_2, \cdots, X_n$, given $T_n = t$, where $T_n = \sum_1^n X_i$, is multinomial:

$$P(X_1 = x_1, \cdots, X_n = x_n \mid T_n = t) = \frac{t!}{x_1! \cdots x_n!}\left(\frac{1}{n}\right)^t \quad \text{if} \quad \sum_1^n x_i = t,$$

where the x_i are nonnegative integers.
(b) Consider the stopping rule φ, which stops at the first n for which $T_n \geq 2$. Find an estimate $\mathbf{d}^0$ based on $\{T_j\}$ which is as good as $\mathbf{d}$, where $d_j(x_1, \cdots, x_j) = x_1$, regardless of the loss function $L(\theta, a)$, provided that it is convex in $a \in \mathcal{A}$ for each $\theta \in \Theta$.

Ans: $d_n{}^0(t_n) = t_n/((n - 1)t_n + 1)$.

8. Let $X_1, X_2, \cdots$ be a sample from the geometric distribution, $\mathcal{NB}(1, \theta)$.
(a) Show that the conditional distribution of $X_1, X_2, \cdots, X_n$, given $T_n = t$, where $T_n = \sum_1^n X_i$, gives equal weight to the $\binom{n+t-1}{n}$ points, $(x_1, \cdots, x_n)$, where the x_i are nonnegative integers and $\sum_1^n x_i = t$.

(b) Consider the stopping rule φ, which stops at the first n for which $X_n = 0$. Find $d_n^0(t) = E_\varphi(X_1 \mid N = n, T_n = t)$.

Ans: $d_n^0(t) = t/(n-1)$ for $n > 1$, $d_1^0(t) = 0$.

(c) Find the stopping rule φ^0 of (7.24).

Ans: $\varphi_0^0 = \varphi_0$, $\varphi_1^0 = \varphi_1$, $\varphi_n^0(t) = (n-1)/t$, $n > 1$.

9. Let $X_1, X_2, \cdots, X_M$ represent Bernoulli trials which arise from sampling a population of M items without replacement, θ of which are "defective" and $M - \theta$ "nondefective." Thus X_i is one or zero according as the ith item sampled is defective or nondefective. We know that $\{T_j\}$, where $T_j = \sum_1^j X_i$, is a sufficient sequence for θ. Show that $\{T_j\}$ is transitive.

10. Let $X_1, X_2, \cdots, X_M$ be as in Exercise 9 and let φ be a fixed stopping rule, truncated at M. Show that the conditional distribution of $X_1, \cdots, X_n$ given $N = n$ and $T_n = t_n$ is identical to the same conditional distribution when $X_1, \cdots, X_n$ are independent Bernoulli trials. Thus Examples 1, 2, and 4 and Exercises 5 and 6 would give identical results for the distribution of Exercise 9.

11. Give an example of independent random variables $X_1, X_2, \cdots$ and a sufficient sequence $\{T_j\}$ which is not transitive.

12. Prove the converse of Lemma 2. *Hint*. For a given bounded integrable function $f_0(x_1, \cdots, x_j)$ put $f(X_1, \cdots, X_j) = \exp(iuT_j) f_0(X_1, \cdots, X_j)$ in (7.21) and use the unicity of the Fourier transform.

7.4 Invariant Sequential Decision Problems

In this section we take a brief look at the use of the invariance principle in sequential decision problems.

We are given $(\Theta, \mathcal{A}, L)$, the random variables $X_1, X_2, \cdots$ available for observation, and the sequence of cost functions $c_j(\theta, x_1, \cdots, x_j)$. It is assumed that there is a group $\mathcal{G}$ of transformations from the real line onto the real line such that the fixed sample size problem involving $X_1, \cdots, X_j$ is invariant under the group $\mathcal{G}_j$ of transformations $(X_1, \cdots, X_j) \to (gX_1, \cdots, gX_j)$ for $g \in \mathcal{G}$. Although various generalizations are possible and necessary for certain important problems, we treat this rather simple structure because it contains the main points of interest.

Definition 1. A sequential decision problem is said to be *invariant* under a group $\mathcal{G}$ of measurable transformations on the real line, if

(a) (invariance of distributions) there is a group $\bar{\mathcal{G}}$ of transformations on Θ and a mapping $g \in \mathcal{G} \to \bar{g} \in \bar{\mathcal{G}}$ such that for all j the distribution of $(gX_1, \cdots, gX_j)$, given θ, is identical to the distribution of $(X_1, \cdots, X_j)$, given $\bar{g}\theta$, for all $g \in \mathcal{G}$ and $\theta \in \Theta$;

(b) (invariance of loss) there is a group $\tilde{\mathcal{G}}$ of transformations on $\mathcal{A}$ and a mapping $g \in \mathcal{G} \to \tilde{g} \in \tilde{\mathcal{G}}$ such that $L(\bar{g}\theta, \tilde{g}a) = L(\theta, a)$ for all $g \in \mathcal{G}$, $\theta \in \Theta$, and $a \in \mathcal{A}$;

(c) (invariance of cost) for all j, $c_j(\bar{g}\theta, gx_1, \cdots, gx_j) = c_j(\theta, x_1, \cdots, x_j)$ for all $g \in \mathcal{G}$, $\theta \in \Theta$, and $(x_1, \cdots, x_j) \in \mathfrak{X}_1 \times \cdots \times \mathfrak{X}_j$.

In short, for a sequential decision problem to be invariant, every fixed sample decision problem must be invariant under the same group $\mathcal{G}$ acting componentwise on the observations. Each of these fixed-sample decision problems must have the same associated group $\bar{\mathcal{G}}$ acting on Θ and the same associated group $\tilde{\mathcal{G}}$ acting on $\mathcal{A}$. Furthermore, the cost functions must be invariant—a point that did not arise in the fixed sample problem. The invariance of the cost is obviously satisfied in the important case in which the cost is independent of θ and the observations.

Definition 2. A decision rule $(\boldsymbol{\varphi}, \boldsymbol{\delta})$ is said to be *invariant* if

(a) (invariance of the stopping rule) for each j, $\varphi_j(gx_1, \cdots, gx_j) = \varphi_j(x_1, \cdots, x_j)$ for all $g \in \mathcal{G}$ and $(x_1, \cdots, x_j) \in \mathfrak{X}_1 \times \cdots \times \mathfrak{X}_j$;

(b) (invariance of the terminal decision rule) for each j, if

$$\psi_j(x_1, \cdots, x_j) > 0,$$

then

$$\tilde{g}\delta_j(x_1, \cdots, x_j) = \delta_j(gx_1, \cdots, gx_j) \tag{7.27}$$

(where $\tilde{g}\delta$ represents the distribution of $\tilde{g}Z$ when Z is a random variable with distribution δ).

Remark 1. Condition (b) in Definition 2 implies if $\psi_0 > 0$ that, considering δ_0 as a distribution on the real line, $\delta_0(A) = \delta_0(\tilde{g}A)$ for all $g \in \mathcal{G}$ and all Borel sets A (for nonrandomized rules, $d_0 = \tilde{g}d_0$). This is satisfied, for example, when for all $g \in \mathcal{G}$, $\tilde{g}$ is the identity transformation on $\mathcal{A}$, as in hypothesis-testing problems. In some problems, for example, when $\tilde{\mathcal{G}}$ is the group of translations on the real line, there exists no such distribution δ_0, in which case every invariant rule must take at least one observation, that is, $\varphi_0 = \psi_0 = 0$.

Remark 2. If $\boldsymbol{\varphi}$ is invariant, so is $\boldsymbol{\psi}$ [that is, Definition 2(a) implies, for each j, $\psi_j(gx_1, \cdots, gx_j) = \psi_j(x_1, \cdots, x_j)$ for all $g \in \mathcal{G}$ and all $(x_1, \cdots, x_j) \in \mathfrak{X}_1 \times \cdots \times \mathfrak{X}_j$].

The main property of invariant rules follows.

Theorem 1. The risk function of an invariant rule is constant on orbits of $\bar{\mathcal{G}}$ [that is, $R(\bar{g}\theta, (\boldsymbol{\varphi}, \boldsymbol{\delta})) = R(\theta, (\boldsymbol{\varphi}, \boldsymbol{\delta}))$ for all $\bar{g} \in \bar{\mathcal{G}}$].

Proof. Let Z be a random variable on $\mathfrak{A}$, whose distribution is given by δ. The invariance of the loss [Definition 1 (b)] implies

$$L(\bar{g}\theta, \tilde{g}\delta) = L(\theta, \delta), \tag{7.28}$$

for $L(\bar{g}\theta, \tilde{g}\delta) = EL(\bar{g}\theta, \tilde{g}Z) = EL(\theta, Z) = L(\theta, \delta)$. Consequently,

$$
\begin{aligned}
R(\theta, (\boldsymbol{\varphi}, \boldsymbol{\delta}))
&= \sum_{j=0}^{\infty} E_\theta\, \psi_j(X_1, \cdots, X_j)[L(\theta, \delta_j\,(X_1, \cdots, X_j)) \\
&\qquad\qquad + c_j(\theta, X_1, \cdots, X_j)] \\
&= \sum_{j=0}^{\infty} E_\theta\, \psi_j(X_1, \cdots, X_j)[L(\bar{g}\theta, \tilde{g}\delta_j(X_1, \cdots, X_j)) \\
&\qquad\qquad + c_j(\bar{g}\theta, gX_1, \cdots, gX_j)] \\
&= \sum_{j=0}^{\infty} E_\theta\, \psi_j(gX_1, \cdots, gX_j)[L(\bar{g}\theta, \delta_j(gX_1, \cdots, gX_j)) \\
&\qquad\qquad + c_j(\bar{g}\theta, gX_1, \cdots, gX_j)] \\
&= \sum_{j=0}^{\infty} E_{\bar{g}\theta}\, \psi_j(X_1, \cdots, X_j)[L(\bar{g}\theta, \delta_j(X_1, \cdots, X_j)) \\
&\qquad\qquad + c_j(\bar{g}\theta, X_1, \cdots, X_j)] \\
&= R(\bar{g}\theta, (\boldsymbol{\varphi}, \boldsymbol{\delta})),
\end{aligned} \tag{7.29}
$$

using for the central three equalities the invariance of the loss and cost, of the decision rule, and of the distributions, respectively, thus completing the proof.

In the case in which all invariant rules have constant risk, for example, when $\bar{\mathcal{G}}$ is transitive on Θ, we may search for a best invariant sequential decision rule. Theorem 2 simplifies the search and does for invariant rules what Theorem 7.2.1 does for Bayes rules.

For a given invariant rule $(\boldsymbol{\varphi}, \boldsymbol{\delta})$, the functions φ_j and ψ_j are invariant under the group $\mathcal{G}_j$ of transformations $(x_1, \cdots, x_j) \to (gx_1, \cdots, gx_j)$, and hence are functions of the maximal invariant $\mathbf{Y}_j$ under $\mathcal{G}_j$. We use the notation $\varphi_j(\mathbf{Y}_j)$ and $\psi_j(\mathbf{Y}_j)$ to emphasize this dependence.

Theorem 2. Suppose that $\bar{\mathcal{G}}$ is transitive over Θ and let $\mathbf{Y}_j$ be a maximal invariant under $\mathcal{G}_j$. Suppose that $(\boldsymbol{\varphi}, \boldsymbol{\delta}^0)$ is an invariant decision rule such that δ_j^0 minimize among invariant rules δ_j

$$E_{\theta_0}(L(\theta_0, \delta_j(X_1, \cdots, X_j)) \mid \mathbf{Y}_j = \mathbf{y}_j) \tag{7.30}$$

for some fixed θ_0 and for all y_j for which $\psi_j(\mathbf{y}_j) > 0$. Then $R(\theta, (\boldsymbol{\varphi}, \boldsymbol{\delta}^0)) \leq R(\theta, (\boldsymbol{\varphi}, \boldsymbol{\delta}))$ for any invariant decision rule $(\boldsymbol{\varphi}, \boldsymbol{\delta})$ which uses the same stopping rule $\boldsymbol{\varphi}$.

Proof. Fix θ_0 and $\boldsymbol{\varphi}$ invariant. Then

$$\begin{aligned} R(\theta, (\boldsymbol{\varphi}, \boldsymbol{\delta})) &= R(\theta_0, (\boldsymbol{\varphi}, \boldsymbol{\delta})) \\ &= \sum_{j=0}^{\infty} E_{\theta_0}\{\psi_j(\mathbf{Y}_j)[L(\theta_0, \delta_j(X_1, \cdots, X_j)) \\ &\qquad\qquad + c_j(\theta_0, X_1, \cdots, X_j)]\} \\ &= \sum_{j=0}^{\infty} E\{\psi_j(\mathbf{Y}_j) E_{\theta_0}[L(\theta_0, \delta_j(X_1, \cdots, X_j)) \\ &\qquad\qquad + c_j(\theta_0, X_1, \cdots, X_j) \mid \mathbf{Y}_j]\}. \end{aligned}$$

This expression is clearly minimized by choosing δ_j to minimize the expression (7.30), thus completing the proof.

It is interesting to note (a) that δ^0 is independent of the invariant stopping rule $\boldsymbol{\varphi}$ and (b) that δ_j^0 is a best invariant rule for the fixed sample size problem involving $X_1, \cdots, X_j$. Therefore, in finding best invariant sequential decision rules in such problems, we may use the same method used to find Bayes rules, namely, by solving the problem truncated at J and letting $J \to \infty$. To emphasize the similarity we use a notation parallel to that of Section 7.2.

We assume that the group $\bar{\mathcal{G}}$ is transitive over Θ and that Theorem 2 applies. Fix θ_0 and let $\mathbf{Y}_j$ be a maximal invariant under $\mathcal{G}_j$. Let the minimum of the conditional expected loss given $\mathbf{Y}_j = \mathbf{y}_j$ over the invariant rules be denoted by $\rho_j(\mathbf{y}_j)$:

$$\rho_j(\mathbf{y}_j) = \inf_{\delta_j \text{ invariant}} E_{\theta_0}\{L(\theta_0, \delta_j(X_1, \cdots, X_j)) \mid \mathbf{Y}_j = \mathbf{y}_j\}.$$

The minimum invariant conditional expected loss plus cost of stopping after j observations, given $\mathbf{Y}_j = \mathbf{y}_j$, is denoted by $U_j(\mathbf{y}_j)$:

$$U_j(\mathbf{y}_j) = \rho_j(\mathbf{y}_j) + E_{\theta_0}\{c_j(\theta_0, X_1, \cdots, X_j) \mid \mathbf{Y}_j = \mathbf{y}_j\}.$$

Suppose the decision problem is truncated at J, $\varphi_J \equiv 1$. If the statistician has observed $X_1 = x_1, \cdots, X_{J-1} = x_{J-1}$, the total risk if he stops is $U_{J-1}(\mathbf{y}_{J-1})$. He may decide to continue only on the basis of $\mathbf{Y}_{J-1}$, so that heuristically his risk, if he continues, is $E_{\theta_0}\{U_J(\mathbf{Y}_J) \mid \mathbf{Y}_{J-1} = \mathbf{y}_{J-1}\}$. Thus he should stop or continue, depending on whether $U_{J-1}(\mathbf{y}_{J-1})$ is larger or smaller than $E_{\theta_0}\{U_J(\mathbf{Y}_J) \mid \mathbf{Y}_{J-1} = \mathbf{y}_{J-1}\}$. The minimum invariant conditional risk given $\mathbf{Y}_{j-1} = \mathbf{y}_{j-1}$, denoted by $V_{j-1}^{(J)}(\mathbf{y}_{j-1})$, may be defined inductively as $V_J^{(J)} = U_J$ and

$$V_{j-1}^{(J)}(\mathbf{y}_{j-1}) = \min\left[U_{j-1}(\mathbf{y}_{j-1}), E_{\theta_0}\{V_j^{(J)}(\mathbf{Y}_j) \mid \mathbf{Y}_{j-1} = \mathbf{y}_{j-1}\}\right] \quad (7.31)$$

for $j = 1, \cdots, J$. The constant $V_0^{(J)}$ denotes the minimum invariant risk for the truncated problem. An optimal invariant stopping rule is then $\boldsymbol{\varphi}^0 = (\varphi_0{}^0, \cdots, \varphi_J{}^0)$, where

$$\varphi_{j-1}^0(\mathbf{y}_{j-1}) = \begin{cases} 1 & \text{if } U_{j-1}(\mathbf{y}_{j-1}) < E_{\theta_0}\{V_j^{(J)}(\mathbf{Y}_j) \mid \mathbf{Y}_{j-1} = \mathbf{y}_{j-1}\}, \\ \text{any} & \qquad = \\ 0 & \qquad > \end{cases} \quad (7.32)$$

for $j = 1, \cdots, J$ (of course, $\varphi_J{}^0(\mathbf{y}_J) \equiv 1$).

Theorem 3. Let $\boldsymbol{\delta}^0$ be as in Theorem 2 and let $\boldsymbol{\varphi}^0$ be defined by (7.32). Then $(\boldsymbol{\varphi}^0, \boldsymbol{\delta}^0)$ is a best invariant decision rule for the problem truncated at J.

The proof is left to the reader.

In analogy to Theorem 7.2.3, we may define $V_0^{(\infty)}$ as the minimum invariant risk for the general problem

$$V_0^{(\infty)} = \inf_{(\boldsymbol{\varphi}, \boldsymbol{\delta}) \text{ invariant}} R(\theta, (\boldsymbol{\varphi}, \boldsymbol{\delta}))$$

and prove Theorem 4.

Theorem 4. If $E\rho_J(\mathbf{Y}_J) \to 0$ as $J \to \infty$ or if L is bounded, then $V_0^{(J)} \to V_0^{(\infty)}$ as $J \to \infty$.

The proof is completely analogous to Theorem 7.2.3. Corresponding to Theorems 7.2.4 and 7.2.5, we have Theorems 5 and 6.

Theorem 5. If $U_j(\mathbf{y}_j)$ is a constant independent of $\mathbf{y}_j$, then the best invariant sequential decision rule for the truncated problem is a fixed sample size rule. If, in addition, $V_0^{(J)} \to V_0^{(\infty)}$ as $J \to \infty$, the best invariant sequential decision rule for the general problem is a fixed sample size rule.

Theorem 6. If $V_0^{(J)} \to V_0^{(\infty)}$ as $J \to \infty$ and for all $j > J_0$

$$\rho_{j-1}(\mathbf{y}_{j-1}) - E_{\theta_0}\{\rho_j(\mathbf{Y}_j) \mid \mathbf{Y}_{j-1} = \mathbf{y}_{j-1}\}$$
$$\leq E_{\theta_0}\{c_j(\theta_0, X_1, \cdots, X_j) - c_{j-1}(\theta_0, X_1, \cdots, X_{j-1}) \mid \mathbf{Y}_{j-1} = \mathbf{y}_{j-1}\} \tag{7.33}$$

for almost all $\mathbf{y}_{j-1}$, then $V_0^{(J_0)} = V_0^{(\infty)}$.

Example 1. *Invariant sequential estimates of a location parameter.* Consider the estimation of a location parameter θ for a sequence of random variables $X_1, X_2, \cdots$. By this we mean that the joint density of $X_1, \cdots, X_j$, given θ, is in the form of $f_{X_1,\cdots,X_j}(x_1, \cdots, x_j \mid \theta) = f_j(x_1 - \theta, \cdots, x_j - \theta)$ for some function f_j. We assume that the loss is a function of the difference $a - \theta$, [that is, $L(\theta, a) = W(a - \theta)$] and that the cost of each observation is a constant c (that is, $c_j(\theta, x_1, \cdots, x_j) = jc$). This problem is invariant under the translations $g_b(x) = x + b$, $\bar{g}_b(\theta) = \theta + b$, $\tilde{g}_b(a) = a + b$: the group $\bar{\mathcal{G}}$ is transitive for this problem, and all invariant rules have constant risk. The group $\tilde{\mathcal{G}}$ is transitive also, so that there exists no invariant distribution δ_0 over $\mathcal{A}$ and every invariant rule must take at least one observation. For $j = 2, 3, \cdots$, the maximal invariant under the group $\mathcal{G}_j$ is $\mathbf{Y}_j = (X_2 - X_1, X_3 - X_1, \cdots, X_j - X_1)$. From Theorem 2 an optimal terminal decision rule d_j^0 is the same as the fixed sample size rule found in Section 4.7, namely, $d_j^0(\mathbf{X}) = X_1 - b_j^0(\mathbf{Y}_j)$, where $b_j^0(\mathbf{Y}_j)$ minimizes $E_0(W(X_1 - b_j(\mathbf{Y}_j)) \mid \mathbf{Y}_j)$. In three important cases this minimum value is independent of $\mathbf{Y}_j$, and the best invariant sequential rule is a fixed sample size rule: the normal distribution and Exercises 5 and 6.

If $X_1, X_2, \cdots$ are independent random variables, normally distributed with mean θ and variance 1, then the search for a best invariant estimate, d_j^0, for the fixed sample size problem may be restricted to the functions of the sufficient statistic $T_j = \sum_1^j X_i$. Furthermore, the maximal invariant $\mathbf{Y}_j = (X_2 - X_1, \cdots, X_j - X_1)$ is independent of T_j, as noted in Section 3.3. Hence $E_0(W(d_j^0(T_j)) \mid \mathbf{Y}_j) = E_0(W(d_j^0(T_j)))$ is independent of $\mathbf{Y}_j$. By Theorems 4 and 5, if W is bounded or

$$E_0(W(d_j^0(T_j))) \to 0 \quad \text{as} \quad j \to \infty,$$

the best invariant sequential estimate of θ is a fixed size estimate. If $W(x) = x^2$, then

$$d_j^0(T_j) = \bar{X}_j = \frac{T_j}{j} \quad \text{and} \quad U_j = \operatorname{Var} \bar{X}_j + jc = \frac{1}{j} + jc.$$

The best invariant estimate in this case turns out to be exactly the minimax estimate found in Exercise 7.2.6.

Example 2. Let $X_1, X_2, \cdots$ be a sample from the uniform distribution $\mathcal{U}(\theta - 1/2, \theta + 1/2)$ and suppose that we wish to estimate θ sequentially, using squared error loss $L(\theta, a) = (\theta - a)^2$ and constant cost c per observation. Because θ is a location parameter, we may search for a best invariant estimate. The optimal invariant terminal decision rule d_j^0 is the best invariant rule for the fixed sample size problem, namely $d_j^0(X_1, \cdots, X_j) = M_j$, where M_j is the mid-range,

$$M_j = (\max(X_1, \cdots, X_j) + \min(X_1, \cdots, X_j))/2$$

(see Exercise 4.7.2(a)). When $\theta = 0$, the distribution of M_j, given $\mathbf{Y}_j$, is uniform over the interval $(-(1 - R_j)/2, (1 - R_j)/2)$, where R_j is the range, $R_j = \max(X_1, \cdots, X_j) - \min(X_1, \cdots, X_j)$ (Exercise 9). Hence

$$\rho_j(\mathbf{Y}_j) = E_0(M_j^2 \mid \mathbf{Y}_j) = \frac{(1 - R_j)^2}{12}. \tag{7.34}$$

Because $E_0\rho_J(\mathbf{Y}_J) = (2(J + 1)(J + 2))^{-1} \to 0$ as $J \to \infty$ (Exercise 9), it follows from Theorem 4 that $V_0^{(J)} \to V_0^{(\infty)}$ as $J \to \infty$, so that the general problem may be approximated by truncated problems. We show below that an optimal invariant stopping rule for the problem truncated at J is $\varphi_0 = 0$ and for $j = 1, 2, \cdots, J - 1$

$$\varphi_j(\mathbf{Y}_j) = \begin{cases} 1 & \text{if} \quad R_j \geq 1 - \sqrt[3]{24c}, \\ 0 & \text{if} \quad R_j < 1 - \sqrt[3]{24c}. \end{cases} \tag{7.35}$$

Because the rule of (7.35) is independent of J, it must therefore be an optimal invariant rule for the nontruncated problem as well.

To use (7.31) we need to compute

$$E_0\{\rho_j(\mathbf{Y}_j) \mid \mathbf{Y}_{j-1}\} = \frac{E_0\{(1 - R_j)^2 \mid \mathbf{Y}_{j-1}\}}{12}.$$

When $\theta = 0$, the distribution of R_j, given $\mathbf{Y}_{j-1}$, has the distribution function

$$P\{R_j \leq z \mid \mathbf{Y}_{j-1}\} = \begin{cases} 0 & \text{if } z < R_{j-1} \\ R_{j-1} + (z - R_{j-1})(2 - z - R_{j-1})(1 - R_{j-1})^{-1} & \text{if } R_{j-1} \leq z \leq 1. \end{cases} \tag{7.36}$$

From this it follows that

$$E_0\{(1 - R_j)^2 \mid \mathbf{Y}_{j-1}\} = R_{j-1}(1 - R_{j-1})^2 + \int_{R_{j-1}}^{1} (1 - z)^2 \frac{2(1 - z)}{(1 - R_{j-1})}\, dz$$

$$= \frac{(1 - R_{j-1})^2(1 + R_{j-1})}{2}. \tag{7.37}$$

Hence

$$V_{J-1}^{(J)}(\mathbf{Y}_{J-1})$$

$$= \min\left[\frac{(1 - R_{J-1})^2}{12} + (J - 1)c,\ \frac{(1 - R_{J-1})^2(1 + R_{J-1})}{24} + Jc\right]$$

$$= \begin{cases} \dfrac{(1 - R_{J-1})^2}{12} + (J - 1)c & \text{if } R_{J-1} \geq 1 - \sqrt[3]{24c}, \\[2ex] \dfrac{(1 - R_{J-1})^2(1 + R_{J-1})}{24} + Jc & \text{if } R_{J-1} < 1 - \sqrt[3]{24c}, \end{cases}$$

and (7.35) is valid for $j = J - 1$. It is easy to see by induction that

$$V_{j-1}^{(J)}(\mathbf{Y}_{j-1}) = \tfrac{1}{12}(1 - R_{j-1})^2 + (j - 1)c$$

if $R_{j-1} \geq 1 - \sqrt[3]{24c}$ and that

$$E\{V_j^{(J)}(\mathbf{Y}_j) \mid \mathbf{Y}_{j-1}\} < \tfrac{1}{12}(1 - R_{j-1})^2 + (j - 1)c$$

if $R_{j-1} < 1 - \sqrt[3]{24c}$. This implies that (a) is valid for $j = 1, 2, \cdots, J - 1$, as asserted.

This rule, in fact, is minimax, as shown by Wald (1950), using Bayes methods described in Section 7.2. Blyth (1951) studied this problem, using a more general loss function, $L(\theta, a) = W(a - \theta)$, where W is a nondecreasing function of $|a - \theta|$. He shows that Wald's solution (7.35) with $\sqrt[3]{24c}$ replaced by a general function of W and c is still minimax, provided there does not exist a rule which takes no observations and has smaller risk.

EXAMPLE 3. ESTIMATION OF A DISTRIBUTION FUNCTION. Let Θ be the set of all continuous distribution functions on the real line, let $\mathcal{A}$ be the set of all distribution functions on the real line, and let

$$L(F, \hat{F}) = \int (F(x) - \hat{F}(x))^2 F(x)^{-1}(1 - F(x))^{-1}\, dF(x)$$

for all $F \in \Theta$ and $\hat{F} \in \mathcal{A}$. Let $X_1, X_2, \cdots$ be a sample from some distribution $F \in \Theta$ and let the cost of each observation be the same constant c, $c_j(\theta, x_1, \cdots, x_j) = jc$. This sequential problem is invariant under the group $\mathcal{G}$ of continuous strictly increasing transformations g from the real line onto the real line, with $\tilde{g}\, F(x) = F(g^{-1}x)$ and $\tilde{g}\, \hat{F}(x) = \hat{F}(g^{-1}x)$. The optimal invariant terminal decision rule d_j^0 was found in Section 4.8 to be the sample distribution function

$$d_j^0 = \hat{F}_j(x) = \sum_{i=0}^{j} \frac{i}{j} I_{[Z_i^{(j)}, Z_{i+1}^{(j)})}(x),$$

where $Z_0^{(j)} = -\infty$, $Z_{j+1}^{(j)} = +\infty$, and $Z_1^{(j)} \leq \cdots \leq Z_j^{(j)}$ are the order statistics of $X_1, \cdots, X_j$. The maximal invariant under $\mathcal{G}_j$ is the set of ranks $\mathbf{Y}_j = (Y_1, \cdots, Y_j)$ of $(X_1, \cdots, X_j)$. Because $\mathbf{Y}_j$ is independent of $(Z_1^{(j)}, \cdots, Z_j^{(j)})$, ρ_j is independent of $\mathbf{y}_j$.

$$\rho_j(\mathbf{y}_j) = \sum_{i=0}^{j} \int_0^1 \left(t - \frac{i}{j}\right)^2 \binom{j}{i} t^{i-1}(1 - t)^{j-i-1}\, dt = \frac{1}{j}.$$

The best invariant sequential decision rule is thus a fixed sample size rule: take a sample of size j, where j minimizes $jc + 1/j$, and estimate F to be the sample distribution function.

Exercises

1. Prove Theorem 3.

2. Prove Theorem 4.

3. Prove Theorem 5.

4. Prove Theorem 6.

5. Let $X_1, X_2, \cdots$ be a sample of an exponential distribution with unknown location parameter θ and density

$$f(x \mid \theta) = \exp[-(x - \theta)] I_{(\theta,\infty)}(x).$$

Let the loss function be a function of the difference

$$L(\theta, a) = W(a - \theta),$$

and the cost of each observation a constant, $c_j(\theta, x_1, \cdots, x_j) = jc$.
(a) Show that the search for an optimal invariant terminal decision rule, d_j, may be restricted to the functions of the sufficient statistic $T_j = \min(X_1, \cdots, X_j)$.
(b) Show that the maximal invariant $\mathbf{Y}_j$ is independent of T_j and conclude that a best invariant sequential rule is a fixed sample size rule, provided the hypotheses of Theorem 4 are satisfied.
(c) Find a best invariant decision rule when the loss is $L(\theta, a) = (e^{\theta-a} - 1)^2$ and $c = 1/12$. (This is the same problem as estimating the scale parameter $\varphi = e^{-\theta}$ of a uniform distribution $u(0, \varphi)$ with loss $L(\varphi, a') = (\varphi - a')^2/\varphi^2$.)

6. Let $X_1, X_2, \cdots$ be a sample from the gamma distribution $\mathcal{G}(\alpha, \theta)$ with α known. Let the loss be a function of a/θ, $L(\theta, a) = W(a/\theta)$ and let the cost of each observation be constant, $c_j(\theta, x_1, \cdots, x_j) = jc$. Because θ is a scale parameter, this problem is invariant under scale changes.
(a) Show that the search for an optimal invariant terminal decision rule, d_j, may be restricted to the functions of the sufficient statistic $T_j = \sum_1^j X_i$.
(b) Show that the maximal invariant $\mathbf{Y}_j = (X_2/X_1, \cdots, X_j/X_1)$ is independent of T_j and conclude that a best invariant sequential rule is a fixed sample size rule, provided the hypotheses of Theorem 4 are satisfied.
(c) Find a best invariant decision rule when the loss is $L(\theta, a) = (\theta^2 - a^2)^2/\theta^4$ and $c = 1/60$.

7. *Basu's theorem.* Basu (1955). Let $\mathbf{X}$ be a random vector whose distribution depends on θ. Suppose that $\mathbf{T}$ based on $\mathbf{X}$, is a complete sufficient statistic for θ and that $\mathbf{Y}$, based on $\mathbf{X}$, has a distribution which does not depend on θ. Then $\mathbf{T}$ and $\mathbf{Y}$ are stochastically independent. *Hint.* Let $g(\mathbf{Y})$ be an arbitrary function of $\mathbf{Y}$. Show that $E_\theta[E\{g(\mathbf{Y}) \mid \mathbf{T}\} - Eg(\mathbf{Y})] = 0$ for all θ and use the definition of completeness.

8. (a) Prove Exercises 5(b) and 6(b) using Basu's theorem.
(b) Let $X_1, X_2, \cdots, X_n$ be a sample of size $n(n > 2)$ from a normal distribution. Show that $\bar{X}$ and any function of the differences $(X_2 - X_1, \cdots, X_n - X_1)$ are independent. Show that $(\bar{X}, s^2)$ is independent of any function of the ratio of the differences $((X_2 - X_1)/(X_n - X_1), \cdots, (X_{n-1} - X_1)/(X_n - X_1))$.

9. Let $X_1, \cdots, X_j$ be a sample from the uniform distribution on the interval $(-\frac{1}{2}, \frac{1}{2})$, let $M_j =$ the mid-range $= (\max(X_1, \cdots, X_j) + \min(X_1, \cdots, X_j))/2$, let $R_j =$ the range $= \max(X_1, \cdots, X_j) - \min(X_1, \cdots, X_j)$, and let $\mathbf{Y}_j =$ the differences $= (X_2 - X_1, \cdots, X_j - X_1)$.
(a) Show that (M_j, R_j) and the ratios of the differences $((X_2 - X_1)/(X_j - X_1), \cdots, (X_{j-1} - X_1)/(X_j - X_1))$ are independent (use Basu's theorem).
(b) Show that the conditional distribution of M_j, given $\mathbf{Y}_j$, is the same as the conditional distribution of M_j, given R_j.
(c) Show that the conditional distribution of M_j, given R_j, is uniform on the interval $[-(1 - R_j)/2, (1 - R_j)/2]$.
(d) Verify (7.34) and show $E(1 - R_j)^2/12 = (2(j + 1)(j + 2))^{-1}$.
(e) Find the conditional distribution of (M_{j-1}, X_j), given R_{j-1}, and verify (7.36).

10. *Estimation of a median.* Let Θ be the set of all continuous distribution functions on the real line, let $\mathcal{A}$ be the real line, and let $L(F, a) = (F(a) - \frac{1}{2})^2$ for $F \in \Theta$ and $a \in \mathcal{A}$ (see Exercise 4.8.3). Let the cost of each observation be constant, $c_j(\theta, x_1, \cdots, x_j) = jc$. Show that the best invariant sequential decision rule is a fixed sample size rule and find ρ_j.

7.5 Sequential Tests of a Simple Hypothesis Against a Simple Alternative

When the random observables $X_1, X_2, \cdots$ are independent and identically distributed, there is a particularly simple way of viewing the posterior distributions of θ, X_{j+1}, X_{j+2}, $\cdots$, given $X_1, \cdots, X_j$, $j = 0, 1, 2, \cdots$ for a given prior distribution τ of θ. This viewpoint is stated in Lemma 1.

We state and prove this lemma by assuming densities for the distributions of the variables in question, although it is valid quite generally. We let $f(x \mid \theta)$ denote the common density of the variables $X_1, X_2, \cdots$, given θ, and $\tau(\theta)$ represents the density of θ. The posterior density of θ, given $X_1 = x_1, \cdots, X_j = x_j$, is denoted by $\tau(\theta \mid x_1, \cdots, x_j)$, so that

$$\tau(\theta \mid x_1, \cdots, x_j) = \frac{\tau(\theta)\left(\prod_{i=1}^{j} f(x_i \mid \theta)\right)}{\int \tau(\theta)\left(\prod_{i=1}^{j} f(x_i \mid \theta)\right) d\theta}. \tag{7.38}$$

Lemma 1. If $X_1, \cdots, X_n$ are independent identically distributed with density $f(x \mid \theta)$ and θ has prior distribution with density $\tau(\theta)$, the distribution of θ, X_{j+1}, X_{j+2}, $\cdots$, X_n, given $X_1 = x_1, \cdots, X_j = x_j$, can be described by saying that $X_{j+1}, \cdots, X_n$ are independent identically distributed with density $f(x \mid \theta)$ and θ has density $\tau(\theta \mid x_1, \cdots, x_j)$.

Proof. The joint density of $\theta, X_1, \cdots, X_n$ is $\tau(\theta) \prod_1^n f(x_i \mid \theta)$. The marginal density of $X_1, \cdots, X_j$ is $\int \tau(\theta) \prod_1^j f(x_i \mid \theta)\, d\theta$. Hence the conditional density of $\theta, X_{j+1}, \cdots, X_n$, given $X_1 = x_1, \cdots, X_j = x_j$,

$$\frac{\tau(\theta)\left(\prod_1^n f(x_i \mid \theta)\right)}{\int \tau(\theta)\left(\prod_1^j f(x_i \mid \theta)\right) d\theta} = \tau(\theta \mid x_1, \cdots, x_j) \prod_{j+1}^{n} f(x_i \mid \theta), \tag{7.39}$$

as was to be proved.

This lemma entails certain general implications. Suppose that $X_1, X_2, \cdots$ are independent identically distributed, with density $f(x \mid \theta)$, and that θ has prior distribution with density $\tau(\theta)$. If we observe $X_1 = x_1, \cdots, X_j = x_j$, the only thing that has changed about our view of the future (as far as the distribution of θ and the future observables is concerned) is that the prior density of θ has now changed to $\tau(\theta \mid x_1, \cdots, x_j)$.

Suppose now, in addition to the assumption that $X_1, X_2, \cdots$ are independent and identically distributed, that there is a constant cost per observation, that is, $c_j = jc$, where $c > 0$. Then in the Bayes sequential decision problem the only thing that has changed about our view of the future after observing $X_1 = x_1, \cdots, X_j = x_j$ is that the prior distribution of θ has changed from $\tau(\theta)$ to $\tau(\theta \mid x_1, \cdots, x_j)$. Therefore it seems reasonable to expect that in such Bayes sequential decision problems (nontruncated) the decision whether to stop and what terminal decision to take if it is decided to stop should depend only on the distribution $\tau(\theta \mid x_1, \cdots, x_j)$ and not otherwise on $x_1, \cdots, x_j$ or j. In other words, we expect that an optimal stopping rule can be found among the class of rules which stop for the first j for which $\tau(\theta \mid x_1, \cdots, x_j)$ is in some fixed subset Θ_∞^* of the space Θ^* of all prior distributions on Θ. This is, in fact, true rather generally. However, except for the case of finite Θ, I know of no good applications of this approach; that is to say, I know of no problems other than those in which Θ is finite for which the optimal subset Θ_∞^* of Θ^* is characterizable in some useful way. In problems of testing a simple hypothesis against a simple alternative, in which Θ consists of two elements so that Θ^* can be identified with the unit interval on the real line, it turns out that the optimal subset is easily described. It is to this problem that we now turn our attention.

Consider a sequential decision problem in which $\Theta = \{\theta_0, \theta_1\}$, $\mathfrak{A} = \{a_0, a_1\}$, and

$$\begin{aligned} L(\theta_0, a_0) &= L(\theta_1, a_1) = 0, \\ L(\theta_0, a_1) = w_{01} \quad &\text{and} \quad L(\theta_1, a_0) = w_{10}, \end{aligned} \tag{7.40}$$

where $w_{01} > 0$ and $w_{10} > 0$. For $i = 0$ or 1, a_i may be considered as the action of accepting the hypothesis H_i that θ_i is the true state of nature. We assume that the random observables $X_1, X_2, \cdots$ are independent and identically distributed with common density (mass function) $f_i(x) = f(x \mid \theta_i)$ if θ_i is the true state of nature, $i = 0, 1$, $f_0(x) \not\equiv f_1(x)$. Furthermore, we assume that there is a constant cost c per observation, so that $c_j = jc$. Let π denote the prior probability that θ_1 is the true state of nature. The posterior probability that θ_1 is the true state of nature, given $X_1 = x_1, \cdots, X_j = x_j$, is denoted by $\pi_{x_1, \cdots, x_j}$.

$$\pi_{x_1, \cdots, x_j} = \frac{\pi \prod_1^j f_1(x_i)}{\pi \prod_1^j f_1(x_i) + (1 - \pi) \prod_1^j f_0(x_i)}. \tag{7.41}$$

For the fixed sample size problem based on $X_1, \cdots, X_j$ the Bayes terminal decision rule is to take that action with the smaller conditional expected loss, given $X_1, \cdots, X_j$. Thus the terminal decision rule $\boldsymbol{\delta}^\pi$, where

$$\delta_j^\pi(x_1, \cdots, x_j) = \begin{cases} 0 & \text{if} \quad w_{10}\pi_{x_1,\cdots,x_j} < w_{01}(1 - \pi_{x_1,\cdots,x_j}), \\ \text{any} & \text{if} \quad = \\ 1 & \text{if} \quad > \end{cases} \tag{7.42}$$

is Bayes with respect to π in the sequential problem.

It is necessary to extend the notation of Section 7.2 to show dependence on the prior probability π. Instead of $\rho_j(x_1, \cdots, x_j)$ and $U_j(x_1, \cdots, x_j)$, we write $\rho_j(x_1, \cdots, x_j; \pi)$ and $U_j(x_1, \cdots, x_j; \pi)$. For ρ_0 and U_0 we write $\rho_0(\pi)$ and $U_0(\pi)$. The minimum posterior Bayes expected loss for the fixed sample size problem is

$$\rho_j(x_1, \cdots, x_j; \pi) = \begin{cases} w_{10}\pi_{x_1,\cdots,x_j} & \text{if} \quad \pi_{x_1,\cdots,x_j} \leq \dfrac{w_{01}}{w_{01} + w_{10}}, \\ w_{01}(1 - \pi_{x_1,\cdots,x_j}) & \text{if} \quad \geq \end{cases} \tag{7.43}$$

The minimum posterior Bayes expected loss plus cost is

$$U_j(x_1, \cdots, x_j; \pi) = \rho_j(x_1, \cdots, x_j; \pi) + jc. \tag{7.44}$$

In particular,

$$U_0(\pi) = \rho_0(\pi) = \begin{cases} w_{10}\pi & \text{if} \quad \pi \leq \dfrac{w_{01}}{w_{01} + w_{10}}, \\ w_{01}(1 - \pi) & \text{if} \quad \geq \end{cases} \tag{7.45}$$

Instead of $V_j^{(J)}(x_1, \cdots, x_j)$, we write $V_j^{(J)}(x_1, \cdots, x_j; \pi)$, so that $V_j^{(J)}$ is defined inductively in j for fixed J by $V_J^{(J)}(x_1, \cdots, x_J; \pi) = U_J(x_1, \cdots, x_J; \pi)$ and for $0 \leq j < J$

$$\begin{aligned} V_j^{(J)}(x_1, \cdots, x_j; \pi) = \min [&U_j(x_1, \cdots, x_j; \pi), \\ &E\{V_{j+1}^{(J)}(x_1, \cdots, x_j, X_{j+1}; \pi) \mid X_1 = x_1, \cdots, X_j = x_j, \pi\}], \end{aligned} \tag{7.46}$$

where from now on we write $E\{\ |\ \pi\}$ as the expectation when π is the prior probability of θ_1 and $E\{\ |\ X_1 = x_1, \cdots, X_j = x_j; \pi\}$ as the conditional expectation, given $X_1 = x_1, \cdots, X_j = x_j$ when π is the prior probability of θ_1.

The implications of Lemma 1 may be expressed more concretely in terms of $\pi_{x_1,\cdots,x_j}$, U_j and $V_j^{(J)}$.

Lemma 2. (a) $(\pi_{x_1,\cdots,x_j})_{x_j+1,\cdots,x_n} = \pi_{x_1,\cdots,x_j,x_j+1,\cdots,x_n}$.

$$\text{(b)}\quad U_j(x_1, \cdots, x_j; \pi) = U_{j-1}(x_2, \cdots, x_j; \pi_{x_1}) + c = \cdots$$

$$= U_0(\pi_{x_1,\cdots,x_j}) + jc,$$

$$\text{(c)}\quad V_j^{(J)}(x_1, \cdots, x_j; \pi) = V_{j-1}^{(J-1)}(x_2, \cdots, x_j; \pi_{x_1}) + c = \cdots$$

$$= V_0^{(J-j)}(\pi_{x_1,\cdots,x_j}) + jc.$$

Proof. Part (a) is easily checked from (7.41). Part (b) follows immediately from (7.43), (7.44), (7.45), and (a)

To prove (c) we proceed by induction on the difference $J - j$. For $J - j = 0$, (c) is equivalent to (b). Because, from Lemma 1, the distribution of X_{j+1}, given $X_1 = x_1, \cdots, X_j = x_j$, when π is the probability that θ_1 is the true state of nature, is the same as the distribution of X_{j+1} when $\pi_{x_1,\cdots,x_j}$ is the probability that θ_1 is the true state of nature, (7.46) may be written

$$V_j^{(J)}(x_1, \cdots, x_j; \pi) = \min [U_j(x_1, \cdots, x_j; \pi),$$
$$E\{V_{j+1}^{(J)}(x_1, \cdots, x_j, X_{j+1}; \pi) \mid \pi_{x_1,\cdots,x_j}\}]. \quad (7.47)$$

Now, if all equalities in (c) are true when $J - j = k$, then if $J - j = k + 1$, (7.47) may be written

$$V_j^{(J)}(x_1, \cdots, x_j; \pi) = \min [U_0(\pi_{x_1,\cdots,x_j}),$$
$$E\{V_1^{(J-j)}(X_{j+1}; \pi_{x_1,\cdots,x_j}) \mid \pi_{x_1,\cdots,x_j}\}] + jc$$
$$= V_0^{(J-j)}(\pi_{x_1,\cdots,x_j}) + jc.$$

The rest of the equalities in (c) follow from this using (a), which completes the induction.

For a fixed stopping rule ϕ we denote the minimum Bayes risk by

$r(\pi, \boldsymbol{\phi})$ (please excuse the abuse of notation):

$$r(\pi, \boldsymbol{\phi}) = \inf_{\boldsymbol{\delta}} r(\pi, (\boldsymbol{\phi}, \boldsymbol{\delta}))$$

$$= \sum_{j=0}^{\infty} E\{\psi_j(X_1, \cdots, X_j)\, U_j(X_1, \cdots, X_j; \pi) \mid \pi\}. \qquad (7.48)$$

The functions $V_0^{(\infty)}(\pi)$, $V_0^{(J)}(\pi)$, and $r(\pi, \boldsymbol{\phi})$ are all well-behaved functions of π, as Lemma 3 indicates.

Lemma 3. The functions $V_0^{(\infty)}(\pi)$, $V_0^{(J)}(\pi)$ for each J and $r(\pi, \boldsymbol{\phi})$ for all $\boldsymbol{\phi}$, are concave functions of π in $[0, 1]$.

Proof. The proof for all these functions uses the same idea and depends only on the facts that $r(\pi, (\boldsymbol{\phi}, \boldsymbol{\delta}))$ is linear in π and that each of the functions above is an infimum of $r(\pi, (\boldsymbol{\phi}, \boldsymbol{\delta}))$ over a suitable space of decision rules. We give the proof for $V_0^{(\infty)}(\pi)$ as an example. Let $0 < \alpha < 1$ and $0 \le \pi_i \le 1$, $i = 1, 2$. Then

$$\begin{aligned} V_0^{(\infty)}(\alpha\pi_1 + (1-\alpha)\pi_2) &= \inf_{(\boldsymbol{\phi},\boldsymbol{\delta})} r(\alpha\pi_1 + (1-\alpha)\pi_2, (\boldsymbol{\phi}, \boldsymbol{\delta})) \\ &= \inf_{(\boldsymbol{\phi},\boldsymbol{\delta})} [\alpha r(\pi_1, (\boldsymbol{\phi}, \boldsymbol{\delta})) + (1-\alpha)\, r(\pi_2, (\boldsymbol{\phi}, \boldsymbol{\delta})] \\ &\ge \alpha \inf_{(\boldsymbol{\phi},\boldsymbol{\delta})} r(\pi_1, (\boldsymbol{\phi}, \boldsymbol{\delta})) + (1-\alpha) \inf_{(\boldsymbol{\phi},\boldsymbol{\delta})} r(\pi_2, (\boldsymbol{\phi}, \boldsymbol{\delta})) \\ &= \alpha\, V_0^{(\infty)}(\pi_1) + (1-\alpha)\, V_0^{(\infty)}(\pi_2), \end{aligned}$$

as was to be proved.

A method of computing the $V_0^{(J)}(\pi)$ inductively is given in Theorem 1.

Theorem 1. $V_0^{(J+1)}(\pi) = \min[U_0(\pi), c + E\{V_0^{(J)}(\pi_{X_1}) \mid \pi\}]$.

The proof is an exercise.

An optimal stopping rule for the truncated problem is easily described. Let Θ_J^* denote the set of all prior distributions $\pi \in [0, 1] = \Theta^*$ such that in the problem truncated at J stopping without taking any observations is Bayes with respect to π.

$$\Theta_J^* = \{\pi : V_0^{(J)}(\pi) = U_0(\pi)\}. \tag{7.49}$$

Because the $V_0^{(J)}(\pi)$ are nonincreasing in J for fixed π, these sets are nested

$$\Theta^* = \Theta_0^* \supset \Theta_1^* \supset \Theta_2^* \supset \cdots.$$

These sets may be used to describe the Bayes stopping rules for the truncated problem.

Theorem 2. In a sequential decision problem truncated at J with independent identically distributed observations $X_1, X_2, \cdots$ and constant cost c per observation the stopping rule $\boldsymbol{\phi}^{J,\pi}$, where

$$\phi_j^{J,\pi}(x_1, \cdots, x_j) = \begin{cases} 1 & \text{if} \quad \pi_{x_1,\cdots,x_j} \in \Theta_{J-j}^*, \\ 0 & \text{otherwise.} \end{cases} \tag{7.50}$$

is Bayes with respect to π.

Proof. From (7.11) the stopping rule

$$\phi_j(x_1, \cdots, x_j) = \begin{cases} 1 & \text{if} \quad U_j(x_1, \cdots, x_j; \pi) = V_j^{(J)}(x_1, \cdots, x_j; \pi), \\ 0 & \text{otherwise,} \end{cases}$$

is Bayes with respect to π. From Lemma 2 the equality

$$U_j(x_1, \cdots, x_j; \pi) = V_j^{(J)}(x_1, \cdots, x_j; \pi) \tag{7.51}$$

is equivalent to the equality $U_0(\pi_{x_1,\cdots,x_j}) = V_0^{(J-j)}(\pi_{x_1,\cdots,x_j})$ from which the theorem follows.

It will be seen that the limit $\boldsymbol{\phi}^\pi$ of the rules $\boldsymbol{\phi}^{J,\pi}$ of (7.50) is Bayes in the nontruncated problem. To prove this result we first demonstrate that Theorem 1 is valid in the limit as $J \to \infty$. The resulting equation is the fundamental functional equation of dynamic programming for this problem.

Theorem 3. $V_0^{(\infty)}(\pi) = \min [U_0(\pi), c + E\{V_0^{(\infty)}(\pi_{X_1}) \mid \pi\}]$.

Proof. Because the loss is bounded, we have from Theorem 7.2.3 and monotone convergence (or alternatively, from Lemma 4 below) that $E(V_0^{(J)}(\pi_{X_1}) \mid \pi) \to E(V_0^{(\infty)}(\pi_{X_1}) \mid \pi)$ for each π. The result follows immediately from Theorem 1.

Lemma 4. $V_0^{(J)}(\pi) \to V_0^{(\infty)}(\pi)$ uniformly in π, $0 \leq \pi \leq 1$; that is, $\sup_{0 \leq \pi \leq 1} (V_0^{(J)}(\pi) - V_0^{(\infty)}(\pi)) \to 0$ as $J \to \infty$.

This lemma may be proved using only the monotone pointwise convergence of $V_0^{(J)}(\pi)$ to $V_0^{(\infty)}(\pi)$ and the concavity of the functions involved. Alternatively, one may prove that the functions involved are continuous on $[0, 1]$ and apply Dini's theorem (see any recent text on real analysis). The proof is left to the student.

It is easy to make a guess at an optimal stopping rule for the nontruncated problem. If we define, analogous to (7.49),

$$\Theta_\infty^* = \{\pi : V_0^{(\infty)}(\pi) = U_0(\pi)\}, \tag{7.52}$$

it is easily checked that the sets Θ_J^* decrease to the limit Θ_∞^* (Exercise 2). Therefore we would expect that the limit of the rules $\boldsymbol{\phi}^{J,\pi}$ of Theorem 2 would be optimal in the nontruncated problem. This limit is denoted by $\boldsymbol{\phi}^\pi$, where

$$\phi_j^\pi(x_1, \cdots, x_j) = \begin{cases} 1 & \text{if } \pi_{x_1,\cdots,x_j} \in \Theta_\infty^*, \\ 0 & \text{otherwise.} \end{cases} \tag{7.53}$$

We intend to prove that $\boldsymbol{\phi}^\pi$ is optimal, that is, $r(\pi, \boldsymbol{\phi}^\pi) = V_0^{(\infty)}(\pi)$. For this purpose we introduce the rules ${}^n\boldsymbol{\phi}^{J,\pi}$. For $j = 0, 1, \cdots, n-1$

$${}^n\phi_j^{J,\pi}(x_1, \cdots, x_j) = \phi_j^\pi(x_1, \cdots, x_j)$$

and for $j = n, n+1, \cdots$

$$\begin{aligned} {}^n\phi_j^{J,\pi}(x_1, \cdots, x_j) &= \phi_j^{J,\pi_{x_1,\cdots,x_n}}(x_{n+1}, \cdots, x_j) \\ &= \begin{cases} 1 & \text{if } \pi_{x_1,\cdots,x_j} \in \Theta_{n+J-j}^*, \\ 0 & \text{otherwise.} \end{cases} \end{aligned}$$

Thus ${}^n\boldsymbol{\phi}^{J,\pi}$ represents the rule that uses $\boldsymbol{\phi}^\pi$ for n stages followed by $\boldsymbol{\phi}^{J,\pi_{x_1,\cdots,x_n}}$. It is easy to see, using Lemma 2(a), that for $n \geq 1$,

$${}^n\phi_j^{J,\pi}(x_1, \cdots, x_j) = {}^{n-1}\phi_{j-1}^{J,\pi_{x_1}}(x_2, \cdots, x_j),$$

hence that for $n \geq 1$, using (7.1),

$$^{n}\psi_j^{J,\pi}(x_1, \cdots, x_j) = {}^{n-1}\psi_{j-1}^{J,\pi_{x_1}}(x_2, \cdots, x_j)(1 - \psi_0^{\pi}). \tag{7.54}$$

Lemma 5. Let $\epsilon > 0$; and let J be so large that $V_0^{(J)}(\pi) \leq V_0^{(\infty)}(\pi) + \epsilon$ for all π. (Such a J exists from Lemma 4.) Then for all π and $n = 0, 1, 2, \cdots,$

$$r(\pi, {}^{n}\boldsymbol{\phi}^{J,\pi}) \leq V_0^{(\infty)}(\pi) + \epsilon. \tag{7.55}$$

Proof. Lemma 5 is obviously true for $n = 0$, for ${}^{0}\boldsymbol{\phi}^{J,\pi} = \boldsymbol{\phi}^{J,\pi}$ and $r(\pi, \boldsymbol{\phi}^{J,\pi}) = V_0^{(J)}(\pi)$. We proceed by induction on n. Suppose that (7.55) is valid for $n = k$. From Lemma 2(b), (7.48), and (7.54),

$$\begin{aligned} r(\pi, {}^{k+1}\boldsymbol{\phi}^{J,\pi}) &= \sum_{j=0}^{\infty} E\{{}^{k+1}\psi_j^{J,\pi}(X_1, \cdots, X_j)\, U_j(X_1, \cdots, X_j; \pi) \mid \pi\} \\ &= \psi_0^{\pi}\, U_0(\pi) + (1 - \psi_0^{\pi}) \sum_{j=1}^{\infty} E\{{}^{k}\psi_{j-1}^{J,\pi_{X_1}}(X_2, \cdots, X_j) \\ &\qquad \times [U_{j-1}(X_2, \cdots, X_j; \pi_{X_1}) + c] \mid \pi\}, \end{aligned} \tag{7.56}$$

but because the distribution of $X_2, \cdots, X_j$, given π and X_1, is the same as the distribution of $X_2, \cdots, X_j$, given π_{X_1}, the summation in the expression on the right side of (7.56) may be written

$$\begin{aligned} &E\{\sum_{j=1}^{\infty} {}^{k}\psi_{j-1}^{J,\pi_{X_1}}(X_2, \cdots, X_j)[U_{j-1}(X_2, \cdots, X_j; \pi_{X_1}) + c] \mid \pi\} \\ &= E\{E\{\sum_{j=1}^{\infty} {}^{k}\psi_{j-1}^{J,\pi_{X_1}}(X_2, \cdots, X_j)[U_{j-1}(X_2, \cdots, X_j; \pi_{X_1}) + c] \mid \pi, X_1\} \mid \pi\} \\ &= E\{r(\pi_{X_1}, {}^{k}\boldsymbol{\phi}^{J,\pi_{X_1}}) \mid \pi\} + c. \end{aligned}$$

Hence by using the induction hypothesis, Theorem 3, and the definition of ψ_0^{π}, we obtain

$$\begin{aligned} r(\pi, {}^{k+1}\boldsymbol{\phi}^{J,\pi}) &= \psi_0^{\pi}\, U_0(\pi) + (1 - \psi_0^{\pi})[c + E\{r(\pi_{X_1}, {}^{k}\boldsymbol{\phi}^{J,\pi_{X_1}}) \mid \pi\}] \\ &\leq \psi_0^{\pi}\, U_0(\pi) + (1 - \psi_0^{\pi})[c + E\{V_0^{(\infty)}(\pi_{X_1}) + \epsilon \mid \pi\}] \\ &\leq V_0^{(\infty)}(\pi) + \epsilon, \end{aligned}$$

as was to be proved.

Theorem 4. $r(\pi, \boldsymbol{\phi}^\pi) = V_0^{(\infty)}(\pi)$.

Proof. Note first that $\boldsymbol{\phi}^\pi$ stops with probability one. If $\pi = 0$ or 1, then $\boldsymbol{\phi}^\pi$ takes no observations. If $0 < \pi < 1$, and $P_\theta, \boldsymbol{\phi}^\pi\{N = \infty\} = \Delta > 0$ for $\theta = \theta_0$ or θ_1, then $E_\theta, {}^n\boldsymbol{\phi}^{\pi,J}\{N\} \geq n\Delta \rightarrow \infty$ as $n \rightarrow \infty$, contradicting Lemma 5.

It is clear that $r(\pi, \boldsymbol{\phi}^\pi) \geq V_0^{(\infty)}(\pi)$ from the definition of $V_0^{(\infty)}(\pi)$. To show $r(\pi, \boldsymbol{\phi}^\pi) \leq V_0^{(\infty)}(\pi)$, define for arbitrary $\boldsymbol{\phi}$

$$r_n(\pi, \boldsymbol{\phi}) = \sum_{j=0}^{n-1} E\{\psi_j(X_1, \cdots, X_j)\, U_j(X_1, \cdots, X_j; \pi) \mid \pi\},$$

so that $r_n(\pi, \boldsymbol{\phi})$ is nondecreasing in n and converges to $r(\pi, \boldsymbol{\phi})$ as n increases to infinity. Therefore, for a given $\epsilon > 0$ there is an n large enough so that

$$\begin{aligned} r(\pi, \boldsymbol{\phi}^\pi) - \epsilon &\leq r_n(\pi, \boldsymbol{\phi}^\pi) \\ &= r_n((\pi, {}^n\boldsymbol{\phi}^{J,\pi}) \\ &\leq r(\pi, {}^n\boldsymbol{\phi}^{J,\pi}) \\ &\leq V_0^{(\infty)}(\pi) + \epsilon, \end{aligned}$$

where J is chosen as in Lemma 5. Because $\epsilon > 0$ is arbitrary, the proof is complete.

Exercises

1. (a) Prove Theorem 1. (b) Prove Lemma 4.

2. Let Θ_∞^* be defined by (7.52) and let Ω denote $\lim_{J\to\infty} \Theta_J{}^*$, where $\Theta_J{}^*$ is defined by (7.49). Prove $\Omega = \Theta_\infty^*$.

3. Show that $\Theta_\infty^* = \Theta^*$ if, and only if, $c \geq c_0$, where

$$c_0 = \frac{w_{01}w_{10}}{w_{01} + w_{10}} [P_{\theta_1}(f_1(X_1) > f_0(X_1)) - P_{\theta_0}(f_1(X_1) > f_0(X_1))].$$

Hint. Show that $\Theta_\infty^* = \Theta^*$ if, and only if, $U_0(\hat{\pi}) \leq c + E\{U_0(\hat{\pi}_{X_1}) \mid \hat{\pi}\}$ where $\hat{\pi} = w_{01}/(w_{01} + w_{10})$.

4. If $f_i(x)$ is the mass function of the binomial distribution $\mathcal{B}(1, p_i)$ for $i = 0, 1$, where $p_0 < p_1$, the number c_0 of Exercise 3 is $c_0 = (p_1 - p_0)w_{01}w_{10}/(w_{01} + w_{10})$.

5. Let $X_1, X_2, \cdots$ be independent Bernoulli trials, that is, $\mathcal{B}(1, p)$, and consider the problem of testing $H_0: p = 1/3$ against $H_1: p = 2/3$, with $w_{01} = w_{10} = 1$ and $c = 1/12$. Show that $\Theta_1^* = \Theta_2^* = \cdots = \Theta_\infty^* = (5/12, 7/12)^c$ and

$$V_0^{(\infty)}(\pi) = \begin{cases} \pi & \text{if} \quad 0 \le \pi \le 5/12, \\ 5/12 & \text{if} \quad 5/12 \le \pi \le 7/12, \\ 1 - \pi & \text{if} \quad 7/12 \le \pi \le 1. \end{cases}$$

Hint. In fact,

$$E\{V_0^{(1)}(\pi_{X_1}) \mid \pi\} = \begin{cases} \pi & \text{if} \quad 0 \le \pi \le 5/19, \\ \dfrac{5 + 17\pi}{36} & \text{if} \quad 5/19 \le \pi \le 7/17, \\ 1/3 & \text{if} \quad 7/17 \le \pi \le 10/17, \\ \dfrac{22 - 17\pi}{36} & \text{if} \quad 10/17 \le \pi \le 14/19, \\ 1 - \pi & \text{if} \quad 14/19 \le \pi \le 1. \end{cases}$$

6. Let $f_0(x) = (1/2)I_{(0,2)}(x)$ and $f_1(x) = (1/2)I_{(1,3)}(x)$; the problem is to test one uniform distribution against another. If $w_{01} = w_{10} = 1$, show that c_0 of Exercise 3 is 1/4. If $c < 1/4$, show that $\Theta_1^* = \Theta_2^* = \cdots = \Theta_\infty^* = (2c, 1 - 2c)^c$ and

$$V_0^{(J)}(\pi) = \begin{cases} \pi & \text{if} \quad 0 \le \pi \le 2c, \\ \dfrac{\pi + 2c(2^J - 1)}{2^J} & \text{if} \quad 2c \le \pi \le 1/2, \\ \dfrac{1 - \pi + 2c(2^J - 1)}{2^J} & \text{if} \quad 1/2 \le \pi \le 1 - 2c, \\ 1 - \pi & \text{if} \quad 1 - 2c \le \pi \le 1. \end{cases}$$

7.6 The Sequential Probability Ratio Test

In this section we define Wald's sequential probability ratio test for testing a simple hypothesis against a simple alternative and prove the optimum property of this test first conjectured by Wald (1947) and proved by Wald and Wolfowitz (1948).

Let $X_1, \cdots, X_n, \cdots$ be independent identically distributed random variables and consider testing the hypothesis H_0 that the common density (mass function) of the X_i is $f_0(x)$ against the alternative hypothesis H_1 that the common density (mass function) of the X_i is $f_1(x)$. *The probability ratio* or *likelihood ratio* based on $X_1, \cdots, X_j$ is defined as

$$\lambda_j(X_1, \cdots, X_j) = \left[\prod_{i=1}^{j} \frac{f_1(X_i)}{f_0(X_i)} \right] \tag{7.57}$$

for $j = 1, 2, \cdots$. For $j = 0$, the probability ratio is arbitrarily defined as one, that is, $\lambda_0 = 1$. Unusually small values of λ_j tend to indicate that H_0 is true, whereas unusually large values of λ_j tend to indicate that H_1 is true.

The sequential probability ratio test with boundaries A and B, denoted by SPRT(A, B), where $0 < A \le 1 \le B < \infty$, is defined by the stopping rule $\boldsymbol{\phi}$ and terminal decision rule $\boldsymbol{\delta}$:

$$\phi_j(x_1, \cdots, x_j) = \begin{cases} 0 & \text{if} \quad A < \lambda_j(x_1, \cdots, x_j) < B, \\ 1 & \text{otherwise,} \end{cases} \tag{7.58}$$

$$\delta_j(x_1, \cdots, x_j) = \begin{cases} 1 & \text{if} \quad \lambda_j(x_1, \cdots, x_j) \ge B > A, \\ 0 & \text{if} \quad \lambda_j(x_1, \cdots, x_j) \le A < B, \\ \text{any} & \text{if} \quad A = B = 1. \end{cases} \tag{7.59}$$

(where δ_j represents the probability of accepting H_1).

In other words, the SPRT(A, B) continues sampling as long as λ_j stays strictly between A and B. Sampling is stopped at the first j for which $\lambda_j \le A$ or $\lambda_j \ge B$ and, if $\lambda_j \le A$, hypothesis H_0 is accepted, whereas, if $\lambda_j \ge B$, hypothesis H_1 is accepted.

Note that if $A = 1 < B$, the SPRT(A, B) accepts H_0 without sampling (regardless of B). Similarly, if $B = 1 > A$, the SPRT(A, B) accepts H_1 without sampling. In the special case $A = B = 1$, the SPRT(A, B) takes no samples and decides at random whether to accept hypothesis

H_0 or hypothesis H_1. If this special case is ruled out, the SPRT(A, B) is exactly determined by the numbers A and B.

More generally, we may define a larger family of tests called *extended sequential probability ratio tests.* An extended SPRT(A, B) is defined as above with the exception that in the stopping rule randomization is permitted on the boundaries; that is, (7.58) becomes

$$\phi_j(x_1, \cdots, x_j) = \begin{cases} 0 & \text{if} \quad A < \lambda_j(x_1, \cdots, x_j) < B, \\ \text{any} & \text{if} \quad \lambda_j(x_1, \cdots, x_j) = A \text{ or } B, \\ 1 & \text{otherwise.} \end{cases} \tag{7.60}$$

These rules extend the sequential probability ratio tests in the same sense that in nonsequential problems of testing hypotheses the randomized tests extend the nonrandomized tests. These rules are, in fact, Bayes rules in certain related sequential decision problems, as studied in Section 7.5.

Our first objective is to show that the Bayes rule of (7.42) and (7.53) is a SPRT(A, B) for some A and B. For this and later results it is useful to know something about the function

$$V^*(\pi) = c + E\{V_0^{(\infty)}(\pi_{X_1}) \mid \pi\} \tag{7.61}$$

which represents the minimum Bayes risk over all rules that require at least one observation. The stopping rule (7.53) requires stopping if the expected loss incurred by stopping without taking further observations $(U_0(\pi_{x_1, \cdots, x_j}))$ is less than or equal to the minimum Bayes risk of continuing with at least one more observation $(V^*(\pi_{x_1, \cdots, x_j}))$, since Θ_∞^* may be written as $\{\pi^*: U_0(\pi) \leq V^*(\pi)\}$. The first lemma gives the main properties of $V^*(\pi)$.

Lemma 1. The function $V^*(\pi)$ is continuous and concave in the interval $0 \leq \pi \leq 1$, with $V^*(0) = V^*(1) = c$.

Proof. The fact that $V^*(\pi)$ is concave follows as in Lemma 7.5.3 from the fact that $V^*(\pi)$ is the infimum of a class of linear functions of π, or alternatively, from the fact that $V_0^{(\infty)}(\pi)$ is concave (Exercise 1). Continuity of $V^*(\pi)$ in the open interval $0 < \pi < 1$ follows automatically (see Exercise 2.8.2). Finally, because $0 \leq V_0^{(\infty)}(\pi) \leq w_{10}\pi$, we have

$$0 \leq E\{V_0^{(\infty)}(\pi_{X_1}) \mid \pi\} \leq w_{10}E\{\pi_{X_1} \mid \pi\} = w_{10}\pi,$$

so that $V^*(0) = c$ and $V^*(\pi) \to c$ as $\pi \to 0$. By symmetry $V^*(1) = c$ and $V^*(\pi) \to c$ as $\pi \to 1$, thus completing the proof.

The next lemma says that the set Θ_∞^*, defined by (7.52) or equivalently by

$$\Theta_\infty^* = \{\pi : U_0(\pi) \leq V^*(\pi)\}, \tag{7.62}$$

is the complement of an open interval, perhaps empty. The function $U_0(\pi)$ is made up of two linear parts; on $0 \leq \pi \leq w_{01}/(w_{01} + w_{10})$, $U_0(\pi) = w_{10}\pi$, whereas on $w_{01}/(w_{01} + w_{10}) \leq \pi \leq 1$, $U_0(\pi) = w_{01}(1 - \pi)$. Because $V^*(\pi)$ is concave and $V^*(0) > U_0(0)$, there is at most one root of the equation $V^*(\pi) = w_{01}\pi$ in the interval $0 \leq \pi \leq w_{01}/(w_{01} + w_{10})$.

We use π_L (π lower) to denote this root if it exists (see Fig. 7.1). If this root does not exist, we define π_L as $w_{01}/(w_{01} + w_{10})$. Hence

$$\pi_L = lub \left\{\pi : V^*(\pi) \geq w_{10}\pi,\ 0 \leq \pi \leq \frac{w_{01}}{w_{01} + w_{10}}\right\}. \tag{7.63}$$

Similarly, π_U (π upper) is defined as the root, if it exists, of $V^*(\pi) =$

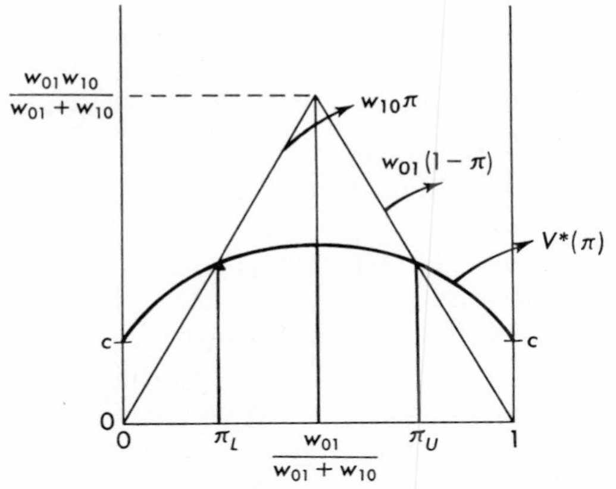

Fig. 7-1

$w_{01}(1 - \pi)$ in the interval $w_{01}/(w_{01} + w_{10}) \leq \pi \leq 1$; if this root does not exist, π_U is defined as $w_{01}/(w_{01} + w_{10})$.

$$\pi_U = \text{glb}\left\{\pi : V^*(\pi) \geq w_{01}(1 - \pi), \frac{w_{01}}{(w_{01} + w_{10})} \leq \pi \leq 1\right\}. \tag{7.64}$$

From these definitions we see that

$$0 < \pi_L \leq \frac{w_{01}}{w_{01} + w_{10}} \leq \pi_U < 1. \tag{7.65}$$

Furthermore, it is easy to see that $\pi_L = w_{01}/(w_{01} + w_{10})$ if, and only if, $\pi_U = w_{01}/(w_{01} + w_{10})$. In particular, we have Lemma 2:

Lemma 2

$$\Theta_\infty^* = [0, \pi_L] \cup [\pi_U, 1].$$

Now we may prove Theorem 1, which, complemented by Exercise 2, states that the Bayes rule studied in Section 7.5 is a SPRT(A, B).

Theorem 1. If $\pi_L \leq \pi \leq \pi_U$, then the Bayes rule $(\boldsymbol{\phi}^\pi, \boldsymbol{\delta}^\pi)$ of equations (7.42) and (7.53) is a SPRT(A, B), where

$$A = \frac{1-\pi}{\pi} \cdot \frac{\pi_L}{1 - \pi_L} \quad \text{and} \quad B = \frac{1 - \pi}{\pi} \cdot \frac{\pi_U}{1 - \pi_U}. \tag{7.66}$$

Proof. If $\pi_L \leq \pi \leq \pi_U$, then A and B defined by (7.66) satisfy $0 < A \leq 1 \leq B < \infty$. From Lemma 2 the stopping rule (7.53) may be written

$$\phi_j^\pi(x_1, \cdots, x_j) = \begin{cases} 0 & \text{if} \quad \pi_L < \pi_{x_1, \cdots, x_j} < \pi_U, \\ 1 & \text{otherwise;} \end{cases}$$

but because

$$\pi_{x_1, \cdots, x_j} = \frac{\pi \lambda_j(x_1, \cdots, x_j)}{\pi \lambda_j(x_1, \cdots, x_j) + (1 - \pi)},$$

$\pi_L < \pi_{x_1, \cdots, x_j} < \pi_U$ is equivalent to $A < \lambda_j(x_1, \cdots, x_j) < B$ with A

and B defined by (7.66). Thus ϕ^π is identical to the stopping rule (7.58). Similarly, the terminal decision rule (7.42) may be written

$$\delta_j^\pi(x_1, \cdots, x_j) = \begin{cases} 0 & \text{if} \quad \pi_{x_1,\cdots,x_j} < \dfrac{w_{01}}{w_{01} + w_{10}}, \\ \text{any} & \qquad\qquad = \\ 1 & \qquad\qquad > \end{cases}$$

But using inequalities (7.65), we see that this terminal decision rule is essentially the same as rule (7.59), for the only differences occur at points $x_1, \cdots, x_j$ where stopping never occurs. This completes the proof.

If $\pi < \pi_L$ or $\pi > \pi_U$, the Bayes rule (ϕ^π, δ^π) is still a SPRT(A, B), although (7.66) no longer holds (see Exercise 2). As a matter of fact, it is not difficult to show that every Bayes rule is an extended SPRT(A, B).

The risk function for this sequential hypothesis testing problem may be written

$$R(\theta_0, (\phi, \delta)) = w_{01}\alpha_0(\phi, \delta) + c\,E\{N \mid H_0, \phi\},$$

$$R(\theta_1, (\phi, \delta)) = w_{10}\alpha_1(\phi, \delta) + c\,E\{N \mid H_1, \phi\},$$

where

$$\alpha_0(\phi, \delta) = P\{\text{accept } H_1 \mid H_0, (\phi, \delta)\},$$

$$\alpha_1(\phi, \delta) = P\{\text{accept } H_0 \mid H_1, (\phi, \delta)\}.$$

The general (non-Bayesian) approach to testing H_0 against H_1 therefore involves comparing decision rules on the basis of these four numbers: the two error probabilities, $\alpha_0(\phi, \delta)$ and $\alpha_1(\phi, \delta)$, and the two expected sample sizes, $E\{N \mid H_0, \phi\}$ and $E\{N \mid H_1, \phi\}$. The Wald-Wolfowitz optimum property of the SPRT is stated in terms of these four numbers. This optimum property is merely a restatement of the Bayesian optimality of these tests, but in a surprisingly strong form, in which the SPRT simultaneously minimizes both expected sample sizes when the probabilities of error are fixed.

Theorem 2. Let (ϕ, δ) be the SPRT(A, B) and let (ϕ', δ') be any other decision rule for which

$$\alpha_0(\phi', \delta') \le \alpha_0(\phi, \delta) \quad \text{and} \quad \alpha_1(\phi', \delta') \le \alpha_1(\phi, \delta).$$

Then

$$E\{N \mid H_0, \phi\} \leq E\{N \mid H_0, \phi'\}$$

and

$$E\{N \mid H_1, \phi\} \leq E\{N \mid H_1, \phi'\}.$$

The proof of this theorem may be based on the following converse of Theorem 1. Given A, B, and π, with $0 < A \leq 1 \leq B < \infty$ and $0 < \pi < 1$, there exists a sequential decision problem (that is, a choice of $c > 0$, $w_{01} > 0$ and $w_{10} > 0$) such that the SPRT(A, B) is Bayes with respect to π (that is, such that equations (7.66) hold). Proofs may be found in Wald and Wolfowitz (1948), in Lehmann (1959) (due to Le Cam), in Burkholder and Wijsman (1963), and in Matthes (1963). Here we essentially follow Burkholder and Wijsman, who noticed that a weaker result (Lemma 4) is somewhat easier to prove and suffices for the proof of Theorem 2. This weaker result states that the converse to Theorem 1 is valid at least for certain values of π arbitrarily close to zero or one.

It is necessary to extend the notation π_L and π_U to show dependence on c, w_{01}, and w_{10}. However, because multiplying c, w_{01}, w_{10} by the same positive constant merely multiplies U_0 and V^* by that constant, we arbitrarily put $w_{01} = w$ and $w_{10} = 1 - w$, where $0 < w < 1$. This amounts merely to choosing the scale so that $w_{01} + w_{10} = 1$. Thus we use $\pi_L(c, w)$ and $\pi_U(c, w)$ to denote the lower and upper boundaries (7.63) and (7.64) when c is the cost, $w_{01} = w$, and $w_{10} = 1 - w$. Similarly, $V^*(\pi; c, w)$ replaces $V^*(\pi)$, given c is the cost, $w_{01} = w$, and $w_{10} = 1 - w$.

Lemma 3. For fixed w, $\pi_L(c, w)$ and $\pi_U(c, w)$ are continuous functions of c and $\pi_L(c, w) \to 0$ and $\pi_U(c, w) \to 1$ as $c \to 0$.

Proof. Let $c_0(w)$ be the smallest value of c such that $\pi_L(c, w) = w = \pi_U(c, w)$. (See Exercise 7.5.3.) Then for $c \geq c_0(w)$, $\pi_L(c, w) = w$. For $0 < c \leq c_0(w)$, $\pi_L(c, w)$ is defined by the properties

$$\begin{aligned} (1 - w)\pi &< V^*(\pi; c, w) \quad \text{if} \quad \pi < \pi_L(c, w), \\ &= \qquad\qquad\qquad \text{if} \quad = \\ &> \qquad\qquad\qquad \text{if} \quad > \end{aligned} \tag{7.67}$$

(a) For fixed π and w, $V^*(\pi; c, w)$ is continuous and nondecreasing in $c > 0$. This follows because $V^*(\pi; c, w)$, being the infimum (over the class of rules which take at least one observation) of functions $r(\pi, (\boldsymbol{\phi}, \boldsymbol{\delta}))$ which are linear and nondecreasing in c, is concave (as in the proof of Lemma 7.5.3) hence continuous (see Exercise 2.8.2) and nondecreasing.

(b) For fixed w, $\pi_L(c, w)$ is nondecreasing. If $c' \geq c$, then $(1 - w)\pi(c', w) = V^*(\pi(c', w); c', w) \geq V^*(\pi(c', w); c, w)$, which implies that $\pi(c', w) \geq \pi(c, w)$, using (7.67).

(c) For fixed w, $\pi_L(c, w)$ is continuous. To show continuity from the right suppose that c_n are decreasing and $c_n \to c$ as $n \to \infty$. Suppose, contrary to continuity, that $\pi_L(c_n, w) \to \pi' > \pi_L(c, w)$. Then $V^*(\pi'; c, w) < (1 - w)\pi' \leq V^*(\pi'; c_n, w) \to V^*(\pi'; c, w)$, providing a contradiction. Similarly, $\pi_L(c, w)$ is continuous from the left.

(d) For fixed w, $\pi_L(c, w) \to 0$ as $c \to 0$. Because there exists a test of fixed sample size n for which the error probabilities are arbitrarily small if n is sufficiently large, we have $V^*(\pi; c, w) \to 0$ as $c \to 0$. Hence $\pi_L(c, w) \to 0$ as $c \to 0$.

(e) Finally, for fixed w, $\pi_U(c, w)$ is continuous in c and $\pi_U(c, w) \to 1$ as $c \to 0$, by symmetry, thus completing the proof.

Lemma 4. For given $\epsilon > 0$, and $0 < A \leq 1 \leq B < \infty$, there exist (a) π, c and w, with $0 < \pi < \epsilon$ satisfying (7.66) and (b) π', c', and w', with $1 - \epsilon < \pi' < 1$ satisfying (7.66).

Proof. We prove only part (a); part (b) follows by symmetry. Fix w so that $w < A\epsilon$ and consider the function

$$\frac{\pi_L(c, w)}{1 - \pi_L(c, w)} \cdot \frac{1 - \pi_U(c, w)}{\pi_U(c, w)}.$$

This is a continuous function of c, which for sufficiently large values of c takes the value 1 (since $\pi_L = \pi_U$) and tends to zero as c tends to zero. Hence there exists a value of c such that

$$\frac{\pi_L(c, w)}{1 - \pi_L(c, w)} \cdot \frac{1 - \pi_U(c, w)}{\pi_U(c, w)} = \frac{A}{B}.$$

If we choose $\pi = \pi_L(c, w)/(A + (1 - A)\pi_L(c, w))$, it follows that

$$\frac{1 - \pi}{\pi} = A \frac{1 - \pi_L(c, w)}{\pi_L(c, w)} = B \frac{1 - \pi_U(c, w)}{\pi_U(c, w)},$$

exactly (7.66). Furthermore, using (7.65) in the form $\pi_L \le w$, we see that

$$\pi = \frac{\pi_L(c, w)}{A + (1 - A)\,\pi_L(c, w)} \le \frac{\pi_L(c, w)}{A} \le \frac{w}{A} < \epsilon,$$

thus completing the proof.

Proof of Theorem 2. From Lemma 4(a) find c, $w_{01} = w$, and $w_{10} = 1 - w$ so that (7.66) is satisfied and π is less than a preassigned small number $\epsilon > 0$. Then, from Theorem 1, the SPRT(A, B), namely, $(\boldsymbol{\phi}, \boldsymbol{\delta})$, is Bayes with respect to π. Hence

$$\begin{aligned}
0 &\le r\,(\pi, (\boldsymbol{\phi}', \boldsymbol{\delta}')) - r(\pi, (\boldsymbol{\phi}, \boldsymbol{\delta}))\\
&= \pi w(\alpha_1(\boldsymbol{\phi}', \boldsymbol{\delta}') - \alpha_1(\boldsymbol{\phi}, \boldsymbol{\delta})) + (1 - \pi)(1 - w)(\alpha_0(\boldsymbol{\phi}', \boldsymbol{\delta}')\\
&\qquad\qquad - \alpha_0(\boldsymbol{\phi}, \boldsymbol{\delta}))\\
&\quad + \pi c(E\{N \mid H_0, \boldsymbol{\phi}'\} - E\{N \mid H_0, \boldsymbol{\phi}\}) + (1 - \pi)\, c(E\{N \mid H_1\boldsymbol{\phi}'\}\\
&\qquad\qquad - E\{N \mid H_1, \boldsymbol{\phi}\})\\
&\le \pi c(E\{N \mid H_0, \boldsymbol{\phi}'\} - E\{N \mid H_0, \boldsymbol{\phi}\}) + (1 - \pi)\, c(E\{N \mid H_1, \boldsymbol{\phi}'\}\\
&\qquad\qquad - E\{N \mid H_1, \boldsymbol{\phi}\}).
\end{aligned}$$

Because this is valid for certain values of π arbitrarily close to zero, this implies $0 \le E\{N \mid H_1, \boldsymbol{\phi}'\} - E\{N \mid H_1, \boldsymbol{\phi}\}$. Using Lemma 4(b) or invoking symmetry, we must also have $0 \le E\{N \mid H_0, \boldsymbol{\phi}'\} - E\{N \mid H_0, \boldsymbol{\phi}\}$ to complete the proof.

Exercises

1. Let $g(\pi)$ be a concave function of $\pi \in [0, 1]$. Let $\pi(x) = \pi f_1(x)/(\pi f_1(x) + (1 - \pi) f_0(x))$, where $f_0(x)$ and $f_1(x)$ are densities (probability mass functions). Let $h(\pi) = E\{g(\pi(X)) \mid \pi\}$, where $E\{\ \mid \pi\}$ represents expectation with respect to the distribution with density (mass function) $\pi f_1(x) + (1 - \pi) f_0(x)$. Show that $h(\pi)$ is concave in $\pi \in [0, 1]$.
2. (a) If $\pi \le \pi_L$, then the Bayes rule $(\boldsymbol{\phi}^\pi, \boldsymbol{\delta}^\pi)$ of (7.42) and (7.53) is a SPRT(A, B) with $A = 1$.

(b) If $\pi \geq \pi_U$, the Bayes rule $(\boldsymbol{\phi}^\pi, \boldsymbol{\delta}^\pi)$ of (7.42) and (7.53) is a SPRT(A, B) with $B = 1$.

3. *An alternative optimum property of the* SPRT(A, B). Let $(\boldsymbol{\phi}, \boldsymbol{\delta})$ be the SPRT(A, B) and let $(\boldsymbol{\phi}', \boldsymbol{\delta}')$ be any other decision rule for which

$$E\{N \mid H_0, \boldsymbol{\phi}'\} \leq E\{N \mid H_0, \boldsymbol{\phi}\}$$

and

$$E\{N \mid H_1, \boldsymbol{\phi}'\} \leq E\{N \mid H_1, \boldsymbol{\phi}\}.$$

Then

$$\alpha_0(\boldsymbol{\phi}, \boldsymbol{\delta}) \leq \alpha_0(\boldsymbol{\phi}', \boldsymbol{\delta}') \quad \text{and} \quad \alpha_1(\boldsymbol{\phi}, \boldsymbol{\delta}) \leq \alpha_1(\boldsymbol{\phi}', \boldsymbol{\delta}').$$

4. Show that

$$\pi_L(w, c) \geq \min\,(w, c/(1 - w))$$

and

$$\pi_U(c, w) \leq \max\,(w, 1 - (c/w)).$$

5. (a) Show that the SPRT(A, B) for testing $f_0(x) = \frac{1}{2}I_{(0,2)}(x)$ against $f_1(x) = \frac{1}{2}I_{(1,3)}(x)$ is either $(\boldsymbol{\phi}', \boldsymbol{\delta}')$: continue sampling as long as $1 \leq x_j \leq 2$; as soon as $x_j < 1$, stop and accept H_0; as soon as $x_j > 2$, stop and accept H_1, or $(\boldsymbol{\phi}'', \boldsymbol{\delta}'')$: without sampling, accept H_1 with probability δ_0'' and accept H_0 with probability $1 - \delta_0''$.
(b) Show $\alpha_0(\boldsymbol{\phi}', \boldsymbol{\delta}') = \alpha_1(\boldsymbol{\phi}', \boldsymbol{\delta}') = 0$ and $E\{N \mid H_0, \boldsymbol{\phi}'\} = E\{N \mid H_1, \boldsymbol{\phi}'\} = 2$. Show $\alpha_0(\boldsymbol{\phi}'', \boldsymbol{\delta}'') = w_{01}\delta_0''$, $\alpha_1(\boldsymbol{\phi}'', \boldsymbol{\delta}'') = w_{10}(1 - \delta_0'')$, and $E\{N \mid H_0, \boldsymbol{\phi}''\} = E\{N \mid H_1, \boldsymbol{\phi}''\} = 0$.
(c) Using (b), show that $\pi_L(c, w) = 2c/(1 - w)$ and $\pi_U(c, w) = 1 - (2c/w)$, provided $c \leq w(1 - w)/2$, and that $\pi_L = \pi_U = w$ otherwise.

6. Show that the SPRT(A, B) for testing $f_0(x) = \frac{1}{2}I_{(0,2)}(x)$ against $f_1(x) = I_{(0,1)}(x)$ with $0 < A < 1 < B$ has the following form for some integer $J \geq 1$: $\phi_0 = 0$, $\phi_J \equiv 1$, and for $j = 1, \cdots, J - 1$,

$$\phi_j(x_1, \cdots, x_j) = \begin{cases} 0 & \text{if } x_j < 1, \\ 1 & \text{if } x_j \geq 1. \end{cases}$$

$$\delta_j(x_1, \cdots, x_j) = \begin{cases} 0 & \text{if } x_j \geq 1, \\ 1 & \text{if } x_j < 1. \end{cases}$$

(b) For the test of (a), show that $\alpha_0 = 2^{-J}$, $\alpha_1 = 0$, $E\{N \mid H_0, \boldsymbol{\phi}\} = 2(1 - 2^{-J})$ and $E\{N \mid H_1, \boldsymbol{\phi}\} = J$.

(c) Let $w_{01} = 1$, $w_{10} = 2$, $c = 1/15$; and $\pi = 1/3$. Show that the test of part (a) with $J = 4$ is Bayes with respect to π. (Why is it not Bayes with respect to π if w_{10} is changed to $1/6$?)

7.7 The Fundamental Identity of Sequential Analysis

The problem of computing the two probabilities of error and the two expected sample sizes of a given SPRT(A, B) is numerically tedious of solution. Fortunately, simple approximations for these quantities exist which are often sufficiently accurate for statistical purposes. These approximations, due to Wald, also provide an approximate solution to the inverse problem of finding numbers A and B for which the SPRT(A, B) has given probabilities of error. In this section we describe Wald's approximations and give indications concerning their accuracy.

The Wald approximations to the two probabilities of error are easy to derive. Consider a given SPRT(A, B) of (7.58) and (7.59), with $A < 1 < B$, and let $\alpha_0 = P\{\text{accept } H_1 \mid H_0\}$ and $\alpha_1 = P\{\text{accept } H_0 \mid H_1\}$ denote the two error probabilities. Then

$$\alpha_1 = P\{\lambda_N \le A \mid H_1\} = \sum_{n=1}^{\infty} \int_{Q_n} \cdots \int \prod_1^n f_1(x_i)\, dx_i, \tag{7.68}$$

where Q_n is the set $\{N = n, \lambda_N \le A\}$. For $(x_1, \cdots, x_n) \in Q_n$, $\lambda_n(x_1, \cdots, x_n) \le A$ or equivalently $\prod_1^n f_1(x_i) \le A \prod_1^n f_0(x_i)$ with "approximate" equality, for $\prod_1^{n-1} f_1(x_i) > A \prod_1^{n-1} f_0(x_i)$ in Q_n. Hence

$$\alpha_1 \le \sum_{n=1}^{\infty} \int_{Q_n} \cdots \int A \prod_1^n f_0(x_i)\, dx_i$$

$$= AP\{\lambda_N \le A \mid H_0\} = A(1 - \alpha_0), \tag{7.69}$$

with "approximate" equality. A similar argument gives

$$\alpha_0 = P\{\lambda_N \ge B \mid H_0\} \le \frac{1}{B} P\{\lambda_N \ge B \mid H_1\} = \frac{1 - \alpha_1}{B}. \tag{7.70}$$

Thus we have the inequalities

$$A \ge \frac{\alpha_1}{1 - \alpha_0}, \qquad B \le \frac{1 - \alpha_1}{\alpha_0}. \tag{7.71}$$

The Wald approximations consist in replacing the inequalities in (7.71) with equalities. This immediately gives an approximate solution to the inverse problem of finding A and B to achieve specified error probabilities. Approximate error probabilities of a given SPRT(A, B) may be found by solving (7.71) (with inequalities replaced by approximate equalities)

$$\alpha_0 \doteq \frac{1 - A}{B - A}, \qquad \alpha_1 \doteq \frac{1 - B^{-1}}{A^{-1} - B^{-1}} = \frac{A(B - 1)}{B - A}. \tag{7.72}$$

It is important to notice one striking feature of these formulas; namely, these approximations to the error probabilities of a given SPRT(A, B) are independent of the distributions under H_0 and H_1. Up to this approximation the distribution problems involved in sequential probability ratio tests are much simpler than those involved in the non-sequential counterparts.

The general problem may be considered as a topic in the theory of random walks, as may be seen by reformulating the stopping rule (7.58) in terms of the random variables Z_i defined for $i = 1, 2, \cdots$ by

$$Z_i = \log \frac{f_1(X_i)}{f_0(X_i)}. \tag{7.73}$$

It is assumed as part of definition (7.73) that $Z_i = +\infty$ if $f_1(X_i) > 0$ and $f_0(X_i) = 0$ and that $Z_i = -\infty$ if $f_1(X_i) = 0$ and $f_0(X_i) > 0$; thus, under H_0 or H_1, $Z_1, Z_2, \cdots$ is a sequence of independent identically distributed random variables, possibly taking infinite values with positive probability. If we let

$$S_j = \sum_{i=1}^{j} Z_i, \tag{7.74}$$

then the likelihood ratio (7.57) takes the form $\lambda_j = \exp\{S_j\}$ (where S_0 is defined as zero). Hence the stopping rule (7.58) may be written

$$\phi_j(S_j) = \begin{cases} 0 & \text{if} \quad a < S_j < b, \\ 1 & \text{otherwise,} \end{cases} \tag{7.75}$$

where $a = \log A \leq 0 \leq \log B = b$. No difficulties arise in the degenerate cases $a = 0$ or $b = 0$, and we assume that $a < 0 < b$. Stated in this form, this problem is seen to be equivalent to the gambler's ruin problem which runs as follows. Two gamblers, I and II, with respective resources $-a > 0$ and $b > 0$, play a sequence of games. The ith game results in the payment of the random amount Z_i by gambler II to gambler I, the quantities $Z_1, Z_2, \cdots$ being independent and identically distributed. (If Z_i is negative, II receives $-Z_i$ from I.) Playing continues until one of the players is required to pay all he has or more. The problem is to find the probability that a given player will be ruined and the expected number of games to be played.

We approach this problem assuming an arbitrary (possibly infinite valued) distribution as the common distribution of $Z_1, Z_2, \cdots$. This leads to finding the probability of rejecting H_0 and the expected sample size of a given SPRT under an arbitrary distribution (not just H_0 and H_1). We first give a theorem which implies that the SPRT will terminate with probability one under H_0 or H_1, provided H_0 and H_1 are distinguishable.

Let $Z_1, Z_2, \cdots$ be independent and identically distributed random variables, let $S_n = \sum_1^n Z_i$, and define N as the first n for which it is not true that $a < S_n < b$ where $a < 0 < b$. If $a < S_n < b$ for all n, then N is defined as $+\infty$.

Theorem 1. If $P\{Z_i = 0\} \neq 1$, then $P\{N < \infty\} = 1$; in fact, there exist numbers r and c, $0 < r < 1$, $c > 0$, such that $P\{N \geq n\} \leq cr^n$.

Proof. If $P\{Z_i = 0\} \neq 1$, then there exists an $\epsilon > 0$ such that either $P\{Z_i > \epsilon\} > 0$ or $P\{Z_i < -\epsilon\} > 0$. We assume $P\{Z_i > \epsilon\} = \delta > 0$ because the case $P\{Z_i < -\epsilon\} > 0$ may be treated by symmetry. Let m be any integer for which $m\epsilon > (b - a)$. Then

$$P\{S_{k+m} - S_k > (b - a) \mid S_1, \cdots, S_k\}$$
$$= P\{\sum_{k+1}^{k+m} Z_i > (b - a)\} \geq \delta^m \qquad \text{for all } k.$$

Hence for all k

$$P\{N \geq mk + 1\} = P\{a < S_i < b \quad \text{for} \quad i = 1, \cdots, mk\} \leq (1 - \delta^m)^k.$$

Let $c = (1 - \delta^m)^{-1}$ and $r = (1 - \delta^m)^{1/m}$. For a given n find that k for which $km < n \leq (k + 1)m$. Then

$$P\{N \geq n\} \leq P\{N \geq km + 1\} \leq (1 - \delta^m)^k = cr^{(k+1)m} \leq cr^n,$$

thus completing the proof.

This theorem implies that when $P\{Z_i = 0\} \neq 1$ all moments of the distribution of N are finite. In fact, the probability generating function $E\rho^N$ is finite for $0 \leq \rho < 1/r$, where r satisfies the conclusion of Theorem 1 (see Exercise 1).

Next, we prove the fundamental identity of sequential analysis due to Wald (1947). It is assumed for this identity that the independent and identically distributed random variables $Z_1, Z_2, \cdots$ are finite-valued. For subsequent applications to the sequential probability ratio test, this entails no real loss of generality, for if each $Z_i > (b - a)$ is replaced by $Z_i = (b - a)$ and each $Z_i < -(b - a)$ is replaced by $Z_i = -(b - a)$ neither the stopping time N nor the boundary crossed (upper or lower) is changed. In other words, the distribution of N and the probability of exiting via the upper (lower) boundary are not changed if the distribution of the Z_i is truncated at $\pm(b - a)$. (It may even be assumed that the common distribution of the Z_i is bounded in absolute value by $b - a$, but the fundamental identity is not affected if this assumption is omitted.)

Therefore let $Z_1, Z_2, \cdots$ be independent, identically distributed, finite-valued, random variables and let S_n and N be defined as before. Let $M(t)$ denote the moment generating function of the common distribution of the Z_i.

$$M(t) = E \exp(tZ_i). \tag{7.76}$$

The set of real t for which $M(t)$ is finite is an interval (possibly degenerate) containing the origin.

Theorem 2. (*Fundamental Identity of Sequential Analysis*). If

$$P\{Z_i = 0\} \neq 1 \quad \text{and} \quad P\{|Z_i| < \infty\} = 1,$$

then

$$E\{\exp(tS_N)\, M(t)^{-N}\} \equiv 1 \tag{7.77}$$

for all real t for which $M(t)$ is finite.

Proof. [Following Bahadur (1958)]. If $p(z)$ denotes the density (mass function) of Z_i, we define a one-parameter exponential family of distributions having the density

$$p(z \mid t) = \exp(tz)\, M(t)^{-1} p(z) \tag{7.78}$$

for those t for which $M(t)$ is finite. Because $M(0) = 1$, $p(z \mid 0) = p(z)$. If Z is not degenerate at zero when $t = 0$, it is not degenerate at zero for any t. Hence for all t for which $M(t)$ is finite

$$P_t\{N < \infty\} \equiv 1 \tag{7.79}$$

from Lemma 1. This is equivalent to (7.77), since

$$E\{\exp(tS_N)\, M(t)^{-N}\} = \sum_{j=0}^{\infty} E\{\psi_j(Z_1, \cdots, Z_j) \exp(tS_j)\, M(t)^{-j}\}$$

$$= \sum_{j=0}^{\infty} E_t\{\psi_j(Z_1, \cdots, Z_j)\} = P_t\{N < \infty\}, \tag{7.80}$$

where ψ is derived from the stopping rule ϕ of (7.75), thus completing the proof.

It should be noted that the identity (7.80) is valid for a general stopping rule ϕ and that therefore the fundamental identity is valid for all stopping rules ϕ for which (7.79) is satisfied.

It may be shown (see Wald (1947) or Bahadur (1958)) that derivatives of all orders may be passed beneath the expectation sign in the fundamental identity at all points interior to the set on which $M(t)$ exists. Assuming such a result, we prove Theorem 3 as an application of the fundamental identity. An alternate proof of this theorem, valid under less restrictive conditions, is sketched in Exercise 2.

Theorem 3. Assume that $P\{Z_i = 0\} \neq 1$, $P\{|Z_i| < \infty\} = 1$, and that $M(t)$ exists in a neighborhood of the origin. Then

(a) $ES_N = \mu EN$ and

(b) $E(S_N - N\mu)^2 = \sigma^2 EN$,

where $\mu = EZ_i$ and $\sigma^2 = \text{Var } Z_i$.

Proof. One derivative beneath the expectation sign in (7.77) yields

$$E\left\{\exp[tS_N - N\log M(t)]\left(S_n - N\frac{d}{dt}\log M(t)\right)\right\} = 0,$$

which evaluated at $t = 0$ is exactly (a), for $d/dt \log M(t)\,|_{t=0} = \mu$. A second derivative beneath the expectation sign yields

$$E\left\{\exp\,[tS_N - N\log M(t)]\left((S_N - N\frac{d}{dt}\log M(t))^2 - N\frac{d^2}{dt^2}\log M(t)\right)\right\} = 0,$$

which evaluated at $t = 0$ gives (b), for $d^2/dt^2 \log M(t)\,|_{t=0} = \sigma^2$.

As a first application of these theorems we derive the Wald approximations of $P\{S_N \geq b\}$ and EN under an arbitrary distribution of the independent identically distributed sequence $Z_1, Z_2, \cdots$. The idea involved consists of two parts.

First, we find a nonzero real number t_0 for which $M(t_0) = 1$, if possible. For such a number the fundamental identity implies

$$E\exp\,(t_0 S_N) = 1. \tag{7.81}$$

Second, the distribution of S_N is approximated by the distribution giving mass $P\{S_N \leq a\}$ to a and mass $P\{S_N \geq b\} = 1 - P\{S_N \leq a\}$ to b. This is essentially the same approximation involved in deriving the approximations (7.72); we pretend that whenever S_n crosses the lower (or upper) boundary it does so by hitting a (or b) exactly. Such an approximation turns (7.81) into

$$\exp\,(t_0 a)P\{S_N \leq a\} + \exp\,(t_0 b)P\{S_N \geq b\} \doteq 1, \tag{7.82}$$

from which we may solve for $P\{S_N \geq b\}$,

$$P\{S_N \geq b\} \doteq \frac{1 - \exp\,(t_0 a)}{\exp\,(t_0 b) - \exp\,(t_0 a)}\,. \tag{7.83}$$

Because a moment generating function is strictly convex in the domain of its existence, ($M''(t) = Ee^{tZ}Z^2 > 0$; it is assumed that Z is not degenerate at zero) and because $M(0) = 1$ it is easy to see that there is at most one nonzero real t_0 for which $M(t_0) = 1$. If $\mu = EZ = 0$, then, because $\mu = M'(0)$ represents the slope of $M(t)$ at the origin, there is no nonzero real t_0 for which $M(t_0) = 1$. Barring this case, that is to say, if $\mu \neq 0$, then in general we can find a nonzero real t_0 for which $M(t_0) = 1$ (Exercise 3). "In general" here means provided $M(t)$ exists for all t (which may be taken to be the case by truncating Z at $\pm(b - a)$) and provided $P(Z > 0) > 0$ and $P(Z < 0) > 0$ (which, if violated, would lead to a trivial $P\{S_N \geq b\}$). Hence it is reasonable to assume the existence of a nonzero real t_0 for which $M(t_0) = 1$, provided $\mu \neq 0$.

The case $\mu = 0$ can be treated in a similar manner by using Theorem 3(a), which states that $ES_N = 0$. The same approximation that yielded (7.82) from (7.81) gives $aP\{S_N \leq a\} + bP\{S_N \geq b\} \doteq 0$. Thus, corresponding to (7.83), we have for $\mu = 0$

$$P\{S_N \geq b\} \doteq \frac{-a}{b - a} \tag{7.84}$$

This may also be obtained from (7.83) by letting $t_0 \to 0$, which is the case if $\mu \to 0$.

The approximations for EN now follow easily from (7.83) and (7.84), using Theorem 3. If $\mu \neq 0$, Theorem 3(a) implies that

$$EN = \frac{1}{\mu} ES_N \doteq \frac{1}{\mu} (aP\{S_N \leq a\} + bP\{S_N \geq b\})$$

$$\doteq \frac{1}{\mu} \frac{a[\exp(t_0 b) - 1] + b[1 - \exp(t_0 a)]}{\exp(t_0 b) - \exp(t_0 a)}. \tag{7.85}$$

If $\mu = 0$, Theorem 3(b) shows that

$$EN = \frac{1}{\sigma^2} ES_N^2 \doteq \frac{1}{\sigma^2} (a^2 P\{S_N \leq a\} + b^2 P\{S_N \geq b\})$$

$$\doteq \frac{1}{\sigma^2} \frac{a^2 b - b^2 a}{b - a} = \frac{-ab}{\sigma^2}. \tag{7.86}$$

These results are easy to apply to obtain approximations to the probabilities of error and expected sample sizes of a given SPRT(A, B) when the probability that Z_i assumes infinite values is zero under H_0 and H_1, because the t_0 satisfying $M(t_0) = 1$ is easily seen to be $+1$ under H_0 and -1 under H_1 (Exercise 5). In this case the approximations (7.83) reduce to (7.72). In fact, when the distribution of the Z_i is restricted to a small interval about the origin, useful upper and lower bounds of the probabilities of error may be obtained that generalize the bounds (7.71) (Exercise 6).

In the situation in which Z_i takes just three values, $-z$, 0, and $+z$, these approximations (7.83) – (7.86) are exact when a and b are taken as integer multiples of z, for then there is no overlap over the boundaries.

This situation arises in sequential probability ratio tests of symmetric binomial hypotheses treated in Example 1. (See also Exercise 7.)

EXAMPLE 1. Let $X_1, X_2, \cdots$ be independent Bernoulli trials with probability p of success, $0 < p < 1$, and consider testing the hypothesis $H_0{:}p = p_0$ against the alternative $H_1{:}p = 1 - p_0$, where $0 < p_0 < 1/2$. Then $f_0(x) = p_0^x(1 - p_0)^{1-x}$ for $x = 0, 1$ and $f_1(x) = (1 - p_0)^x p_0^{1-x}$ for $x = 0, 1$, so that

$$Z_i = \log \frac{f_1(X_i)}{f_0(X_i)} = (2X_i - 1) \log \frac{(1 - p_0)}{p_0}$$

Hence, if p is the true value of the parameter, then Z_i takes the values $+z$ and $-z$ with respective probabilities p and $1 - p$, where

$$z = \log\left(\frac{1 - p_0}{p_0}\right) > 0.$$

In using the SPRT(A, B), we may as well define $a = \log A$ and $b = \log B$ as integer multiples of z, say,

$$a = -jz \quad \text{and} \quad b = kz,$$

where j and k are positive integers. In this case the Wald approximations (7.72) and (7.83) through (7.86) are exact. Because

$$M(t) = Ee^{tZ} = pe^{tz} + (1 - p)e^{-tz},$$

the nonzero root t_0 of equation $M(t) = 1$ is easily found, provided $\mu = EZ = pz - (1 - p)z \neq 0$, that is, provided $p \neq \frac{1}{2}$:

$$t_0 = z^{-1} \log\left(\frac{1 - p}{p}\right).$$

Because $\{S_N \geq b\}$ is the event $\{\text{accept } H_1\}$, (7.83) and (7.84) now give the exact power function

$$P\{\text{accept } H_1 \mid p\} = \begin{cases} \dfrac{1 - [p/(1 - p)]^j}{[(1 - p)/p]^k - [p/(1 - p)]^j} & \text{if } p \neq \frac{1}{2}, \\[2ex] \dfrac{j}{k + j} & \text{if } p = \frac{1}{2}. \end{cases}$$

The expected sample size in (7.85) and (7.86) (no longer approximations) becomes

$$E\{N \mid p\} = \begin{cases} \dfrac{1}{2p-1} \dfrac{k\{1 - [p/(1-p)]^j\} - j\{[(1-p)/p]^k - 1\}}{[(1-p)/p]^k - [p/(1-p)]^j} & \text{if} \quad p \neq \frac{1}{2}, \\ jk & \text{if} \quad p = \frac{1}{2}. \end{cases}$$

These are the formulas for the probability of ruin and expected duration of the classical gambler's ruin problem found, for example, in Feller (1957), Chapter 14.

When the random variables Z_i can assume only a finite number of integral multiples of a constant, exact formulas may be derived for the probability of accepting H_1 and for the expected sample size. See Wald (1947), Appendix A4, for details. Another family of distributions of the Z_i for which $P\{S_N \geq b\}$ and EN are capable of being expressed by exact formulas has been described by J. H. B. Kemperman (1961). A very special case of this family is treated in the following example, in which the exact values of $P\{S_N \geq b\}$ and EN are compared with the Wald approximations.

EXAMPLE 2. Let the common distribution of $X_1, X_2, \cdots$ have a density of the form

$$f(x \mid \theta) = \frac{1 - \theta^2}{2} \exp\left(-\mid x \mid + \theta x\right), \tag{7.87}$$

where $\mid \theta \mid < 1$. Consider the SPRT(A, B) of the hypothesis $H_0:\theta = -1/2$ against the alternative $H_1:\theta = +1/2$. From (7.73) we see that $Z_i = \log\left(f(X_i \mid 1/2)/f(X_i \mid -1/2)\right) = X_i$. The moment generating function of the Z_i under an arbitrary distribution in the form of (7.87),

$$M_\theta(t) = \frac{1-\theta^2}{2} \int_{-\infty}^{\infty} \exp\left[-\mid z \mid + (\theta + t)z\right] dz = \frac{1 - \theta^2}{1 - (\theta + t)^2},$$

exists for all t in the interval $-1 - \theta < t < 1 - \theta$. Following Kemperman, we make the computation

$$\begin{aligned}
E_\theta\{&\exp(tS_n)\textstyle\prod_1^{n-1}(1-\varphi_j(Z_1,\cdots,Z_j))I_{(b,\infty)}(S_n)\}\\
&= E_\theta\{\exp(tS_{n-1})\textstyle\prod_1^{n-1}(1-\varphi_j(Z_1,\cdots,Z_j))\\
&\qquad\qquad \times E_\theta\{\exp(tZ_n)\, I_{(b-S_{n-1},\infty)}(Z_n)\mid Z_1,\cdots,Z_{n-1}\}\}\\
&= E_\theta\{\exp(tS_{n-1})\textstyle\prod_1^{n-1}(1-\varphi_j(Z_1,\cdots,Z_j))\\
&\qquad\qquad \times \displaystyle\int_{b-S_{n-1}}^{\infty}\frac{(1-\theta^2)}{2}\exp(-z+\theta z+tz)\,dz\}\\
&= E_\theta\{\exp(tS_{n-1})\textstyle\prod_1^{n-1}(1-\varphi_j(Z_1,\cdots,Z_j))\displaystyle\frac{(1-\theta^2)}{2(1-\theta-t)}\\
&\qquad\qquad \times \exp[-(1-\theta-t)(b-S_{n-1})]\}\\
&= \frac{(1-\theta)e^{tb}}{(1-\theta-t)}B_n(\theta),
\end{aligned}\tag{7.88}$$

where

$$\begin{aligned}
B_n(\theta) = \frac{(1+\theta)}{2}&\exp[-b(1-\theta)]\\
&\times E_\theta\{\exp[(1-\theta)S_{n-1}]\textstyle\prod_1^{n-1}(1-\varphi_j(Z_1,\cdots,Z_j))\}.
\end{aligned}$$

Putting $t=0$ in (7.88) shows that $B_n(\theta) = P_\theta\{N=n,\ S_n\geq b\}$. Similarly,

$$\begin{aligned}
E_\theta\{\exp(tS_N)\textstyle\prod_1^{n-1}(1-\varphi_j(Z_1,\cdots,Z_j))I_{(-\infty,a)}(S_n)\}&\\
= \frac{(1+\theta)e^{ta}}{(1+\theta+t)}A_n(\theta)&,
\end{aligned}$$

where $A_n(\theta) = P_\theta\{N=n,\ S_N\leq a\}$. Hence, from the fundamental identity,

$$\begin{aligned}
1 &\equiv E_\theta\{\exp(tS_N)\,M_\theta(t)^{-N}\}\\
&= \sum_{n=1}^{\infty}E_\theta\{\exp(tS_n)\textstyle\prod_1^{n-1}(1-\varphi_j(Z_1,\cdots,Z_j))[I_{(b,\infty)}(S_n)\\
&\qquad\qquad + I_{(-\infty,a)}(S_n)]\}\,M_\theta(t)^{-n}\\
&= \frac{(1-\theta)e^{tb}}{(1-\theta-t)}\sum_{n=1}^{\infty}B_n(\theta)\,M_\theta(t)^{-n} + \frac{(1+\theta)e^{ta}}{(1+\theta+t)}\sum_{n=1}^{\infty}A_n(\theta)\,M_\theta(t)^{-n}.
\end{aligned}\tag{7.89}$$

Because

$$P_\theta\{S_N \geq b\} = \sum_{n=1}^{\infty} B_n(\theta) = 1 - \sum_{n=1}^{\infty} A_n(\theta),$$

we can solve (7.89) for $P_\theta\{S_N \geq b\}$ after replacing t with that nonzero real value, t_θ, for which $M_\theta(t_\theta) = 1$; namely, $t_\theta = -2\theta$. This calculation yields

$$P_\theta\{S_N \geq b\} = \begin{cases} \dfrac{(1-\theta)^2 - (1+\theta)^2 e^{-2\theta a}}{(1-\theta)^2 e^{-2\theta b} - (1+\theta)^2 e^{-2\theta a}} & \text{for} \quad \theta \neq 0, \\[2ex] \dfrac{1-a}{2+b-a} & \text{for} \quad \theta = 0, \end{cases} \tag{7.90}$$

where $P_\theta\{S_N \geq b\}$ may be determined at $\theta = 0$ by taking the limit as $\theta \to 0$. $P_\theta\{S_N \geq b\}$ at $\theta = 0$ may also be found by differentiating the identity (7.89) with respect to t and putting $t = 0$. This computation gives the expected sample size $E_\theta N$ for $\theta \neq 0$ as follows:

$$0 = \left(\frac{1}{1-\theta} + b\right) \sum_{n=1}^{\infty} B_n(\theta) - M_\theta'(0) \sum_{n=1}^{\infty} n B_n(\theta)$$
$$- \left(\frac{1}{1+\theta} - a\right) \sum_{n=1}^{\infty} A_n(\theta) - M_\theta'(0) \sum_{n=1}^{\infty} n A_n(\theta).$$

Because $E_\theta N = \sum_1^\infty n(B_n(\theta) + A_n(\theta))$ and $M_\theta'(0) = 2\theta/(1-\theta^2)$, we find

$$E_\theta N = \begin{cases} \dfrac{1-\theta^2}{2\theta}\left[\left(\dfrac{1}{1-\theta} + b\right) P_\theta(S_N \geq b) - \left(\dfrac{1}{1+\theta} - a\right) \times (1 - P_\theta(S_N \geq b))\right] & \text{for} \quad \theta \neq 0, \\[2ex] \dfrac{(1+b)(1-a)+1}{2} & \text{for} \quad \theta = 0, \end{cases} \tag{7.91}$$

where $E_\theta N$ is determined at $\theta = 0$ by continuity or by taking a second derivative of (7.89) at $t = 0$ (Exercise 8).

The Wald approximations (7.83) and (7.84) give

$$P_\theta(S_N \geq b) \doteq \begin{cases} \dfrac{1 - \exp(-2\theta a)}{\exp(-2\theta b) - \exp(-2\theta a)} & \text{for } \theta \neq 0, \\[2ex] \dfrac{-a}{b - a} & \text{for } \theta = 0. \end{cases} \tag{7.92}$$

For $\theta = 0$ the Wald approximation is close to the true value (7.90) if $-a$ and b are large; if $a = b$, it is exact. However, as $\theta \to -1$, the true value of $P_\theta\{S_N \geq b\}$ tends to zero whereas its approximation converges to $(1 - e^{2a})/(e^{2b} - e^{2a}) > 0$; and, as $\theta \to +1$, $P_\theta\{S_N \geq b\} \to 1$ while its approximation converges to $(e^{-2a} - 1)/(e^{-2a} - e^{-2b}) < 1$. We investigate in detail the accuracy of the approximations when $\theta = \pm\frac{1}{2}$.

Because $\alpha_0 = P_{-1/2}(S_N \geq b)$ and $\alpha_1 = 1 - P_{1/2}(S_N \geq b)$, the exact values of α_0 and α_1, computed from (7.90), are

$$\alpha_0 = \frac{3 - e^a}{9e^b - e^a} = \frac{3 - A}{9B - A},$$

$$\alpha_1 = \frac{3 - e^{-b}}{9e^{-a} - e^{-b}} = \frac{3 - B^{-1}}{9A^{-1} - B^{-1}}.$$

From (7.72) it is seen that the Wald approximations of α_0 and α_1 are about three times too large when B and A^{-1} are large.

A person using the Wald approximations and desiring error probabilities α_0 and α_1 would choose A and B to satisfy (7.71) with equality, whereas the correct values of A and B are found in solving the above equations for A and B, namely, $A = 3\alpha_1/(1 - \alpha_0)$ and $B = (1 - \alpha_1)/3\alpha_0$ (provided $A < 1 < B$). This person would use an A one third the correct value and a B three times the correct value. The effect would be to give him error probabilities about one third as large as prescribed. This, being on the conservative side, is excusable. However, it entails a corresponding increase in the expected sample sizes. It is important to see how large the increase is. From (7.91)

$$E\{N \mid H_0\} = E_{-1/2}N = (3/4)[(2 - a) - \alpha_0(8/3 + b - a)].$$

Changing A to one third its value and B to three times its value changes

$$-a \text{ to } -a + \log 3 \quad \text{and} \quad b \text{ to } b + \log 3,$$

whereas α_0 changes from $(3 - A)/(9B - A)$ to $(9 - A)/(81B - A) = \alpha_0^*$. The increase in this expected value is $(3/4)[(1 - 2\alpha_0^*) \log 3 + (8/3 + b - a)(\alpha_0 - \alpha_0^*)]$. Because α_0 stays bounded away from zero if b is fixed, this increase in expected value tends to $+\infty$ as $a \to -\infty$; but because $E_{-1/2}N$ tends to $+\infty$ at the same rate, the percentage increase is at least bounded. On the other hand, in the symmetric case $a = -b$, this increase in expected value is bounded by

$$(3/4)[\log 3 + (8/3 - 2b)\alpha_0] = (3/4)[\log 3 + (8/3 + 2 \log B) \times (3B + 1)^{-1}] \leq (3/4)[\log 3 + (8/3)(1/4)] \leq 3/2.$$

This is in line with the conclusions of the more general study of Wald (1947): that the use of these approximations for A and B in general decreases the probabilities of error from those desired and that, in general, although the expected sample size increases, this increase is not very serious.

Although sufficiently accurate for most statistical applications, the Wald approximations to the probabilities of error may be rather poor, as we have seen. Even more care must be exercised in using the Wald approximations to the expected sample sizes. If, for example, μ in (7.85) is much larger than b, then so is ES_N, from Theorem 3(a); but when ES_N is replaced by $aP\{S_N \leq a\} + bP\{S_N \geq b\} < b$, the ratio $EN = ES_N/\mu$ is approximated by something much less than one. In general, the Wald approximations to EN are underestimates.

As a numerical example, suppose it is desired to have $\alpha_0 = \alpha_1 = 0.01$. The correct values of A and B to use are $A = 1/33$ and $B = 33$. If these values are used,

$$E_{-1/2}N = E_{1/2}N = (3/4)[(2 + \log 33) - \alpha_0(8/3 + 2 \log 33)] = 4.11\cdots;$$

if the Wald approximations $A = 1/99$ and $B = 99$ are used, then α_0 and α_1 reduce to $0.0034\cdots$, and $E_{-1/2}N$ and $E_{1/2}N$ increase to $4.92\cdots$, an increase of only $0.81\cdots$. On the other hand, the Wald approximations give $E_{-1/2}N = E_{1/2}N = 3.38\cdots$, too low by $1.54\cdots$.

If $\alpha_0 = 0.001$ and $\alpha_1 = 0.1$ (about the least symmetric case that might occur), the correct values are $A = 0.300\cdots$ and $B = 300$. With these values, $E_{-1/2}N = 2.39\cdots$ and $E_{1/2}N = 5.06\cdots$. The Wald approximations give $A = 0.100\cdots$ and $B = 900$. If these values are used, then $\alpha_0 = 0.00036\cdots$, $\alpha_1 = 0.033\cdots$, $E_{-1/2}N = 3.22\cdots$, and $E_{1/2}N = 6.31\cdots$, the increases in the expected sample sizes being $0.83\cdots$ and $1.25\cdots$. On the other hand, the Wald approximations give $E_{-1/2}N = 1.72\cdots$ and $E_{1/2}N = 4.42\cdots$, too low by $1.50\cdots$ and $1.89\cdots$.

Exercises

1. Suppose that a random variable N taking nonnegative integer values satisfies the conclusion of Theorem 1. Show that $E\rho^N$ is finite for $0 \leq \rho < 1/r$ and conclude that all moments of N are finite.
2. Let $X_1, X_2, \cdots$ be independent identically distributed random variables, let $S_n = \sum_1^n X_j$, and let ϕ be any stopping rule for which $E\phi N < \infty$.
(a) Show that if $E|X_1| < \infty$ and $\mu = EX_1$, then $E\phi S_N = \mu E\phi N$.
(b) Show that if $\operatorname{var} X_1 = \sigma^2 < \infty$, then $E\phi(S_N - N\mu)^2 = \sigma^2 E\phi N$.
Hint. It is sufficient to give the proof for $\mu = 0$.
(a) Show

$$\sum_{n=1}^{\infty} E\{\psi_n(X_1, \cdots, X_n) \sum_{j=1}^{n} X_j\} = \sum_{j=1}^{\infty} E\{X_j \sum_{n=j}^{\infty} \psi_n(X_1, \cdots, X_n)\}$$

$$= \sum_{j=1}^{\infty} E(X_j) E(\sum_{n=j}^{\infty} \psi_n(X_1, \cdots, X_n))$$

$$= 0,$$

since X_j is independent of

$$\sum_{n=j}^{\infty} \psi_n(X_1, \cdots, X_n) = 1 - \sum_{n=0}^{j-1} \psi_n(X_1, \cdots, X_n) \text{ a.s.}$$

(b) (Chow, Robbins, and Teicher (1965)) In the proof analogous to (a) it is difficult to justify the interchange of summations. However, if $\phi^{(J)}$ denotes ϕ truncated at J, show as in (a)

$$E\phi^{(J)} S_N^2 = \sum_{i=1}^{J} \sum_{j=1}^{J} E\{X_i X_j \sum_{n=\max(i,j)}^{J} \psi_n^{(J)}(X_1, \cdots, X_n)\}$$

$$= \sigma^2 E\phi^{(J)} N \to \sigma^2 E\phi N \qquad \text{as} \quad J \to \infty.$$

Since $E\phi^{(J)}S_N^2 \geq \sum_1^J E\psi_n(X_1, \cdots, X_n)S_n^2 \to E\phi S_N^2$ as $J \to \infty$, it is sufficient to show $E\phi^{(J)}S_N^2 \leq E\phi S_N^2$. By an application of (a), $E\{\sum_{n=J+1}^{\infty} \psi_n(x_1, \cdots, x_J, X_{J+1}, \cdots, X_n) \sum_{j=J+1}^{n} X_j\} = 0$ for fixed $(x_1, \cdots, x_J)$, so that

$$E\phi S_N^2 - E\phi^{(J)}S_N^2 = \sum_{n=J+1}^{\infty} E\psi_n(X_1, \cdots, X_n)(S_n^2 - S_J^2)$$

$$= \sum_{n=J+1}^{\infty} E\psi_n(X_1, \cdots, X_n)(S_n - S_J)^2 \geq 0.$$

3. Let $M(t) = Ee^{tZ}$ and suppose that $M(t)$ exists for all t. If

$$P\{Z < 0\} > 0, \qquad P\{Z > 0\} > 0,$$

and $EZ \neq 0$, then there exists a unique nonzero real number t_0 such that $M(t_0) = 1$. *Hint.* Show that $M(t)$ is convex, $M(0) = 1$, $M'(0) \neq 0$, and $\lim_{t\to\pm\infty} M(t) = +\infty$.

4. (a) Show that $\mu = EZ$ and t_0 in Exercise 3 are of opposite sign.
(b) Show that the approximation (7.85) always gives a positive value for EN (but, unfortunately, does not always approximate EN to be at least one).

5. Let the distribution of a random variable X have density $f_0(x)$ under hypothesis H_0 and density $f_1(x)$ under hypothesis H_1. Let $Z = \log(f_1(X)/f_0(X))$ and suppose that the distributions under H_0 and H_1 are mutually absolutely continuous, so that

$$P\{|Z| < \infty \mid H_i\} = 1 \quad \text{for} \quad i = 0, 1.$$

Let

$$M_i(t) = E\{e^{tZ} \mid H_i\}, \qquad i = 0, 1.$$

Show that $M_0(+1) = 1$ and $M_1(-1) = 1$.

6. (a) Suppose that $P\{\epsilon_1 \leq Z_i \leq \epsilon_2\} = 1$, where $\epsilon_1 < 0 < \epsilon_2$. Show that $P\{S_N \geq b\}$ of (7.83) satisfies the inequalities

$$\frac{1 - \exp(t_0 a)}{\exp[t_0(b + \epsilon_2)] - \exp(t_0 a)} \leq P\{S_N \geq b\}$$

$$\leq \frac{1 - \exp[t_0(a + \epsilon_1)]}{\exp(t_0 b) - \exp[t_0(a + \epsilon_1)]}$$

(b) Suppose that $P\{\epsilon_1 \le Z_i \le \epsilon_2\} = 1$ is satisfied under H_0 and H_1. Using $t_0 = \pm 1$ in 6(a) show that (7.71) may be extended to

$$\frac{\alpha_1}{1-\alpha_0} \le A \le \frac{\alpha_1 e^{-\epsilon_1}}{1-\alpha_0}, \quad \frac{e^{-\epsilon_2}(1-\alpha_1)}{\alpha_0} \le B \le \frac{1-\alpha_1}{\alpha_0}.$$

(c) If $X_1, X_2, \cdots$ are as in Example 1, find the bounds on the probabilities of error given by 6(a) for the SPRT(A, B) of the hypothesis $H_0{:}p = 1/4$ against the hypothesis $H_1{:}p = 1/2$, when $A = 1/9$ and $B = 9$.

Ans. $8/161 \le \alpha_0 \le 25/241$, $16/241 \le \alpha_1 \le 17/161$.

7. Of two given coins, it is known that one is fair and the other has probability 2/3 of heads, but it is not known which is the fair coin. The coins, tossed simultaneously and independently, produce vector observations (X_1, Y_1), (X_2, Y_2), $\cdots$, where X_i, respectively Y_i, is 1 or 0, depending on whether the ith toss of the first coin, respectively the second coin, is heads or tails. Find for a given SPRT(A, B) of the hypothesis $H_0{:}(X_i, Y_i) \in (\mathcal{B}(1, 2/3), \mathcal{B}(1, 1/2))$ against the hypothesis $H_1{:}(X_i, Y_i) \in (\mathcal{B}(1, 1/2), \mathcal{B}(1, 2/3))$, the exact probabilities of error and expected sample sizes.

Ans. If $A = 2^{-j}$ and $B = 2^k$, with j and k integers, then $E\{N \mid H_0\} = 6(j(B-1) - k(1-A))/B - A)$.

8. Verify (7.91) for $E_\theta N$ at $\theta = 0$.

9. In Example 2 find the Bayes rule with respect to the distribution, giving probability $\frac{1}{2}$ to both H_0 and H_1, when $w_{01} = w_{10}$ and $c = 1/100$. Is this rule minimax? *Ans.* $A^{-1} = B = 41.08\cdots$.

10. In Example 2 test the hypothesis $H_0{:}\theta = -\theta_0$ against the alternative $H_1{:}\theta = +\theta_0$ $(0 < \theta_0 < 1)$ and determine how the Wald approximations to the probabilities of error fare as a function of θ_0.

11. Let $X_1, X_2, \cdots$ be independent, identically distributed random variables with common probability mass function

$$f(x \mid \theta) = \begin{cases} \theta^x & x = 1, 2, \cdots, \\ \dfrac{1-2\theta}{1-\theta} & x = 0, \end{cases}$$

where $0 < \theta < 1/2$, and consider the SPRT(A, B) of the hypothesis $H_0{:}\theta = 1/5$ against the alternative $H_1{:}\theta = 3/8$. The exact values of the probability of rejecting H_0 and the expected sample size under an arbitrary value of θ can be found by a method similar to that used in Example 2.

(a) Show

$$M_\theta(t) = (1 - 2\theta)(1 - \theta)^{-1} \exp(-t\Delta) + \theta(\exp(-t\Delta) - \theta)^{-1},$$

where $\Delta = \log(15/8)$.

(b) Show, corresponding to (7.89),

$$1 \equiv e^{ta} \sum_{n=1}^{\infty} A_n(\theta)\, M_\theta(t)^{-n} + (1 - \theta)e^{tb}(1 - e^{t\Delta}\theta)^{-1} \sum_{n=1}^{\infty} B_n(\theta)\, M_\theta(t)^{-n}$$

where $A_n(\theta) = P_\theta\{N = n, S_N \le a\}$ and $B_n(\theta) = P_\theta\{N = n, S_N \ge b\}$.

(c) Find $P_\theta\{S_N \ge b\}$ and $E_\theta N$. *Ans.* If $a = -j\Delta$, $b = k\Delta$, j and k positive integers, then for $\theta \ne (2 - \sqrt{2})/2$, $P_\theta\{S_N \ge b\} = (1 - \rho^{-j})/(2(1 - \theta^2)\rho^k - \rho^{-j})$ where $\rho = (1 - 2\theta)/(2\theta(1 - \theta))$, and

$$E_\theta N = (1 - \theta)^2(2\theta^2 - 4\theta + 1)^{-1} \times [j - (j + k + \theta(1 - \theta)^{-1})\, P_\theta\{S_N \ge b\}].$$

12. *Wald approximation for the probability generating function of N.* Assume that for a given ρ between zero and one there exist two distinct real roots t_1 and t_2 of the equation $M(t) = \rho^{-1}$.

(a) Using the fundamental identity and the approximation

$$E\{\exp(t_i S_N) M(t_i)^{-N}\} \doteq \exp(t_i a) E\{\rho^N I_{(-\infty,a)}(S_N)\} + \exp(t_i b) E\{\rho^N I_{(b,\infty)}(S_N)\}$$

for $i = 1, 2$, solve for $E\{\rho^N I_{(-\infty,a)}(S_N)\}$ and $E\{\rho^N I_{(b,\infty)}(S_N)\}$ and then find $E\rho^N$ as the sum of these two.

(b) Find $E\{\rho^N \mid H_0\}$ for the distributions of Example 1.

13. Find $E_\theta\{\rho^N\}$ exactly for the distributions of Example 2.

If the distribution of Z gives positive probability to $\pm\infty$, the Wald approximations to the expected sample size may not be found directly from (7.85) and (7.86). Instead, it is suggested that t_0, μ, and σ^2 be found for the distribution of Z truncated at $\pm(b - a)$ and used in (7.83)–(7.86). This not only gives a method for getting the expected sample size, but also provides approximations for the error probabilities which are generally somewhat better than approximations (7.72). An alternative method of dealing with the problem when Z assumes infinite values with positive probability, presented in the next exercise, is based on the ability to find certain expectations for the conditional distribution of Z, given Z is finite. This method gives exact results when exact results are available for this conditional distribution.

14. Let $\hat{Z}_1, \hat{Z}_2, \cdots$ be independent identically distributed finite-valued random variables. Let $\epsilon_1, \epsilon_2, \cdots$ be independent identically distributed and independent of $\hat{Z}_1, \hat{Z}_2, \cdots$, such that $P\{\epsilon_1 = -\infty\} = p_-$, $P\{\epsilon_1 = +\infty\} = p_+$ and $P\{\epsilon_1 = 0\} = p_0$, where $p_- + p_0 + p_+ = 1$. Define $Z_i = \hat{Z}_i + \epsilon_i$. Let $\hat{S}_n = \sum_1^n \hat{Z}_i (S_n = \sum_1^n Z_i)$ and let $\hat{N}(N)$ be the first n for which it is not true that $a < \hat{S}_n < b (a < S_n < b)$. Thus $N \leq \hat{N}$. Prove the following formulas.

(a) $(1 - p_0)EN = 1 - Ep_0^{\hat{N}}$.

(b) $(1 - p_-)P\{S_N = +\infty\} = p_+(1 - Ep_0^{\hat{N}})$

(c) $P\{b \leq S_N < +\infty\} = E\{p_0^{\hat{N}} I_{(b,\infty)}(\hat{S}_{\hat{N}})\}$.

15. Consider the SPRT(A, B) of the hypothesis $H_0{:}f_0(x)$ against the alternative $H_1{:}f_1(x)$, where

$$f_0(x) = \begin{cases} \frac{4}{5} & x = 0, \\ \\ \frac{1}{5} & x = 1. \end{cases} \qquad f_1(x) = \begin{cases} \frac{2}{5} & x = 0, \\ \frac{2}{5} & x = 1, \\ \frac{1}{5} & x = 2. \end{cases}$$

Let $A = 2^{-j}$ and $B = 2^k$, with j and k positive integers. Find the exact values of α_0, α_1, $E\{N \mid H_0\}$ and $E\{N \mid H_1\}$.

Ans. $\alpha_1 = (2^k - 2^{-k})/(2^{j+k} - 2^{-(j+k)})$,

$$E\{N \mid H_1\} = \frac{5(2^k - 1)(2^j - 1)}{(2^{j+k} - 1)}.$$

References

Agarwal, O. P. (1955). "Some minimax invariant procedures for estimating a cumulative distribution function." *Ann. Math. Statist.* **26,** 450–463.

Anscombe, F. J., and R. J. Aumann, (1963). "A definition of subjective probability." *Ann. Math. Statist.* **34,** 199–205.

Arrow, K. J., D. Blackwell, and M. A. Girshick (1949). "Bayes and minimax solutions of sequential decision problems." *Econometrica* **17,** 213–244.

Bahadur, R. R. (1954). "Sufficiency and statistical decision functions." *Ann. Math. Statist.* **25,** 423–462.

Bahadur, R. R. (1958). "A note on the fundamental identity of sequential analysis." *Ann. Math. Statist.* **29,** 534–543.

Baños, A. (1967). *On Pseudo-Games.* Ph.D. thesis, UCLA.

Basu, D. (1955). "On statistics independent of a complete sufficient statistic." *Sankhya* **15,** 377–380.

Berkson, J. (1955). "Maximum likelihood and minimum χ^2 estimates of the logistic function." *J. Amer. Statist. Assoc.* **50,** 130–162.

Blackwell, D. (1947). "Conditional expectation and unbiased sequential estimation." *Ann. Math. Statist.* **18,** 105–110.

Blackwell, D. (1951). "On the translation parameter problem for discrete variables." *Ann. Math. Statist.* **22,** 393–399.

Blackwell, D., and M. A. Girshick, (1954). *Theory of Games and Statistical Decisions.* Wiley, New York.

Blyth, C. R. (1951). "On minimax statistical decision procedures and their admissibility." *Ann. Math. Statist.* **52,** 22–42.

Burkholder, D. L., and R. A. Wijsman, (1963). "Optimum properties and admissibility of sequential tests." *Ann. Math. Statist.* **34,** 1–17.

Chernoff, H., and Moses, L. E. (1959). *Elementary Decision Theory.* Wiley, New York.

Chow, Y. S., and H. Robbins (1965). "On optimal stopping rules for S_n/n." *Illinois J. Math.* **9,** 444–454.

Chow, Y. S., H. Robbins, and H. Teicher (1965). "Moments of randomly stopped sums." *Ann. Math. Statist.* **36,** 789–799.

DeGroot, M. H. (1959). "Unbiased sequential estimation for binomial populations." *Ann. Math. Statist.* **30,** 80–101.

Dynkin, E. B. (1951). "Necessary and sufficient statistics for a family of probability distributions." *Uspekhi Mat. Nauk* (N.S.) **6,** No. 1 (41), 68–90. (Also *Selected Transl. Math. Stat. Prob.*, 17–40.

Dvoretzky, A., A. Wald, and J. Wolfowitz (1951). "Elimination of randomization in certain statistical decision procedures and zero-sum two-person games." *Ann. Math. Statist.* **22,** 1–21.

Feller, W. (1957). *An Introduction to Probability Theory and Its Applications,* Vol. 1, 2nd ed. Wiley, New York.

Ferguson, T. S. (1961). "On the rejection of outliers." *Proc. 4th Berkeley Symp. on Math. Statist. and Prob.* **1,** 253–287. University of California Press, Berkeley.

Fix, E. (1949). "Tables of the non-central χ^2. *Univ. of Calif. Pub. in Statist.* **1,** 15–19.

Fix. E., J. L. Hodges, and E. L. Lehmann (1959). "The restricted χ^2 test." In *Studies in Probability and Statistics Dedicated to Harald Cramér.* Almquist and Wiksell, Stockholm.

Fox, M. (1956). "Charts of the power of the *F*-test." *Ann. Math. Statist.* **27,** 484–497.

Girshick, M. A., F. Mosteller, and L. J. Savage (1946). "Unbiased estimates for certain binomial sampling problems with applications." *Ann. Math. Statist.* **17,** 13–23.

Hall, W. J., R. A. Wijsman, and J. K. Ghosh (1965). "The relationship between sufficiency and invariance with applications in sequential analysis." *Ann. Math. Statist.* **36,** 575–614.

Halmos, P. R., and L. J. Savage (1948). "Application of the Radon-Nikodym theorem to the theory of sufficient statistics." *Ann. Math. Statist.* **20,** 225–241.

Hoeffding, W. (1951). "Optimum non-parametric tests." *Proc. 2nd Berkeley Symp. on Math. Statist. and Prob.,* 83–92. University of California Press, Berkeley.

Hoeffding, W. (1960). "Lower bounds for the expected sample size and the average risk of a sequential procedure." *Ann. Math. Statist.* **31,** 352–368.

Hoel, P. G. (1945). "Testing the homogeneity of Poisson frequencies." *Ann. Math. Statist.* **16,** 362–368.

Hoel, P. G., and R. P. Peterson (1949). "A solution to the problem of optimum classification." *Ann. Math. Statist.* **20,** 433–438.

Jahnke, E., and F. Emde (1945). *Tables of Functions with Formulae and Curves.* 4th ed. Dover, New York.

James, W., and C. Stein (1961). "Estimation with quadratic loss " *Proc. 4th Berkeley Symp. on Math. Statist. and Prob.* **1,** 361–380. University of California Press, Berkeley.

Jensen, J. L. W. V. (1906). "Sur les functions convexes et les inégalités entre les valeurs moyennes." *Acta Math.* **30,** 175–193.

Karlin, S. (1957a). "Pólya type distributions II." *Ann. Math. Statist.* **28,** 281–308.

Karlin, S. (1957b). "Pólya type distributions III: Admissibility for multi-action problems." *Ann. Math. Statist.* **28,** 839–860.

Karlin, S. (1959). *Mathematical Methods and Theory in Games, Programming, and Economics,* Vols. 1 and 2. Addison-Wesley, Reading, Mass.

Karlin, S., and H. Rubin (1956). "The theory of decision procedures for distributions with monotone likelihood ratio." *Ann. Math. Statist.* **27,** 272–299.

Karlin, S. and D. Truax (1960). "Slippage problems." *Ann. Math. Statist.* **31,** 296–323.

Kemperman, J. H. B. (1961). "The Passage Problem for a Stationary Markov Chain," *Statistical Research Monographs,* Vol. 1. University of Chicago Press, Chicago.

Kiefer, J. (1957). "Invariance, minimax sequential estimation and continuous time processes." *Ann. Math. Statist.* **28,** 573–601.

Kudo, H. (1955). "On minimax invariant estimates of the transformation parameter." *Natural Science Report*, **6,** 31–73. Ochanomizu University.

Le Cam, L. (1955). "An extension of Wald's theory of statistical decision functions." *Ann. Math. Statist.* **26,** 69–81.

Lehmann, E. L. (1959). *Testing Statistical Hypotheses.* Wiley, New York.

Lehmann, E. L., and H. Scheffé, (1950, 1955). "Completeness, similar regions and unbiased estimation." *Sankhya* **10,** 305–340, and **15,** 219–236.

Lehmann, E. L. and C. Stein (1950). "Completeness in the sequential case." *Ann. Math. Statist.* **21,** 376–385.

Locks, M. O., M. J. Alexander, and B. J. Byars (1963). "New Tables of the Non-Central t Distribution," ARL 63–19. Aeronautical Research Laboratories, Officer of Aerospace Research, USAF.

Loève, M. (1963). *Probability Theory,* 3rd ed. Van Nostrand, Princeton, N. J.

Luce, R. D., and H. Raiffa (1957). *Games and Decisions.* Wiley, New York.

Matthes, T. K. (1963). "On the optimality of the sequential probability ratio tests." *Ann. Math. Statist.* **34,** 18–21.

McKinsey, J. C. C. (1952). *Introduction to the Theory of Games.* The Rand Corp., McGraw-Hill, New York.

Neyman, J. (1935). "Su un teorema concernente le cosiddette statistiche sufficienti." *Giornale dell' Instituto degli Attuari* **6,** 320–334.

Neyman, J. (1952). *Lectures and Conferences on Mathematical Statistics and Probability,* 2nd ed. Graduate School, U. S. Department of Agriculture.

Neyman, J., and E. S. Pearson (1933). "On the problem of the most efficient tests of statistical hypotheses." *Philos. Trans. Roy. Soc. London Series A* **231,** 289–337.

Neyman, J., and E. S. Pearson (1936, 1938). "Contributions to the theory of testing statistical hypotheses. I. Unbaised critical regions of type A and type A_1. II. Certain theorems on unbiased critical regions of type A. III. Unbiased tests of simple statistical hypotheses specifying the values of more than one unknown parameter." *Statist. Res. Mem.* **1,** 1–37, and **2,** 25–57.

Owen, D. B. (1963). *Factors for One-Sided Tolerance Limits and for Variables Sampling Plans,* SCR–607. Sandia Corp., Albuquerque, N. M.

Paige, L. J., and J. D. Swift, (1961). *Elements of Linear Algebra.* Ginn, Boston.

Parzen, E. (1960). *Modern Probability Theory and Its Applications.* Wiley, New York.

Paulson, E. (1952). "An optimum solution to the k-sample slippage problem for the normal distribution." *Ann. Math. Statist.* **23,** 610–616.

Pearson, E. S., and H. O. Hartley (1951). "Charts of the power function for analysis of variance tests derived from the noncentral *F*-distribution." *Biometrika* **38,** 112–130.

Pitman, E. J. G. (1939). The estimation of location and scale parameters of a continuous population of any given form." *Biometrika* **30,** 391–421.

Raiffa, H., and R. Schlaifer (1961). *Applied Statistical Decision Theory.* Graduate School of Business Administration, Harvard.

Rao, C. R. (1945). "Information and accuracy attainable in estimation of statistical parameters." *Bull. Calcutta Math. Soc.* **37,** 81–91.

Ray, S. N. (1965). "Bounds on the maximum sample size of a Bayes sequential procedure." *Ann. Math. Statist.* **36,** 859–878.

Resnikoff, G. J., and G. J. Leiberman (1957). *Tables of the Non-Central t-Distribution.* Stanford University Press, Stanford, Calif.

Savage, L. J. (1947). "A uniqueness theorem for unbiased sequential estimation." *Ann. Math. Statist.* **18,** 295–297.

Savage, L. J. (1954). *The Foundations of Statistics.* Wiley, New York.

Scheffé, H. (1953). "A method for judging all contrasts in the analysis of variance." *Biometrika* **40,** 87–104.

Scheffé, H. (1959). *The Analysis of Variance.* Wiley, New York.

Stein, C. (1956). "Inadmissibility of the usual estimator for the mean of a multivariate normal distribution." *Proc. 3rd Berkeley Symp. on Math. Statist. and Prob.* **1,** 197–206. University of California Press, Berkeley.

Steinhaus, H. (1957). "The problem of estimation." *Ann. Math. Statist.* **28,** 633–648.

Tucker, H. G. (1962). *An Introduction to Probability and Mathematical Statistics.* Academic Press, New York.

Tukey, J. (1953). "Some selected quick and easy methods of statistical analysis." *Trans. N. Y. Acad. Sci.,* Series II, **16,** 88–97.

Valentine, F. A. (1964). *Convex Sets.* McGraw-Hill, New York.

von Neumann, J., and O. Morgenstern (1944). *Theory of Games and Economic Behavior,* 3rd. ed. Princeton University Press, 1953, Princeton, N. J.

Wald, A. (1947). *Sequential Analysis.* Wiley, New York.

Wald, A. (1950). *Statistical Decision Functions.* Wiley, New York.

Wald, A., and J. Wolfowitz (1948). "Optimum character of the sequential probability ratio test." *Ann. Math. Statist.* **19,** 326–339.

Wald, A., and J. Wolfowitz (1951). "Two methods of randomization in statistics and the theory of games." *Ann. Math.* **53,** 581–586.

Widder, D. V. (1946). *The Laplace Transform.* Princeton University Press, Princeton, N.J.

Wilks, S. S. (1962). *Mathematical Statistics.* Wiley, New York.

Williams, J. D. (1954). *The Compleat Strategyst.* McGraw-Hill, New York.

Wolfowitz, J. (1946). "On sequential binomial estimation." *Ann. Math. Statist.* **17,** 489–493.

Wolfowitz, J. (1950). Minimax estimates of the mean of a normal distribution with known variance." *Ann. Math. Statist.* **21,** 218–230.

Index